The New Aspects of Subnuclear Physics

THE SUBNUCLEAR SERIES

Series Editor: **ANTONINO ZICHICHI**
European Physical Society
Geneva, Switzerland

1. 1963 STRONG, ELECTROMAGNETIC, AND WEAK INTERACTIONS
2. 1964 SYMMETRIES IN ELEMENTARY PARTICLE PHYSICS
3. 1965 RECENT DEVELOPMENTS IN PARTICLE SYMMETRIES
4. 1966 STRONG AND WEAK INTERACTIONS
5. 1967 HADRONS AND THEIR INTERACTIONS
6. 1968 THEORY AND PHENOMENOLOGY IN PARTICLE PHYSICS
7. 1969 SUBNUCLEAR PHENOMENA
8. 1970 ELEMENTARY PROCESSES AT HIGH ENERGY
9. 1971 PROPERTIES OF THE FUNDAMENTAL INTERACTIONS
10. 1972 HIGHLIGHTS IN PARTICLE PHYSICS
11. 1973 LAWS OF HADRONIC MATTER
12. 1974 LEPTON AND HADRON STRUCTURE
13. 1975 NEW PHENOMENA IN SUBNUCLEAR PHYSICS
14. 1976 UNDERSTANDING THE FUNDAMENTAL CONSTITUENTS OF MATTER
15. 1977 THE WHYS OF SUBNUCLEAR PHYSICS
16. 1978 THE NEW ASPECTS OF SUBNUCLEAR PHYSICS
17. 1979 POINTLIKE STRUCTURES INSIDE AND OUTSIDE HADRONS

Volume 1 was published by W. A. Benjamin, Inc., New York; 2-8 and 11-12 by Academic Press, New York and London; 9-10 by Editrice Compositori, Bologna; 13-17 by Plenum Press, New York and London.

The New Aspects of Subnuclear Physics

Edited by

Antonino Zichichi
European Physical Society
Geneva, Switzerland

SPRINGER SCIENCE+BUSINESS MEDIA, LLC

Library of Congress Cataloging In Publication Data

International School of Subnuclear Physics, 16th, Erice, Italy, 1978.
The new aspects of subnuclear physics.

(The subnuclear series; 16)
"Proceedings of the sixteenth International School of Subnuclear Physics, held in Erice, Italy, July 31–August 11, 1978."
Includes index.
1. Particles (Nuclear physics)–Congress. I. Zichichi, Antonino. II. Title. III. Title: Subnuclear physics. IV. Series: Subnuclear series; 16.
QC793.I555 1978 539.7'21 80-15088

ISBN 978-1-4615-9172-6 ISBN 978-1-4615-9170-2 (eBook)
DOI 10.1007/978-1-4615-9170-2

Proceedings of the Sixteenth International School of Subnuclear Physics, held in Erice, Sicily, Italy, July 31–August 11, 1978.

Preface

In August 1978 a group of 80 physicists from 51 laboratories of 15 countries met in Erice to attend the 16th Course of the International School of Subnuclear Physics.

The countries represented at the School were: Austria, Denmark, Federal Republic of Germany, Finland, France, Israel, Italy, the Netherlands, Sweden, Switzerland, South Africa, Turkey, the United Kingdom, The United States of America, and Yugoslavia.

The School was sponsored by the Italian Ministry of Public Education (MPI), the Italian Ministry of Scientific and Technological Research (MRSI), the North Atlantic Treaty Organization (NATO), the Sicilian Regional Government, and the Weizmann Institute of Science.

As usual, the Course was devoted to a review of the most outstanding problems and results in Subnuclear Physics, with particular emphasis on the new aspects; there were mainly two: supersymmetry and electroweak interactions. In his famous lecture at Erice in 1967, Sid Coleman reviewed "All possible symmetries of the S matrix." All but one, namely that which tells you: if you have a fermion you must have a boson. This is supersymmetry, and this produces the superspace, i.e. an entity which has not only the Einstein-"bosonic" coordinates, but also "fermionic" coordinates. From superspace we get supergravity; and this means that one day we should be able to detect not only the graviton (with spin 2) but also the gravitino (spin 3/2). If we add "flavour", "colour", and "family" as other intrinsic degrees of freedom, we get extended supergravity. It is the grand unified theory of all the forces of nature. This is the dream. In real life we have the Glashow-Salam-Weinberg unified theory of the electromagnetic and weak interactions, which is enjoying powerful experimental support. On the "classical" front, new puzzles arise. The field of subnuclear physics could not be more exciting.

I hope the reader will enjoy the book as much as the students have enjoyed the School and the most attractive part of it: the informal, free, and unconventional Discussion Sessions. Thanks to the Scientific Secretaries, these Discussions have been reproduced as closely as possible to the real happening.

At various stages of my work I have enjoyed the collaboration of many friends in Erice, Bologna, and Geneva, whose contributions have been extremely important for the School and are highly appreciated.

Antonino Zichichi
December 1979
Geneva

Contents

LECTURES

SPECIALIZED SEMINARS

SUPERSYMMETRIC THEORIES OF FUNDAMENTAL INTERACTIONS

S. Ferrara

CERN

Geneva, Switzerland

1. INTRODUCTION

Supersymmetry[1)] is a new symmetry of local quantum field theory which extends in a non-trivial way the relativity group of space-time. The new symmetry operations are carried by spinorial charges which, according to the spin-statistic theorem, obey anticommutation relations. Particles are classified, in any supersymmetric field theory, according to representations of the symmetry group, and irreducible representations contain particle states of different spin, both integer and half-integer. Supersymmetry is therefore the first example of a genuine relativistic spin containing symmetry.

Previous no-go theorems which prevented possible relativistic generalizations of $SU_w(6)$ are now circumvented.

The key point is that the algebraic structure related to supersymmetry is not an ordinary Lie algebra, to which the above-mentioned theorems applied, but rather a graded Lie algebra (GLA).

Graded Lie algebras contain generators obeying both commutation and anticommutation relations. The spinorial charges obeying anti-commutation relations are called the odd elements of the GLA, while even generators obeying commutation relations form an ordinary Lie algebra which contains, as part, the Poincaré Lie algebra.

Investigations of several model field theories in the past years and the detailed study of their renormalizability properties have shown that not only is supersymmetry a consistent symmetry at the quantum level but usually it improves the divergences of perturbation theory. This general feature of any supersymmetric model gave us the great hope that a supersymmetric version[2)] of the Einstein theory of gravitation, now called supergravity, could once and for all overcome the well-known difficulties related to the quantization of gravity.

This program[3)] has been partially successful, and there are existing models which have finite one- and two-loop quantum corrections in scattering processes due to gravitational forces. It is at present an open problem whether this trend will persist at higher loops and will lead to a renormalizable theory of gravity.

One of the main goals of supersymmetric theories of particles is to provide a new framework for unifications of fundamental interactions. The two main directions that have been pursued have been, from one side, a supersymmetric extension of chromodynamics unified with weak and electromagnetic interactions[4)]; from the other side, in a more ambitious way, the construction of the so-called extended supergravity theories in which all fundamental particles of nature sit in the same irreducible multiplet[2)].

A third scheme that is worth while mentioning is the construction of superconformal gravity theories which can be regarded as supersymmetric extensions of Weyl gravity[3)].

The first class of theories[4)] seem at present to give the most realistic examples because they are flexible enough to accommodate present particle phenomenology. The exciting feature of these supersymmetric extensions of strong, weak, and electromagnetic interactions is that they imply the existence of a class of new particles having a new quantum number R, the old known particles having R = 0. The spinorial charge of the supersymmetry algebra is the carrier (one unity) of this new quantum number.

In these theories there are two colourless, massless spin ½ particles beyond the usual electron and muon neutrino; they are the photino (the spin ½ partner of the photon) and the goldstino, the Goldstone fermion of spontaneously broken supersymmetry. This last massless excitation is eaten by the gravitino, the spin 3/2 partner of the graviton, when supersymmetry is promoted to a local invariance and gravity is switched on.

As a consequence of the super-Higgs effect, an intimate mixing between weak and gravitational interactions is expected through the massive gravitino. Needless to say, possible detectable effects[5)] of the new particles implied by these supersymmetric unified theories are crucial in order to give further support to this scheme of unification. The unsatisfactory point of this approach is that it does not avoid the old renormalizability problems of gravitation. This is mainly due to the fact that the generators of the local gauge group of strong, weak and electromagnetic interactions commute with the spinorial charge, and therefore a true unification is lacking. The second scheme of unification, based on extended supergravity theories[6)], looks the most appealing because all the fundamental particles of nature sit in the same irreducible multiplet of the underlying symmetry. It is clear that the spinorial charges no longer commute with the generators of the gauge group because, for example, they make the transition between coloured (gluons, quarks) and colourless (leptons, weak bosons, Higgs scalars, graviton) states. Symmetry breaking in these theories is expected to occur dynamically, with possible elimination of the cosmological term which is induced, at the classical level, by the Yang-Mills coupling. These models have finite one- and two-loop quantum corrections[3)]. They seem to be the best candidates for renormalizable theories of gravitation. The drawback of these unified theories is that they are so constrained that both the local gauge group and the representations to which the lower spin particles belong are entirely fixed by the symmetry. The Yang-Mills invariance is restricted to be SO(N), and N cannot exceed 8 under the requirement that all fundamental particles have maximum spin 2, which is the case for the graviton. The SO(8) model contains exact colour and e.m. symmetry but accommodates only a U(1) subgroup of the weak isospin group SU(2), which is believed to be the minimal group of weak interactions.

The third scheme, relying on conformal theories[3,7] in order to have contact with Nature, must exhibit both conformal and supersymmetry breaking. It has the nice property of having one coupling constant only and a vanishing cosmological constant, but it is plagued with the problems of instability and non-unitarity of higher derivative Lagrangian field theories.

To summarize, although at present all the unification schemes based on supersymmetric interactions seem to have their drawbacks, they nevertheless seem fascinating. At present they appear to offer an almost unique possibility of having a unified theory of all particle interactions including gravitation. They seem also to offer a solution to the challenging problem of a consistent formulation of a quantum theory of gravitation.

Finally, it is worth while mentioning that supersymmetric theories of particles provide a theoretical explanation for the existence of fermions and bosons, the building blocks of the universe (matter and radiation) through a symmetry principle which is ultimately related to the underlying geometrical structure of space and time.

2. SUPERSYMMETRY ALGEBRAS AND PARTICLE SUPERMULTIPLETS

Supersymmetry algebras[8] provide an enlargement of the basic space-time symmetry algebras through an operation which is often called the graduation of a given Lie algebra.

The basic supersymmetry algebra consists of N spinorial charges Q^i_α, $\alpha = 1, \ldots, 4$, $i = 1, \ldots, N$, which transform according to the fundamental self-conjugate representation of the Lorentz group

$$i[M_{\mu\nu}, Q^i_\alpha] = (\sigma_{\mu\nu})^\beta_\alpha Q^i_\beta \tag{2.1}$$

These 4N charges are the odd elements or the grading representation of the graded Poincaré algebra. The even part consists of the Poincaré generators P_μ, $M_{\mu\nu}$, and of possible central charges Z^{ij}, Z'^{ij}.

The basic commutators are

$$\begin{aligned}
&[Q^i_\alpha, P_\mu] = 0 \\
&\{Q^i_\alpha, \bar{Q}^j_\beta\} = -2\gamma^\mu_{\alpha\beta} P_\mu \delta^{ij} + \delta_{\alpha\beta} Z^{ij} + \gamma^5_{\alpha\beta} Z'^{ij} \\
&[Z^{ij}, Q^\ell_\alpha] = [Z'^{ij}, Q^\ell_\alpha] = 0 \\
&[Z^{ij}, Z^{\ell m}] = [Z'^{ij}, Z'^{\ell m}] = [Z^{ij}, Z'^{\ell m}] = 0
\end{aligned} \tag{2.2}$$

The operators Z,Z′ are real and antisymmetric. In Eq. (2.2) we have used a Majorana (real) representation for the γ-metrics with metric (-+++). Moreover, $\gamma_5 = \gamma_0\gamma_1\gamma_2\gamma_3$, $\sigma_{\mu\nu} = \frac{1}{4}[\gamma_\mu, \gamma_\nu]$.

The algebra given by Eq. (2.2) is called extended supersymmetry for N > 1 and simple supersymmetry for N = 1. In the latter case, Z = Z′ = 0.

All irreducible representations of the algebra given by Eq. (2.2) acting on single-particle states of given momentum have been classified[9].

Each irreducible representation decomposes into a finite direct sum of irreducible Poincaré representations, all degenerate in mass and with spin shifted by one-half unity. Of particular relevance are the zero mass representations. These representations are suitable for describing unified gauge theories in which gauge particles of different helicity, such as the graviton and the photon, sit in the same irreducible representation. The multiplicity and the helicity content of one-particle states of a massless irreducible multiplet of N-extended supersymmetry can easily be obtained by means of the Wigner method of induced representations. Starting from a singlet state of helicity $|\lambda|$ and then successively applying those spinorial charges Q_i^- (i = 1, ..., N) which act as ½ unit helicity-lowering operators, one gets 2^N states of helicity $|\lambda|$, $|\lambda| - \frac{1}{2}$, ..., $|\lambda| - (K/2)$, ..., up to $|\lambda| - (N/2)$, the multiplicity of a state of given helicity $|\lambda| - (K/2)$ being $N!/[K!(N-K)!]$.

To get acceptable representations consistent with local field theory, CPT conjugate states of reversed helicities must be added. These are obtained by starting with a singlet state of helicity $-|\lambda|$ and by repeatedly applying those spinorial charges Q_i^+ (i = 1, ..., N) which act as helicity-raising operators. Then one gets a doubling of the states, which are therefore 2^{N+1} with the exception of those representations for which $|-\lambda| = |\lambda| - (N/2)$, i.e. $|\lambda| = N/4$. These representations are self-conjugate and contain only 2^N states.

From the previous analysis, it is very easy to make the complete list of all basic massless representations of N-extended supersymmetry with a given helicity content. In the construction of supersymmetric gauge-field theories, the interest is in those multiplets which correspond to gauge fields and matter fields of fundamental interactions. This physical requirement limits us to considering representations with $|\lambda_{max}| = 2$. With this restriction in mind it is easy to show that only a limited number of extended supersymmetries, precisely up to N = 8, can be realized in conventional Lagrangian field theories. In particular let us call, by convention, scalar, vector, spinor, and tensor multiplets those representations having $|\lambda|_{max} = \frac{1}{2}, 1, \frac{3}{2}$, and 2, respectively. They

Table 1

Basic massless irreducible representations of extended supersymmetry

Isospin \ Helicity	$\lambda = 2$	$\lambda = 3/2$	$\lambda = 1$	$\lambda = \frac{1}{2}$	$\lambda = 0$
N = 1	1	1			
		1	1		
			1	1	
				1	2
SO(2)	1	2	1		
		1	2	1	
			1	2	2
				2	4
SO(3)	1	3	3	1	
		1	3	3	$2 = 1 \oplus 1$
			1	$3 \oplus 1$	$3 \oplus 3$
SO(4)	1	4	6	4	$2 = 1 \oplus 1$
		1	4	$6 \oplus 1$	$4 \oplus 4$
			1	4	$6 = 3 \oplus 3$
SO(5)	1	5	10	$10 \oplus 1$	$5 \oplus 5$
		1	$5 \oplus 1$	$10 \oplus 5$	$10 \oplus 10$
SO(6)	1	6	$15 \oplus 1$	$20 \oplus 6$	$15 \oplus 15$
		1	6	15	20
SO(7)	1	$7 \oplus 1$	$21 \oplus 7$	$35 \oplus 21$	$35 \oplus 35$
SO(8)	1	8	28	56	$70 = 35 \oplus 35$

exist in extended supersymmetric theories up to N = 2, 4, 6, and 8 spinor charges, respectively. Because of the anticommuting properties of the spinor charges, it is easily seen that particle states of helicity $|\lambda_{max}| - (K/2)$ transform according to the K rank antisymmetric tensor representation of the SO(N) orthogonal group. In Table 1 we have listed all possible representations of extended supersymmetry which can describe basic particles of fundamental interactions.

Massive representations exist as well. Basic irreducible representations contain 2^{2N}-particle states. More general representations have dimension $2^{2N} \cdot (2J + 1) \cdot d_N$ where J, d_N is the spin and the dimension of the "isospin representation" to which the ground state belongs. The latter state is, by definition, that state of the representation which is annihilated by the 2N spinorial charges which act as destruction operators in the 2N dimensional Clifford algebra spanned by the 4N Majorana charges Q^i_α. In Table 2 we list the massive representations for basic multiplets which can occur in a given Lagrangian field theory. So far we have considered the graded version of the Poincaré algebra. Also of physical interest are the gradings of the O(4, 2) conformal and O(3, 2) de Sitter algebras. The former is relevant for Yang-Mills theories and for

Table 2

Basic massive irreducible representations of extended supersymmetry

Mass M ≠ 0 \ Spin	J = 1	J = ½	J = 0
N = 1	1	2	1
		1	2
N = 2	1	4	5
		2	4

supersymmetric versions of Weyl gravity, while the latter is appropriate for supergravity theories with a cosmological constant. The grading of the O(4, 2) ∿ SU(2, 2) conformal algebra is obtained as follows: for a given N there are 8N odd generators: Q^i_α, S^i_α (α = 1, ..., 4; i = 1, ..., N). They behave as 2N self-conjugate SL(2, C) 4-spinors or as N self-conjugate SU(2, 2) 8-spinors. The even part of the algebra contains the 15 generators of the conformal group $M_{\mu\nu}$, P_ν, D, K_ν, and the N^2 generators of the unitary group U(N). The de Sitter grading can be obtained trivially as a graded subalgebra of the conformal algebra. The odd elements are the N spinors $R^i_\alpha = \frac{1}{2}(Q^i_\alpha + S^i_\alpha)$, while the even elements are the 10 generators $M_{\mu\nu}$, $L_\mu = \frac{1}{2}(P_\mu - K_\mu)$ of $O(3, 2) \subset O(4, 2)$ and the $\frac{1}{2}N(N - 1)$ internal symmetry generators of $SO(N) \subset U(N)$.

3. SUPERSYMMETRIC LAGRANGIAN FIELD THEORIES

A great performance of supersymmetry in the physics of particles and fields has certainly been its successful application to Lagrangian field theories. In the past years it has been shown that all renormalizable theories, such as the Yukawa theory, the ϕ^4 theory, and the Yang-Mills theories, can be suitably extended in order to manifest supersymmetry. Supersymmetry usually severely restricts the couplings and the masses in renormalizable interactions. With great surprise it has been realized that the Ward identities related to this new symmetry not only are preserved by the renormalization procedure but usually improve renormalization. In fact quantum divergences in supersymmetric field theories have the tendency to cancel. In many model field theories the quantum infinities which are absorbed in the renormalization constants are less than one would expect from naive power counting arguments. In spite of this success, no example of non-renormalizable field theory which becomes renormalizable after imposing supersymmetry has yet been found. It is at present an open question whether supergravity could be just the first example of this kind of theory.

In order to apply supersymmetry in Lagrangian field theory, representations of this symmetry on local fields must be described.

The simplest of these representations was found by Wess and Zumino[8] in the following form:

$$\begin{aligned}
\delta A(x) &= i\bar{\epsilon}\chi(x)\\
\delta B(x) &= i\bar{\epsilon}\gamma_5\chi(x)\\
\delta\chi(x) &= \not{\partial}(A(x)+\gamma_5 B(x))\varepsilon + (F(x)+\gamma_5 G(x))\varepsilon\\
\delta F(x) &= i\bar{\epsilon}\not{\partial}\chi(x)\\
\delta G(x) &= i\bar{\epsilon}\gamma_5\not{\partial}\chi(x)
\end{aligned} \tag{3.1}$$

This multiplet is often called the scalar multiplet. Here A(x), F(x) are scalars, B(x), G(x) pseudoscalars, and $\chi(x)$ a Majorana spinor; $\delta A(x)$ stands for $[\bar{\varepsilon}Q, A(x)]$, etc. ε is a constant anti-commuting spinorial parameter which may be regarded as an odd element of a Grassmann algebra, the group manifold of a graded Lie algebra.

One can indeed verify that (3.1) is a representation of the algebra given by Eq. (2.2) with N = 1. In fact the commutator of two such variations gives, when applied to any field, an infinitesimal translation:

$$[\delta_1, \delta_2]A(x) = -2i\,\bar{\epsilon}_1\gamma^\mu\epsilon_2\,\partial_\mu A(x)\ , \text{ etc.} \tag{3.2}$$

We note, *en passant*, that only a finite number of local fields are needed to close the representation.

If we impose the mass-shell conditions,

$$(\Box - m^2)A(x) = (\Box - m^2)B(x) = (\not{\partial} + m)\chi(x) = 0 \tag{3.3}$$

They in turn imply

$$\delta(mA+F) = \delta(mB+G) = 0$$

which have the solution F = −mA, G = −mB.

For the first time a new phenomenon happens in the representation space of the symmetry.

The number of fields which are needed to close the algebra off-mass shell is greater than the number of fields which realize the symmetry on particle states. The fields F and G play the role of auxiliary fields and they can be eliminated using their field equations. Note that the particle fields A, B, χ just give rise to a representation on single-particle states, which can be identified in Tables 1 and 2 as the lowest spin representation of supersymmetry with N = 1.

It is worth while mentioning that the auxiliary fields, like F and G for the example under consideration, play a crucial role in supersymmetric theories. Although they can be eliminated from any given Lagrangian because the equations of motion are purely algebraic, it is much more convenient to retain them in order to have linear transformation laws and closure of the algebra. This last property is very important in order to have a tensor calculus of general validity, as in any other symmetry, and then simple rules for constructing invariants.

In order to have all the ingredients for the formulation of supersymmetric interaction, we need another representation, called the vector multiplet, which is realized as follows:

$$\begin{aligned}
\delta C &= i\bar{\epsilon}\gamma_5\zeta \\
\delta\zeta &= \gamma^\mu v_\mu \epsilon - \partial_\mu C\gamma_5\gamma^\mu\epsilon + (M+\gamma_5 N)\epsilon \\
\delta M &= i\bar{\epsilon}\lambda + i\bar{\epsilon}\partial\!\!\!/\zeta \\
\delta N &= i\bar{\epsilon}\gamma_5\lambda + i\bar{\epsilon}\gamma_5\partial\!\!\!/\zeta \\
\delta v_\mu &= i\bar{\epsilon}\gamma_\mu\lambda + i\bar{\epsilon}\partial_\mu\zeta \\
\delta\lambda &= -\tfrac{1}{2}(\partial_\mu v_\nu - \partial_\nu v_\mu)\gamma^\mu\gamma^\nu\epsilon + \gamma_5 D\epsilon \\
\delta D &= i\bar{\epsilon}\gamma_5\partial\!\!\!/\lambda
\end{aligned} \tag{3.4}$$

It consists of three pseudoscalars (scalars) D, C, N, a scalar (pseudoscalar) M, an (axial) vector V_μ, and two Majorana spinors ζ, λ.

The multiplets defined by the transformation laws in (3.1) and (3.4) have been first obtained by looking at the minimal set of fields needed to close the algebra.

A complete development of the tensor calculus of global supersymmetry has been possible only after the introduction of the superfield concept[10]. This technique proved extremely powerful for treating supersymmetric theories in a simple way, and at the same time provided a geometrical meaning to supersymmetry transformations by enlarging the Minkowski space-time to a superspace whose points are labelled by extra anticommuting coordinates.

The rules of tensor calculus can be summarized as follows: Two scalar multiplets S_1, S_2 can be multiplied in two different ways in order to give either a new scalar multiplet $S = S_1S_2$ or a (complex) vector multiplet $S_1\bar{S}_2$. The real and imaginary part of the latter multiplet $\frac{1}{2}(S_1\bar{S}_2 - \bar{S}_1S_2)$, $\frac{1}{2}i(S_2\bar{S}_1 - S_1\bar{S}_2)$ are two real vector multiplets which are respectively symmetric and antisymmetric in the exchange $1 \leftrightarrow 2$. Two vector multiplets V_1, V_2 can be multiplied to give a new vector multiplet $V = V_1 \cdot V_2$. To any scalar multiplet S of components A, B, χ, F, G, one can associate a new scalar multiplet of components F, G, $\not\partial\chi$, $\Box A$, $\Box B$, to be called the kinetic multiplet T(S). To any vector multiplet V of components C, ζ, M, N, V_μ, λ, D, one can associate a scalar multiplet (with a spinor index) $W_\alpha(V)$ of components λ, $\sigma^{\mu\nu}F_{\mu\nu} + \gamma_5 D$, $\not\partial\lambda$, where $F_{\mu\nu} = \partial_\mu V_\nu - \partial_\nu V_\mu$.

From the transformation laws in Eqs. (3.1) and (3.4) we further observe that the F (G) component of a scalar multiplet and the D component of a vector multiplet transform as total divergences under supersymmetry. This provides the rule for the construction of invariant actions! The F (G) component of any (composite) scalar multiplet or the D component of any (composite) vector multiplet is the candidate for globally supersymmetric Lagrangian densities.

The simplest non-trivial model field theory with global supersymmetry is the self-interaction of a scalar multiplet S, the so-called Wess-Zumino model[8]. This is obtained by adding the following three supersymmetric invariant terms:

$$\mathcal{L}_{KIN} = (S\bar{S})_D = -\frac{1}{2}(\partial_\mu A)^2 - \frac{1}{2}(\partial_\mu B)^2 - \frac{i}{2}\bar{\chi}\partial\!\!\!/\chi + \frac{1}{2}F^2 + \frac{1}{2}G^2$$

$$\mathcal{L}_{MASS} = m(S^2)_F = m\left(FA + GB - \frac{i}{2}\bar{\chi}\chi\right) \tag{3.5}$$

$$\mathcal{L}_{INT} = g(S^3)_F = g\left(FA^2 - FB^2 + 2GAB - i\bar{\chi}(A - \gamma_5 B)\chi\right)$$

The total Lagrangian is given by

$$\mathcal{L} = \mathcal{L}_{KIN} + \mathcal{L}_{MASS} + \mathcal{L}_{INT} \tag{3.6}$$

The current related to the invariance under supersymmetry of the theory can be derived via the standard Noether procedure:

$$J_\mu = \partial\!\!\!/(A - \gamma_5 B)\gamma_\mu\chi - (F + \gamma_5 G)\gamma_\mu\chi$$

and is conserved as a consequence of the field equations.

After elimination of the auxiliary fields F and G by means of their field equations,

$$\begin{aligned} F + mA + g(A^2 - B^2) &= 0 \\ G + mB + 2gAB &= 0 \end{aligned} \tag{3.7}$$

we get

$$\begin{aligned} \mathcal{L} = &-\frac{1}{2}(\partial_\mu A)^2 - \frac{1}{2}(\partial_\mu B)^2 - \frac{1}{2}m^2(A^2+B^2) - \frac{i}{2}\bar{\chi}(\partial\!\!\!/ + m)\chi \\ &- gmA(A^2+B^2) - \frac{g^2}{2}(A^2+B^2)^2 - ig\bar{\chi}(A - \gamma_5 B)\chi \end{aligned} \tag{3.8}$$

This Lagrangian could be equally well called the supersymmetric extension of the ϕ^3, ϕ^4 or Yukawa theory. The resulting supersymmetry of the Lagrangian not only implies equality of masses for the particle fields A, B, and χ but also a precise relationship between the interaction terms.

The amazing fact of this theory is that not only are the relations between masses and couplings preserved by renormalization, but they allow the theory to be renormalized with only one basic infinite renormalization constant, a (logarithmically divergent) wave function renormalization Z common to all fields[11]. For instance, the renormalized mass and coupling constant are given, to each order of perturbation theory, by the following relations:

$$g_r = Z^{3/2} g_0 \quad , \quad m_r = Z m_0$$

These relations are reflected in an important relation between the functions $\beta(g)$, $\gamma(g)$ which appear in the renormalization equations of the theory.

The greatest achievement of supersymmetric Lagrangian field theories is perhaps the construction of supersymmetric extensions of Yang-Mills theories[12]. These theories are the building blocks of any supersymmetric unified theory of weak, strong, and electromagnetic interactions which at present have a chance of being realistic.

The gauge multiplet, which is the multiplet to which the gauge mesons belong, is given by a vector multiplet as described by the transformation laws (3.4). The gauge function, in a supersymmetric theory, is also enlarged to an entire multiplet. The additional gauge freedom coming from this bigger gauge invariance, present in any supersymmetric gauge theory, can be used to reduce the vector multiplet to have $C = \zeta = M = N = 0$. This is the so-called Wess-Zumino gauge. In this gauge there is still the freedom of making ordinary gauge transformations on the vector potential V_μ. In the

Wess-Zumino gauge the remaining fields λ, V_μ, and D transform as follows under supersymmetry transformations:

$$\begin{aligned} \delta\lambda &= -G_{\mu\nu}\sigma^{\mu\nu}\varepsilon + D\gamma_5\varepsilon \\ \delta V_\mu &= i\bar{\varepsilon}\gamma_\mu\lambda \\ \delta D &= i\bar{\varepsilon}\not{D}\lambda \end{aligned} \tag{3.9}$$

The fields λ, V_μ, D are Lie algebra valued fields, $G_{\mu\nu}$ is the Yang-Mills field strength $G_{\mu\nu} = \partial_\mu V_\nu - \partial_\nu V_\mu + ig[V_\mu, V_\nu]$, and D_μ is the covariant derivative $D_\mu\lambda = \partial_\mu\lambda + ig[V_\mu, \lambda]$. The commutator of two transformations as in (3.9) gives a combined translation and Yang-Mills gauge transformation with field-dependent parameter $\omega = 2i\,\bar{\varepsilon}_1\gamma^\mu\varepsilon_2 V_\mu$.

The invariant Yang-Mills Lagrangian is given by $\mathrm{Tr}(W_\alpha W^\alpha)_F$, in which W_α is the scalar multiplet previously introduced. It reduces to

$$\mathrm{Tr}\left(-\frac{1}{4}G_{\mu\nu}^2 - \frac{i}{2}\bar{\lambda}\not{D}\lambda + \frac{1}{2}D^2\right) \tag{3.10}$$

For pure supersymmetric Yang-Mills theories the equation of motion of the auxiliary field D is simply D = 0 and one gets the ordinary Yang-Mills coupling of a massless Majorana spinor in the adjoint representation of the gauge group. The conserved Noether current of supersymmetry is given by

$$J_\mu = -\frac{1}{2}\mathrm{Tr}\left(G_{\nu\rho}\gamma^\nu\gamma^\rho\gamma_\mu\lambda\right)$$

Because of superconformal invariance (see Section 2) there exists in this theory a second conserved current $I_\mu = -\gamma\cdot xJ_\mu$ related to the second charge S_α of the graded conformal algebra. One could add to the previous Lagrangian a supersymmetric and gauge-invariant interaction with n flavour scalar multiplets S^i. The resulting field equations for the auxiliary field D are no longer trivial. For instance, if one adds a scalar multiplet of mass m in the adjoint representation, one gets the additional term:

$$\mathcal{L}_{MATTER} = \mathrm{Tr}\Big(-\tfrac{1}{2}(\mathcal{D}_\mu A)^2 - \tfrac{1}{2}(\mathcal{D}_\mu B)^2 - \tfrac{i}{2}\bar{\chi}\not{D}\chi + \tfrac{1}{2}(F^2+G^2)$$
$$+ g\bar{\lambda}[A+\gamma_5 B, \chi] + ig\,D[A,B] + m(FA + GB - \tfrac{i}{2}\bar{\chi}\chi)\Big) \tag{3.11}$$

Elimination of the auxiliary fields F, G, and D gives the final Lagrangian

$$\mathcal{L}_{MATTER} = \mathrm{Tr}\Big(-\tfrac{1}{4}G_{\mu\nu}^2 - \tfrac{i}{2}\bar{\lambda}\not{D}\lambda - \tfrac{i}{2}\bar{\chi}\not{D}\chi - \tfrac{1}{2}m^2(A^2+B^2)$$
$$- \tfrac{i}{2}m\bar{\chi}\chi + g\bar{\lambda}[A+\gamma_5 B, \chi] + \tfrac{g^2}{2}[A,B]^2\Big) \tag{3.12}$$

Supersymmetric Yang-Mills theories offer examples of theories with scalars which are asymptotically free. For the previous theory with a gauge group SU(N) and n flavours, the Callan-Symanzik function is given, in the one-loop approximation, by

$$\beta(g) = -\frac{g^3}{16\pi^2}(3-n)N \tag{3.13}$$

The theory is asymptotically free if $n < 3$. If $n = 3$, $\beta(g) = 0$. A remarkable phenomenon happens if, for $n = 3$, one adds an additional self-coupling to the scalar multiplet to make the theory SU(4) symmetric and therefore invariant under an extended $N = 4$ supersymmetry algebra[13]. For that theory, $\beta(g)$ has been shown to vanish also at the second loop[14], and it is not impossible that such a theory could provide the first example of an exactly scale (conformal) invariant interaction in four dimensions.

4. SUPERGRAVITY AND TENSOR CALCULUS FOR LOCAL SUPERSYMMETRY

Supergravity[2)] is the supersymmetric extension of general relativity; it enlarges the gauge group of Einstein transformations to a graded gauge group in which a new fermionic invariance is present. This new fermionic symmetry can be regarded as the curve-space generalization of the flat-space symmetry of the massless Rarita-Schwinger equation. It is the appropriate gauge principle for describing, in a consistent way, interacting massless spin $3/2$ particles. From a more general point of view, supergravity can be regarded as the gauge theory of a graded Lie algebra, the supersymmetry algebra, in much the same way as the Einstein theory can be viewed as the gauge theory of the Poincaré group, and Yang-Mills theories as gauge theories of an internal symmetry group. From a more physical point of view, the basic quantum mechanical concepts of fermionic spin and the connection between spin and statistics enter into the formulation of gravitational theories for the first time. A tensor calculus which provides the construction of multiplets and invariant actions in local symmetry is now available. An obvious, but invaluable, consequence is that one can now obtain actions for complicated systems by simply adding invariant subactions, without the tiresome order-by-order Noether constructions which were needed previously. Heart-warming calculations involving complicated Fierz identities, Bianchi identities, high-precision γ-matrix algebra, possible only to supergravity practitioners, are finally circumvented.

The simplest version of a supergravity theory is the gauging of the N = 1 supersymmetry algebra which is given by Eq. (2.2) for i = 1.

The fundamental Lagrangian of pure supergravity (without coupling to sources), as was originally obtained[15)], is

$$\mathcal{L}_{SG} = -\frac{e}{2K^2} R(e,\hat{\omega}) - \frac{1}{2}\bar{\Psi}_\mu \gamma_5 \gamma_\nu D_\rho \psi_\sigma \, \varepsilon^{\mu\nu\rho\sigma} \qquad (4.1)$$

The covariant derivative is

$$D_\rho \psi_\sigma = \partial_\rho \psi_\sigma + \frac{1}{2} \hat{\omega}_{\rho ab}(e,\psi) \sigma^{ab} \psi_\sigma - \Gamma^{\nu}_{\rho\sigma}(g) \psi_\nu \tag{4.2}$$

$\hat{\omega}_{\rho ab}$ is the spin-connection with spin $3/2$ torsion:

$$\hat{\omega}_{\rho ab}(e,\psi) = \omega_{\rho ab}(e) + K_{\rho ab}(e,\psi) \tag{4.3}$$

where

$$\omega_{\rho ab}(e) = \frac{1}{2}\left[e^{\sigma}_{a}(\partial_\rho e_{b\sigma} - \partial_\sigma e_{b\rho}) + e^{\mu}_{a} e^{\nu}_{b} (\partial_\nu e_{c\mu}) e^{c}_{\rho}\right] - (a \leftrightarrow b)$$

$$K_{\rho ab}(e,\psi) = \frac{\kappa^2}{4}\left(\bar{\psi}_\rho \gamma_a \psi_b - \bar{\psi}_\rho \gamma_b \psi_a + \bar{\psi}_a \gamma_\rho \psi_b\right) \tag{4.4}$$

$K_{\rho ab}$ being the contorsion tensor.

$\mathcal{L}_{SG}$ describes the (locally) supersymmetric interaction of the massless graviton of helicity states $\lambda = \pm 2$ with the massless gravitino of helicity states $\lambda = \pm 3/2$. From the point of view of the symmetry, the Lagrangian (4.1) describes the self-interaction of the $(2, 3/2)$ massless multiplet of global supersymmetry (see Table 1).

Coordinate invariance and local supersymmetry are necessary requirements in order that the fields $e_{a\mu}$ and ψ_μ describe the interaction of physical degrees of freedom only. The Lagrangian $\mathcal{L}_{SG}$ in Eq. (4.1) is invariant (up to a 4-divergence) under the following field variations*):

---

*) From now on we use a different γ-matrix convention than in the previous sections, namely $g_{\mu\mu} = (1, 1, 1, 1)$ and 0 otherwise, $\gamma_\mu^2 = 1$, $\gamma_5^2 = 1$.

$$\delta e_{a\mu} = \kappa \bar{\epsilon} \gamma_a \psi_\mu$$

$$\delta \psi_\mu = \frac{2}{\kappa} D_\mu \epsilon \qquad \left(D_\mu = \partial_\mu + \frac{1}{2} \hat{\omega}_{\mu ab} \sigma^{ab} \right) \tag{4.5}$$

This is unsatisfactory because one would then expect that

$$[\delta_{\epsilon_1}, \delta_{\epsilon_2}] = \delta_G + \delta_L + \delta_S \tag{4.6}$$

where δ_G, δ_L, and δ_S are general coordinate transformations, local Lorentz transformations, and local supersymmetry transformations. Instead, from (4.5) one gets that the commutator of the local transformations of parameters $\varepsilon_1(x)$, $\varepsilon_2(x)$ gives, when acting on ψ_μ,

$$[\delta_{\epsilon_1}, \delta_{\epsilon_2}] \psi_\mu = (\delta_G + \delta_L + \delta_S) \psi_\mu + \text{"eqs. of motion terms"} , \tag{4.7}$$

where "eqs. of motion terms" vanish only on the spin-$3/2$ shell, i.e. if

$$\varepsilon^{\mu\nu\rho\sigma} \gamma_5 \gamma_\nu D_\rho \psi_\sigma = 0 \tag{4.8}$$

This phenomenon already occurred in global supersymmetry when fields such as F and G in Eq. (3.1) and D in Eq. (3.9) played the role of auxiliary fields. *"La seule raison d'être"* of these fields is to make the algebra closed. The same phenomenon takes place in supergravity. The complete set of potentials needed to close the local algebra is given by the following multiplet[16]:

$$(e_{a\mu}, \psi_\mu, S, P, A_\mu) \tag{4.9}$$

where S, P, A_μ are a scalar, a pseudoscalar, and an axial vector field, respectively. These fields are the auxiliary fields of supergravity.

The local supersymmetry variations for the gauge potentials with a closed algebra are the following ($R^\mu = e^{-1}\varepsilon^{\mu\nu\rho\sigma}\gamma_5\gamma_\nu D_\rho\psi_\sigma$)

$$\begin{aligned}
\delta e_{a\mu} &= \kappa\bar{\epsilon}\gamma_a\psi_\mu \\
\delta\psi_\mu &= \frac{2}{\kappa}D_\mu\epsilon + \frac{1}{3}\gamma_\mu(S - i\gamma_5 P)\epsilon + i\left(A_\mu - \frac{1}{3}\gamma_\mu \not{A}\right)\gamma_5\epsilon \\
\delta P &= -\frac{i}{2}\bar{\epsilon}\gamma_5\gamma\cdot R - \frac{1}{2}\kappa\bar{\epsilon}\gamma\cdot\psi P + \frac{i}{2}\kappa\bar{\epsilon}\gamma_5\gamma\psi S + \frac{1}{2}\kappa\bar{\epsilon}A\cdot\psi \\
\delta S &= \frac{1}{2}\bar{\epsilon}\gamma\cdot R - \frac{1}{2}\kappa\bar{\epsilon}\gamma\cdot\psi S - \frac{i}{2}\kappa\bar{\epsilon}\gamma_5\gamma\cdot\psi P + \frac{i}{2}\kappa\bar{\epsilon}\gamma_5 A\cdot\psi \\
\delta A_\mu &= \frac{3}{2}i\bar{\epsilon}\gamma_5\left(R_\mu - \frac{1}{3}\gamma_\mu\gamma\cdot R\right) + \frac{\kappa}{2}\bar{\epsilon}\psi_\mu P + \frac{i}{2}\kappa\bar{\epsilon}\gamma_5\psi_\mu S - \\
&\quad - \frac{\kappa}{2}\bar{\epsilon}\gamma\cdot\psi A_\mu + \kappa\bar{\epsilon}\not{A}\psi_\mu - \frac{\kappa}{4}e\,\varepsilon_{\mu\nu\rho\sigma}A^\nu\bar{\epsilon}\gamma_5\gamma^\rho\psi^\sigma
\end{aligned} \tag{4.10}$$

The Lagrangian which is invariant under (4.10) is

$$\begin{aligned}
\mathcal{L}_{SG} &= -\frac{e}{2\kappa^2}R(e,\hat{\omega}) - \frac{1}{2}\bar{\psi}_\mu\gamma_5\gamma_\nu D_\rho\psi_\sigma\varepsilon^{\mu\nu\rho\sigma} + \\
&\quad + \frac{e}{3}\left(g^{\mu\nu}A_\mu A_\nu - S^2 - P^2\right)
\end{aligned} \tag{4.11}$$

From (4.10) one indeed gets a closed algebra

$$[\delta_{\epsilon_1}, \delta_{\epsilon_2}] = \delta_S(\epsilon') + \delta_G(\xi^\mu) + \delta_L(\omega^{ab}) \tag{4.12}$$

with

$$\begin{aligned}
&\epsilon' = -\kappa\bar{\epsilon}_2\gamma^\mu\epsilon_1\psi_\mu\ , \quad \xi^\mu = 2\bar{\epsilon}_2\gamma^\mu\epsilon_1\ , \quad \omega^{ab} = 2\bar{\epsilon}_2\gamma^\rho\epsilon_1\hat{\omega}_\rho^{ab} - \\
&- \frac{2}{3}i\kappa\bar{\epsilon}_2\gamma_c\epsilon_1 A_d\varepsilon^{abcd} - \frac{4}{3}\kappa\bar{\epsilon}_1\sigma^{ab}\epsilon_2 S - \frac{4}{3}i\kappa\bar{\epsilon}_2\sigma^{ab}\gamma_5\epsilon_1 P
\end{aligned}$$

In pure supergravity the field equations for the auxiliary fields are $S = P = A_d = 0$, and one re-obtains the original Lagrangian (4.1) and the "Stony Brook algebra", which no longer closes unless $R_\mu = 0$.

We make the following observation. Inspection of the basic group property (4.12) shows that the commutator algebra contains both vector $\bar{\varepsilon}_1\gamma^\mu\varepsilon_2$ and tensor $\bar{\varepsilon}_1\sigma^{ab}\varepsilon_2$ structures. Tensor terms $\bar{\varepsilon}_1\sigma^{ab}\varepsilon_2$ are proportional to the S, P fields. If S and P cannot be eliminated, the commutator algebra that we have introduced seems not to be equivalent to the Breitenlohner commutator algebra[17] (at least in the present formulation) because of the lack, in the latter, of any tensor terms whatsoever.

It is useful to make a comparison between the different algebraic schemes that have been constructed in order to give a geometrical structure to supergravity.

In the Breitenlohner scheme there are 126 = 9 × 14 potentials in the basic multiplet. These potentials are just the field content of a Lie algebra valued vector multiplet

$$V = V_a P^a + \bar{V}_\alpha Q_\alpha + V_{ab} M^{ab} \tag{4.13}$$

over the 14-dimensional graded Poincaré algebra. It is clear that (4.13) implies a first-order formulation of supergravity because the spin-connection is treated as an independent potential. De Wit and Grisaru[18] have been able to reduce the 126 potentials to 54 by making a second-order formulation of the Breitenlohner scheme. There are still 22 auxiliary fields in this formulation. The formulation that we have been discussing in this section makes use of only 38 potentials (4.9), of which only 6 (S, P, A_μ) are auxiliary. Of course our formulation is also of the second order in the sense that the spin-connection is not an independent dynamical variable.

It is worth while mentioning that the Breitenlohner potentials are related to the superspace formulation that has been pursued by Brink, Gell-Mann, Ramond and Schwarz[19], while the present formulation is related to the superspace formulation of Wess and Zumino[20] and Volkov, Akulov and Soroka[21]. These formulations certainly overlap whenever the over-all theory is superconformal, because in that case S and P can be gauged away, or if supergravity is coupled to conformal matter, because in that case S = P = 0 are consistent constraints.

With the completed gauge potential multiplet, one can now formulate a tensor calculus for local supersymmetry[22] in much the same way as it was formulated in global supersymmetry (see Section 3).

The transformation laws of a scalar multiplet with components $Z = A + iB$, $\chi_L = \frac{1}{2}(1 + \gamma_5)\chi$, $H = F + iG$, becomes (we put $k = 1$ from now on):

$$\begin{aligned} \delta Z &= 2\bar{\epsilon}_L \chi_L \\ \delta \chi_L &= \not{\hat{\partial}} Z \epsilon_R + H \epsilon_L \\ \delta H &= 2\bar{\epsilon}_R \not{\hat{\partial}} \chi_L - \tfrac{2}{3}\bar{\epsilon}_L \chi_L u - \tfrac{i}{3}\bar{\epsilon}_R \not{A} \chi_L \end{aligned} \tag{4.14}$$

For a vector multiplet in the "Wess-Zumino" gauge we have:

$$\begin{aligned} \delta V_\mu &= -\bar{\epsilon}\gamma_\mu \lambda \\ \delta \lambda &= \sigma_{\mu\nu} \hat{F}^{\mu\nu} \epsilon + i\gamma_5 D \epsilon \\ \delta D &= i\bar{\epsilon}\gamma_5 \not{\hat{\partial}} \lambda - \tfrac{i}{2}\bar{\epsilon}\not{A}\lambda \end{aligned} \tag{4.15}$$

where $u = S - iP$ and curly derivatives $\mathcal{D}_\mu$ are supercovariant derivatives:

$$\begin{aligned}
\mathcal{D}_\mu z &= \partial_\mu z - \bar{\psi}_{\mu L} \chi_L \\
\mathcal{D}_\mu \chi_L &= D_\mu \chi_L - \frac{1}{2} \not{\mathcal{D}} z \, \psi_{\mu R} - \frac{1}{2} H \psi_{\mu L} \\
\hat{F}_{\mu\nu} &= \partial_\mu V_\nu - \partial_\nu V_\mu + \frac{1}{2} \bar{\psi}_\mu \gamma_\nu \lambda - \frac{1}{2} \bar{\psi}_\nu \gamma_\mu \lambda \\
\mathcal{D}_\mu \lambda &= D_\mu \lambda - \frac{1}{2} \hat{F}_{ab} \sigma^{ab} \psi_\mu - \frac{i}{2} \gamma_5 D \psi_\mu
\end{aligned} \tag{4.16}$$

Equations (4.14) and (4.15) reduce to the global case (3.1) and (3.9) for vanishing potentials.

Supercovariant derivatives are constructed in the same way as ordinary covariant derivatives in Yang-Mills theories or in any other gauge theory. Moreover, if the vector multiplet is a Yang-Mills multiplet, the supercovariant derivatives contain also Yang-Mills connections.

Scalar and pseudoscalar densities are given by

$$\begin{aligned}
& eF + \frac{e}{2} \bar{\psi} \cdot \gamma \chi + \frac{e}{2} \bar{\psi}_\mu \sigma^{\mu\nu} (A - i\gamma_5 B) \psi_\nu + e(AS + PB) \\
& eG - i\frac{e}{2} \bar{\psi} \cdot \gamma \gamma_5 \chi + i\frac{e}{2} \bar{\psi}_\mu \sigma^{\mu\nu} (B + i\gamma_5 A) \psi_\nu + e(SB - PA)
\end{aligned} \tag{4.17}$$

Again Eq. (4.17) reduces to the flat density F(G) for vanishing potential.

The kinetic multiplet T(A, B, χ, F, G) has components

$$A_T = F + \frac{1}{3}(SA + BP), \quad B_T = -G - \frac{1}{3}(BS - AP)$$

$$\chi_T = \not{\mathcal{D}}\chi - \frac{i}{6}\not{A}\gamma_5\chi - \frac{1}{3}(A - i\gamma_5 B)\eta$$
$$F_T = \tilde{\Box} A \qquad (4.18)$$
$$G_T = -\tilde{\Box} B$$

where $\tilde{\Box}$ is a supercovariant d'Alembertian and η is defined below.

One can also construct scalar multiplets which are functions of the gauge potentials only.

These multiplets vanish in the flat limit but they are crucial for describing the pure supergravity part of the action. Of particular relevance is the scalar curvature multiplet $\tilde{R}$ with components

$$A_{\tilde{R}} = S \,, \quad B_{\tilde{R}} = P$$
$$\chi_{\tilde{R}} = \eta = \frac{1}{2}(\gamma\cdot R - \gamma\cdot\psi\, S - i\gamma_5\gamma\cdot\psi\, P + i\gamma_5 A\cdot\psi)$$
$$F_{\tilde{R}} = \frac{1}{2}R + \frac{1}{2}\bar{\psi}_\mu\gamma^\nu\gamma^\mu R_\nu + \frac{1}{4}\bar{\psi}_\mu(S - i\gamma_5 P)\psi^\mu \qquad (4.19)$$
$$+ \frac{i}{4}\bar{\psi}\cdot A\gamma_5\gamma\cdot\psi - \frac{2}{3}(S^2 + P^2 + \frac{1}{2}A_\mu^2)$$
$$G_{\tilde{R}} = -e^{-1}\partial_\mu(eA^\mu) + \frac{i}{2}\bar{\psi}_\mu\gamma_5 R^\mu - \frac{i}{2}\bar{\psi}_\mu\sigma^{\mu\nu}\gamma_5 R_\nu$$
$$+ \frac{i}{4}\bar{\psi}_\mu\gamma_5(S - i\gamma_5 P)\psi^\mu + \frac{1}{4}\bar{\psi}\cdot A\gamma\cdot\psi$$

The supergravity Lagrangian $\mathcal{L}_{SG}$ given by Eq. (4.11) is the scalar density constructed with $-\tilde{R}$. The locally supersymmetric action for a scalar multiplet in the improved form is the scalar density constructed with $\frac{1}{2}\Sigma \otimes T(\Sigma)$.

The Yang-Mills locally supersymmetric action is the scalar density constructed with $\bar{W}^\alpha(V) \otimes W_\alpha(V)$. A useful identity is $T(I) = {}^1\!/_3\tilde{R}$. Here $I = (1, 0, 0, 0, 0)$ is the identity multiplet.

5. SPONTANEOUS SYMMETRY-BREAKING AND SUPER-HIGGS MECHANISM

Spontaneous symmetry-breaking is one of the most outstanding problems encountered in building models for unified interactions in the framework of supersymmetry invariance.

Exact supersymmetry assigns equal masses to bosons and fermions and, apart from some massless particles such as the photon, graviton, and neutrino, Nature does not look at all supersymmetric. In order not to spoil the nice renormalizability properties of supersymmetric theories, the breaking mechanism has to be spontaneous. When global supersymmetry is spontaneously broken, according to general arguments, a massless Goldstone fermion arises in the mass spectrum[11,23]. For spontaneously broken N-extended supersymmetry, one would expect N massless fermions.

In a unified theory with global supersymmetry a necessary condition for spontaneous breaking is that one of the auxiliary fields F, G, D of the basic scalar and vector multiplets takes a non-vanishing value at the minimum of the potential[23]. This is because then there is a spin ½ Majorana field for which

$$\delta\lambda = \frac{1}{a}\epsilon + \text{q-number terms} . \tag{5.1}$$

This transformation law in turn implies that the symmetry charge Q_α does not annihilate the vacuum.

An example of Lagrangian with spontaneous breaking of supersymmetry is the Volkov-Akulov[8] Lagrangian

$$\mathcal{L}_{VA} = -\frac{1}{2a^2}\det\left(\delta_\mu^\nu + i a^2 \bar{\lambda}\gamma^\nu\partial_\mu\lambda\right) \tag{5.2}$$

in which the Goldstone fermion λ transforms non-linearly under supersymmetry

$$\delta\lambda = \frac{1}{a}\epsilon + ia\bar{\epsilon}\gamma^{\mu}\lambda\partial_{\mu}\lambda \tag{5.3}$$

If supersymmetry is promoted to a local invariance a super-Higgs effect can occur, namely the goldstino (Goldstone fermion) can be eaten by the massless gravitino, the gauge fermion of local supersymmetry, after which banquet it becomes massive. This is nothing but the supersymmetric version of the Higgs-Kibble mechanism of ordinary gauge theories.

Deser and Zumino[24] have studied the coupling of the Volkov-Akulov Lagrangian to de Sitter supergravity and have shown that, under the assumption of a vanishing cosmological term, there is a relationship between the mass of the gravitino and the constant a which measures the breaking of the symmetry:

$$m^2_{\psi} = \frac{1}{6a^2} \tag{5.4}$$

Using the established tensor calculus[22] for supergravity, the super-Higgs effect has recently been investigated[25] for general interactions of the scalar multiplet $\Sigma = (A, B, \chi, F, G)$ coupled to supergravity.

The most general interaction for a scalar multiplet Σ with second-order derivative Lagrangian is given by the following expression:

$$\sum_{nm} a_{nm}\Sigma^{n}\otimes T(\Sigma^{m}) + \sum_{n} b_n \Sigma^{n} \tag{5.5}$$

If we denote by $\Phi = \sum_{nm} a_{nm} z^n\bar{z}^m$, $g = \sum_n b_n z^n$ $(z = A + iB)$, this interaction depends on two arbitrary functions $\Phi(z, \bar{z})$ and $g(z)$. If $\mathcal{L}$ is assumed to be parity conserving, then a_{nm}, b_n are real and, moreover, a_{nm} are symmetric.

The Lagrangian, in its form (5.5), has the following starting terms:

$$\mathcal{L} = \frac{e}{6}\Phi R - e\Phi_{,z\bar{z}}\partial_\mu z \partial_\mu \bar{z} - e\Phi_{,z\bar{z}}\bar{\chi}\not{D}\chi$$

$$+\frac{1}{6}\Phi\bar{\psi}_\mu\gamma_5\gamma_\nu D_\rho\psi_\sigma\varepsilon^{\mu\nu\rho\sigma} + \ldots \quad (5.6)$$

Under certain regularity conditions, namely

$$\Phi(z,\bar{z}) < 0\ ,\quad (\log - \Phi)_{,z\bar{z}} < 0 \quad (5.7)$$

it is possible to perform a Weyl rescaling on the vierbein field $e_{a\mu}$ and a field redefinition on the fermion fields ψ_μ, χ such that the final Lagrangian takes a simple and compact form.

These field redefinitions are

$$-\Phi/3 = e^{-2\lambda}\ ,\quad \mathcal{G} = \log(-\Phi/3)^3/|g|^2\ ,\ g \neq 0$$

$$e^N_{a\mu} = e_{a\mu}\, e^{-\lambda}$$

$$z^N = z$$

$$\chi^N_L = \left(\frac{g}{g^*}\right)^{\frac{1}{4}}\chi_L\sqrt{-2\mathcal{G}_{,z\bar{z}}}\; e^{\lambda/2} \quad (5.8)$$

$$\psi^N_{\mu L} = \left(\frac{g}{g^*}\right)^{-\frac{1}{4}}\psi_{\mu L}\, e^{-\lambda/2} - \frac{2}{\sqrt{-2\mathcal{G}_{,z\bar{z}}}}\gamma^N_\mu\chi^N_L\lambda_{,\bar{z}}$$

In terms of the new variables and after elimination of the auxiliary fields S, P, A_μ, F, G, the Lagrangian (5.5) takes the form

$$\mathcal{L} = \mathcal{L}_{SG} + \mathcal{L}_{MATTER} \quad (5.9)$$

$\mathcal{L}_{SG}$ is the supergravity Lagrangian (4.1) and

$$\mathcal{L}_{MATTER} = \mathcal{L}_B + \mathcal{L}_F$$

where

$$\mathcal{L}_B = e\, G_{,z\bar{z}}\, \partial_\mu z \partial_\mu \bar{z} + e^{-G}\left(3 + \frac{|G_{,z}|^2}{G_{,z\bar{z}}}\right) \qquad (5.10)$$

$$\begin{aligned} \mathcal{L}_F = & -\frac{e}{2}\bar{\chi}\,\not{D}\,\chi + e\, e^{-G/2}\Big[\bar{\psi}_\mu \sigma^{\mu\nu}\psi_\nu - \\ & - (-2G_{,z\bar{z}})^{-1/2}\,\bar{\psi}\cdot\gamma\,\hat{G}_{,z}\,\chi + \\ & + (2G_{,z\bar{z}})^{-1}\,\bar{\chi}\left(\frac{\hat{G}_{,zz\bar{z}}\hat{G}_{,z}}{G_{,z\bar{z}}} + \hat{G}_{,z}^2 - \hat{G}_{,zz}\right)\chi \end{aligned} \qquad (5.11)$$

+ four-fermion terms

+ two-fermion terms containing $\partial_\mu z$, $\partial_\mu \bar{z}$.

In Eq. (5.11) the terms which have not been written explicitly are not relevant for the super-Higgs effect.

The Lagrangian (5.9) as well as the transformation laws of the new fields contain only the function G. This means that if we redefine the two functions ϕ and g by means of the transformation

$$g(z) \to g(z)h(z) \quad , \quad \Phi(z,\bar{z}) \to \Phi(z,\bar{z})\left(h(z)h(\bar{z})\right)^{1/3}$$

we do not change the Lagrangian.

The conditions for the occurrence of spontaneous symmetry breaking are

$$
\begin{aligned}
&V_{,z}(z_0) = 0 && \text{in which } z_0 \text{ is an absolute minimum} \\
&V(z_0) = 0 && \\
&e^{-\mathcal{G}/2}\,\mathcal{G}_{,\bar z} \neq 0 && \text{at } z = z_0 .
\end{aligned}
\tag{5.12}
$$

The first condition implies the existence of a minimum, the second condition the absence of a cosmological term, the third condition the presence of a constant term in the transformation law of χ. By observing that

$$
-V(z,\bar z) = e^{-\mathcal{G}}\left(3 + \frac{|\mathcal{G}_{,z}|^2}{\mathcal{G}_{,z\bar z}}\right)
\tag{5.13}
$$

from the first two conditions we get

$$
\begin{aligned}
&|\mathcal{G}_{,z}|^2 = -3\,\mathcal{G}_{,z\bar z} \\
&\mathcal{G}_{,z\bar z}\left(\mathcal{G}_{,zz}\mathcal{G}_{,\bar z} + \mathcal{G}_{,z\bar z}\mathcal{G}_{,z}\right) - \mathcal{G}_{,z}\mathcal{G}_{,\bar z}\,\mathcal{G}_{,zz\bar z} = 0 \quad \text{at } z = z_0 .
\end{aligned}
\tag{5.14}
$$

It is easy to see that at $z = z_0$, after a chiral transformation

$$
\tilde\chi = \left(\frac{\hat{\mathcal{G}}_{,z}}{\hat{\mathcal{G}}_{,\bar z}}\right)^{1/2}\chi
$$

the bilinear fermion terms appear in a universal form

$$\mathcal{L}_F^{MASS} = e^{-\mathcal{G}_0/2}\left[\bar{\psi}_\mu \sigma^{\mu\nu}\psi_\nu - \sqrt{\tfrac{3}{2}}\,\bar{\psi}\cdot\gamma\,\tilde{\chi} - \bar{\tilde{\chi}}\,\tilde{\chi}\right] \tag{5.15}$$

Finally the substitution

$$\hat{\psi}_\mu = \psi_\mu - \frac{1}{\sqrt{6}}\gamma_\mu \tilde{\chi} - \sqrt{\tfrac{2}{3}}\, e^{\mathcal{G}_0/2}\,\partial_\mu \tilde{\chi} \tag{5.16}$$

eliminates $\tilde{\chi}$ altogether and leaves only the kinetic and mass term for $\hat{\psi}_\mu$ with mass

$$m_\psi = e^{-\mathcal{G}_0/2} \qquad\qquad \mathcal{G}_0 = \mathcal{G}\Big|_{z=z_0}$$

In the class of models for which $G_{z\bar{z}} = -\frac{1}{2}$ everywhere [$\phi = -3e^{-\frac{1}{6}z\bar{z}}$, g(z) arbitrary], one gets a universal mass formula

$$m_A^2 + m_B^2 = 4m_\psi^2 \tag{5.17}$$

Formula (5.17) is independent of the input potential g(z) which originated the spontaneous symmetry breaking. An explicit example in which the conditions (5.12) are fulfilled is given by $g(z) = \lambda(z + b)$ (λ arbitrary and $b = 2\sqrt{2} - \sqrt{6}$). Then $z_0 = \sqrt{6} - \sqrt{2}$ and $m_A^2 = 2\sqrt{3}\,m_\psi^2$, $m_B^2 = 2(2 - \sqrt{3})m_\psi^2$.

The analysis carried out in this section shows the very important fact that the super-Higgs effect is a real phenomenon which can occur in local supersymmetry. Moreover, the mass formula that we have derived gives a bound on the mass of the gravitino in terms of the mass of the scalar particles.

Formula (5.17) is interesting because it could be of real interest in more complicated models describing unified theories of fundamental interactions.

6. EXTENDED SUPERGRAVITY AND MODELS UNIFYING ELECTRODYNAMICS AND CHROMODYNAMICS WITH GRAVITATION

Extended supergravity theories[2] arise from the gauging of an extended supersymmetry algebra described by Eqs. (2.2). The basic gauge multiplet has a particle content dictated by Table 1. These theories have the property of unifying gravity with interactions of particles with lower spin. If $N \geq 4$, scalar, spin ½, and spin 1 particles appear in the same multiplet of the graviton.

The simplest example of these theories is the gauging of $N = 2$ extended supergravity[26]. This theory can be regarded as the unification of Maxwell theory with gravity. The photon sits in the same multiplet as the graviton, together with two spin $^3\!/_2$ Majorana fields.

It is amazing to observe that this theory improves the usual Maxwell-Einstein theory because it gives finite one- and two-loop corrections to scattering processes. For example, in the non-supersymmetric Maxwell-Einstein system, photon-photon scattering is one-loop divergent, but it turns out to be finite in the supersymmetric version[2,3].

Theories[27] with $N \geq 3$ have a global non-Abelian symmetry SO(N). However, in the presence of the gravitational constant, only, the SO(N) group is not gauged. The $N(N - 1)/2$ spin 1 gauge potentials gauge an Abelian group $U(1)^{N(N-1)/2}$, the group of the central charges. The SO(N) group can be gauged[28] provided one adds to the Lagrangian a mass-like term for the gravitino with mass $m_\psi = \sqrt{2}(g/k)$ and a cosmological term with strength $6(g/k^2)^2$. Here g is the SO(N) Yang-Mills coupling constant.

The reason for this mechanism is easily understood by noting that with a cosmological term present, the graded group is no longer IO(3, 1) but rather the de Sitter group O(3, 2). As discussed in Section 2, the internal symmetry part of an extended graded O(3, 2) algebra is precisely SO(N).

Apart from the problem of interpretation of the cosmological term which one could think to be absent in a realistic situation in which spontaneous symmetry-breaking occurs, these theories have the appearance of truly unified theories of particle interactions in which an underlying Yang-Mills gauge group SO(N) is present. It is not clear at present whether these theories can be used for current particle phenomenology. The main difficulty lies in the fact that the biggest gauge group which can be realized in a Lagrangian field theory is SO(8), which is too small to accommodate the minimal gauge group of unified weak, strong and electromagnetic interactions which is $SU(3) \otimes U(2)$.

Nevertheless, the SO(8) group contains the SU(3) group, and it can be regarded as a unified theory of gravity with quantum chromodynamics.

Let us now take a closer look at the Gell-Mann classification[6)] of the particle states of SO(8) supergravity with respect to their SU(3)-colour content.

The eight spin $^{3}/_{2}$ gravitinos split into $3 \oplus \bar{3} \oplus 1 \oplus 1$ of $SU_c(3)$. Electric charges are assigned to be $Q = -^{1}/_{3}, ^{1}/_{3}, 0, 0$, and the rest of the analysis follows without any further assumption. The adjoint representation of SO(8) to which the spin 1 gauge bosons belong is split into: $8(0) \oplus 1(0) \oplus 1(0) \oplus 3(-^{1}/_{3}) \oplus 3(-^{1}/_{3}) \oplus 3(^{2}/_{3}) \oplus \bar{3}(^{1}/_{3}) \oplus$ $\oplus \bar{3}(^{1}/_{3}) \oplus \bar{3}(-^{2}/_{3})$. The charge assignment is given in brackets.

The particles can be identified with the gluons, photons, Z, heavy, and superheavy vector bosons which already occur in other attempts to unify non-gravitational interactions. The 56 spin $\frac{1}{2}$ Majorana particles break up into Dirac spinors as follows:

$28 = 3(2/3) \oplus 3(-1/3) \oplus 3(-1/3) \oplus 3(2/3) \oplus 6(1/3) \oplus 8(0) \oplus 1(-1) \oplus 1(0)$. These particles can be identified with the u, d, s, c quarks, a fifth quark flavour, a neutral octet (heavy leptons), and the electron and its neutrino. As one can see, the missing particles are at least the charged vector bosons $W^{\pm}$, the muon and its neutrino, and the τ.

7. CONCLUSIONS

In the present survey of supersymmetry and its application to Lagrangian field theory and to fundamental particle interactions, we have tried to cover the classification of particle states according to this new symmetry, the construction of the basic Lagrangians, and the description of a new gauge principle -- local supersymmetry -- which appears to be the first convincing framework for unifying all the interactions, including gravitation.

We have also discussed, in some detail, some of the recent technical developments of the field, such as the tensor calculus for local supersymmetry and the study of the super-Higgs effect. Because of lack of space and time we have not, however, been able to cover other important results and to talk about the progress which has been made in the fields, including the superspace formulation of globally and locally supersymmetric theories and the renormalization situation in supergravity.

Another important development has been in the study of local supersymmetry from the point of view of differential geometry, using the fibre bundle approach[29]. It is worth while mentioning also the study of the anomalies, such as the axial anomaly and the stress tensor anomaly in several supergravity models[30].

Recent progress in the tensor calculus for supergravity, with the extension of the density formulae to vector multiplets, has also been made[31].

Finally, we would like also to mention the construction of superconformal theories[3], which arise from the gauging of the graded conformal algebra and which can be regarded as the supersymmetric extension of Weyl gravity. Very recently, more general (sometimes non-local) supersymmetries have been proposed in the context of source theory[32].

ACKNOWLEDGEMENT

I would like to thank Prof. B. Zumino for many enlightening discussions on supersymmetry.

REFERENCES

1) Reviews of supersymmetry and extensive lists of references can be found in:
B. Zumino, Proc. 17th Internat. Conf. on High Energy Physics, London, 1974 (Science Research Council, Didcot, 1974), p. I-254.
J. Wess, Lecture Notes in Physics (Springer, Berlin, 1974), vol. 37, p. 352.
L. Corwin, Y. Ne'eman and S. Sternberg, Rev. Mod. Phys. 47, 573 (1975).
A. Salam and J. Strathdee, Phys. Rev. D 11, 521 (1975).
L. O'Raifeartaigh, Comm. Dublin Inst. for Advanced Studies, Series A (Theor. Phys.), Nr. 22 (1975).
B. Zumino, Proc. Conf. on Gauge Theories and Modern Field Theory, Boston, 1975 (MIT Press, Cambridge, Mass., 1976) p. 255.
A. Salam and J. Strathdee, ICTP preprint IC/76/22 (1976).
S. Ferrara, Rivista Nuovo Cimento 6, 105 (1976).
P. Fayet and S. Ferrara, Phys. Rep. 32, 251 (1977).

2) Reviews of supergravity and references can be found in:
S. Deser, Lectures given at the Symposium on Differential Geometrical Methods in Mathematical Physics, Bonn, 1977.
D.Z. Freedman, P. van Nieuwenhuizen and S. Ferrara, Proc. 18th Internat. Conf. on High Energy Physics, Tiflis, 1976 (JINR, Dubna, 1977), vol. II, p. T34.
P. van Nieuwenhuizen, Proc. Symposium on Asymptotic Structure of Space-Time, Cincinnati, 1976 (Plenum Press, New York, 1977), p. 407.
B. Zumino, Proc. 17th Scottish Universities Summer School, Edinburgh, 1976 (SUSSP Publications, Edinburgh, 1977), p. 549.

D.Z. Freedman, P. van Nieuwenhuizen and S. Ferrara, Proc. Annual Meeting of the Amer. Phys. Soc. (Dept. of Particles and Fields), Brookhaven Nat. Lab., 1976 (Report BNL 50598, Upton, 1977), p. E1.

In Deeper pathways in high-energy physics, Proc. Orbis Scientiae, Coral Gables, 1977 (eds. A. Perlmutter and L.F. Scott), (Plenum Press, New York, 1977), see contributions of R. Arnowitt (p. 179), D.Z. Freedman (p. 205), M. Grisaru and P. van Nieuwenhuizen (p. 233), and B. Zumino (p. 259).

B. Zumino, Proc. Internat. Conf. on Particle Physics, Budapest, 1977 (Central Research Inst. for Physics, Budapest, 1978), vol. II, p. 1209.

D.Z. Freedman, preprint CalTech-68-625 (1977); Proc. Annual Meeting of the Amer. Phys. Soc. (Dept. of Particles and Fields), Argonne Nat. Lab., 1977, to be published.

J. Wess, Proc. 8th GIFT Internat. Seminar in Theoretical Physics, Salamanca, 1977 (ed. J.A. de Azcárraga), (Springer, Berlin, 1978), p. 81.

R. Arnowitt, Talk given at the 5th Conf. on Experimental Meson Spectroscopy, Boston, 1977, Proc. to be published.

S. Ferrara, Proc. 6th Internat. Colloquium on Group Theoretical Methods in Physics, Tübingen, 1977 (Lecture Notes in Theoretical Physics, vol. 79, Springer, Berlin, 1978), p. 22.

Reviews for amateurs can be found in:

G.B. Lubkin, Physics Today, under "Search & Discovery", 30, No. 6, 17 (June 1977).

Scientific American, under "Science and the Citizen", 237, No. 1, 59 (July 1977).

J. Scherk, La Recherche 8, No. 82, 878 (October 1977).

D.Z. Freedman and P. van Nieuwenhuizen, Scientific American 238, No. 2, 126 (February 1978).

3) See, for instance, P. van Nieuwenhuizen, preprint CERN TH.2473 (1978), invited talk at the Orbis Scientiae, Coral Gables, 1978.

4) P. Fayet, Proc. CNRS National Colloquium, Montpellier, May 1977, to be published.
P. Fayet, preprint CalTech-68-641 (1978), talk given at the Orbis Scientiae, Coral Gables, 1978.

5) G. Farrar and P. Fayet, Phys. Letters 76B, 575 (1978).
T. Goldman, Los Alamos report LA-UR-78-1104 (1978).

6) M. Gell-Mann, invited paper at the Washington Meeting of the Amer. Phys. Soc., April 1977; and to be published.

7) P.G.O. Freund, Univ. Chicago report EFI 77-58, invited talk given at the Conf. on Nonassociative Algebras in Physics, Univ. Virginia (Charlottesville), 1977.

8) Y.A. Gol'fand and E.P. Likhtam, JETP Letters 13, 323 (1971).
D.V. Volkov and V.P. Akulov, Phys. Letters 46B, 109 (1973).
J. Wess and B. Zumino, Nuclear Phys. B70, 39 (1974).
R. Haag, J.T. Lopuszanski and M. Sohnius, Nuclear Phys. B88, 257 (1975).

9) A. Salam and J. Strathdee, Nuclear Phys. B80, 499 (1974) and B84, 127 (1975).
W. Nahm, Nuclear Phys. B135, 149 (1978).

10) A. Salam and J. Strathdee, Nuclear Phys. B76, 477 (1974).
S. Ferrara, J. Wess and B. Zumino, Phys. Let. B 51, 239 (1974).

11) J. Iliopoulos and B. Zumino, Nuclear Phys. B76, 310 (1974).

12) S. Ferrara and B. Zumino, Nuclear Phys. B79, 413 (1974).
A. Salam and J. Strathdee, Phys. Letters 51B, 353 (1974).
B. de Wit and D.Z. Freedman, Phys. Rev. D 12, 2286 (1975).

13) F. Gliozzi, J. Scherk and D. Olive, Nuclear Phys. B122, 253 (1977).
L. Brink, J.H. Schwarz and J. Scherk, Nuclear Phys. B121, 77 (1977).

14) D.R.T. Jones, Phys. Letters 72B, 199 (1977).
E.C. Poggio and H.N. Pendleton, Phys. Letters 72B, 200 (1977).

15) D.Z. Freedman, P. van Nieuwenhuizen and S. Ferrara, Phys. Rev. D 13, 3214 (1976).
S. Deser and B. Zumino, Phys. Letters B62, 335 (1976).

16) S. Ferrara and P. van Nieuwenhuizen, Phys. Letters 74B, 333 (1978).
K.S. Stelle and P.C. West, Phys. Letters 74B, 330 (1978).

17) P. Breitenlohner, Phys. Letters 67B, 49 (1977) and Nuclear Phys. B124, 500 (1977).

18) B. de Wit and M. Grisaru, Phys. Letters 74B, 57 (1978).

19) L. Brink, M. Gell-Mann, P. Ramond and J. Schwarz, Phys. Letters 74B, 336 (1978) and report Phys. Letters 76B, 417 (1978).

20) J. Wess and B. Zumino, Phys. Letters 66B, 361 (1977).
R. Grimm, J. Wess and B. Zumino, Phys. Letters B73, 415 (1978).
J. Wess and B. Zumino, Phys. Letters 74B, 51 (1978).

21) V.P. Akulov, D.V. Volkov and V.A. Soroka, JETP Letters 22, 396 (1975).

22) S. Ferrara and P. van Nieuwenhuizen, Phys. Letters 76B, 404 (1978); and Ecole Normale Sup. report LPTENS 78/14 (1978).
S. Ferrara, P. van Nieuwenhuizen and M. Grisaru, Nuclear Phys. B138, 430 (1978).

23) A. Salam and J. Strathdee, Phys. Letters 49B, 465 (1974).
P. Fayet and J. Iliopoulos, Phys. Letters 51B, 461 (1974).
L. O'Raifeartaigh, Nuclear Phys. B96, 331 (1975).

24) S. Deser and B. Zumino, Phys. Rev. Letters 38, 1433 (1977).

25) E. Cremmer, B. Julia, J. Scherk, P. van Nieuwenhuizen, S. Ferrara and L. Girardello, Ecole Normale Sup. report LPTENS 78/17 (1978).

26) S. Ferrara and P. van Nieuwenhuizen, Phys. Rev. Letters 37, 1669 (1976).

27) D.Z. Freedman, Phys. Rev. Letters 38, 105 (1977).
S. Ferrara, J. Scherk and B. Zumino, Phys. Letters 66B, 35 (1977).
A. Das, Phys. Rev. D 15, 2805 (1977).
E. Cremmer, J. Scherk and S. Ferrara, Phys. Letters 68B, 234 (1977).
E. Cremmer and J. Scherk, Nuclear Phys. B127, 259 (1977).
B. de Wit and D.Z. Freedman, Nuclear Phys. B130, 105 (1977).
E. Cremmer, B. Julia and J. Scherk, Phys. Letters 76B,409 (1978).

28) D.Z. Freedman and A. Das, Nuclear Phys. B120, 221 (1977).

29) S.W. MacDowell and F. Mansouri, Phys. Rev. Letters 38, 739 (1977).
P.K. Townsend and P. van Nieuwenhuizen, Phys. Letters B67, 439 (1977).
Y. Ne'eman and T. Regge, Phys. Letters B76, 54 (1978).

30) See, for instance, N.K. Nielsen, M.T. Grisaru, H. Römer and P. van Nieuwenhuizen, preprint CERN TH.2481 (1978), and references therein.

31) K.S. Stelle and P.C. West, report ICTP/77-78/15, to be published in Phys. Letters B, and ICTP/77-78/24.

32) J. Schwinger, UCLA preprint 78/Tep/11 (May 1978).

DISCUSSION

CHAIRMAN: S. Ferrara

Scientific Secretaries: K.L. Giboni, A.D. Kennedy, S. Templeton

DISCUSSION No. 1

- THIRRING:

A cancellation between boson and fermion contributions is needed in gravitation for the vacuum expectation value of the energy-momentum tensor. There is always a positive zero point energy for bosons and a negative zero point energy for fermions, and one wants these to cancel because otherwise one would get a large cosmological constant - in fact they cancel to a fantastically good accuracy. Is this something that would be guaranteed by supersymmetry?

- FERRARA:

If one has exact supersymmetry, it immediately follows that the vacuum expectation value of the energy-momentum tensor is exactly zero.

- THIRRING:

This is for an exact symmetry; but if it's broken, things are not so good.

- FERRARA:

You don't know whether it is broken. The stress tensor is obtained through a supersymmetry rotation of the spinor current and if the spinor charges annihilate the vacuum, its vacuum expectation value is zero. Consequently, there is no induced cosmological term in any exact supersymmetric theory.

- HA:

Can one extend the idea of supersymmetry to the domain of

nuclei and construct a phenomenological theory of nuclear spectra?

- FERRARA:

In principle, supersymmetry can apply to any system, there is no reason why it should be a symmetry particularly relevant to particle physics. In any system with bosons and fermions which are almost degenerate in mass, and which seem to fit into a multiplet structure, one could think of applying supersymmetry.

- HA:

That means that fermions and bosons don't have to be fundamental particles, they could be composite ones.

- ZICHICHI:

So the conclusion was that you don't need fundamental particles and fundamental interactions in order to create supersymmetries? This disturbs me very much.

- FARRAR:

There is no reason to expect it to be supersymmetric, but if you look it might be.

- KLEINERT:

If you take a nucleus with a single degenerate unfilled shell, and you proceed to fill that shell, then indeed there is a slightly broken supersymmetry which describes the levels of this nucleus. A supermultiplet consists of the system with 1, 2, ... nucleons in the shell up to a full shell. Pickup and stripping reactions proceed via a fermionic charge just as weak interactions proceed via I^+ or I^-. It has to be decided by comparison with data to what extent supersymmetry can be found also in the interactions.

- ZICHICHI:

It is like asking whether chemistry would obey SU(3) symmetry: it could do it, but why?

- FARRAR:

I think the thing making supersymmetry interesting for particle physics is that the actual Lagrangian is invariant under the symmetry, not just that one can organize the states. I would be astonished if the interaction would have that symmetry in the case of a nucleus.

- HA:

If gauge invariance is consistent with supersymmetry, both spin ½ fermions as well as vector bosons can be gauge particles. What are these spin ½ gauge particles for the strong, weak and electromagnetic interactions?

- FERRARA:

In a sense they are gauge particles because they are the same superparticle - they are two different states of the same supermultiplet. In some sense one can say that one is unifying matter and radiation. For example in supersymmetric Q.C.D. one has an octet of coloured gluons and an octet of massless spin ½ Majorana particles, which are called gluinos. These are only massless if supersymmetry is unbroken; however, if you want to apply supersymmetry to the real world in a realistic way, it must be spontaneously broken. In this case, provided the colour symmetry remains unbroken, the gluons will remain massless but the gluinos may acquire a mass.

- ZICHICHI:

As the gluinos are also coloured, why don't they likewise remain massless due to SU(3) colour symmetry?

(The gluons must remain massless since they are the gauge bosons of the colour symmetry, whereas the gluinos are not. They transform as separate multiplets under colour transformations, but into each other under the supersymmetry transformations which are spontaneously broken.)

- FARRAR:

Another similar example is that for every quark there must be a scalar quark, which must become massive because they are not seen. However, I will be discussing these points in my lectures.

- GROSSE:

In a Yang-type Lagrangian in which you unify spin 0 with spin 2 you get higher than second derivatives. There is the Velo-Zwanziger phenomenon implying that sometimes higher order equations of motion admit wave solutions travelling faster than light. Does this happen in supersymmetric theories?

- FERRARA:

In normal supergravity one finds the usual kinetic energy term having no higher than second derivatives. However, in Weyl invariant (Superconformal) gravity one finds that the spin two particle propagates with a quartic derivative and the spin 3/2 one with a cubic derivative.

- GROSSE:

Why do we always use Majorana spinors?

- FERRARA:

The essential reason for the basic building block being the Majorana spinor is because of the hermiticity property of the anti-commutator. On the right hand side the momentum is hermitian, and you can play around with the γ-matrices to show that the spinors Q on the left hand side must be self-conjugate. Of course this is no restriction, since one can have as many spinors as one wishes.

- BHANOT:

Can you give an intuitive argument why the Coleman-Mandula and O'Raifeartaigh no-go theorems do not apply to Graded Lie Algebras?

- FERRARA:

For a long time people tried to combine internal symmetries with Poincaré invariance, and eventually these no-go theorems were proved as purely mathematical theorems, and their physical content is unclear.

- BLASI:

The no-go theorems state that one can only combine internal symmetries with Poincaré invariance trivially (as a direct product) if one tries to do this at one space-time point. This is not the case in supersymmetry as it involves neighbouring points.

- FERRARA:

This is so because one has a derivative in the supersymmetric transformation; it is easy to see that this comes about because the product of two supersymmetry transformations is a translation, and the generator of translations is a derivative.

- KENNEDY:

Why can't you introduce this convenient derivative for a Lie algebra?

- FERRARA:

Many people have tried this, but none have succeeded; the basic problem is that one cannot do this with a finite number of fields, and this is one of the assumptions of the Coleman theorem.

- KLEINERT:

Maybe it is because we never tried to put a derivative on the right hand side; and let us not forget that the no-go theorem was really derived for the purpose of trying to get a multiplet with different masses into a combined representation of Poincaré and internal group, and the proof showed that the masses would have to be equal so that relativistic SU(6) would never work out. In supersymmetric theories the masses are still degenerate.

- FERRARA:

Yes, but the spins are different.

- ROTH:

If certain infinities can be eliminated by generalizing the Lie algebras to Graded Lie Algebras, why not look at more generalized algebras to try to eliminate the remaining infinities?

- FERRARA:

Supersymmetry is a good example of the fact that when you go to a richer algebraic structure you can get unexpected benefits. In principle there is no reason to suppose that there shouldn't be richer structures, but no one has found them yet.

- SCHELLEKENS:

Do supersymmetry cancellations also occur in supersymmetric φ^3 theory in such a way that it is finite, and if so to what order has this been checked?

- FERRARA:

It is not finite, but it is renormalizable. It also preserves the supersymmetry to all orders.

- THIRRING:

Is it finite in one space one time dimension?

- FERRARA:

Some people in Dublin have proved it to be.

- BENGTSSON:

Would you care to elaborate somewhat on the differences between particles under SO(7) and SO(8), since the number of particles with a given helicity seems to be the same in the two representations.

- FERRARA:

Several multiplets are identical, such as those with N = 7 and 8 and $\lambda_{max} = 2$, or with N = 3 and 4 and $\lambda_{max} = 1$. They are exactly the same; if you start with N = 3 you automatically build up a multiplet with higher symmetry.

- NILSSON:

Why is the largest symmetry you consider SO(8), why don't you extend this to SO(10)?

- FERRARA:

First of all one gets 10 gravitons, which we don't like because they are gauge particles and we would have to find an invariance for each of them, and we only have one momentum to gauge. Thus this would be very difficult, unless you believe in the strong gravity of Salam and Strathdee. No one has been able to construct a consistent theory containing particles of spin greater than 2, and we would need a spin 5/2 particle for SO(10). Spin 1 particles are usually associated with gauging internal symmetries, spin 3/2 with gauging spin ½ supersymmetry generators, and spin 2 with gauging the Poincaré group. A spin 5/2 particle would have to be associated with gauging a spin 3/2 supersymmetric generator, and this has been ruled out in a paper by Haag, Lopuszanski and Sohnius.

- ZICHICHI:

There is only one charge that we really understand in nature and that is the gravitational charge, which corresponds to the curvature of space-time. In these supersymmetric theories how are we to understand the role of the other charges which are important in nature, can one obtain these charges as some kind of generalized curvature in the space on which supersymmetry is based? What should our physical intuition be? Einstein died without understanding the connection between electric charge and mass because he could not believe there to be something absolutely irreconcilable with the concept of space-time.

- FERRARA:

If one takes the view advocated by Gell-Mann that there is only one fundamental multiplet in nature which contains all fundamental particles, then the maximum symmetry one can have is an SO(8) gauge symmetry. In this one has a collection of particles, one has the graviton and 8 spin 3/2 gravitinos. The graviton is the gauge particle related to the Poincaré group. The gravitinos are gauging the 8 supersymmetric charges. You have 28 spin 1 bosons which are the gauge particles of the 28 dimensional SO(8) internal symmetry group. This theory has only 2 coupling constants, the gravitational charge and the dimensionless gauge coupling constant which only arises in curved space with cosmological term.

- ZICHICHI:

What if I want to work out the strong, weak and other coupling constants?

- FERRARA:

These come from the 2 coupling constants through spontaneous symmetry breaking. This theory is not sufficient for this unification because SO(8) does not contain $SU(3)_{colour} \otimes SU(2)_{weak} \otimes U(1)_{em}$. If you do a reduction of SO(8) into SU(3) as Gell-Mann did, you find that you can predict the charges of the quarks; this is the first model in which you can theoretically predict that the quarks possess fractional charges of the correct values.

- ZICHICHI:

Without cheating? Genuine?

- FERRARA:

Yes.

- KENNEDY:

I want to ask a question about the connection between Supergroups and ordinary Lie groups. I find it hard to visualize a manifold in which some of the variables correspond to anticommuting elements of a Grassmann algebra; how do the group elements correspond to finite transformations when one can only expand to some finite order in the anticommuting parameters? To what extent do Lie's theorems and their converses hold for supergroups?

- FERRARA:

One cannot visualize any physical transformations which realize the supersymmetry transformations because of the fermionic nature of the charges. Nevertheless, a supergroup is a mathematically well defined object. Although the fermionic charges are not observable due to their spinorial nature, they give rise to physically observable effects in amplitudes for scattering processes such as

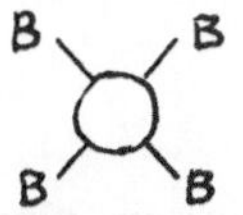

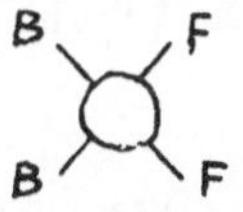

Almost all the theorems that hold for Lie algebras can be extended to Graded Lie algebras; there has been an extensive investigation

of the structure of Graded Lie groups by Sternberg and Konstant at M.I.T. and Harvard. There is an analogue of the Cartan classification of all compact semi-simple Lie algebras for Graded Lie algebras.

- FELTESSE:

Do you violate the principle of causality with this space-time symmetry which transforms fields at different points into each other?

- FERRARA:

No, because the transformation involves only a finite number of derivatives.

The following result holds in any supersymmetric theory and has not been proved for any other type of theory: by taking the relation

$$\{Q_\alpha, \bar{Q}_\beta\} = -2\gamma^\mu_{\alpha\beta} P_\mu$$

and multiplying both sides by γ and taking the trace one can also show that the energy is positive definite as

$$H = P^0 = \sum_\alpha Q^2_\alpha \geqslant 0$$

or at least non-negative. Therefore the supersymmetric theory is a local theory which looks better than the usual one.

- NILSSON:

There is a supergravity theory based on the conformal group, can you get the Einstein theory out of this?

- FERRARA:

Conformal supergravity is completely invariant under local conformal transformations and it has a dimensionless gravitational constant. Thus the theory is completely different from Einstein gravity which has a dimensional coupling constant. There is one connection which is that any solution of the vacuum Einstein equations $R_{\mu\nu} = 0$ is also a solution of the conformally invariant Weyl theory. The action for the Weyl theory is

$$\int d^4x \left(R^2_{\mu\nu} - \tfrac{1}{2} R^2\right)\sqrt{|g|}$$

- KENNEDY:

That looks rather like the Yang-Stephenson theory which was introduced to make gravitation look more like the Yang-Mills theory.

- FERRARA:

Since $\int (R_{\mu\nu\rho\lambda})^2$ can be written as a linear combination of $\int R^2$ and $\int R_{\mu\nu}^2$, all of these theories are similar.

- ROTH:

Is supergravity two loop finite or just two loop renormalizable?

- FERRARA:

It is two loop finite.

- ROTH:

Is there hope that it's going to be finite to all orders or renormalizable to all orders?

- FERRARA:

The hope is that it will prove to be finite to all orders - it is not renormalizable in the sense that the Green's functions are infinite. What happens is that when you go onto mass shell all the infinities cancel, it is finite in the sense of S-matrix elements just like the ordinary Einstein theory without matter. This comes about, because the coefficients of the counter-terms vanish when you use the field equations.

- ROTH

I don't understand. If you have counter terms then you are doing renormalization, aren't you?

- FERRARA:

The theory is not renormalizable as you get an infinite number of counter terms, but as these all vanish on mass shell (at least to the two loop level) all this is unimportant.

- KENNEDY:

The meaning of renormalizability is unclear for such Lagrangians, in any case, as we have a non-polynomial interaction due to the presence of the $\sqrt{|g|}$ in the tensor density.

- FERRARA:

One only has a finite number of terms up to the two loop level, but the theory certainly is not renormalizable.

- ANASTAZE:

If we decompose the 56-plet of Majorana spinors into Dirac spinors you miss some particles such as the muon, the muon neutrino and the tau. Is it possible to explain this, and is this consistent with what we know about elementary particles?

- FERRARA:

If you insist on a scheme in which all particles sit in a single SO(8) multiplet and you believe in Q.C.D. then there are missing particles. This scheme is unsatisfactory from this point of view. At present I don't know of any way out other than putting them into more than one multiplet.

DISCUSSION No. 2 (Scientific Secretaries: K.L. Giboni, A.D. Kennedy, S. Templeton)

- BHANOT:

I would like to ask you about your model with gluons and gluinos. Could you please comment on the independent degrees of freedom and on the ghosts that are needed?

- FERRARA:

All the Lagrangians we have written are classical Lagrangians, if you want to quantize them you must add Fadeev-Popov ghosts; and then for supersymmetric gauge theories follow the usual renormalization procedure.

- BHANOT:

I thought these were not renormalizable but finite.

- FERRARA:

No, the supersymmetric Yang-Mills theory is renormalizable but not finite; only supergravity is one and two loop finite but not renormalizable. It is worth noticing that for example in the supersymmetric Yukawa theory supersymmetry leads us to expect that the wave function, mass and coupling constant renormalizations are each common to all members of the multiplet. Furthermore, the dynamics leads to further relations between the renormalization constants $M_R = Z\, M_o$, $g_R = Z^{3/2} g_o$, where Z is the wave function renormalization. We see that there is only one basic divergence in the theory.

- BHANOT:

Is this a consequence of the generalized Ward identities?

- FERRARA:

No, they are a consequence of the dynamical structure of the theory and the topology of the Feynman diagrams corresponding to the particular form of the Lagrangian chosen.

- JACOB:

You gave a beautiful review of supersymmetry as it now stands after several years of development. Could you please itemize those topics which are particularly "hot" at present and give us an idea of where the field is presently progressing?

- FERRARA:

The prospects for supersymmetry are of two different kinds, firstly technical developments and secondly there are developments which are endeavouring to construct realistic models. In the latter case I refer you to Glennys Farrar's lectures. Putting aside supergravity for the moment, one would like to construct supersymmetric versions of Q.C.D., and of the weak and electromagnetic interactions: in doing this the problem is having a proliferation of particles. All the common particles have partners of different spin, for example the quarks have scalar partners and the gluons have fermionic partners, the gluinos. The problem is to find definite predictions which can be tested experimentally.

The most ambitious aim is to make a finite theory of gravitation, which requires putting matter particles all into the same multiplet as the graviton. There is a programme at present trying to get around the problem that the largest internal symmetry allowable is SO(8), as discussed before.

- BLASI:

What is the meaning of "on the mass shell" when applied to the counter terms of supergravity?

- FERRARA:

The situation is the same as in the ordinary Einstein theory of gravity, where in the absence of matter the theory is one loop renormalizable due to some miraculous cancellations (as was first shown by 't Hooft and Veltman). At one loop order the graviton interaction of pure gravity is given by the following diagram

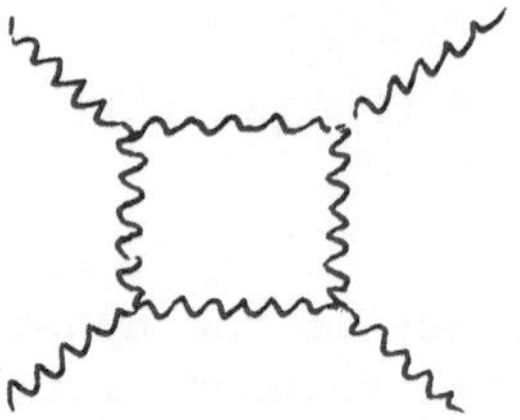

for which the counter terms are of the form R^2, $R^2_{\mu\nu}$, $R^2_{\mu\nu\rho\sigma}$. By "on the mass shell" we mean that we use the Einstein equations of motion to show that these must vanish. In the absence of matter the field equations state that the Einstein tensor $G_{\mu\nu} = R_{\mu\nu} - \frac{1}{2}g_{\mu\nu} R$ vanishes, hence R and $R_{\mu\nu}$ also vanish. Further, by means of the Gauss-Bonnet theorem we find that $R^2_{\mu\nu\rho\sigma}$ is a total divergence and hence gives no contribution.

The argument generalizes to the case of supergravity, where one can have matter present for example in the form of gravitinos. To summarize, by "on mass shell" we mean that the counter terms vanish when the field equations of the theory are used - this is all we mean when saying that the theory is finite.

- BLASI:

Are the two loop counter terms invariant under supersymmetry transformations?

- FERRARA:

Yes, this is necessary to ensure the finiteness at the two loop level. In pure gravity the counter term, which is the product of three Riemann tensors, occurs which causes the theory to be non-finite, however, in supergravity supersymmetry makes sure that no such counter terms can arise. We assume the quantized theory to remain supersymmetric, but this is true only if we have a supersymmetric invariant regularization procedure: however, it is necessary to make also the assumption that the divergence can be written as a local counter term.

- GERHOLD:

Supersymmetry seems to allow one to construct consistent spin 2 theories which was not possible before. Is it possible to extend this to spins higher than 2?

- FERRARA:

In order to construct a theory with a definite spin greater than $\frac{1}{2}$ one must have some gauge principle; for spin 1 this comes from internal symmetries, as one always gets one Lorentz index more than is present on the generators of the symmetry. For spin 3/2 this is exactly the case, which stems from the fermionic supersymmetry transformations, and for the spin 2 case from the Poincaré symmetry. There are no corresponding symmetries for the higher spin cases.

- FARRAR:

A few years ago nobody knew about supersymmetry. Isn't it possible that someone might find a higher symmetry by, for example, generalizing the supersymmetry generators to behave as vector spinors?

- FERRARA:

Unfortunately such a generalization doesn't work. This was proved by Haag, Lopuszanski and Sohnius assuming the usual axioms of quantum field theory. They proved that one cannot have more than spin ½ fermionic generators, because the anticommutator of two pure spin 3/2 parts of the supersymmetry generators must vanish as one only has the momentum to put on the right hand side, and nothing with two Lorentz indices.

- FARRAR:

What if you have a product of momenta or something like that on the right hand side?

- FERRARA:

One cannot do this, because the anticommutators must be linear in the generators in order to preserve the algebraic structure. (One could include products but these would be new generators and the algebra would not close.)

- GERHOLD:

A similar problem arises when one considers high energy scattering mediated by a high spin particle. In this case the problem was circumvented by Reggeizing the exchange particle. Is it possible to Reggeize supersymmetric theories in an analogous fashion?

- FERRARA:

I can't make any general statements, but if you want a supersymmetric theory to Reggeize you must have supersymmetric Regge trajectories. This means that for each value of the mass the multiplicity and spins must go as a supersymmetry multiplet.

It is interesting to note that this is exactly what happens in the Ramond-Schwartz dual model; indeed it was just this model which was generalized to give us the concept of supersymmetry.

- LACKNER:

My question concerns the superspace introduced by Salam and Strathdee. There are both commuting and anticommuting coordinates, and by supersymmetry transformations I can add products of anti-commuting numbers - which are commuting but nilpotent - to the ordinary coordinates x^μ. How does one interpret this?

- FERRARA:

I think it would be useful to give a brief explanation of what superspace is.

Supersymmetry tells us that we do not live in Minkowski space but in a "graded" space $Y = (x_\mu, \theta^i_\alpha)$, which is at least 8 dimensional and in general 4 + 4N dimensional. If we consider how the operators of the algebra

$$[P_\mu, Q_\alpha] = 0$$

$$\{Q_\alpha, \bar{Q}_\beta\} = -2\gamma^\mu_{\alpha\beta} P_\mu$$

act on this space, we find for normal translations with parameter a_μ

$$x'_\mu = x_\mu + a_\mu$$

$$\theta'_\alpha = \theta_\alpha ,$$

whilst for supertranslations parameterized by ϵ we obtain

$$x'_\mu = x_\mu + i\bar{\epsilon}\gamma_\mu \theta$$

$$\theta'_\alpha = \theta_\alpha + \epsilon_\alpha .$$

For the commutator of two such supertranslations we find

$$[\delta_1, \delta_2]\, x_\mu = 2i\bar{\epsilon}_1\gamma_\mu\epsilon_2$$

$$[\delta_1, \delta_2]\, \theta_\alpha = 0,$$

where the quantity $a_\mu = 2i\bar{\epsilon}_1\gamma_\mu\epsilon_2$ is nilpotent. In this approach we construct superfields on superspace rather than consider multiplets of ordinary fields on Minkowski space. It is convenient to replace the Majorana spinor coordinates by two Weyl spinors:

$$\theta_\alpha = \begin{pmatrix} \theta_\alpha \\ \bar{\theta}^{\dot{\alpha}} \end{pmatrix} \qquad \begin{matrix} \alpha = 1,2 \\ \dot{\alpha} = 1,2 \end{matrix}$$

where $\bar{\theta}_{\dot{\alpha}} = \epsilon_{\dot{\alpha}\dot{\beta}} \bar{\theta}^{\dot{\beta}}$.

We now define a superfield $\phi(Y) = \phi(x,\theta_\alpha) = \phi(x,\theta_\alpha,\bar{\theta}^{\dot{\alpha}})$ as a field defined over all the coordinates of superspace. We can easily relate this to fields defined over Minkowski space by performing a formal Taylor expansion in the Weyl components, and because these coordinates belong to an anticommuting Grassmann algebra the expansion terminates after a finite number of terms - for the case of N = 1 we end up with 2^4 fields as follows:

$$\begin{aligned}\phi(x,\theta_\alpha,\bar{\theta}^{\dot{\alpha}}) = {} & C(x) + \chi^\alpha(x)\theta_\alpha + S_{\dot{\alpha}}(x)\bar{\theta}^{\dot{\alpha}} + \\ & + \theta^\alpha\theta_\alpha M(x) + \bar{\theta}_{\dot{\alpha}}\bar{\theta}^{\dot{\alpha}} N(x) + \theta_\alpha \sigma^{\mu\alpha\dot{\alpha}} \bar{\theta}_{\dot{\alpha}} V_\mu(x) + \\ & + \theta^\alpha\theta_\alpha\bar{\theta}^{\dot{\alpha}}\lambda_{\dot{\alpha}}(x) + \bar{\theta}_{\dot{\alpha}}\bar{\theta}^{\dot{\alpha}}\theta_\alpha\xi^\alpha(x) + \theta_\alpha\theta^\alpha\bar{\theta}_{\dot{\alpha}}\bar{\theta}^{\dot{\alpha}} D(x) = \sum_n \phi^{\alpha_1\ldots\alpha_n}(x)\,\theta_{\alpha_1}\ldots\theta_{\alpha_n}\end{aligned}$$

Unfortunately we have too many fields here, they do not correspond to a single supermultiplet, but we can reduce the number by restricting the coordinate $\bar{\theta}_\alpha$ to be zero. In this case only three terms survive:

$$\phi(x,\theta_\alpha,0) = C(x) + \chi^\alpha(x)\theta_\alpha + \theta^\alpha\theta_\alpha M(x)$$

This is called a chiral superfield, and we see that if we impose the condition $\bar{D}_{\dot{\alpha}}\phi = 0$ where $\bar{D}_{\dot{\alpha}}$ is the usual way of writing the derivative in the spinorial basis, then

$$\phi_1(x,\theta,\bar{\theta}) = \phi_1(x,\theta)$$

$$\phi_{chiral}(x,\theta,\bar{\theta}) = \phi_1(x + i\theta\sigma\bar{\theta},\theta),$$

and for the superfield S:

$$S(x,\theta,\bar{\theta}) = S(x + i\theta\sigma\bar{\theta},\theta)$$

we obtain the three non-vanishing terms

$$S(x,\theta) = \underset{\downarrow \atop A + iB}{\mathcal{A}(x)} - \underset{\downarrow \atop \frac{1}{2}(1+i\gamma_5)\chi}{\chi_\alpha(x)}\theta^\alpha + \underset{\downarrow \atop F - iG}{\mathcal{F}(x)}\theta^\alpha\theta_\alpha$$

which we identify with the complex fields indicated. Actually we still have too many fields, but we find that the fields F and G do not correspond to physical degrees of freedom, but merely serve as auxiliary fields; it would be possible to eliminate them from the theory, but it is easier to keep them so that we can work in a linear representation of supersymmetry. The situation is therefore somewhat analogous to that of Einstein gravitation, where we can work either with the physical fields themselves as in the original Hilbert formulation or with the auxiliary field Γ , the connection, as in the Palatini method which makes the variational problem linear.

- NILSSON:

Has one been able to eliminate all the auxiliary fields that are automatically introduced in the superfield formalism?

- FERRARA:

Yes, this has indeed been done. It is easy to see how this happens, if we consider the scalar chiral multiplet (A, B, χ, F, G) which transforms under a supersymmetry transformation ϵ as:

$$\delta A(x) = i\bar{\epsilon}\chi(x)$$
$$\delta B(x) = i\bar{\epsilon}\gamma_5\chi(x)$$
$$\delta\chi(x) = \partial_\mu[A(x) - \gamma_5 B(x)]\gamma^\mu\epsilon + [F(x) + \gamma_5 G(x)]\epsilon$$
$$\delta F(x) = i\bar{\epsilon}\not{\partial}\chi(x)$$
$$\delta G(x) = i\bar{\epsilon}\gamma_5\not{\partial}\chi(x).$$

We can immediately check that these fields do indeed form a representation of the supersymmetry algebra by evaluating the commutators

$$(\delta_1\delta_2 - \delta_2\delta_1)A(x) = [\delta_1, \delta_2]A(x) =$$
$$-2i\bar{\epsilon}_1\gamma^\mu\epsilon_2\partial_\mu A(x) \qquad \text{etc.}$$

We see therefore that (A, B, χ, F, G) is the superfield of the particle multiplet $(0^\pm, \frac{1}{2})$. If we ask how the fields $(\mathcal{A}, \chi_L, \mathcal{F})$ we had before transform, we find that

$$\delta\mathcal{A} = 2i\bar{\epsilon}_L\chi_L$$
$$\delta\chi_L = \not{\partial}\mathcal{A}\epsilon_R + \mathcal{F}\epsilon_L$$
$$\delta\mathcal{F} = 2i\bar{\epsilon}_R\not{\partial}\chi_L$$

so $\mathcal{F}$ is not a particle field, but, as we mentioned before, just an auxiliary field. "La seule raison d'être" of the field $\mathcal{F}$ is to close the algebra. It is worth mentioning that the auxiliary fields, like F, G in the example under consideration, play a crucial role in supersymmetric theories. Although they can be eliminated from any given Lagrangian because their field equations are purely algebraic, it is much more convenient to retain them in order to have linear transformations and closure of the algebra. This last property is decisive in order to have a tensor calculus of general validity, as in any other symmetry, and then simple rules for the construction of invariants. For example we evaluate the tensor product

$$(A_1, B_1, \chi_1, F_1, G_1) \otimes (A_2, B_2, \chi_2, F_2, G_2) = (A_3\ B_3\ \chi_3\ F_3\ G_3)$$

and we find

$$\begin{aligned}
A_3 &= A_1 A_2 - B_1 B_2 \\
B_3 &= A_1 B_2 + A_2 B_1 \\
\chi_3 &= (A_1 - \gamma_5 B_1)\chi_2 + (A_2 - \gamma_5 B_2)\chi_1 \\
F_3 &= F_1 A_2 + F_2 A_1 + G_1 B_2 + G_2 B_1 - i\bar{\chi}_1 \chi_2 \\
G_3 &= G_1 A_2 + G_2 A_1 - F_1 B_2 - F_2 B_1 + i\bar{\chi}_1 \gamma_5 \chi_2
\end{aligned}$$

It has been possible over the last few years to develop a complete tensor calculus for supersymmetry; general rules for the multiplication of multiplets exist and enable us to construct invariant interactions.

- LAURSEN:

Are there infrared problems in supergravity?

- FERRARA:

No, gravitational theories are non-renormalizable because the coupling constant has the dimensions of inverse mass, and such theories behave well in the infrared limit. For the Yang-Mills theory you have infrared problems, because the theory is renormalizable.

- GERHOLD:

Do infra-red divergences remain even in supersymmetric Yang-Mills theory?

- FERRARA:

In supersymmetric theories you have all the usual infra-red problems of ordinary gauge theories.

- ROTH:

Can you state what relation exists between the gravitational and electromagnetic coupling constants in the supergravity unification of gravity and electromagnetism?

- FERRARA:

There is no relation between them. You hope that in a truly unified theory there will be some mass scale to relate them, like the relation between the Fermi constant G for weak interactions and the fine structure constant which is provided by the unification of weak interactions and electromagnetism. For gravity this mass must be of the order of 10^{18} GeV - in fact there are some supergravity theories in which you really encounter this mass. You also come across a mass scale of this size when you try to unify $SU(3)_{colour}$ with $SU(2) \otimes U(1)$.

- ZICHICHI:

But the coupling constant of $SU(2) \otimes U(1)$ is much closer to that of SU(3) than it is to that of gravity.

- FERRARA:

Because the SU(3) and the $SU(2) \otimes U(1)$ theories are renormalizable, the coupling constants only approach each other logarithmically with energy. Hence you must have a very large mass scale to unify them with, but as this mass scale is of the order of the Planck mass it is probably only meaningful to unify them together with gravity.

- ROTH:

You stated that the two loop finiteness of supergravity occurs because there are no counter terms that are supersymmetric to this order. To higher orders you would expect there to be supersymmetric counter terms whose coefficients don't vanish, otherwise it should be simple to show finiteness.

- FERRARA:

In general there will be supersymmetric counter terms to higher orders. For example the three loop terms have been constructed. One would not expect the infinities to cancel due to supersymmetry alone, one must employ the dynamics in addition, just as in the super-Yukawa theory the divergences turned out to be less than expected.

- ROTH:

But don't the extra cancellations occur because of the Ward identities?

- FERRARA:

Yes, but they are Ward identities that come not only from the supersymmetry but also from the particular form of the Lagrangian and the topology of the Feynman diagrams.

- KLEINERT:

The ordinary ϕ^3 is superrenormalizable, so surely the super-ϕ^3 has one infinite renormalization constant less.

- FERRARA:

Although this theory is only ϕ^3 in superspace and so appears to be superrenormalizable it is not really superrenormalizable, since on the expansion of the superfield quartic terms arise.

- GROSSE:

Are there special difficulties due to the Adler-Bell-Jackiw-Bardeen anomalies in supersymmetric models?

- FERRARA:

There are models in which one gets spin 3/2 anomalies in addition to the spin ½ anomaly. Previous calculations disagreed on the coefficients of the anomalous terms, but this was due to different treatments of the quantization schemes.

- KENNEDY:

What is known about the cosmology of supergravity? For example does one get super black holes or things like that?

- FERRARA:

The only comment I can make is about supergravity theories with torsion. To explain torsion, consider what happens when a vector is parallelly transported around an infinitesimal rectangle: the change in a vector can be read from the commutation of two covariant derivatives which symbolically can be written as:

$$[D, D] = R \cdot M + T \cdot D$$

in which R, M and T denote the curvature, Lorentz transformation and torsion, respectively. Thus we have a rotation of the vector proportional to the Riemann curvature tensor and in addition a displacement proportional to the torsion. In such theories there are differing conclusions about black holes and singularities. However, in general this is an interesting area which ought to be studied.

- KENNEDY:

Trautman has suggested that singularities do not occur in non-supersymmetric theories with torsion, such as the Einstein-Cartan theory. However, there has indeed been some disagreement over this.

- FERRARA:

In the usual theory with torsion it is assumed that the torsion is generated by spin ½ particles; if this is so then the torsion is the dual of an axial field (essentially the axial spin ½ current). In supergravity the torsion is due to the spin 3/2 particles, so all the components of the torsion are independent, and it would be interesting to study how this affects the usual results. Certainly the problem ought to be studied.

SUPERSYMMETRY IN NATURE

Glennys R. Farrar

California Institute of Technology

Pasadena, California 91125

I. INTRODUCTION AND MOTIVATION

Nature is certainly Poincaré invariant. However it is possible that her forces possess a much higher degree of space-time invariance than we have heretofore realized, such as invariance under transformations which relate fermions and bosons, so-called supersymmetry transformations[(1)]. As will be explained below, exact supersymmetry of the forces (i.e., of the Lagrangian) as well as of the physical states requires bosons and fermions to come in mass-degenerate multiplets; evidently, nature does not exhibit such behavior. However it is nonetheless possible that nature is supersymmetric but that this is not manifest through degeneracy between bosons and fermions. Such a state of affairs is called spontaneously broken supersymmetry and occurs when the lowest energy state (the vacuum) is not supersymmetric even though the forces are[(2)].

In order to study the question of whether nature does possess this higher symmetry, it is necessary first to demonstrate that supersymmetry can be broken spontaneously[(3,4)], then demonstrate that realistic models with spontaneously broken supersymmetry can be constructed[(5)]— in particular having the correct spectrum of "observed" particles (photon, leptons, gluons, quarks, $W^{\pm}$, Z^{o}), no unobserved particles such as light spin-0 leptons, and correct weak interaction phenomenology, e.g., V & A currents. Finally, we can use such a model to determine where and how to detect evidence for the supersymmetry of the fundamental forces of nature(6,7).

These lectures are organized as follows. Section II shows in a simple but interesting example due to Fayet(4), how spontaneous supersymmetry breaking can arise. This example has the property that, depending on certain parameters, it can either have spontaneously broken supersymmetry with gauge invariance unbroken, or it can have spontaneously broken gauge invariance with supersymmetry unbroken — the supersymmetric Higgs model. Thus while in a realistic model both supersymmetry and weak interaction gauge invariance will be spontaneously broken, we can see here the inherent differences in the spontaneous symmetry breaking phenomenon for these two types of symmetries. A concluding subsection (II.c) explains that the supersymmetric self-interaction of a massive vector multiplet is precisely the supersymmetric Higgs model(4). Although this is not central to our discussion of spontaneous supersymmetry breaking, it is included because the fact that in a supersymmetric theory there is no need for a contrived "Higgs mechanism" in order to have a renormalizable theory of a massive vector particle provides an additional theoretical motivation for finding supersymmetry attractive.

Section III is concerned with the realistic model constructed by Fayet(5). Since these lectures do not assume any expertise in the technical aspects of supersymmetry, the student desiring full details is referred to the original literature. The focus here is on the problems which appear in constructing a realistic model and how they are overcome (Section III.a), and on those physical consequences of the supersymmetry which are most accessible experimentally. Section III.b discusses a new phase invariance, R(8,9), which arises naturally in many cases and which leads to a new conserved additive quantum number, R, or to a conserved R-parity. Section III.c describes the properties and behavior of a new class of hadrons which appear in supersymmetric theories, called R-hadrons because their R-quantum number differs from that of ordinary hadrons[9,6]. Section III.d briefly discusses the properties of the goldstino — the goldstone fermion associated with spontaneous supersymmetry breaking, and the possibilities for detecting the super-Higgs effect which eliminates the goldstino in favor of a massive spin-3/2 particle when supergravity is included[10].

Given this picture of a possible supersymmetric world which corresponds to the real world insofar as has been observed, we show (Section IV) how several experiments already performed give additional constraints on the theory[6,7]. There is so far no positive experimental evidence that nature is supersymmetric. We close (Section V) with a discussion of the implications of the absence of low mass R-hadrons with the properties expected in the models discussed in Section III.

II. SPONTANEOUS SUPERSYMMETRY (OR GAUGE INVARIANCE) BREAKING

When supersymmetry is unbroken, all the particles in each supermultiplet must have equal masses. This follows from the basic structure of the supersymmetry algebra[1]: Q_α, the supersymmetry generator which transforms boson fields into fermion fields and vice versa, is a Majorana[11] spinor satisfying

$$[Q_\alpha, P^\mu] = 0$$
$$\{Q_\alpha, \bar{Q}_\beta\} = -2(\gamma\cdot P)_{\alpha\beta} \quad , \tag{1}$$

where P^μ are the generators of translations. The fundamental interactions of nature are said to be "supersymmetric" if, under a transformation generated by the Q_α, the Lagrangian density changes only by a total divergence. P^2 commutes with all the operators of the algebra so a supersymmetry multiplet is made up of fermions and bosons which have the same mass in the absence of spontaneous supersymmetry breaking. In order to understand how spontaneous supersymmetry breaking works, it is useful to consider a simple model due to P. Fayet[4]. It describes a single massless "gauge" multiplet, V (containing a vector, V_μ, and a Majorana spinor λ), interacting with a left-handed (massless) chiral multiplet, S (containing a left-handed Dirac spinor, ψ_L, and a complex scalar ϕ). In the absence of symmetry breaking all these particles would be massless and the Lagrangian is by construction supersymmetric and gauge invariant. Depending on the relative signs of two parameters, this model exhibits either spontaneous supersymmetry breaking (Section II.a) — giving a gauge invariant model with mass splittings among fermions and bosons — or spontaneous gauge invariance breaking (Section II.b) — giving a Higgs model which is supersymmetric. The ultimate model must exhibit simultaneous supersymmetry and gauge invariance breaking, however it is pedagogically useful to study this simple model in which only one or the other arises. An added bonus is that in this model we can see (Section II.c) how the whole Higgs phenomenon is especially natural from the standpoint of supersymmetry[4].

It is particularly easy to see the supersymmetry and gauge invariance of this abelian gauge model[4] in superfield notation[12]. (If this eludes you, simply take on faith the gauge invariance and supersymmetry of the Lagrangian written below in terms of physical fields (eq. 5).)

$$\mathcal{L} = \mathcal{L}_0 + \xi D + [S^*(\exp 2\ eV)S]_{D\text{-component}} \quad , \tag{2}$$

where $\mathcal{L}_0$ is the free Lagrangian density for the massless superfield V, and e can be taken > 0 without loss of generality. The action from this Lagrangian is invariant under generalized gauge transformations[13], and since it is the "D component" of another superfield, it is also manifestly invariant under supersymmetry transformations[13].

Wess and Zumino showed[13] how to make a partial specification of generalized gauge so as to eliminate all but the V_μ, λ and D components of V, making eq. (2) polynomial in e:

$$\mathcal{L} = [-\frac{1}{4} V_{\mu\nu}V^{\mu\nu} - \frac{i}{2}\bar{\lambda}\not{\partial}\lambda + \frac{D^2}{2}] + \xi D - i\bar{\psi}_L\not{D}\psi_L - (D_\mu\phi)^\dagger D^\mu\phi$$

$$+ \frac{1}{2}(F+iG)^\dagger(F+iG) + ie\sqrt{2}\,[\bar{\psi}_L\lambda\phi+\phi^\dagger\bar{\lambda}\psi_L] + eD\phi^\dagger\phi \quad , (3)$$

where $V_{\mu\nu} \equiv \partial_\mu V_\nu - \partial_\nu V_\mu$ and $D_\mu \equiv \partial_\mu + ieA_\mu$. Here D, F, and G are auxiliary fields(12); requiring the action to be stationary with respect to D, F, and G allows us to eliminate them, e.g., replace

$$D = -(\xi + e\phi^\dagger\phi) \quad , \quad (4)$$

etc., in eq. (3). Then

$$\mathcal{L} = -\frac{1}{4} V_{\mu\nu}V^{\mu\nu} - \frac{i}{2}\bar{\lambda}\not{\partial}\lambda - i\bar{\psi}_L\not{D}\psi_L - (D_\mu\phi)^\dagger D^\mu\phi + ie\sqrt{2}[\bar{\psi}_L\lambda\phi + \phi^\dagger\bar{\lambda}\psi_L]$$

$$-\frac{1}{2}[\xi + e\phi^\dagger\phi]^2 \quad . \quad (5)$$

Section II.a

We want to know whether the supersymmetry of this Lagrangian is spontaneously broken, i.e., whether the minimum of the potential of the scalar field ϕ corresponds to a supersymmetric state or not. How can we tell? One way is simply to solve for the physical states and see if fermion-boson pairs are degenerate or not. (Since the Lagrangian itself is supersymmetric, we know any supersymmetry breaking is spontaneous.) However before doing that we can look for a more general criterion for spontaneous supersymmetry breaking. A necessary condition that supersymmetry is spontaneously broken is that the supersymmetry generator, Q, does not annihilate the vacuum: $Q|0\rangle \neq 0$.

Since $\{Q,\bar{\lambda}\} = \ldots. + \gamma_5 D$, a sufficient condition for supersymmetry breaking is

$$\langle 0|\{Q,\bar{\lambda}\}|0\rangle = \gamma_5 \langle 0|D|0\rangle \neq 0 \quad .$$

Hence, supersymmetry will be broken in this model if and only if the auxiliary field D has a non-vanishing vacuum expectation value (v.e.v.) $\langle D \rangle$. (In more complicated models involving more multiplets, other auxiliary fields may take on non-vanishing v.e.v.'s when supersymmetry is spontaneously broken.) More generally, in global supersymmetry, the anticommutation relation $\{Q_\alpha, \bar{Q}_\beta\} = -2\,(\gamma_\mu)_{\alpha\beta}P^\mu$ implies

$$H = \frac{1}{4} \sum_{\alpha=1}^{4} Q_\alpha^2$$

so that $Q_\alpha |0\rangle \neq 0$ for some α requires that $\langle H \rangle \neq 0$.

Now in this model $\langle D \rangle \neq 0$ (using eq. (4)) requires that $[\xi + e\phi^\dagger\phi]$ must have a non-vanishing v.e.v. at the minimum of the potential. This will happen if $\xi > 0$, in which case the potential is as shown in Fig. 1; and is a minimum for $\phi = 0$. Evidently $\langle \mathcal{U} \rangle = 1/2\,\langle D^2 \rangle \neq 0$ for this configuration. Thus when $\xi > 0$ the Lagrangian (eq. (5)), describes a massless vector V_μ, two massless left-handed Dirac spinors ψ_L and λ_L, and a complex scalar ϕ of mass $\sqrt{e\xi}$: hence we see explicitly the supersymmetry breaking, and the fact that gauge invariance is conserved.

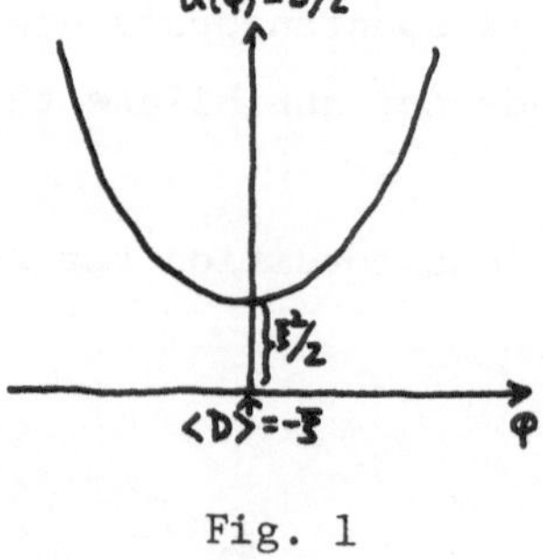

Fig. 1

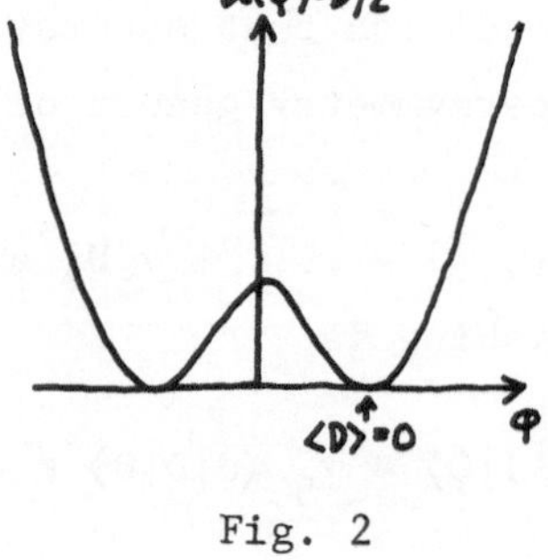

Fig. 2

Section II.b

Compare this to the case when $\xi < 0$, shown in Fig. 2. The potential is minimum when $|\phi| = \sqrt{|\xi|/e}$, so that gauge invariance is spontaneously broken and $\langle D\rangle = -[\xi + e\phi^{\dagger}\phi] = 0$, leading us to expect that supersymmetry will be conserved. To obtain the physical particle content of eq. (5) in this case, we must expand ϕ about its value at the potential minimum. We can rewrite the Lagrangian in order to make its physical content more evident. Let $\phi = -\rho/\sqrt{2}\, e^{i\theta}$ with ρ real and $\langle\rho\rangle \equiv v$ a constant, satisfying $\xi + 1/2\, ev^2 = 0$; now

$$|D_\mu\phi|^2 = \frac{1}{2}\partial_\mu\rho\partial^\mu\rho + \frac{\rho^2}{2}|ev_\mu + \partial_\mu\theta|^2 \quad .$$

However since the Lagrangian is gauge invariant we can rename $eV_\mu + \partial_\mu\theta \equiv eW_\mu$ without changing any physics. Note also that the second to last term generates a mass term of the form $m\bar{E}E$ where $E = \psi_L + \lambda_R$. Finally make the shift $\phi = -(v+B_t)/\sqrt{2}$ so that Lagrangian is written in terms of fields having zero v.e.v. Then

$$\begin{aligned}\mathcal{L} = [&-\frac{1}{4}W_{\mu\nu}W^{\mu\nu} - \frac{e^2v^2}{2}W_\mu W^\mu - i\bar{E}\not{\partial}E - iev\,\bar{E}E - \frac{1}{2}\partial_\mu B_t\partial^\mu B_t \\ &- \frac{1}{2}e^2v^2B_t^{\,2}] - e^2vB_tW_\mu W^\mu - \frac{e^2}{2}B_t^{\,2}W_\mu W^\mu + e\bar{E}_L\not{W}E_L \\ &- ieB_t\bar{E}E - \frac{1}{2}e^2vB_t^{\,3} - \frac{e^2}{8}B_t^{\,4} \quad . \end{aligned} \tag{6}$$

This represents a vector boson W_μ, a spinor E, and a scalar B_t, all of mass m = ev, and their interactions. Hence supersymmetry is conserved whereas gauge invariance is not. This latter is evident both from the absence of a massless vector particle, and from the fact that the physical Dirac particle E, as a superposition of ψ and λ, does not have a definite charge: charge is no longer a good quantum number since gauge invariance is broken. This model is precisely the supersymmetric extension of the Higgs model: with supersymmetry requiring the existence of a fermion partner for the W, and the equality of the W, Higgs (B_t) and fermion (E) masses. Furthermore, the usual

Higgs model has three free parameters whereas its supersymmetric extension has only two, e and m = ev, reflecting the degeneracy of the masses in the supersymmetric case.

Section II.c

This abelian Higgs model can be obtained in a different way[4]. Consider a massive vector supermultiplet W which contains as physical fields the vector W_μ, the Dirac spinor E and the real scalar B_t, all of mass m. The free Lagrangian of this multiplet is

$$\mathcal{L}_0 = -\frac{1}{4} W_{\mu\nu}W^{\mu\nu} - i\bar{E}\partial\!\!\!/ E - \frac{1}{2}\partial_\mu B_t \partial^\mu B_t - \frac{m^2}{2} W_\mu W^\mu - im\bar{E}E - \frac{m^2}{2} B_t{}^2 . \tag{7}$$

If this massive vector multiplet is allowed to interact with itself in a way which does not introduce a four-fermi interaction[4],

$$\mathcal{L} = \mathcal{L}_0 + [\xi W + \frac{v^2}{4} \exp(2eW)]_{D\text{-component}} . \tag{8}$$

Again e can be taken > 0 without loss of generality. Fayet[4] shows that for $\xi < 0$ supersymmetry is conserved and this is equivalent in content to the Higgs model of Section II.b, eq. (6) above.

Restating this result: If supersymmetry had been known before the Higgs mechanism, we still could have arrived at a renormalizable description of weak interactions mediated by massive vector particles. We simply would have considered the most general supersymmetric Lagrangian for a self-interacting massive vector (excluding four-fermi interactions) coupled, if we had wished, to any number of chiral multiplets. We would not have been surprised that such a model was renormalizable, since supersymmetry is known to improve the convergence of field theories[1]. Later some clever person would have rediscovered the underlying gauge invariance of the model; how this emerges is explained in section 5 of Ref. 4.

III. A REALISTIC MODEL

There may be many different ways of realizing supersymmetry in a model which is consistent with nature. For instance, a very ambitious program would be to unify gravity and all other interactions in such a way that all particles are contained in a single supermultiplet. This would involve "extended supersymmetry," where there are N supersymmetry generators Q_α^i, i=1...N, and the supersymmetry transformation changes not only the spin of a field but also its internal quantum numbers. Although such a scheme would be extremely attractive, if it could be made to work out, the only examples of spontaneously broken supersymmetric Lagrangians which are known are for N=1 and N=2, not enough for unification between gravity and internal symmetries.

A more modest approach, taken by Fayet[5,9], is to construct a supersymmetric model which gives an acceptable description of particle physics: weak, electromagnetic and strong interactions, leaving gravity to be coupled later. It is plausible that such a model could give an accurate approximation, in the particle physics sector, of a fully unified theory including gravity. This is the approach we discuss here.

Section III.a

The problems which arise in constructing a more realistic model are substantial, partly because nature has many particles which must be incorporated: gluons, photon, weak gauge bosons, quarks, and leptons. Including their supersymmetry partners leads to the minimal array of particles shown in Table 1, arranged according to the multiplets existing before the breaking of supersymmetry. The major problems to be surmounted in constructing such a realistic model are the following:

i) Nature involves Dirac spinors whereas the algebraic formalism of supersymmetry naturally involves Majorana spinors. How can compatibility be assured? This is intimately connected to the

Table 1: Particle content of the theory with the assignment of R-number indicated in parentheses.

Multiplets	Vectors	Spinors	Scalars
Massless gauge multiplets	Photon (0) Gluons (0)	Photino (1) Gluinos (1)	
Massive gauge multiplets	Intermediate Vector Bosons (0)	Heavy Leptons (± 1)	Higgs scalars (0)
Matter multiplets		Quarks (0) ℓ, ν_ℓ (0) ⋮ ζ (± 1)	Scalar quarks (± 1) Scalar leptons (± 1) ⋮ a,b (± 2)

problem of defining a conserved fermion number in a supersymmetric theory. Although from a pair of Majorana spinors one can always form a ψ_L and ψ_R, for them to form together a single massive Dirac ψ, ψ_L and ψ_R must carry the same quantum numbers!

ii) There are many scalar and pseudoscalar particles in the theory; friends of the W's, quarks, and leptons, which could potentially induce non-V,A "weak" interactions. How can such undesirable exchanges be avoided?

iii) Supersymmetry breaking is much more difficult here than in the simple abelian model discussed in Section II, since the Lagrangian now involves many more multiplets and their auxiliary fields. Instead of a simple expression such as $D = -(\xi + e\phi^\dagger\phi)$ which must be forced to be non-zero to assure spontaneous supersymmetry breaking, one has a set of equations giving each auxiliary field as a constant plus a polynomial of degree two in the physical spin-0 fields, so that there is in general a solution for the $\langle\phi_i\rangle$ for which all the auxiliary fields have vanishing v.e.v.'s and consequently there is no spontaneous supersymmetry breaking.

iv) Among the terms in this polynomial will be one or more involving scalars carrying color or e.m. charge. In general there is no reason the potential should have its minimum for $\langle\phi_{colored}\rangle = \langle\phi_{charged}\rangle = 0$, hence it is nontrivial to avoid spontaneous breaking of color and QED gauge symmetry.

This list is formidable, and to some extent the construction of a model which works is a matter of trial and error. I cannot give a general prescription for overcoming these difficulties, particularly the last two. It is interesting that the existence of a suitable phase invariance (called R-invariance) solves problems i) and ii) and contributes to the solution of iii).

In particular i) requires the existence of a system of phase transformations on spinors of the form $\lambda \rightarrow e^{\gamma_5\alpha}\lambda$ for Majorana spinors, corresponding to $\psi_L \rightarrow e^{i\alpha}\psi_L$. This is guaranteed for chiral multiplets with a conserved "external" (i.e., not carried by the supersymmetric generator) quantum number for the fermion member, e.g., lepton or baryon number. However if one wants there to be a fermion number associated with the fermi friends of the vector bosons (so that they are not produced, e.g., as $pp \rightarrow EE + X$, but rather as $pp \rightarrow E\bar{E} + X$) such a method cannot be used (since the vector bosons cannot carry a conserved "fermion" number) so that an additional phase invariance for these fermions is necessary and is provided by R-invariance.

Section III.b

A phase invariance such that the supersymmetry generator carries one unit of "charge" is called "R-invariance." The existence of conserved lepton or baryon number for the chiral multiplets, plus R-invariance of the theory, means that an R quantum number $0, \pm 1, \pm 2$ can be assigned to every particle, such that "new" particles (photino, gluinos, goldstino, scalar quarks and leptons, and the fermi friends of the W's) carry R = ± 1, while the "old" particles (photon, gluons, quarks, leptons, W's and Higgs) all carry R = 0. Table 1 shows the R quantum number assignment in Fayet's model[5]. The student can easily convince himself that in the limit $m_s \approx m_W >>$ masses of ordinary objects, R-conservation and ordinary lepton and baryon number conservation guarantees the absence of scalar or pseudoscalar exchanges of a strength comparable to the weak interaction.

Moreover, the existence of R-invariance may eliminate certain terms in the equations giving the auxiliary fields in terms of physical fields. This may mean that the auxiliary fields cannot all vanish simultaneously and consequently lead to the desired spontaneous supersymmetry breaking.

In many theories R-invariance does not need to be specially imposed. E.g., this is the case for theories describing only the gauge interactions of chiral superfields with no direct mass term. For instance the simple "abelian Higgs model" is automatically R-invariant. Finding this is a good exercise for the reader. Remember: it is defined so that a supersymmetry transformation changes R by one unit. The actual value of R assigned to a given particle may be arbitrary (as opposed to ΔR between members of the same supersymmetry multiplet). In this case it is usually convenient to define R as above: so the "ordinary" particles all have R = 0.

R-invariance has further consequences. It is important phenomenologically in that it may suppress low energy production of exotica. E.g., goldstinos and photinos can only be produced in pairs: hadrons → hadrons + ($\lambda_G\bar{\lambda}_G$ or $\lambda_G\bar{\lambda}_\gamma$ or $\lambda_\gamma\bar{\lambda}_\gamma$, etc.). R-invariance may be broken spontaneously by giving a non-zero v.e.v. to a field carrying $|R| = 2$. In this case, or if R-invariance is broken explicitly by a term in the Lagrangian transforming like $|R| = 2$, R is still conserved modulo-2, leaving a conserved R-parity with "old" particles having even R-parity and most new ones having odd R-parity. Such a conserved R-parity is all that is required for the phenomenology discussed below.

Section III.c

The gauge group of Fayet's model[5] is $SU_c(3) \times SU(2) \times U(1) \times U(1)$. After $SU(2) \times U(1) \times U(1)$ is spontaneously broken leaving electromagnetism and QCD exact, the particles are in the supersymmetry multiplets shown in Table 1. Supersymmetry is spontaneously broken by giving a v.e.v. to a linear combination of auxiliary fields associated with the spinor ζ and the neutral, colorless vectors (γ, Z^o, Z'). The resulting Goldstone spinor (goldstino) is then a linear combination of the ζ and the fermi friends of the neutral vectors. The spin-0 quarks and leptons and the remaining fermi friends of the colorless vectors all become heavy, with masses $\sim m_W$. However as shown below, the photino and gluinos (fermi partners of the photon and gluons, respectively) stay massless at the classical level (i.e., in tree approximation).

According to the Goldstone theorem, if supersymmetry is a global symmetry (it becomes local only when gravity is included) the goldstino is exactly massless. As for the photino and gluinos, a direct mass term in the Lagrangian cannot be responsible for their mass without breaking supersymmetry explicitly, since the photon and gluons must be precisely massless. Furthermore they cannot get masses through the Higgs mechanism - i.e., by virtue of being coupled to a field which develops a non-zero vacuum expectation value at the classical minimum of the potential. This is because the photino and gluinos couple to charge and color respectively, so that they would become massive only if a charged or colored field had a non-zero vacuum expectation value, breaking spontaneously e.m. or color gauge invariance which is not supposed to happen. Hence for the question of experimental detection of evidence for supersymmetry, the most important fields are the goldstino, photino, and gluinos which are all massless apart from possible quantum and gravitational corrections.

Because the gluinos are a color octet, they presumably combine with other colored fields, quarks, spin-0 quarks, gluons or gluinos, to form color-singlet hadrons. The most interesting such hadrons, from the point of view of present energies, are "R-mesons," made from $q\bar{q}$ + gluino, "R-baryons," made from qqq + gluino, and "R-glueballs," made from gluon + gluino. These can be expected to be rather light ($\lesssim 1\frac{1}{2}$ GeV/c^2) in the case that gluinos are massless. Since gluinos couple to gluons in pairs, with the ordinary QCD coupling, we expect R-hadron pairs to be produced at "normal" hadronic rates in hadron collisions. Would we have detected this phenomenon if it were occurring?

To answer this we must first ask how R-hadrons behave once produced. Since we assume that R or R-parity is conserved, the R-hadron either rescatters strongly, with the final hadronic state preserving its R quantum number, or it decays conserving R. That is, if it is an R-hadron resonance it will decay via the strong interactions to ordinary hadrons and the lightest R-hadron consistent with strong interaction selection rules. If it cannot decay strongly or electromagnetically to a lighter R-hadron it will decay to ordinary hadrons by the emission of a photino or goldstino, (collectively called nuinos) conserving R. Typical tree approximation diagrams involved in this decay are shown in Fig. 3. The gluino-quark-spin-0-quark vertex has strength $\sim g_s$, while the quark-spin-0-quark-nuino vertex has strength $\sim$ e (for a photino). The goldstino coupling is largely arbitrary and can be taken much smaller, however for definiteness we consider here the case that the goldstino

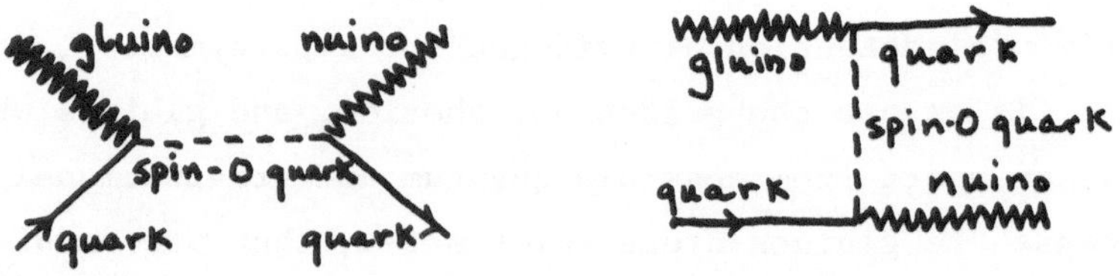

Fig. 3

coupling is comparable to the photino's. The reader can work out for himself variants on this. Taking also for definiteness $m_s \sim m_W$ we see that in tree approximation the amplitude for these processes is $\sim (g_s/e)(1/\sin\theta_c)$ times that for a strangeness changing weak decay. Considering that a typical R-meson or R-glueball decay is to $\pi\pi$ nuino, and taking the R-hadron mass to be 1-1½ GeV/c^2, gives a lifetime estimate 10^{-12} - 10^{-15} sec(6,14). The uncertainty in this estimate reflects the uncertainty in the gluino and nuino couplings (mixing angles) and in m_s, the variation due to phase space for initial masses ranging between 1 and 1½ GeV/c^2, and the uncertainty in the weak interaction dynamics of decays used for comparison(14).

The detection of R-hadrons becomes a question of detecting the production of a pair of particles each of which very quickly decays to hadrons + nuino. The subsequent interaction of the nuinos is governed in tree approximation by the diagrams of Fig. 3, reading backwards. Hence we expect

$$\sigma(\text{nuino} + \text{hadrons} \rightarrow \text{R-hadrons} + \text{hadrons}) \sim 10\text{-}100\ \sigma_{NC}^{\nu N}$$

or $$\sigma(\text{nuino} + \text{hadrons} \rightarrow \text{nuino} + \text{hadrons}) \sim \sigma_{NC}^{\nu N}$$

I.e., the nuinos are neutral and weakly interacting, so generally will go undetected. Therefore R-hadrons would have been detected neither in missing mass nor invariant-mass distributions. If their lifetime were $\sim 10^{-14}$ sec they could be seen in emulsion short-track searches; otherwise they could be observed in missing energy searches, or in experiments designed to detect the interaction of a produced nuino. These experiments will be discussed in Section IV.

Section III.d

Another distinctive feature of supersymmetry is the presence of the goldstino and photino which are light or massless. What are the possibilities for producing and detecting them? For instance one could have $q\bar{q}$ or $e^+e^- \rightarrow$ nuino + antinuino as shown in Fig. 4.

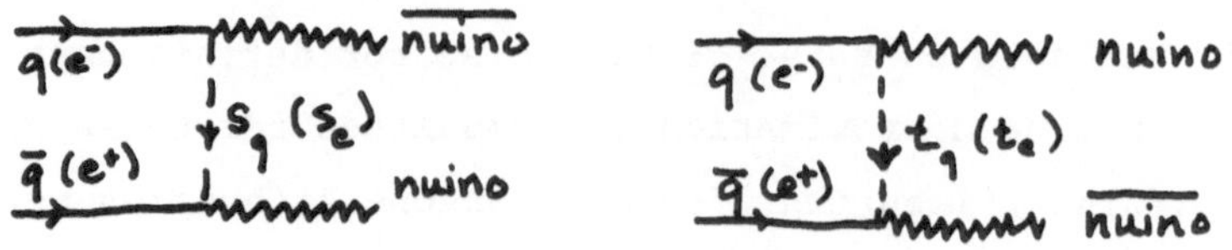

Fig. 4

If the mass of the scalar quark or lepton is $\sim m_W$ as we have been assuming, these processes would be at the ordinary weak level, and consequently hardly detectable.

In fact the photino could be massive through quantum effects, further reducing these rates. While the goldstino is supposed to be strictly massless in the absence of gravitational effects, it's coupling is not fixed, as for the photino, and might be much smaller than e. Thus we cannot give a theoretical lower limit on goldstino-antigoldstino production, either. Furthermore, for many processes involving goldstinos, decoupling theorems apply in the limit $k_{goldstino} \to 0$[15].

We have been focusing on the limit in which gravitational effects can be ignored, so let us briefly consider how gravity affects the phenomena we have been discussing, and whether there is any possibility of detecting evidence for supergravity. The principal effect of incorporating gravity, i.e., of making supersymmetry a local rather than merely global symmetry, is to induce the "super-Higgs" phenomenon[10]. Recall that when ordinary local gauge invariance is spontaneously broken, there is a massive vector meson rather than a massless goldstone boson and a massless gauge particle. Similarly, when supersymmetry is a local invariance so that it describes supergravity including at least one spin-2 and one spin-3/2 gravitino, its spontaneous breaking generates a massive gravitino rather than the massless goldstino present when the symmetry is only global. Fayet has shown[16] that when supersymmetry is broken so that ${m_s}^2 \sim {m_w}^2$, the gravitino mass is $\sim 10^{-5} - 10^{-6}$ eV/c^2. To the extent that the gravitino coupling is very weak, the massive spin-3/2 gravitino with its four helicity states behaves like a superposition of a massless gravitino with helicity $\pm$ 3/2 and an essentially massless goldstino responsible for the helicity $\pm$ 1/2 parts. One might suppose that if goldstinos can be pair produced at a weak interaction level, so could gravitinos. In fact what happens[16] is that the helicity $\pm$ 3/2 component of the massive gravitino decouples in the limit that the gravitational coupling, κ, vanishes, leaving coupled only the helicity $\pm$ 1/2 component, which becomes the goldstino in this limit. Thus while in principle one can produce gravitino-antigravitino pairs with weak interaction cross-sections, the helicity $\pm$ 3/2 components will essentially never be produced so that the special feature characteristic of supergravity will not be detectable.

IV. EXPERIMENTAL SITUATION

Section IV.a

At present there are three types of experiments which may be capable of detecting R-hadrons if their properties are as expected under the considerations described above: R-hadrons with masses $\sim$ 1-1½ GeV/c^2 pair-produced hadronically, decaying in 10^{-12} sec or less to hadrons and a neutral, weakly interacting nuino.

If the R-lifetime is near 10^{-14} sec, the limit[17] on charm production by 300 GeV/c protons, $\sigma_{pN \to \text{charm} + X} < 1.5\ \mu b$, obtained by searching in an emulsion for short-lived particles applies directly to $R\bar{R}$ production[6]. This is not very useful since the R-lifetime need not lie near 10^{-14} sec.

The most distinctive characteristic of an $R\bar{R}$ production event is its apparent non-conservation of energy-momentum (due to the escape of the nuinos) and simultaneous absence of charged leptons (so that the "non-conservation" of energy and momentum cannot be blamed on a missing neutrino). A recent experiment using the Caltech-Stanford calorimeter[18] at Fermilab to measure the hadron energy in muonless events induced by 400 GeV/c protons can be used to place an upper limit on $R\bar{R}$ production[7]. The distribution of measured hadron energy is not perfectly gaussian for energies less than or equal to 340 GeV. This may be simply a calorimetric phenomenon, or may be due to charm production followed by decay into hadrons + $e\nu_e$, or may be due to $R\bar{R}$ production followed by the decay of each into hadrons + nuino. Therefore only an upper limit can be given on $R\bar{R}$ production, and this limit depends on the nuino energy spectrum. That is, if R's are produced with little energy in the pN collision and are light so that the Q-value of their decay into nuinos is small, the experiment is much less sensitive to their detection than if they carry a lot of energy and can give a lot of energy to the nuinos. Also the limit depends on the assumptions made in going from the p-Fe to p-N. Assuming $\sigma_{pN \to R\bar{R}+X}/\sigma_{pA \to R\bar{R}+X} \sim \sigma^{tot}_{pN}/\sigma^{tot}_{pA}$ we obtained[7]

$$\sigma_{pN \to R\bar{R}+X} \lesssim 100 \ \mu b \text{ for } m_R = 1 \ \mathrm{GeV}/c^2$$

or

$$\sigma_{pN \to R\bar{R}+X} \lesssim 20 \ \mu b \text{ for } m_R = 3 \ \mathrm{GeV}/c^2 ,$$

at $\sqrt{s} = 27$ GeV. This limit is independent of the R-lifetime, as long as $\tau_R \lesssim 10^{-11}$ sec, and independent of the nuino re-interaction cross-section as long as that is less than ~ one mb so that the nuinos escape the detector without interacting.

These limits on R-hadron production could be further reduced if one were willing to make assumptions about nuino interactions. In that case the CERN beam dump experiments[19] can be brought to bear on the question[6]. There, 400 GeV/c protons interact with a beam dump, producing a beam of ν_e, $\bar{\nu}_e$, ν_μ, $\bar{\nu}_\mu$ and possibly nuino and $\overline{\text{nuino}}$ downstream. By detecting the neutrino charged current interactions with BEBC, Gargamelle, or the CDHS detector, the flux of neutrinos can be measured, giving a prediction for the number of lepton-less (neutral-current-like) events to be expected. If there were an excess of neutral-current-like events, they could be attributed to nuinos. Thus BEBC[19] can give a limit on the product of $R\bar{R}$ production and nuino re-interaction cross-sections in their set up:

$$2\ \sigma_{pN \to R\bar{R} + X} \cdot \sigma_i \lesssim 2 \times 10^{-66}\ \text{cm}^4$$

Assuming $\sigma_i \gtrsim 10\ \sigma^{\nu}_{NC}$, as suggested in Section III.c, this leads to the limit at $\sqrt{s} \approx 27$ GeV:

$$\sigma_{pN \to R\bar{R} + X} \lesssim 2\ \mu\text{b}$$

Section IV.b

Although the estimate $\tau_R < 10^{-12}$ sec is probably correct for the models discussed in Section III, one can ask what limits can be placed on R-hadron production if R-hadrons happened to be more stable than that. If relatively stable charged R-hadrons exist they could "contaminate" hyperon beams if $\tau_R \gtrsim 10^{-9}$ sec. Unless their masses were close to those of known hadrons they would be detected, giving limits on their production cross-section. Ref. (20) gives limits obtained this way on low energy charged R-hadron production, but because of the important threshold effect for production of a pair with masses $\gtrsim 1$ GeV/c^2, high energy hyperon beam results will be the most useful[21].

Another case is that the only R-hadrons stable to strong and e.m. decays are neutral - perhaps the flavor singlet R-glueball. That $R\bar{R}$ pairs are being produced at a "hadronic" rate cannot in general be ruled out in this case[21] if the "stable" neutrals have lifetimes $\gtrsim 10^{-11}$ sec.

V. CONCLUSIONS FROM THE PRESENT EXPERIMENTAL SITUATION

The least model-dependent experimental limit on $R\bar{R}$ production, the calorimeter experiment described in Section IV.a, indicates that $\sigma_{pN \to R\bar{R} + X} \lesssim 100\ \mu b$ for $m_R \sim 1\ GeV/c^2$ or $\lesssim 20\ \mu b$, for $m_R \sim 3\ GeV/c^2$. For $m_R \sim 1\ GeV/c^2$ a cross-section $\lesssim 100\ \mu b$ at $\sqrt{s} = 27$ GeV is small: compare it to $p\bar{p}$ production which at the same energy is several mb. However $\sigma \lesssim 20\ \mu b$ for $m_R \sim 3\ GeV/c^2$ is to be expected(22). One can probably conclude that R-hadrons are either heavier than $\sim 1\frac{1}{2}\ GeV/c^2$ or their production is anomalously small by the criteria of conventional naive ideas regarding hadron production in QCD.

This may simply mean that we underestimated the R-hadron masses to be expected when gluinos are massless, or may mean that gluinos are massive. A model in which the latter occurs has been constructed by Fayet(23), so that there seems in principle to be no deep reason that gluinos cannot have a mass. It is encouraging, however, for the experimentalist and phenomenologist, that gluinos with masses $\lesssim$ a few GeV/c^2 seem particularly natural.

What, then, is the prospect that the forces of nature are supersymmetric? Up to now we have no experimental evidence that this is the case. However considering the c.m. energy of the experiments in question, the limits on R-hadron production are not sufficiently stringent that they can be regarded as evidence against the supersymmetry of the fundamental forces of nature, even in those models in which supersymmetry is most apparent. Clearly the search for R-hadrons and nuinos should be continued at higher energies and with greater sensitivity because they may be lurking just beyond the range of present experiments.

Acknowledgement

Most of the work described here on the phenomenology of R-hadrons and nuinos was done in collaboration with P. Fayet. In addition I am indebted to him for helping me understand his and others' earlier work on the theory of spontaneous supersymmetry breaking.

References and Footnotes

1. For an introduction to supersymmetry and a more complete list of references, see, e.g., S. Ferrara, Erice Lecture Notes, Aug. 1978 (CERN preprint TH2514) and P. Fayet and S. Ferrara, Phys. Reports 32C, 249 (1977).
2. For example, nature — in particular electromagnetism — is rotation invariant; however if one were living inside a huge ferromagnet, the essential rotation invariance of the fundamental interactions would not be trivially apparent: we would say that the "vacuum" (here the interior of the ferromagnet) did not possess the rotation invariance, even though the fundamental interactions do.
3. P. Fayet and J. Iliopoulos, Phys. Lett. 51B, 461 (1974).
4. P. Fayet, Nuovo Cim. 31A, 626 (1976).
5. P. Fayet, Phys. Lett. 69B, 489 (1977).
6. G. R. Farrar and P. Fayet, Phys. Lett. 76B, 575 (1978).
7. G. R. Farrar and P. Fayet, Caltech Preprint CALT-68-669. To be published in Phys. Lett. B.
8. P. Fayet, Nucl. Phys. B90, 104 (1975); Nucl. Phys. B78, 14 (1974).
9. P. Fayet, Proc. of the Orbis Scientiae, Coral Gables (Florida, USA), Jan. 1978, New Frontiers in H. E. Physics (Plenum Pub. Corp., New York) p. 413 (and Caltech preprint CALT-68-641).
10. S. Deser and B. Zumino, Phys. Rev. Lett. 38, 1433 (1977).
11. In the Majorana representation, in which the Dirac matrices are real, a Majorana spinor is simply a 4-component real spinor.
12. The superfield formalism (A. Salam and G. Strathdee, Nucl. Phys. B76, 477 (1974); S. Ferrara, J. Wess, and B. Zumino, Phys. Lett. 51B, 239 (1974)) allows the fields in a super-multiplet to be treated as a unit, by virtue of the introduction of anticommuting variables, θ. For instance here,

the superfield S contains ψ_L and ϕ, and V contains V_μ and λ. This requires the use of so-called auxiliary fields, which can be eliminated in favor of physical fields. See also (1).

13. J. Wess and B. Zumino, Nucl. Phys. B78, 1 (1974).

14. These estimates can be obtained several ways. (i) Starting with the rates for $K_S^o \to \pi^+\pi^-\gamma$ and $K_L \to \pi^+\pi^-\pi^o$, taking into account differences arising in the amplitudes from factors of $\theta_{Cabibbo}$, g_s, e, etc., one concludes that a 1/2 GeV/c^2 R^o decaying to $\pi\pi$ nuino would have a lifetime $\sim 10^{-11\pm1\frac{1}{2}}$s. Using three-body phase space this can be scaled to $m_R = 1$ GeV/c^2 ($\tau \simeq 10^{-13\pm1\frac{1}{2}}$sec) or $m_R = 1.5$ GeV/c^2 ($\tau \approx 10^{-14\pm1\frac{1}{2}}$sec). (ii) The decay of an R-π via annihilation of a quark and gluino $\to$ spin-0 quark $\to$ quark + nuino can be related to the formula for ordinary $\pi \to \mu\nu$, however without the helicity suppression of $\pi\to\mu\nu$ from the V-A coupling. Taking into account g_s vs. e and using $f_{R-\pi} = f_\pi$ gives, for $m_{R-\pi} = 1.2$ GeV/c^2, $\tau \sim 10^{-12}$–10^{-13}s. (iii) Comparing with $\Sigma \to n\pi\gamma$, accounting for different factors of θ_C, e, and g_s, then accounting for the different Q values would give for $m_R \sim 1.2$ GeV/c^2, $\tau \sim 10^{-13}$ – 10^{-14} s.

15. B. de Wit and D. Z. Freedman, Phys. Rev. Lett. 35. 827 (1975).

16. P. Fayet, Phys. Lett. 70B, 461 (1977).

17. G. Coremans - Bertrand et al., Phys. Lett. 65B, 480 (1976).

18. B. Barish et al., Caltech preprint CALT-68-655.

19. P. C. Bosetti et al., Phys. Lett. 74B, 143 (1978); P. Alibran et al., Phys. Lett. 74B, 134 (1978); T. Hansl et al., Phys. Lett. 74B, 139 (1978).

20. T. Goldman, Phys. Lett. B, to be published.

21. G. R. Farrar, in preparation.

22. Bourquin and J. M. Gaillard, Nucl. Phys. B114, 334 (1976).

23. P. Fayet, Phys. Lett., to be published (Caltech preprint CALT-68-663).

D I S C U S S I O N

CHAIRMAN : G.R. Farrar

DISCUSSION

- ROTH:

Could the goldstino be a neutrino?

- FARRAR:

Bardeen has pointed out that the ν_e cannot be a goldstino since the amplitude for $n \to pe\bar{\nu}_e$ does not vanish in the limit $k_{\nu_e} \to 0$. While the limits are poorer regarding ν_μ, it seems quite unattractive to take the ν_μ to be a goldstino and thus put the ν_e and ν_μ on a very different footing.

- BLASI:

Why is N=4 supersymmetry harder to break than N=1 and 2?

- FARRAR:

The only model known which is invariant under N=3 or 4 extended supersymmetry is pure Yang-Mills, which has no free parameters and no spontaneous supersymmetry breaking. Unless N=4 models describing other supermultiplets are found which could be coupled to the N=4 gauge superfield in order to complicate the potential, there is little hope of spontaneously breaking the N=4 supersymmetry at least in a conventional way.

- WHITAKER:

Suppose the scalar quarks were pair-produced in e^+e^- collisions. Could the characteristic $\sin^2\theta$ angular distribution of scalars be used to distinguish their jets from ordinary quark jets?

- FARRAR:

Since a scalar quark will quickly decay to a quark and gluino, giving 1/2 its energy on the average to the gluino, the events will not have a 2-jet structure. Thus the events will be distinctive, although not as a result of a $\sin^2\theta$ angular distribution. However recall that the cross-section for scalar production in e^+e^- annihilation is only 1/4 as large as for production of a spin-1/2 particle with the same charge. Of course, each quark has two spin-0 friends.

- ISGUR:

Can limits on the branching ratio for $\psi \to \nu\bar{\nu}$ be used to put a lower limit on m_{s_q}?

- FARRAR:

If we assume that C and P are conserved the charge conjugation of a photino-antiphotino system is the same as for a two-photon system. Hence the decay $\psi \to \overline{\text{photino}}$ + photino is forbidden. While $\psi \to \overline{\text{photino}}$ + goldstino is not forbidden since the goldstino need not have the same C as the photino, we have no lower limit on the goldstino coupling so that an absence of this decay does not give any lower limit on the s_c and t_c mass.

SUPERCONDUCTIVITY AND QUARK CONFINEMENT:

MAGNETIC AND ELECTRIC ORDER*+

Kerson Huang

Center for Theoretical Physics
Laboratory for Nuclear Science and Department of Physics
Massachusetts Institute of Technology
Cambridge, MA 02139

I. ELECTRIC AND MAGNETIC ORDER

The well-known Wilson criterion[1] for quark confinement in quantum chromodynamics (QCD) deals with the vacuum expectation of the gauge-invariant operator

$$A(C) = \mathrm{Tr}[p \exp ig \oint_C dx^\mu A_\mu(x)] \tag{1.1}$$

where C is a directed closed path in 4-dimensional Euclidean space-time $(x^1, x^2, x^3, x^0 = i\tau)$. If, for sufficiently large C,

$$\langle A(C)\rangle \sim e^{-\lambda\Sigma(c)}, \quad (\Sigma(c) = \text{area enclosed by } C) \tag{1.2}$$

then a static quark and antiquark, inserted into the QCD vacuum, will attract each other with a linear potential, for large separations between them. The linearity gives rise to linearly rising Regge trajectories for quark bound states (by qualitative reasoning) and is therefore not inconsistent with observations.

The criterion (1.2) is a statement about vacuum fluctuations in QCD without quarks. Nevertheless, it contains information about

*This work is supported in part through funds provided by the U.S. Department of Energy (DOE) under contract EY-76-C-02-3069.

the response of the system to external quarks, just as the spontaneous fluctuations of a thermodynamic system can tell us how the system would respond to external disturbances. It is a plausible but unproved assumption that the inclusion of dynamical quark fields will not qualitatively alter the picture.

Because of Lorentz invariance, the result (1.2) cannot depend on the choice of coordinate frames in Euclidean 4-space. For a large contour C of gentle curvature, it should also be independent of the shape of C. To understand the physical meaning of the Wilson criterion, let us therefore choose C to be a large rectangle, and examine the statement (1.2) in two different coordinate frames.

First suppose that C lies in 3-space, say the x-y plane, as shown if Fig. 1.1(a). The effect of A(C) on the vacuum state is to create instantaneously an infinitely thin tube of color electric flux along the closed curve C (for only the transverse part of $\vec{A}$ contributes to the contour integral, and $\vec{A}_T$ is conjugate to $\vec{E}$). The behavior of <A(C)> depends on what happens to the flux subsequently:

(a) it may spread out over all space, as would be the case in an ordinary vacuum. This would yield $\langle A(C)\rangle \sim 1$.

(b) It may disappear by being locally absorbed into the vacuum, as would be the case in an electron plasma. This would yield $\langle A(C)\rangle \sim e^{-\lambda L(C)}$, when L(C) is the length of C.

(c) It may spread out a little, but remain confined to a thin tube of constant cross section, which eventually shrinks to nothing. This satisfies the Wilson criterion, because the probability that the flux tube will shrink to nothing is proportional to the area swept out.

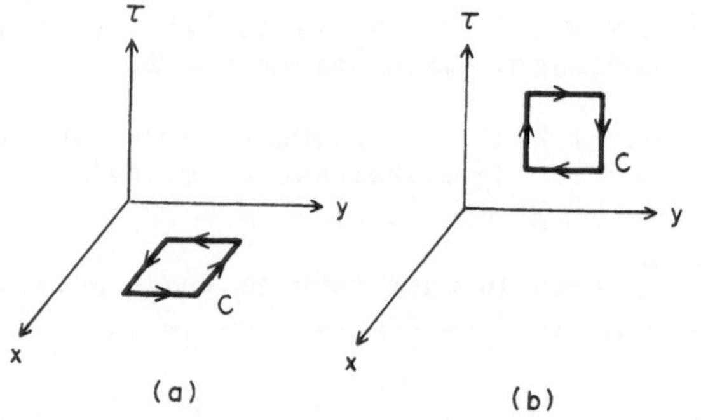

Fig. 1.1. The Wilson contour: (a) in equal-time point of view, (b) in "time-slice" point of view.

From this view, therefore, the Wilson criterion states that color electric flux tends to be confined to a tube. We say that the system has "electric order," and regard $\langle A(C)\rangle$ as an "electric order parameter."

Now let us flip the rectangle C so that its plane becomes parallel to the (imaginary) time axis, as shown in Fig. 1.1(b). The vacuum expectation $\langle A(C)\rangle$ is now a Green's function describing the creation of an electric flux tube of length L, lasting for (imaginary) time interval T. It should be proportional to $e^{-\lambda TE(L)}$, where E(L) is the energy of the flux tube. The Wilson criterion, in this view, states that $E(L) \propto L$, or the flux tube has finite constant energy per unit length. A flux tube that does not end on itself must end on equal and opposite charges (by definition). These charges are not dynamical objects, but merely points where we allow the gauge field to be singular, i.e., external sources. (If we round off the corners of the rectangle C, these point sources will be smeared out.) Thus the Wilson criterion states that a flux tube of finite constant energy per unit length is the sole carrier of color electric flux from an external quark to an external antiquark, giving rise to the linear attractive potential between them. That is, electric order means linear confinement of electric charges.

Whether or not electric order prevails in the QCD vacuum is a question of detailed dynamics, and we do not yet know the answer. We are, however, reminded of a dual phenomenon in a superconductor--magnetic order. By the Meissner effect, magnetic fields are either expelled from a superconductor, or confined to flux tubes called vortex tubes. Our dynamical understanding of superconductivity tells us that, if a magnetic monopole and an antimonopole were inserted into a superconductor with large separation between them, their magnetic flux would be confined to a vortex tube that connects them, with finite constant energy per unit length. Thus we have linear confinement of magnetic charges, i.e., magnetic order. The analog of the Wilson criterion would read

$$\langle B(c)\rangle \sim e^{-\lambda\Sigma(c)}, \quad (\Sigma(c) = \text{area enclosed by C}), \tag{1.3}$$

where the operator B(C) creates an infinitely thin magnetic flux tube (a vortex tube) coinciding with the closed path C. In contrast, the electric order parameter has the behavior

$$\langle A(c)\rangle \sim e^{-\lambda L(c)}, \quad (L(c) = \text{Length of C}), \tag{1.4}$$

because the superconducting ground state is a "plasma" of Cooper pairs.

In terms of the order parameters <A(C)> and <B(C)>, we say that a system has electric order if <A(C)> is "small," i.e. $O[e^{-\lambda\Sigma(c)}]$; it has magnetic order if <B(C)> is "small." When these parameters are not "small," there is no corresponding order, and we say that there is "disorder." In a general system electric and magnetic order are neither correlated nor mutually exclusive.

In this language, a normal metal is a phase with both electric and magnetic disorder. A vortex tube in a superconductor consists of a tube of disordered phase, surrounded by a phase of magnetic order.

We would have a model of linear quark confinement if we could build a theory like superconductivity with <A(C)> and <B(C)> interchanged--a trivial task if we could simply interchange the names of electric field and magnetic field. This is not possible, however, because color electric charges are defined by the generators of the color gauge group, i.e., they are what quarks carry. The problem is to show the existence of electric order in an electric system (in contradistinction to superconductivity, in which magnetic order exists in an electric system).

Many dynamical approaches have been proposed towards such a goal, either explicitly or implicitly[2]. These are not dissimilar to the early efforts to explain superconductivity, before the BCS synthesis, in which bits and pieces of the complicated electron-lattice problem were rummaged through for clues. The possible relevance of electric and magnetic order as a guide in such rummaging has been pointed out by 'tHooft[3], who showed that the possible phases in a non-Abelian gauge system can be deduced by studying the commutation relation between A(C) and B(C'), without recourse to detailed dynamics. A dynamical approach that emphasizes the parallel between quark confinement and superconductivity has been proposed by Thorn[4].

In what follows, we shall first sketch the dynamical basis of magnetic order in superconductivity, and then turn to the search for electric order, or quark confinement, in QCD.

II. SUPERCONDUCTIVITY[5]: MAGNETIC ORDER

1. Experimental Facts

We summarize some relevant experimental facts by describing the response of a superconducting body to external magnetic and electric fields $\vec{B}$ and $\vec{E}$:

(a) $\vec{B}$ is expelled (Meissner effect):

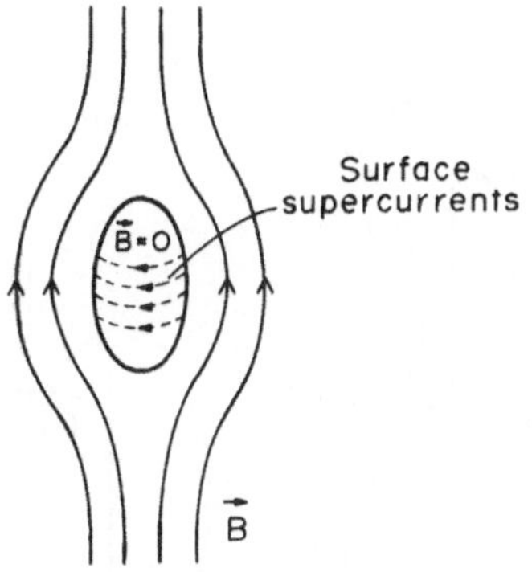

Fig. 2.1. Explusion of Magnetic Field by Type I Superconductor.

An external magnetic field is expelled from the interior of a "type I" superconductor, except in a surface layer whose thickness is called the penetration depth. A non-dissipative current (supercurrent) flows in the surface layer, which can be detected by measuring the magnetic moment of the body. (See Fig. 2.1.)

If we place a superconducting pipe in an external magnetic field, parallel to the axis of the pipe, then supercurrents will be induced, such that the flux Φ trapped inside the pipe is quantized:

$$\Phi = \left(\frac{2\pi\hbar c}{2e}\right) n, \quad (n = \text{integer}). \tag{2.1}$$

If the external field is now turned off, the flux Φ will remain the same, being maintained by a persistent supercurrent. (See Fig. 2.2.)

A related phenomenon is that, in "type II" superconductors, an external magnetic field penetrates the body in the form of quantized flux tubes, with a definite density and geometrical arrangement. (See Fig. 2.3.)

The Meissner effect is an experimental definition of magnetic order. It is destroyed when the external magnetic field strength exceeds a critical value, or if the temperature exceeds a critical value.

(b) <u>$\vec{E}$ is absorbed</u>:

An external electric field is absorbed by induced surface charge densities. (See Fig. 2.4.) These charges can be tapped off by

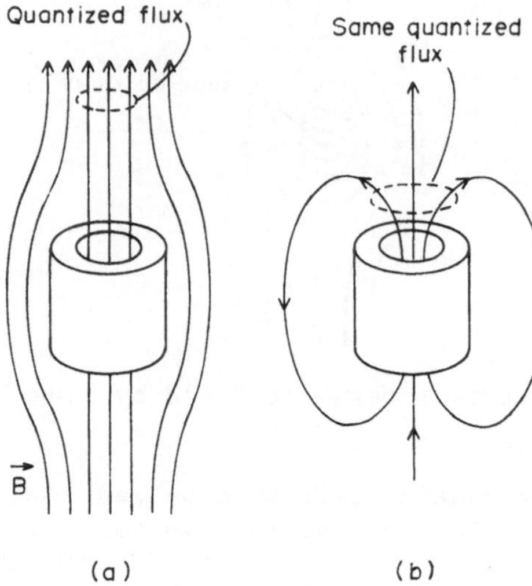

Fig. 2.2. Inducing a persistent supercurrent:
(a) First, place superconducting pipe in external magnetic field.
(b) Then, remove the external magnetic field.

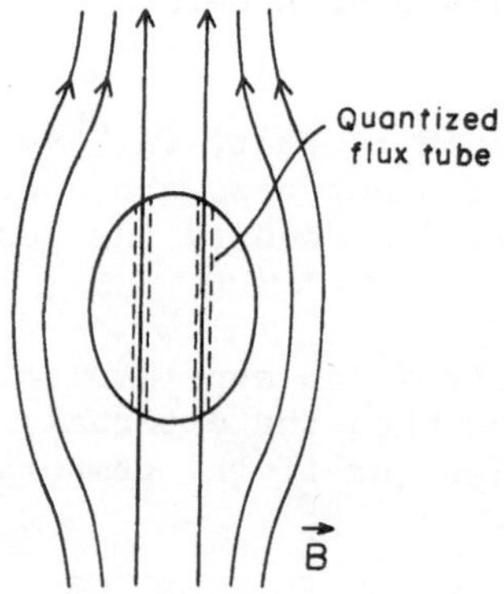

Fig. 2.3. Quantized flux tubes in type II superconductor.

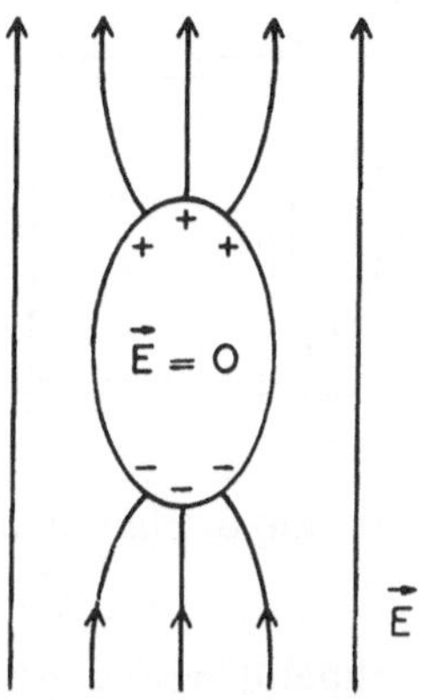

Fig. 2.4. Absorption of electric field by superconductor.

making the superconductor part of an electric circuit. The current flows in the surface layer of the body with no resistance, giving superconductivity its name. (See Fig. 2.5.)

There is no electric order, i.e., no electric Meissner effect.

2. Theory

There seems little doubt that the phenomenon of superconductivity springs from a simple non-relativistic Lagrangian for electrons and heavy ions, coupled solely to the electromagnetic field. When there are few electrons and ions present, the theory can be solved by perturbation theory; but when there are 10^{23} of each, the problem becomes insurmountable. We cannot even deduce the elementary fact that the ions form a crystal lattice!

Our understanding of superconductivity comes not from first principles, but a long historical chain of phenomenology, guided by experiments. It goes roughly as follows:

(a) Start with an electron gas in a crystal lattice of heavy ions.

(b) Understand the plasma oscillation and the screened Coulomb interactions of the electrons, and neglect them.

(c) Suspect the important role of electron-lattice interactions in superconductivity, as suggested by the "isotopic effect."

(d) Realize that the electron-lattice interactions can lead to an effective attraction between electrons, which however small,

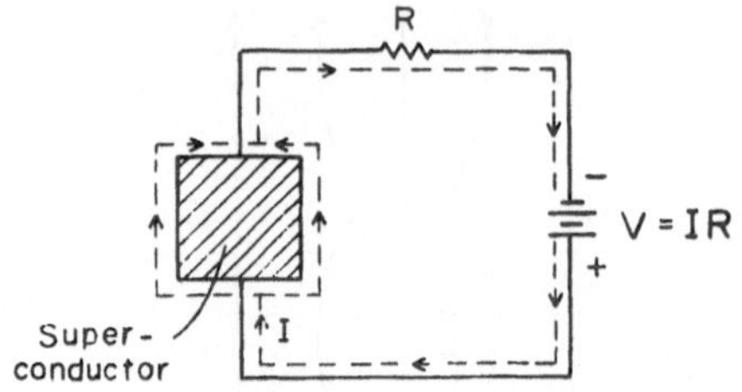

Fig. 2.5. Demonstration of zero resistance of superconductor.

will produce a two-electron bound state of zero angular momentum (Cooper pair), for electrons at the edge of the Fermi distribution.

(e) Adopt an effective electron Hamilton, in which all interactions are ignored except for a simplified form of the electron-electron attraction (even though this represents a small energy, compared to what is omitted). Determine the ground state variationally, leading to a "Bose condensate" of Cooper pairs (BCS theory).

The Bose condensate involves only a small fraction of the electrons, i.e. those near the top of the Fermi distribution. Single-electron excitations are separated from the ground state by an energy gap.

The Meissner effect can be deduced by calculating the ground state expectation of the electron current density, in the presence of a static magnetic field:

$$\langle\vec{j}\rangle = \frac{e\hbar}{2mi} \langle\psi^\dagger \overleftrightarrow{\nabla} \psi\rangle - \frac{e^2}{mc} \langle\psi^\dagger\psi\rangle A. \tag{2.2}$$

The two terms above cancel each other in a free electron gas (apart from a small diamagnetic current). Here the actual calculation is delicate. Effectively one can ignore the normal electrons, which are approximately free. For the superconducting electrons (which are bound in Cooper pairs) the first term vanishes on account of the energy gap. The second term leads to the London equation:

$$\langle\vec{j}\rangle = -\frac{e^2 n_s}{mc} \vec{A}, \quad (\nabla \cdot \vec{A} = 0), \tag{2.3}$$

where n_s is the density of superconducting electrons. (This is to be contrasted with the ordinary diamagnetic response, for which $\langle\vec{j}\rangle \propto \nabla\times\vec{B} = -\nabla^2\vec{A}$.) The Meissner effect comes about as follows. Upon substituting (2.2) into the Maxwell equation $\nabla\times\vec{B} = \langle\vec{j}\rangle/c$, one obtains

$$(\nabla^2 - \mu^2)\vec{B} = 0, \quad \mu = \left(\frac{e^2 n_s}{mc^2}\right)^{1/2} . \tag{2.4}$$

Thus the magnetic field decays in space with decay length μ^{-1}, which is the penetration depth.

The non-dissipative nature of the steady-state supercurrent follows from (2.2), and is therefore traceable to the existence of the energy gap.

We can now understand the dynamics of flux quantization. Suppose we arrange to place an arbitrary amount of magnetic flux through a superconducting pipe, as shown in cross-sectional view in Fig. 2.6. (For example, we could start with a pipe in the normal phase, and cool it to below the transition temperature, after the flux inside has been established by means of a current-carrying solenoid.) How and why will the flux readjust to a quantized value? Inside the superconductor the vector potential is pure gauge, but non-zero:

$$\begin{aligned} &\vec{A} = \nabla\chi , \\ &\oint_C d\vec{x} \cdot \vec{A} = \Phi . \end{aligned} \tag{2.5}$$

By cylindrical symmetry, the solution is

$$\chi = \frac{\Phi\Theta}{2\pi} . \tag{2.6}$$

Hence the Schrödinger equation for a Cooper pair reads

$$-\frac{\hbar^2}{2m^*}\left(\nabla - \frac{i2e}{\hbar c}\nabla\chi\right)^2\psi = E\psi. \tag{2.7}$$

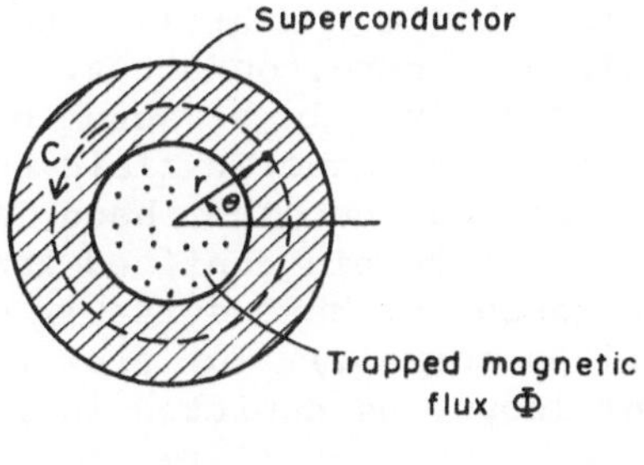

Fig. 2.6. Cross-sectional view of magnetic flux trapped in superconducting pipe.

The wave function has the form $e^{in\Theta}R(r)$, where n must be an integer, by continuity. Using this, we can calculate the energy:

$$E = \frac{\hbar^2}{2m^*}\int d^3r \left|(\nabla - \frac{i2e}{\hbar c}\nabla\chi)\psi\right|^2 = a + b(n - \frac{2e}{\hbar c}\frac{\Phi}{2\pi})^2, \tag{2.8}$$

where a,b are positive definite radial integrals. Thus the energy can assume its minimum possible value only if

$$\Phi = \left(\frac{2\pi\hbar c}{2e}\right) n, \; (n = \text{integer}), \tag{2.9}$$

which is the same as the experimental condition (2.1). When this condition is fulfilled, the Cooper pair does not "know" that there is flux in the pipe.

Because there is a macroscopic number of Cooper pairs, the total energy of the superconductor, as a function of the flux Φ, has extremely sharp minima at the quantized values. If Φ was not quantized initially, a transient flow of Cooper pairs will rapidly readjust it to a quantized value, in order to minimize the energy.[6]

Similarly, if we thrust a superconducting body into an arbitrary magnetic field (below the critical field value), transient supercurrents will bunch the flux lines into quantized flux tubes of finite energy per unit length, with diameters the order of the penetration depth.[7] Whether these flux tubes will remain inside the body (Type II behavior), or be expelled from the body (Type I behavior), is again a matter of energetics, and turns out to depend on the ratio of the penetration depth to a characteristic correlation length.

Finally, we see the dynamical mechanism for monopole confinement. Imagine that a monopole-antimonopole pair was placed at the same point, and then suddenly pulled apart. A magnetic flux tube will always appear between them momentarily, whether the medium is a vacuum, a normal metal, or a superconductor. The question is what happens to the flux subsequently. In a vacuum or a normal metal, the flux will spread, until the magnetic field becomes a dipole field. In a superconductor, on the other hand, our previous discussion shows that the flux will remain confined to a quantized flux tube. (That is, if the monopoles have suitably quantized charges. Otherwise the metal will go normal.) The flux tube is a "solenoid" ringed with supercurrent loops, as depicted in Fig. 2.7(a).

Since a current loop is a magnetic dipole, the response of the superconductor can also be described as a local neutralization of

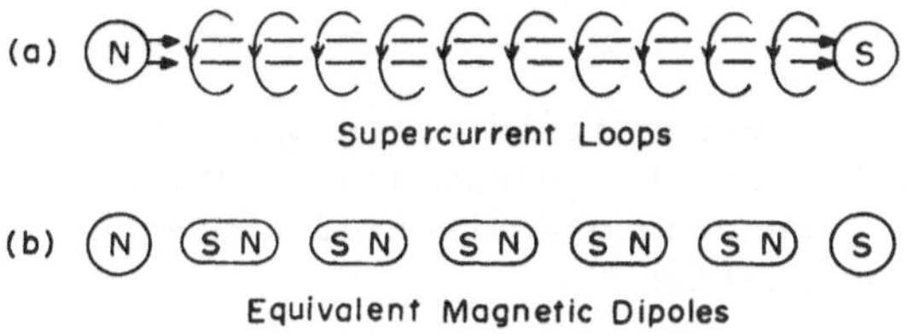

Fig. 2.7. Linear confinement of magnetic monopoles in superconductor:
(a) Formation of flux tube by supercurrent loops.
(b) Equivalent string of induced magnetic dipoles.

magnetic charge: a string of magnetic dipoles "transport" the charge of the monopole to the antimonopole, where it is finally neutralized. (See. Fig. 2.7(b).) Throughout the process of separating the monopole-antimonopole pair, no local magnetic charge density ever appears in the system.

We have given a dynamical description of the response of a superconductor to $A_\mu^{ext}(x)$. In actuality, the system is also coupled to the quantum field $A_\mu(x)$ (which however may be ignored in the determination of the BCS ground state, because the electron-lattice interaction stems from Coulomb interactions). In the superconducting ground state, therefore, there are quantum fluctuations of $A_\mu(x)$, which can be determined from the response to $A_\mu^{ext}(x)$. Indeed, at the beginning of these lectures we have deduced the character of these fluctuations, namely $\langle A(c)\rangle$ and $\langle B(c)\rangle$, on the basis of the known response to external fields. What we have supplied here is our best dynamical understanding of that response.

III. QUARK CONFINEMENT

1. The Order Parameters

We consider quark confinement only in the limited sense of the Wilson criterion, which states that there is electric order in QCD without quarks. (There may, however, be Higgs fields.) We consider a general color gauge group SU(N), N = 3 being the physical value.

The Wilson operator A(C), whose vacuum expectation is the electric order parameter, is defined by

$$A(C) = \mathrm{Tr}\,[p \exp ig \oint_C dx^\mu \, A_\mu(x)], \tag{3.1}$$

where g is the gauge coupling constant, $A_\mu(x)$ is an N×N hermitian matrix whose elements are quantum operators:

$$A_\mu(x) = L_a A_\mu^a(x), \tag{3.2}$$

where $A_\mu^a(x)$ are the gauge field operators, and L_a are the generators of color SU(N), in the fundamental representation. The integral in (3.1) is taken once around a directed closed curve C in 4-dimensional Euclidean space-time

$$x = (x^1, x^2, x^3, x^0 = i\tau). \tag{3.3}$$

The symbol P denotes path ordering, and Tr denotes trace of an N×N matrix.

To define the operator B(C) in QCD, whose vacuum expectation is the magnetic order parameter, let us first define it more carefully in the simpler case of superconductivity. In that case, B(C) creates a thin ring of magnetic flux Φ coinciding with the curve C. Suppose C is a curve in 3-space at a particular time. The flux ring may be characterized by a pure-gauge vector potential $\nabla\chi_C(\vec{x})$, with a multi-valued gauge function $\chi_C(\vec{x})$: If $\vec{x}$ traces out a closed path C' which links C once, then the value of $\chi_C(\vec{x})$ increases by Φ. (See Fig. 3.1.) If we parameterize the path C' by an angle Θ, then $\chi_C(\vec{x})$ is given by (2.5) for points lying on C'. The operator B(C) may be defined by its action on an eigenstate of $\vec{A}(\vec{x})$ with eigenvalue $\vec{a}(\vec{x})$, denoted by $|\vec{a}>$:

$$B(c) \mid \vec{a}> = \mid \vec{a} + \nabla\chi_C \tag{3.4}$$

The electron field, of course, also undergoes a corresponding gauge transformation; but we need not deal with it explicitly.

It is easy to see that, for two closed curves C and C' both in 3-space at equal time,

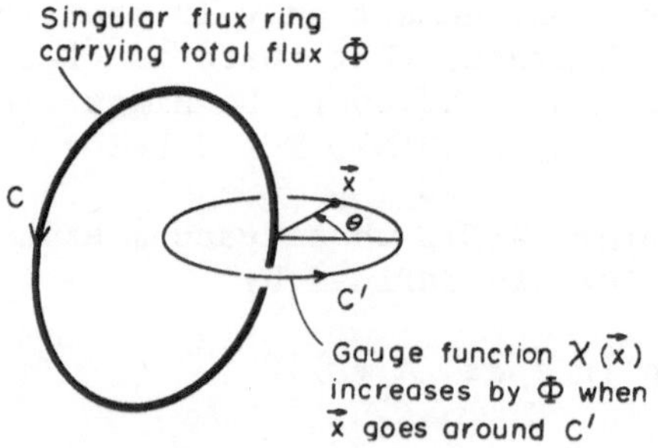

Fig. 3.1. The pure-gauge vector potential $\nabla\chi$ outside of a magnetic flux ring C.

$$A(C')\,B(C) = \begin{cases} B(C)\,A(C') & \text{(if } C \text{ and } C' \text{ are unlinked)} \\ B(C)\,A(C')e^{i2e\Phi} & \text{(if } C \text{ and } C' \text{ are linked once)} \end{cases} \quad (3.5)$$

Hence

$$[A\,(C'),\ B(C)] = 0 \quad (3.6)$$

if the flux Φ is quantized according to (2.9). If a flux ring of arbitrary flux Φ is created in a superconductor, then induced supercurrents will tend to: (a) readjust the value of Φ to a quantized valued, and (b) make the flux ring shrink to nothing. Whether or not it is interesting to consider $B(C)$ with non-quantized values of Φ will depend on the relative time scales involved in (a) and (b). In physical superconductors, the flux will become quantized first. Therefore (3.6) will do for most purposes.

In QCD the term "magnetic flux" has to be made more specific, because $\vec{B} \neq \nabla\times\vec{A}$. With "tHooft[3], we generalize $B(C)$ through the definition in terms of a gauge transformation. Thus $B(C)$ creates a ring of flux of $\nabla\times\vec{A}$, which is divergenceless. Just as in a superconductor, the flux cannot be absorbed.

Let $|\vec{\mathcal{A}}\rangle$ be the eigenstate of $\vec{A}(\vec{x})$ with eigenvalue $\vec{\mathcal{A}}(\vec{x})$, in the $A_0 = 0$ gauge. A local gauge transformation is

$$\vec{\mathcal{A}}\,(\vec{x}) \rightarrow \vec{\mathcal{A}}^{\Omega}(\vec{x})$$

$$\vec{\mathcal{A}}^{\Omega}(\vec{x}) = \Omega(\vec{x})\,\vec{\mathcal{A}}\Omega^{-1}(\vec{x}) + ig\Omega(\vec{x})\nabla\Omega^{-1}(\vec{x}) \quad (3.7)$$

where the $N\times N$ gauge matrix $\Omega(\vec{x})$ (the analog of $e^{ig\chi(\vec{x})}$ in electromagnetism) has the form

$$\Omega(\vec{x}) = e^{iL_a\omega_a(\vec{x})} \quad (3.8)$$

where $\omega_a(\vec{x})$ are arbitrary functions of $\vec{x}$. The transformation laws for $\vec{\mathcal{A}}_a(\vec{x})$ may be obtained from (3.7) by projection, with the help of $Tr(L_aL_b)/Tr(L_a^2) = \delta_{ab}$.

For a closed curve C in 3-space at a fixed time, $B(C)$ is defined by

$$B(C)\,|\vec{\mathcal{A}}\rangle = |\vec{\mathcal{A}}^{\Omega_C}\rangle \quad (3.9)$$

where $\Omega_C(\vec{x})$ is a multi-valued gauge matrix: If $\vec{x}$ traces out a closed path C' that links C once, and is parameterized by an angle θ: $o \rightarrow 2\pi$. (See Fig. 3.1.) Then,

$$\Omega_C(2\pi) = z\Omega_C(o), \quad (3.10)$$

where z is a matrix characterizing the flux in the flux ring. Since $\vec{A}(\vec{x})$ satisfies a differential equation, $\vec{\mathcal{A}}(\vec{x})$ must be continuous in $\vec{x}$. This can be maintained if and only if the matrix z commutes with every matrix of the gauge group. That is, z belongs to the center Z(N) of SU(N), the subgroup consisting of the Nth roots of unity:

$$z = 1_N e^{i2\pi n/N} \tag{3.11}$$

where 1_N is the N×N unit matrix. The corresponding flux is

$$\Phi = 1_N \frac{2\pi n}{gN} \tag{3.12}$$

We also note that (3.11) comes from the fact that the gauge fields transform according to the adjoint representation of SU(N), and are invariant under Z(N). Higgs fields, if present, will not affect (3.11) as long as they are invariant under Z(N). Quarks are not invariant under Z(N), but we do not include them explicitly.

If B(C) creates n quanta of flux (in the sense of (3.11)), we may also think of B(C) as creating one quantum of flux, but reinterpreted the closed path to be C traversed n times. Similarly, upon winding through the n-quantum flux path C once, the change in the gauge function (as given by (3.10)) is the same as that resulting from winding n times through a one-quantum flux path. The commutation relation between A(C') and B(C) may therefore be stated as

$$A(C')\,B(C) = B(C)\,A(C')\,e^{i2\pi n/N} \tag{3.13}$$

where C' and C are two equal-time closed curves that wind through each other n times, and B(C) creates one quantum of flux along C.

Taking the vacuum expectation of both sides of (3.13), we obtain

$$\langle A(C')\,B(C)\rangle = \langle B(C)\,A(C')\rangle\,e^{i2\pi n/N} \tag{3.14}$$

The integer n now loses its originally well-defined geometrical meaning, because the curves C and C' can now be deformed into

Euclidean 4-space (x,y,z,τ), in which linkage has no geometrical meaning: Two curves linked in 3-space can be undone by continuous deformations into the fourth dimension and back.

To see this, consider two plane curves C and C' linked in 3-space at $\tau=0$. Suppose C' initially lies in the x-y plane. We rotate the plane of C' into the fourth dimension, until it lies in the x-τ plane. Then a film sequence of 3-space will look like that depicted in Fig. 3.2. The curve C' intersects 3-space only at two points on the x-axis: where it enters, as marked by ⊙, and where it leaves, as marked by ⊗. These two points move apart from a single point at some negative time, and then come together again and disappear at some positive time. The curve C appears only at the instant $\tau=0$.

We may deform C' continuously by displacing the point ⊗ along the space curve D at the instant $\tau=0$. (See Fig. 3.2.) It is clear that, if we displace it out of the plane of C, even by an infinitesimal amount, then it is possible to go back continuously to an equal-time situation in which C and C' are unlinked, thereby changing n by one unit. The process described does not depend on the shapes of C and C', and goes through even in the limit when the minimum distances among C, C' and D approach infinity.

The only goemetrical way to render n well-defined is to regard some surface spanned by C, or by C', or both of these, as surfaces of discontinuity, with the rule that n changed by one unit whenever the locus of C crosses the C' surface, or vice versa. This means that the phase of $\langle A(C')\, B(C)\rangle$ can be defined up to an additive multiple of $2\pi/N$, which can be fixed by convention. The important thing is that the phase jumps across the surfaces mentioned.

Since the ambiguity of n arises only when we take the vacuum expectation, the remedy must ultimately come from dynamics. The following alternatives are considered by 'tHooft[3].

One possibility is that in the vacuum the curves C and C' can really "feel" each other, no matter how far separated they are. This would require the existence of long-ranged fields in the theory, i.e. "photons." In that case, no dynamical surfaces of discontinuity are needed, and presumably there will be none.

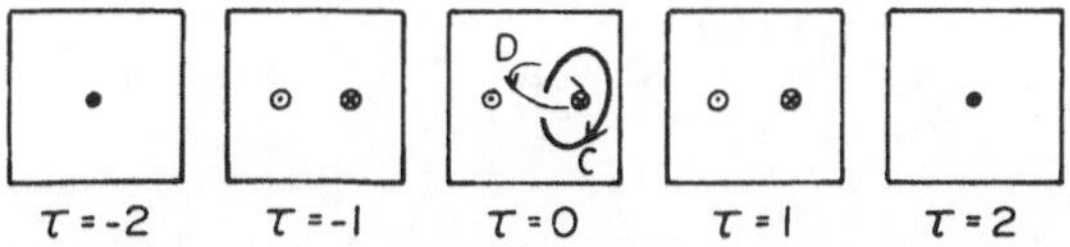

Fig. 3.2. Film sequence showing how to undo a knot by going into the fourth dimension (see text for details).

If there are no "photons" in the theory, then there must be dynamical surfaces of discontinuity with finite energy per unit area. If C acquires such a surface, then one would expect $\langle B(C)\rangle \sim e^{-\lambda\Sigma(C)}$, or magnetic order. If C' does, then one would expect $\langle A(C')\rangle \sim e^{-\lambda\Sigma(C)}$, or electric order. It is also possible that both develop surfaces, in which case there would be simultaneous electric and magnetic order, a case 'tHooft[3] allowed. However, to be consistent with the premise that what is not needed will not happen (because it would cost extra energy), this case should be eliminated.

We conclude that QCD with SU(N) gauge symmetry, with or without Higgs fields, falls into one of the following categories. If there are no "photons" then either quarks or monopoles must be confined; if there are "photons" then neither quarks nor monopoles need be confined. This is summarized in Table I.

TABLE I. CATEGORIES TO WHICH QCD MAY BELONG

(1 = yes, 0 = no)

Category	"Photons"	Magnetic Order: Monopole Confinement	Electric Order: Quark Confinement	Remark
I	1	0	0	Like Weinberg-Salam model
II	0	1	0	Like super-conductivity
III	0	0	1	What we hope for

2. Electric Order?

In general we expect the electric order parameter to have the behavior

$$\langle A(c)\rangle \sim a_o + a_1 e^{-\lambda_1 L(C)} + a_2 e^{-\lambda_2 \Sigma(C)} + \ldots \qquad (3.15)$$

What happens to the electric flux created by A(C) may be correlated with the categories of Table I:

I. If $a_0 \neq 0$, it will disperse, as in the quantum electrodynamic vacuum.

II. If $a_0=0$, but $a_1 \neq 0$, it will be locally absorbed, as in an electric plasma.

III. If $a_0=a_1$, but $a_2 \neq 0$, it will be confined to an electric flux, which then shrinks to nothing. This is the case of electric order.

Electric order means that the vacuum fluctuations are dominated by local fluctuations not of charge density, but of sourceless energy moment density, which must have the form

$$\vec{d} = \nabla \times \vec{k}, \tag{3.16}$$

where $\vec{k}$ is an equivalent magnetic current density. In the presence of external electric flux, magnetic current loops will be induced, confining the flux to an electric flux tube. By the same token, a finite flux tube will appear between well-separated static quark and antiquark. These magnetic current loops, like supercurrent loops in a superconductor, should arise from "magnetic Cooper pairs." What are they, and what could be a mechanism for their formation?

In Abelian electromagnetism there can be no magnetic current density. In QCD, however, it may exist by virture of the self interactions of the gluons (gauge-field quanta). The non-Abelian nature of the gauge group is therefore essential. Let us first clarify the necessary transformation properties of a magnetic current under color SU(3).

There are three "colors" and eight "charges." "Color" refers to an axis in the 3-dimensional vector space of the fundamental representation, and "charge" refers to a generator of the group. In the fundamental representation [3], the eight generators may be chosen to be the hermitian traceless 3x3 Gell-Mann matrices $\frac{1}{2}\lambda_a$, with $\mathrm{Tr}\lambda_a\lambda_b = 2\delta_{ab}$. A group element is represented by a unitary 3x3 matrix

$$\Omega = e^{\frac{i}{2}\lambda_a\omega_a}, \tag{3.17}$$

where ω_a are arbitrary real numbers. It effects the transformation $|3\rangle \to |3\rangle'$, with

$$|3\rangle = \begin{pmatrix} q_1 \\ q_2 \\ q_3 \end{pmatrix}, \quad |3\rangle' = \Omega|3\rangle \tag{3.18}$$

where q_i is called a quark of color index i (i = 1,2,3). There is a distinct conjugate representation $[\bar{3}]$, with vectors

$$<\bar{3}| = (q^1 q^2 q^3), \quad <\bar{3}|' = <\bar{3}|\Omega^{-1} \tag{3.19}$$

where q^i is called an antiquark of color index i (i = 1,2,3). Clearly

$$<\bar{3}|3> = q^i q_i \tag{3.20}$$

is invariant under Ω, i.e. a color singlet. Magnetic current, like $\nabla\times\vec{E}$, transforms according to the adjoint representation [8]:

$$\vec{k}^1 = \Omega\vec{k}\Omega^{-1}, \quad (\vec{k} = \frac{1}{2}\lambda_a \vec{k}_a) \tag{3.21}$$

Thus it transforms like the traceless part of $|3><\bar{3}|$, an expression of the fact that

$$[3] \times [\bar{3}] = [1] + [8]. \tag{3.22}$$

It is now clear that, as far as color transformation properties are concerned, an electric flux tube has the structure

$$|3><\bar{3}|3><\bar{3}|3>\ldots<\bar{3}|3><\bar{3}|. \tag{3.23}$$

It transports color from one end to the other, with no local color density in between. It does for color what the flux tube in superconductivity did for magnetic charge. (See Fig. 2.7.)

The really difficult problem concerns the space-time properties of the flux tube. How do we construct $\vec{k}$ from the gauge fields? What makes the flux tube stable? We can only make some qualitative guesses.

Thorn[4] has suggested that the vacuum of pure QCD (i.e., QCD with neither quarks nor Higgs fields), is a condensate of gluons. The qualitative argument rests on an instability of the bare vacuum, and the direction to which that might lead:

(a) The asymptotic freedom of pure QCD leads to an attraction between static quark and antiquark at small separations. In terms of color channels, this means that $<\bar{3}|$ and $|3>$ attract each other, and thus so would $|3><\bar{3}|$ and $|3><\bar{3}|$. That is, two gluons attract each other in the octet channel.

(b) In perturbation theory, gluons are massless. Therefore any attraction between gluons will lead to an instability, the initial effect of which would be the creation of gluons in order to lower the energy. If there are no other effects to counter this tendency, there will be an avalanche of gluons, making the vacuum a condensate of multi-gluon "bound states." The vacuum fluctuations would be closed gluon strings, whose color structure is represented by the trace of (3.23).

(c) A static quark and antiquark placed in the vacuum, separated by a large distance, will cause one of these closed gluon strings to break (electric field can destroy electric order) and they will attach themselves to opposite ends of the broken string.

To work out the dynamics is an extremely difficult problem, especially because pure QCD contains no free parameters, but only a relation between renormalized coupling constant and an arbitrarily chosen mass scale. One can vary things only by going to a neighboring theory, e.g. by varying the number of colors, or the dimensionality of space-time. Thorn[4] has examined the theory in the limit of an infinite number of colors, using techniques of the infinite momentum frame, and shows that at least in this limit, the gluon-gluon attractions dominate the dynamics, leading to the stability of the gluon string. Table II summarizes the view presented here by drawing parallels from superconductivity.

TABLE II. PARALLELS BETWEEN PURE QCD AND SUPERCONDUCTIVITY

Asymptotic freedom	$\leftrightarrow$	Electron-lattice interaction
Gluon-gluon attraction	$\leftrightarrow$	Electron-electron attraction
Zero mass of bare gluons	$\leftrightarrow$	Existence of bare Fermi surface
Instability of bare vacuum	$\leftrightarrow$	Instability of bare Fermi surface
Multi-gluon bound states	$\leftrightarrow$	Cooper-pair current loops
Electric flux tube	$\leftrightarrow$	Magnetic flux tube

ACKNOWLEDGEMENT

It is a pleasure to thank Charles Thorn for extensive and interesting discussions on the subject matters covered here.

REFERENCES

1. K. G. Wilson, Phys. Rev. D10, 2445 (1974).
2. J. Kogut and L. Susskind, Phys. Rev. D9, 3501 (1974); D16, 395 (1975). S. Mandelstam, Phys. Reports, 23C, 245 (1976). A. M.

Polyakov, Nucl. Phys. B120, 429 (1977). C. Callan, R. Dashen, and D. Gross, Phys. Rev. D17, 2717 (1978).

3. G. 'tHooft, Nucl. Phys. B 138, 1(1978).

4. C. Thorn, Phys. Rev. D19, 639(1979).

5. Some general references on superconductivity are:
L. D. Landau and E. M. Liftschitz, "Electrodynamics of Continuous Media," Pergamon, Oxford (1960), Chapt. VI.
G. Richayzen, "Theory of Superconductivity," Interscience, New York (1965).
P. G. DeGennes, "Superconductivity of Metals and Alloys," Benjamin, New York (1966).

6. N. Byers and C. N. Yang, Phys. Rev. Lett. 7, 46 (1961), contains a more rigorous treatment of flux quantization in a superconductor.

7. H. B. Nielsen and P. Olesen, Nucl. Phys. B61 (1973), discuss the flux tube in a Higgs model, a version of the Landau-Ginsburg mean field theory.

DISCUSSION

CHAIRMAN: K. Huang

Scientific Secretaries: H. Grosse and J.P. Sursock

DISCUSSION

- GROSSE:

In lattice gauge models Wilson's loop criterion has been shown to be true. Is that an example of an electric system having electric order?

- HUANG:

Yes, it is, but somehow this is put in from the beginning, because you de-emphasize the kinetic energy, and therefore confinement comes easier.

- HA:

Superconductivity can only occur when the temperature is below a certain critical value. What would correspond to this parameter in the case of quark confinement?

- HUANG:

It corresponds to the inverse of the coupling constant in the gauge theory. Now, as you know, the coupling constant in pure QCD depends on the momenta involved in the problem. It means that the system you are studying would be either in a phase with electric order or in another phase without electric order, depending on what range of momenta is important for the phenomenon being studied.

- GROSSE:

It is very surprising that neither instantons nor merons play any role in the vacuum structure, although they give optimal contributions to the functional integral. Could you comment on the role of these classical solutions in QCD?

- HUANG:

It is not really known that they give optimal contributions. There is a great variety of notions coming out of QCD: Asymptotic freedom, dimensional transmutation, monopoles, instantons, merons etc. As a parallel one might recall that in superconductivity there are also many notions like photons, phonon-electron interactions, plasma oscillation, screened Coulomb-interactions, Cooper pairs etc.; but the essence of superconductivity lies in the formation of Cooper pairs. We still have to ferret out the right mechanism in QCD.

- BHANOT:

What guarantees that there is no fundamental length in Nature? If there is, then the lattice theory need not have a continuum limit and confinement is trivial?

- HUANG:

Nothing guarantees that. In fact, the lattice theory may even be true. It is just my point of view that one should start with the continuum theory.

- HA:

We have seen a mechanism for quark confinement based on analogy with superconductivity. Whether this mechanism works or not depends very much on the size of the coupling constant of the theory. So it appears that confinement is really a dynamical problem.

- HUANG:

Yes, indeed. Quark confinement in QCD can be demonstrated only by solving the dynamics. An example of this is Thorn's work in the limit of an infinite number of colors.

- SURSOCK:

Are there repulsive forces between gluons?

- HUANG:

There is attraction in the octet channel but in some other channel there may be repulsion.

- GROSSE:

Can you make some comments on how your scheme would apply in two dimensions?

- HUANG:

The scheme based on 't Hooft's commutation relations between A(c) and B(c) is valid only in three spatial dimensions.

- GROSSE:

An analogy of this approach to higher dimension would be to postulate modified commutative relations.

- HUANG:

The concept of linkage of two closed curves exists in three dimensions. One would have to replace curves by surfaces.

- KLEINERT:

Can you explain what you mean by these surfaces spanning the "inside" of the flux curves c and c'?

- HUANG:

This surface expresses the fact that whenever you create a ring the energy involved is proportional to the surface.

- KLEINERT:

Why do you need gluon-gluon attractions to give you a Bose condensate, when you can get a condensate in an ideal Bose gas?

- HUANG:

You need it because there is no number-conservation for gluons, as you have in, say, liquid helium.

- FARRAR:

In Table II quark confinement also means no "photons." This means that there can be no long-ranged force between hadrons. Can one see a mechanism for this?

- HUANG:

It is very hard to see dynamically. The virtue of 't Hooft's method is that you avoid the use of dynamics, but the disadvantage is that you have no idea on how this comes about.

- ROTH:

You stated that Thorn obtained confinement in the limit as the number of colors goes to infinity. Was he able to obtain higher order terms in $1/N_c$?

- HUANG:

Thorn was able to make crude estimates but was not able to make higher order corrections.

- KLEINERT:

In superconductors we understand confinement by considering superconducting charge squeezing magnetic field lines. In QCD we would have gluonic Cooper pairs acting as magnetic monopoles squeezing electric field lines. Can you clarify this point?

- HUANG:

When there is gluon-gluon attraction, you get not only two gluons bound states, but also multi-gluons bound states. In superconductivity you can think of the ground state as a distribution of current loops of Cooper pairs. When a monopole is introduced, these current loops will line up to form a solenoid containing the magnetic flux of the monopole. In QCD, what might happen is that these gluon bound states could be considered as magnetic current loops. They would form a solenoid confining the electric flux when a quark is introduced.

SOFT QCD : LOW ENERGY HADRON PHYSICS WITH CHROMODYNAMICS†

Nathan Isgur*

University of Oxford

Department of Theoretical Physics
1 Keble Road, Oxford OX1 3NP

ABSTRACT

After a brief review of QCD and a discussion of some of the evidence for colour and QCD, the elements of QCD-inspired quark potential models are introduced. They are then applied to the study of hadronic structure with special emphasis on baryons. While a number of open questions remain, there is good evidence that the new ingredients suggested by QCD --- especially the ideas of quark hyperfine interactions and flavour independent confinement potentials --- provide a simple dynamical understanding of much of low energy hadron physics.

---

†Lectures presented at the XVI International School of Subnuclear Physics, Erice in August 1978.

*On leave until September 1978 from the Department of Physics, University of Toronto, Toronto, Ontario M5S 1A7, Canada.

I. Introduction

Quantum chromodynamics, as you all know, is a quantum field theory of strong interactions based on the colour SU(3) gauge group. It is widely held to be the most likely candidate theory for the strong interactions.

Aside from many attractive aesthetic and phenomenological features of this theory, some of which will be dealt with in these lectures, it also has the very important feature of being rigorously testable and therefore of being disproved: while it is a strongly interacting field theory at large distances, at short distances the theory becomes asymptotically free and perturbative techniques can be applied. So although the theory cannot be solved completely (what theory can be?), it can and is presently being confronted in a direct and unambiguous way with experiment. I will mention some of these predictions and tests of hard (high q^2) QCD in this first, introductory lecture.

In the remainder of my lectures, however, I would like to introduce you to some ideas which have sprung from QCD and which have formed a basis for some recent work on low energy hadron dynamics. By using the word "sprung" I mean to imply that these ideas, while inspired by QCD, have in no way yet been rigorously derived from the theory. The phenomena which I will be discussing --- hadron spectroscopy, particle decays, low energy moments, etc. --- involve low q^2 and so can not be reached by a perturbative treatment. By "soft QCD" I therefore mean to indicate both a level of q^2 and of rigour! This is not, of course, an unequivocal disadvantage of this class of models since as a result even should QCD be wrong, it could be that some of the ideas we shall apply under the guise of "soft QCD" will survive; as you will see, they appear to form the basis of at least a successful phenomenology for low energy hadron dynamics.

Finally I should stress that the following account is in the end taken from a rather personal perspective; for other points of view I refer you to the literature and to other recent reviews [1].

II. A Brief Review of Hard QCD

The QCD Lagrangian[2] follows from demanding that the fermionic would be invariant under local SU(3) colour[3] transformations. For the quarks which are assigned to a triplet (3) representation* this requirement leads to the necessity of eight massless coloured gluons as compensating fields analogous to the photon of electrodynamics and to the Lagrangian

$$L_{QCD} = \bar{q}_\alpha [\gamma^\mu (i\partial_\mu - g \frac{\lambda^i_{\alpha\beta}}{2} G_{i\mu}) - m] q_\beta - \frac{1}{4} F^i_{\mu\nu} F^{i\,\mu\nu} \qquad \text{(II.1)}$$

where

$$F^i_{\mu\nu} = \partial_\mu G^i_\nu - \partial^i_\nu G_\mu - igf^{ijk} G^j_\mu G^k_\nu \qquad \text{(II.2)}$$

---

*singlets would be leptons; larger representations than the 3 have been considered: see E. Ma, Phys. Lett. 58B, 442(1975) and Phys. Rev. Lett. 36, 1573(1976); G. Karl, Phys. Rev. D14, 2374(1976); and F. Wilczek and A. Zee, Phys. Rev. D16, 860(1977).

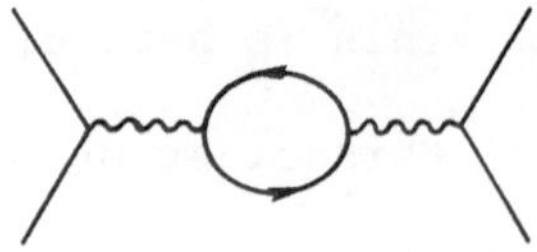

Figure II.1: A vacuum polarization graph in QED.

with the λ^i the 3x3 representations of SU(3), introduced by Gell-Mann for $SU(3)_{flavour}$, which satisfy the relations

$$\left[\frac{\lambda^i}{2}, \frac{\lambda^j}{2}\right] = if^{ijk}\frac{\lambda^k}{2} \qquad \text{(II.3)}$$

$$\left\{\frac{\lambda^i}{2}, \frac{\lambda^j}{2}\right\} = \frac{1}{6}\delta^{ij} + d^{ijk}\frac{\lambda^k}{2} \qquad \text{(II.4)}$$

($\alpha,\beta,.. = 1, 2, 3; i, j, k,.. = 1,.......8$).

Most of the known and supposed consequences of QCD follow from the fact that it is an asymptotically free theory, so we begin by briefly discussing this conclusion.[4] First let's recall the situation in QED. There vacuum polarization diagrams like Figure II.1 lead to the result that

$$\alpha(\vec{q}^{\,2}) = \alpha(\mu^2)\left[1 + \frac{\alpha(\mu^2)}{3\pi}\ln\frac{\vec{q}^{\,2}}{\mu^2} + ...\right] \qquad \text{(II.5)}$$

$$= \frac{\alpha(\mu^2)}{1 - \frac{\alpha(\mu^2)}{3\pi}\ln\vec{q}^{\,2}/\mu^2} + O(\alpha^2) \qquad \text{(II.6)}$$

by summing the leading logarithms (μ is the renormalization point). Thus as $\vec{q}^{\,2}$ increases the effective charge increases corresponding to probing through the shielding of the virtual pairs. In QCD

Figure II.2: A vacuum polarization graph in QCD.

in addition to the quark loop contributions, which also lead to shielding, there are gluon vacuum polarization diagrams like Figure II.2 which give antishielding. In fact one gets

$$\alpha_s(\vec{q}^2) = \alpha_s(\mu^2)\ [1 + \frac{N_f \alpha_s(\mu^2)}{6\pi}\ \ell n\ \frac{\vec{q}^2}{\mu^2} - \frac{11\alpha_s(\mu^2)}{4\pi}\ \ell n\ \frac{\vec{q}^2}{\mu^2} + \ldots] \tag{II.7}$$

where the term proportional to the number of quark flavours N_f comes from the quark loops* and the second term comes from gluon loops so that

$$\alpha_s(\vec{q}^2) = \frac{\alpha_s(\mu^2)}{1 + \frac{33 - 2N_f}{12\pi}\ \alpha_s(\mu^2)\ \ell n\ \frac{\vec{q}^2}{\mu^2}} + O(\alpha_s^2) \tag{II.8}$$

Thus in QCD (so long as $N_f \leq 16$) as $\vec{q}^2$ increases the effective charge decreases logarithmically to zero, and we expect to be able to calculate short distance phenomena perturbatively.

Of course in neither case can the resulting formulae be trusted

---

*Only quarks with mass $m << \vec{q}^2$ contribute, so N_f is actually the the effective number of quarks.

for $\vec{q}^2$ where the effective coupling constant has become large. Thus $\alpha(\vec{q}^2)$ seems to have a pole at large $\vec{q}^2$; for such $\vec{q}^2$ the higher order terms are not negligible, and the result (II.6) breaks down, but we still learn that α gets large in this limit. On the other hand $\alpha_s(\vec{q}^2)$ increases as $\vec{q}^2 \to 0$ and although the result (II.8) cannot be trusted in that region we still learn that α_s gets large in this limit. It is this observation that leads to the hope of confinement, about which we shall have more to say shortly.

There are unfortunately at present very few rigorous tests of QCD (or for that matter for the mere presence of the colour degree of freedom), although a growing number of predictions should be tested in the forseeable future. Some of the most often quoted "classic" tests of the theory are:

1) QCD explains the approximate scaling observed in deep inelastic scattering and also makes definite predictions for the way scaling should be broken. In particular, the theory predicts the q^2 dependence of the moments of the quark and gluon distribution functions. These predictions have been recently tested in neutrino data[5)] where they appear to be in impressive agreement with experiment (see Figure II.3).

2) The ratio $R = \dfrac{\sigma(e^+e^- \to \text{hadrons})}{\sigma(e^+e^- \to \mu^+\mu^-)}$ is in fact approximately constant between new $q\bar{q}$ thresholds and has a value indicating the presence of the factor of three coming from colour as predicted. Although the actual situation is somewhat more complicated, the

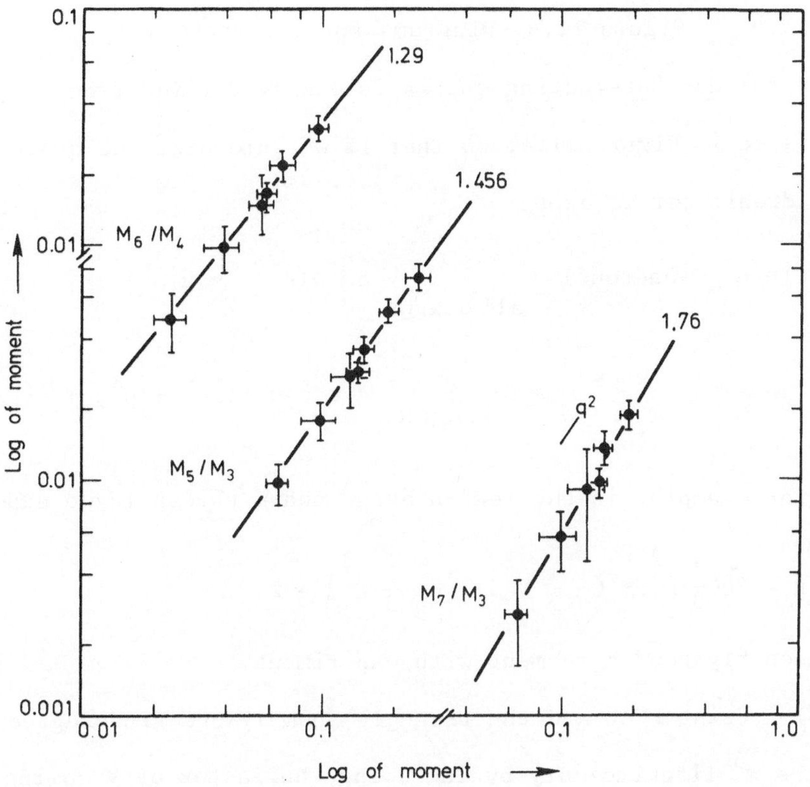

Figure II.3: A test of the QCD predictions for the q^2-dependence of $F_2(x,q^2)$. See reference (5) for details.

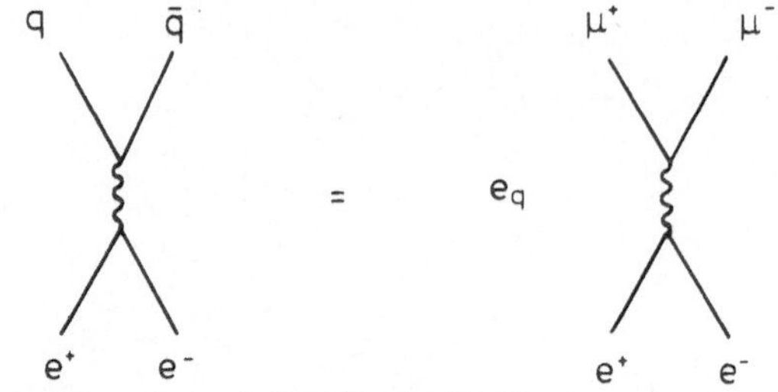

Figure II.4: Diagrams for the ratio R

ratio R for non-interacting quarks is easily derived from the equality illustrated in Figure II.4, so that if we interpret the quark line as a hadronic jet we expect

$$\sigma(e^+e^- \to \text{hadrons}) = \sum_{\text{all quarks}} e_q^2\, \sigma(e^+e^- \to \mu^+\mu^-)$$

$$= 3\Big(\sum_{\text{flavours}} e_q^2\Big)\, \sigma\,(e^+e^- \to \mu^+\mu^-)$$

Thus, for example, in the region below charm threshold we expect

$$R \simeq 3\Big[\Big(\frac{2}{3}\Big)^2 + \Big(-\frac{1}{3}\Big)^2 + \Big(-\frac{1}{3}\Big)^2\Big] = 2$$

in reasonably good agreement with experiment.

3) Using PCAC and the triangle anomaly one finds agreement with the π^0 lifetime only by including the factor of 9 coming from colour.

4) The existence of the colour degree of freedom is essential to our understanding of baryon spectroscopy since the Δ^{++} is totally symmetric in space, spin, and flavour quantum numbers.[2)] This was the first and in some ways remains still the best evidence for the existence of colour as it is completely qualitative.

More recently, diagrammatic methods have been applied to the study of deep inelastic phenomena.[6] From such studies it is possible, under reasonable assumptions about the way in which jets of quarks and gluons are transformed into hadrons, to add to this list several new tests among which are:

5) QCD predicts that the dominant contributions to $\ell N \to \ell' X$ and $e^+e^- \to X$ should consist of two jets; in e^+e^- at least this is certainly the case.

6) The use of the parton model to describe non-classical processes like $pp \to \mu^+\mu^- X$ and $pp \to$ large p_T jets in terms of the same q^2-dependent structure functions seen in deep inelastic lepton scattering has been given a firm, though not yet completely rigorous, theoretical grounding.

In addition to the tested predictions of "hard QCD" which I have listed here, there is now a large body of other predictions awaiting confrontation with experiment. As already mentioned, this is one of the most attractive features of QCD: if it is wrong we should know relatively soon.

III. Soft QCD: An Introduction

Aside from these more or less rigorous conclusions based on QCD there are a number of less well-grounded ideas which have emerged from the theory, and it is these ideas which I principally wish to discuss in these lectures. Here we discuss the background to these ideas in preparation for the detailed applications we will pursue later.

A. A Potential Model for Confinement

The first issue we must address is confinement. This is related to the baryon statistics problem: the colour degree of freedom allows the quarks in the Δ^{++} to be in a totally antisymmetric state, but since they are in a symmetric state of space, spin, and flavour, this is only achieved if they are in a totally antisymmetric colour state. Since

$$3 \times 3 \times 3 = 1 + 8 + 8' + 10$$

this can be done since the singlet is totally antisymmetric in colour. The success of the symmetric quark model for baryon spectroscopy then indicates that only the colour singlet representations exist (at least at low energy). Since $3 \times \bar{3} = 1 + 8$ and since the extra mesons corresponding to fully exploiting the colour degree of freedom have also not been seen it is natural to conclude that only colour singlet mesons have been observed as well. Since single quarks, diquarks, four quark states, etc., cannot be in a colour singlet and have also not been observed, it is natural to elevate this idea to a principle: only colour singlet states exist in nature. Note that this principle does not rule out the existence of combinations like $qq\bar{q}\bar{q}$; we shall return to a brief discussion of such states later.

Obviously, stating this principle, no matter how emphatically, does not solve the confinement problem. Nevertheless, once stated the principle makes QCD seem a likely place to look for a solution. First of all such a principle can obviously only emerge from a theory in which the forces depend on colour! Secondly, we have already seen that in QCD the coupling between colour charges increases with increasing separation (up to the point where perturbation theory breaks down --- after that we don't yet know what happens).

As a simple model of such a force[7)] based on QCD let's consider a non-relativistic colour-symmetric two body potential of the form

$$V_{qq}(r_{12}) = -V(r_{12}) \sum_{i=1}^{8} \left(\frac{\lambda_i}{2}\right)_1 \left(\frac{\lambda_i}{2}\right)_2 \qquad \text{(III.1a)}$$

$$V_{q\bar{q}}(r_{12}) = +V(r_{12}) \sum_{i=1}^{8} \left(\frac{\lambda_i}{2}\right)_1 \left(\frac{\lambda_i^*}{2}\right)_2 \qquad \text{(III.1b)}$$

$$V_{\bar{q}\bar{q}}(r_{12}) = -V(r_{12}) \sum_{i=1}^{8} \left(\frac{\lambda_i^*}{2}\right)_1 \left(\frac{\lambda_i^*}{2}\right)_2 \qquad \text{(III.1c)}$$

where $r_{12} = |\vec{r}_1 - \vec{r}_2|$, $(\lambda_i)_j$ is the ith λ-matrix acting on the colour wave function of the jth quark or antiquark, and $V(r)$ is the effective long-range colour confinement potential(for definiteness one may think of $V(r) = \frac{1}{2}Kr^2$ or br). Now while V_{qq}, $V_{q\bar{q}}$, and $V_{\bar{q}\bar{q}}$ cannot be calculated perturbatively so that they certainly cannot be associated, for example, with one gluon exchange, it is reasonable to assume that they will connect smoothly onto the one gluon exchange chromoelectric potential; this accounts for the signs and complex conjugations that appear in equations (III.1).

I.e. we assume that even at large distances it is the colour charge of the quark or antiquark that counts. It should be stressed at this point that these potentials depend only on the colour and not the flavour of the quark; this assumption, which has support from the study of QCD on a lattice,[8)] means that the eigenstates of the confining potential alone (before the introduction of spin-dependent effects, etc....) have flavour symmetry-breaking only via the explicit appearance of the quark masses in the kinetic energy term of the Hamiltonian.

With these potentials we can calculate the energy of any static configuration of quarks and antiquarks to check for confinement and saturation. We illustrate the way the model works by considering $e^+e^- \to$ hadrons. There are two criteria for a given cluster of quarks and antiquarks to emerge from the interaction region and be observed as a hadron: 1) the cluster must be able to escape from the remaining quarks and antiquarks, and 2) the cluster must be at least temporarily stable. Let us consider the first requirement; specifically, let's imagine that a cluster carrying the quantum numbers of a 3 of $SU(3)_{colour}$ tries to escape from the remaining quarks and antiquarks. Since the overall final state is an SU(3) singlet (the photon is presumably a colour singlet), the remaining debris must be in a $\bar{3}$. However the interaction (III.1) has the property that it confines any representation (whether elementary or composite) to its conjugate representation (whether elementary or composite) with a strength proportional to the colour Casimir quantum number of the representation; thus only colour singlets can escape the region of hadronization. So let's consider an escaping singlet. It will be a stable cluster if its only allowed decay mode is into coloured objects; the only such singlet objects are of course the $q\bar{q}$ and qqq ground states. In addition, of course, one can in principle have two types of semi-stable hadrons: 1) $q\bar{q}$ or qqq excited states which decay into singlets by $q\bar{q}$ pair creation, and 2) multiquark configurations like $qq\bar{q}\bar{q}$ which can decay by either $q\bar{q}$ pair creation (baryonium?)[9) or by rearrangement (cryptoexotics?).[10)

It goes without saying, of course, that the scheme outlined above does not "explain" confinement. But it is useful to know that it is possible to construct simple models based in a sensible way on QCD which have the desired properties. Of course confinement may also be achieved by the use of bags, strings, etc., and our treatment has many features in common with such schemes; we will not, however, deal explicitly with other confinement models here.

Finally, before moving on to the next topic we note that by taking the wave functions

$$|M> = \frac{1}{\sqrt{3}} \delta^{\alpha\beta} |q^{\alpha}\bar{q}^{\beta}> \qquad \text{(III.2)}$$

$$|B> = \frac{1}{\sqrt{6}} \varepsilon^{\alpha\beta\gamma} |q^{\alpha}q^{\beta}q^{\gamma}> \qquad \text{(III.3)}$$

of a meson and a baryon, one can using (II.3), (II.4), (III.1a), and (III.1b) calculate the relevant effective (i.e., colour-averaged) quark-antiquark and quark-quark potentials appropriate to a meson or a baryon in terms of V(r). The result is that

$$V_{q\bar{q}(\text{meson})}(r) = \frac{4}{3}V(r) \qquad \text{(III.4a)}$$

$$V_{qq(\text{baryon})}(r) = \frac{2}{3}V(r) \qquad \text{(III.4b)}$$

so that baryon and meson bound states are governed by related effective potentials.[11)] Incidentally, although it at first sight seems paradoxical that both $q\bar{q}$ and qq should be attractive, in stark contrast to electromagnetism, there is a simple explanation of this result. A quark in a baryon must see the remaining diquark (which could be in either the $\bar{3}$ or 6 representation in 3 x 3) in the $\bar{3}$ state since only $\bar{3}$ can combine with the quark's 3 to make a singlet. But the potentials (III.1) have the property that they are sensitive only to the net colour of an object, so the diquark $\bar{3}$ attracts the quark 3 just as though it were an antiquark. It follows that the effective interaction per quark in the diquark is just half the $q\bar{q}$ interaction, which is the result (III.4).

B. Quark Hyperfine Interactions

While the potentials (III.1) express the hoped-for long range confinement characteristics of QCD, they are of course neither expected or observed to be the whole story. With only these confinement potentials we would, for example, have Δ-N, ρ-π, D^*-D, etc..., degeneracy. Splittings of this type have always been attributed in quark models to the existence of a spin-spin interaction and it is gratifying that such forces have a very natural origin in QCD: chromomagnetic forces analogous to ordinary magnetic dipole-magnetic dipole, or hyperfine, interactions.

In close analogy with QED, in QCD one expects a hyperfine interaction of two quarks i and j which to order α_s is given by[12)]

$$\hat{H}^{ij}_{hyp} = \kappa \frac{\alpha_s}{m_i m_j} \left\{ \frac{8\pi}{3} \vec{S}_i \cdot \vec{S}_j \delta^3(r_{ij}) + \frac{1}{r_{ij}^3} \left[\frac{3\vec{S}_i \cdot \vec{r}_{ij} \vec{S}_j \cdot \vec{r}_{ij}}{r_{ij}^2} - \vec{S}_i \cdot \vec{S}_j \right] \right\} \quad \text{(III.5)}$$

$$\text{where } \kappa = \begin{cases} \left\langle \frac{\vec{\lambda}_i}{2} \cdot \frac{\vec{\lambda}_j}{2} \right\rangle_{qq \text{ in a baryon}} = \frac{2}{3} & \text{(III.6)} \\ \left\langle \frac{\vec{\lambda}_i}{2} \cdot \frac{-\vec{\lambda}_j^*}{2} \right\rangle_{q\bar{q} \text{ in a meson}} = \frac{4}{3} & \text{(III.7)} \end{cases}$$

and where $\alpha_s = g^2/4\pi$ and m_i and m_j are the quark masses. The second term in (III.5), called the tensor term, is just the $\vec{\mu}_i \cdot \vec{B}^j_{external}$ interaction of one colour magnet with the external dipole field of the other (see Figure III.1) and is the familiar force between two macroscopic magnets. The first term is the less familiar Fermi contact term which may (as illustrated in the

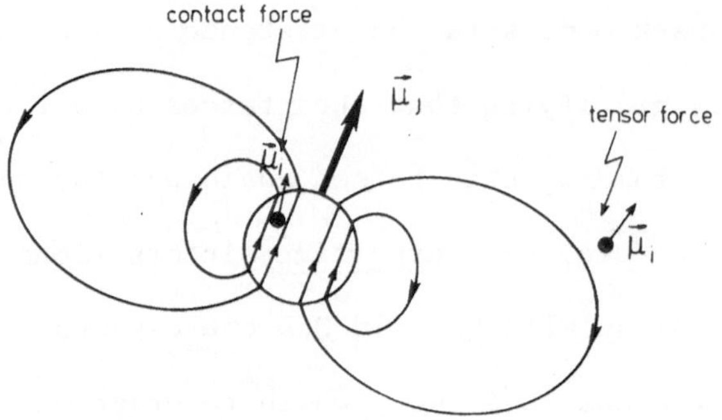

Figure III.1: A representation of the origin of the tensor and contact parts of the hyperfine interaction.

Figure) be visualized as arising from the $\vec{\mu}_i \cdot \vec{B}_{internal}$ interaction of colour magnet i with the colour magnetic field internal to colour magnet j. The Figure makes clear an important fact which we want to immediately stress: if there really are hyperfine interactions between quarks, both the tensor and contact terms must be present and they must be present with the correct relative strength, namely $\frac{8\pi}{3}$*. Incidentally, while the contact term may be a less familiar physical effect, it has one very well-known consequence in electromagnetism: the energy difference between the singlet and triplet spin hydrogen ground states, which is responsible for

*This $\frac{8\pi}{3}$ is just a geometrical factor, namely $2 \times (\frac{4\pi}{3})$, the $\frac{4\pi}{3}$ coming from the volume of the sphere in which $\vec{B}_j$ is parallel to $\vec{\mu}_j$.

the famous and ubiquitous 21 cm line, is due to the contact term (the tensor term has zero expectation value in an s-wave). Thus the $\frac{8\pi}{3}$ in (III.5) has been confirmed experimentally!

Given the close formal analogy between QCD and QED, it would be surprising if they were to give rise to forces of a completely unrelated nature; in fact it is clear that if QCD is correct then at short distances the quark-quark forces must be dominated by one gluon exchange and therefore be very much like electromagnetic forces mediated by one photon exchange. With the long range effects described in terms of the empirical confining potentials (III.1), it is this observation which led to the conjecture[13)] that all other effects may be approximated by one gluon exchange. This proposition should probably be treated with considerable caution; this "two-component" picture of the forces generated by QCD is almost certainly an oversimplification. Nevertheless, it has proven to be very fruitful to consider the possibility that there are analogues to electromagnetic forces in the strong interactions between quarks and in particular there is now, as we shall see, a substantial body of evidence in favour of the existence of the quark hyperfine interactions given by (III.5)

We conclude this section by mentioning one important qualitative feature of colour hyperfine interactions. By analogy with electromagnetism it is obvious that a $q\bar{q}$ pair in a S-wave with parallel spins will have more energy than a similar pair with $S = 0$, so the result $\rho > \pi$ follows automatically showing that this

SU(6)-breaking effect is exactly analogous to the 21 cm line of hydrogen. If $q\bar{q}$ with parallel spins goes up, however, it might have been expected that qq with parallel spins would come down making $\Delta < N$. Here, however, the non-abelian character of the forces comes into play and, just as for the confining force, we know that a q in the Δ sees the remaining two quarks in a $\bar{3}$ so that they have the chromomagnetic fields of an antiquark. Thus the result $\Delta > N$ also follows automatically from this picture and this splitting is also the analogue of the 21 cm line.

C. Other One-Gluon-Exchange-like Effects

In addition to colour magnetic dipole - colour magnetic dipole interactions, it would be natural to expect to see the analogues of other electromagnetic forces.[12)]

In particular, a quark colour magnet in motion in the chromoelectric field of another quark should see a colour magnetic field in its rest frame; the resulting torque will lead to $\vec{L}.\vec{S}$ (spin orbit) terms. It is not possible, however, to unambiguously calculate such effects in terms of one gluon exchange, as was done for hyperfine interactions, because there are inevitably other mechanisms which contribute spin-orbit effects. In particular, an accelerating spin will certainly experience a precession due to relativistic kinematics (Thomas precession) and such a precession appears in a non-relativistic Hamiltonian as an effective dynamical $\vec{L}.\vec{S}$ term. In electromagnetism it is the one photon exchange potential which is also responsible for the Thomas precession and in this case the effect is simply to reduce the "dynamical" $\vec{L}.\vec{S}$ interaction by a famous factor of 2. In our case, however, the quark is being accelerated by both the one gluon exchange potential and the confining potential; the effect will be to more nearly cancel --- or even possibly reverse --- the spin orbit interaction. A reliable calculation of such effects is non-trivial and we shall postpone further consideration of them until later. Fortunately, there is empirical evidence that, at least in baryons, spin-orbit forces are quite small so we can at least partially avoid dealing with this question.

In addition to these spin-dependent terms, the one gluon exchange potential contains a number of spin-independent terms, including of course its Coulomb-like piece. Such terms are difficult to distinguish from the (completely empirical) spin-independent confinement potential V(r) and so we shall generally simply lump them into V(r).

D. Quark Masses

Finally, to conclude this introduction to soft QCD, we need to discuss the values of the quark masses appropriate to this picture. Throughout the following we use the values

$$m_d \simeq m_u \simeq .35 \text{ GeV}$$
$$m_s \simeq .58 \text{ GeV}$$
$$m_c \simeq 1.5 \text{ GeV}$$

Since the mass of a particle is actually a function of the momentum flowing through its propagator, these masses must be considered to be the effective masses of the quarks relevant for this problem; they have no direct relation, for example, to the "PCAC masses".

One feature of these values is that if the quarks have zero anomalous magnetic moment (as indicated, for example, by photo-production amplitudes) then these quark masses lead to predictions for the baryon magnetic moments; for example

$$\mu_p = \frac{M_N}{M_d} \mu_N \simeq 2.8 \mu_N$$

$$\mu_\Lambda = -\frac{1}{3} \frac{m_d}{m_s} \mu_p = -.56\mu_N$$

$$\mu_n = -\frac{2}{3}\mu_p \simeq -1.9 \mu_N$$

in good agreement with the latest experimental values, but in the case of the Λ far from the SU(6) prediction. Note incidentally, this assumption of pointlike behaviour by the quarks is also implicit in the form of the hyperfine interaction (III.5). We shall see that there is evidence that the colour magnetic moments of heavy quarks are suppressed just as indicated by (III.5) in analogy with the above result for the Λ magnetic moment.

Finally, we note that with these masses m_u and m_d are sufficiently small that scaling at $q^2 \sim 1$ GeV^2 is plausible, but sufficiently large that one can understand the general success of the static SU(6) quark model. Since this question often arises, let me be slightly more specific: it is possible to convince yourself that a non-relativistic approximation is reasonable by noting that a harmonic oscillator wave function for the quarks in a proton, if adjusted to reproduce the observed proton charge radius, gives quarks of this mass less than 100 MeV of kinetic energy.

IV. A QCD-based Model for Baryons

We shall stress the baryons as the principle testing ground for the ideas of soft QCD. There are several reasons for this emphasis. Partly there is the fact that the existence of $q\bar{q}$ annihilation channels (via gluons) leads to effects in I = 0 mesons which are not simply treated in terms of a non-relativistic potential model (see Section V-B). Of course such questions can be avoided by concentrating on I = 1 and I = ½ states, but---and this is really the main point---the experimental information on meson spectroscopy is too sparse to allow the luxury of restricting one's attention to a subset of states. The situation may be illustrated by considering the status of the I = 1 P-wave mesons: there are four expected, but only the $A_2(1320)2^{++}$ and $B(1230)1^{++}$ are well established (the mass of the $A_1 1^{++}$ remains very uncertain and it is unclear whether the low-lying scalars like the δ are $q\bar{q}$ or multiquark states[10]). In contrast there are twenty reasonably well established P-wave baryon resonances, each in general having at least several well-measured decay amplitudes.

In addition we shall see that baryons have some special features, absent in mesons, which make them especially suitable for testing the idea of quark hyperfine interactions.

With our reasons for studying baryons established, we now bring together the ideas of the previous sections by explicitly stating our model[14,15]. We assume that the Hamiltonian for baryons is approximately

$$H = \sum_i m_i + H_o + H_{hyp} \qquad \text{(IV.1)}$$

where

$$H_o = \sum_i \frac{p_i^2}{2m_i} + \sum_{i<j} V^{ij}_{conf} \qquad \text{(IV.2)}$$

and

$$H_{hyp} = \sum_{i<j} H^{ij}_{hyp} \qquad \text{(IV.3)}$$

where H^{ij}_{hyp} is given by (III.5) with $\kappa = 2/3$ and V^{ij}_{conf} is the confinement potential assumed to be of the form (III.1a), i.e., a flavour-independent function of the relative qq separation. We shall see that this simple Hamiltonian seems to explain in a reasonably accurate way the properties of all low-lying baryons, including their masses and decay amplitudes! Neglected terms in (IV.1) like relativistic corrections, other one-gluon exchange effects, etc..., seem to be relatively unimportant effects at the level of 10-20%.

I would now like to describe how one goes about discussing

the baryons in this framework. Since we have a three body problem, and since this problem cannot in general be exactly solved, we do perturbation theory around the one soluble case :a harmonic confining potential. This choice turns out to generate a quite reasonable set of basis states; numerical studies have shown that the low lying states of many potentials (for example, linear plus Coulomb) can be approximated by wave functions of the harmonic oscillator form. Thus we assume that

$$V^{ij}_{conf} = \tfrac{1}{2}Kr^2_{ij} + U(r_{ij}) \qquad \text{(IV.4)}$$

where $U(r_{ij})$ is some unknown potential which we expect to incorporate a short-range attractive potential (the Coulomb-like piece) and deviations of the long range part of the potential from the harmonic oscillator form. (Note that a factor of 2/3 from colour averaging has been subsumed into K).

At this stage, of course, our treatment of the Hamiltonian (IV.1) is still general, and consequently insoluble. The approximation we take to solve for the eigenvalues and eigenfunctions of this system is simply to do perturbation theory in U and H_{hyp} in the harmonic oscillator basis. In so doing it turns out to be convenient to separate two cases, depending on the number of strange quarks in the baryon. In the S = 0 (and S = -3) sector the quarks all have equal mass, while in the S = -1 (and S = -2) sector one quark has a mass different from the other two, and these two cases contain some quite different physical effects. We explicitly discuss only the cases S = 0 and S = -1 since the S = -2 and S = -3 sectors follow trivially from these.

A. The Zeroth Order States in the S = 0 Sector

In this case $m_1 = m_2 = m_3 \equiv m$ and the relevant zeroth order Hamiltonian is

$$\tilde{H}_o = \frac{P_\lambda^2}{2m} + \frac{P_\rho^2}{2m} + \frac{3}{2}K(\rho^2 + \lambda^2) \tag{IV.5}$$

where

$$\vec{\rho} \equiv \frac{1}{\sqrt{2}}(\vec{r}_1 - \vec{r}_2) \qquad \vec{P}_\rho \equiv m\frac{d\vec{\rho}}{dt} \tag{IV.6}$$

$$\vec{\lambda} \equiv \frac{1}{\sqrt{6}}(\vec{r}_1 + \vec{r}_2 - 2\vec{r}_3) \qquad \vec{P}_\lambda \equiv m\frac{d\vec{\lambda}}{dt} \tag{IV.7}$$

are two relative coordinates and momenta remaining after separating out the centre-of-mass motion. The low lying eigenstates of this Hamiltonian may be chosen to be[18)]

N=0 $$\psi_{oo}^S \equiv \frac{\alpha^3}{\pi^{3/2}} e^{-\frac{1}{2}\alpha^2(\rho^2+\lambda^2)} \tag{IV.8}$$

1 $$\psi_{11}^{M_\rho} \equiv \frac{\alpha^4}{\pi^{3/2}} \rho_+ e^{-\frac{1}{2}\alpha^2(\rho^2+\lambda^2)} \tag{IV.9a}$$

1 $$\psi_{11}^{M_\lambda} \equiv \frac{\alpha^4}{\pi^{3/2}} \lambda_+ e^{-\frac{1}{2}\alpha^2(\rho^2+\lambda^2)} \tag{IV.9b}$$

2 $$\psi_{oo}^{S'} \equiv \frac{1}{\sqrt{3}}\frac{\alpha^5}{\pi^{3/2}}(\rho^2 + \lambda^2 - 3\alpha^{-2})\, e^{-\frac{1}{2}\alpha^2(\rho^2+\lambda^2)} \tag{IV.10}$$

2 $$\psi_{oo}^{M_\rho} \equiv \frac{2}{\sqrt{3}}\frac{\alpha^5}{\pi^{3/2}}\, \vec{\rho}\cdot\vec{\lambda}\, e^{-\frac{1}{2}\alpha^2(\rho^2+\lambda^2)} \tag{IV.11a}$$

2 $$\psi^{M_\lambda}_{00} \equiv \frac{1}{\sqrt{3}} \frac{\alpha^5}{\pi^{3/2}} (\rho^2 - \lambda^2) e^{-\frac{1}{2}\alpha^2(\rho^2+\lambda^2)} \qquad \text{(IV.11b)}$$

2 $$\psi^{S}_{22} \equiv \frac{1}{2} \frac{\alpha^5}{\pi^{3/2}} (\rho_+^2 + \lambda_+^2) e^{-\frac{1}{2}\alpha^2(\rho^2+\lambda^2)} \qquad \text{(IV.12)}$$

2 $$\psi^{M_\rho}_{22} \equiv \frac{\alpha^5}{\pi^{3/2}} \rho_+\lambda_+ e^{-\frac{1}{2}\alpha^2(\rho^2+\lambda^2)} \qquad \text{(IV.13a)}$$

2 $$\psi^{M_\lambda}_{22} \equiv \frac{1}{2} \frac{\alpha^5}{\pi^{3/2}} (\rho_+^2 - \lambda_+^2)\, e^{-\frac{1}{2}\alpha^2(\rho^2+\lambda^2)} \qquad \text{(IV.13b)}$$

2 $$\psi^{A}_{11} \equiv \frac{\alpha^5}{\pi^{3/2}} (\rho_+\lambda_z - \rho_z\lambda_+) e^{-\frac{1}{2}\alpha^2(\rho^2+\lambda^2)} \qquad \text{(IV.14)}$$

where we have explicitly displayed only the highest state of an orbital angular momentum multiplet and where

$$\alpha = (3Km)^{1/4} \qquad \text{(IV.15)}$$

$$\omega = \left(\frac{3K}{m}\right)^{1/2}. \qquad \text{(IV.16)}$$

By combining these wave functions with quark spin, flavour and colour wave functions to obtain totally antisymmetric wave functions one can construct the non-strange states associated with the first seven supermultiplets of the harmonic oscillator model. These states are discussed in Appendix A where we introduce the notation $|X^{2S+1}L_\pi J^P\rangle$ in which X = N or Δ, S is the quark spin, L = S, P, D,.... is the orbital angular momentum, π = S, M or A is the permutational symmetry (symmetric, mixed, antisymmetric) of the spatial wave function, and J^P is the total angular momentum and parity of the state. Note the degeneracy of the ρ and λ components of a mixed pair of wave functions; this is a consequence of the complete permutational symmetry of this system.[18] In the non-strange sector , but not in the strange, the states $|X^{2S+1}L_{\pi}J^{P}\rangle$ correspond exactly to SU(6) states $|X^{2S+1}[\mu,L^{P}]J^{P}\rangle$ where μ is the multiplicity with μ = 56, 70, and 20 corresponding to π = S, M, and A. Thus the wave functions (IV.8), (IV.10) and (IV.12) are of the types S_S, S'_S and D_S and lead to non-strange states associated with the $[56,0^+]$, $[56',0^+]$ and $[56,2^+]$ supermultiplets, the pairs of wave functions (IV.9a) and (IV.9b), (IV.11a) and (IV.11b), and (IV.13a) and (IV.13b) are of the mixed symmetry types P_M, S_M, and D_M and lead to states associated with the supermultiplets $[70,1^-]$, $[70,0^+]$ and $[70,2^+]$, while the wave function (IV.14), being of the type P_A, leads to the non-strange states associated with the $[20,1^+]$supermultiplet of SU(6).

In the harmonic oscillator model (i.e., if U = 0) these states have energies $E = (3+N)\hbar\omega$, so that all of the supermultiplets $[56',0^+]$, $[70,0^+]$, $[56,2^+]$, $[70,2^+]$ and $[20,1^+]$ are degenerate with a spacing above $[70,1^-]$ equal to the spacing of $[70,1^-]$ above $[56,0^+]$. This is not the situation observed experimentally and this of course means that $U \neq 0$. Significantly, we shall see here that the observed deviations from the harmonic oscillator spectrum are consistent with the action of a short range attractive potential as expected.

B. The Zeroth Order States in the S = -1 Sector

When $m_1 = m_2 = m$ and $m_3 = m'$ as is appropriate to this sector, the harmonic oscillator Hamiltonian becomes

$$\tilde{H}'_o = \frac{P_\rho^2}{2m} + \frac{P_\lambda^2}{2m_\lambda} + \frac{3}{2}K\,(\rho^2 + \lambda^2) \qquad (IV.17)$$

where ρ, λ and P_ρ are as defined above and where

$$\vec{P}_\lambda = m_\lambda \frac{d\vec{\lambda}}{dt} \qquad (IV.18)$$

with

$$m_\lambda = \frac{3mm'}{2m+m'} \qquad (IV.19)$$

The eigenstates of this Hamiltonian are quite distinct from those of the S = 0 sector as the degeneracy between the ρ and λ normal modes has been broken:

$$\omega_\rho = \omega = \left(\frac{3K}{m}\right)^{\frac{1}{2}} \qquad (IV.20)$$

$$\omega_\lambda = \left(\frac{3K}{m_\lambda}\right)^{\frac{1}{2}} \qquad (IV.21)$$

where $\omega_\lambda < \omega_\rho$ here since

$$x \equiv \frac{m_d}{m_s} \simeq 0.6. \qquad (IV.22)$$

The low-lying eigenfunctions of the harmonic oscillator Hamiltonian are now

$$N=0 \qquad \psi_{oo} = \frac{\alpha_\rho^{3/2}\alpha_\lambda^{3/2}}{\pi^{3/2}}\, e^{-\frac{1}{2}\alpha_\rho^2\rho^2}\, e^{-\frac{1}{2}\alpha_\lambda^2\lambda^2} \qquad (IV.23)$$

$$1 \qquad \psi_{11}^{\rho} = \frac{\alpha_\rho^{5/2}\alpha_\lambda^{3/2}}{\pi^{3/2}}\, \rho_+ e^{-\frac{1}{2}\alpha_\rho^2\rho^2}\, e^{-\frac{1}{2}\alpha_\lambda^2\lambda^2} \qquad (IV.24)$$

1 $$\psi_{11}^{\lambda} = \frac{\alpha_\rho^{3/2}\alpha_\lambda^{5/2}}{\pi^{3/2}} \lambda_+ e^{-\frac{1}{2}\alpha_\rho^2\rho^2} e^{-\frac{1}{2}\alpha_\lambda^2\lambda^2} \quad \text{(IV.25)}$$

2 $$\psi_{oo}^{\lambda\lambda} = \sqrt{\frac{2}{3}} \frac{\alpha_\rho^{3/2}\alpha_\lambda^{7/2}}{\pi^{3/2}} (\lambda^2 - \frac{3}{2}\alpha_\lambda^{-2}) e^{-\frac{1}{2}\alpha_\rho^2\rho^2} e^{-\frac{1}{2}\alpha_\lambda^2\lambda^2} \quad \text{(IV.26)}$$

2 $$\psi_{oo}^{\rho\lambda} = \frac{2}{\sqrt{3}} \frac{\alpha_\rho^{5/2}\alpha_\lambda^{5/2}}{\pi^{3/2}} \vec{\rho}.\vec{\lambda}\ e^{-\frac{1}{2}\alpha_\rho^2\rho^2} e^{-\frac{1}{2}\alpha_\lambda^2\lambda^2} \quad \text{(IV.27)}$$

2 $$\psi_{oo}^{\rho\rho} = \sqrt{\frac{2}{3}} \frac{\alpha_\rho^{7/2}\alpha_\lambda^{3/2}}{\pi^{3/2}} (\rho^2 - \frac{3}{2}\alpha_\rho^{-2}) e^{-\frac{1}{2}\alpha_\rho^2\rho^2} e^{-\frac{1}{2}\alpha_\lambda^2\lambda^2} \quad \text{(IV.28)}$$

2 $$\psi_{22}^{\lambda\lambda} = \frac{1}{\sqrt{2}} \frac{\alpha_\rho^{3/2}\alpha_\lambda^{7/2}}{\pi^{3/2}} \lambda_+^2\ e^{-\frac{1}{2}\alpha_\rho^2\rho^2} e^{-\frac{1}{2}\alpha_\lambda^2\lambda^2} \quad \text{(IV.29)}$$

2 $$\psi_{22}^{\rho\lambda} = \frac{\alpha_\rho^{5/2}\alpha_\lambda^{5/2}}{\pi^{3/2}} \rho_+\lambda_+\ e^{-\frac{1}{2}\alpha_\rho^2\rho^2} e^{-\frac{1}{2}\alpha_\lambda^2\lambda^2} \quad \text{(IV.30)}$$

2 $$\psi_{22}^{\rho\rho} = \frac{1}{\sqrt{2}} \frac{\alpha_\rho^{7/2}\alpha_\lambda^{3/2}}{\pi^{3/2}} \rho_+^2\ e^{-\frac{1}{2}\alpha_\rho^2\rho^2} e^{-\frac{1}{2}\alpha_\lambda^2\lambda^2} \quad \text{(IV.31)}$$

2 $$\psi_{11}^{\rho\lambda} = \frac{\alpha_\rho^{5/2}\alpha_\lambda^{5/2}}{\pi^{3/2}} (\rho_+\lambda_z - \rho_z\lambda_+) e^{-\frac{1}{2}\alpha_\rho^2\rho^2} e^{-\frac{1}{2}\alpha_\lambda^2\lambda^2} \quad \text{(IV.32)}$$

where once again we have explicitly shown only the top state of an angular momentum multiplet and where

$$\alpha_\rho \equiv (3Km)^{1/4} = \alpha \quad \text{(IV.33)}$$

$$\alpha_\lambda \equiv (3Km_\lambda)^{1/4} \quad \text{(IV.34)}$$

Notice that much of the degeneracy of the harmonic oscillator levels has disappeared. For example, in the N=1 levels associated with the strange states of the $[70,1^-]$ supermultiplet, the degeneracy between ψ^{ρ}_{11} and ψ^{λ}_{11} has been broken with the ρ-excitation now at higher mass than the λ-excitation. We shall see that this has the consequence that (at least in the absence of hyperfine interactions) SU(3) is in some sense maximally violated in this sector. Similar effects occur in the N=2 levels and in both cases there are interesting consequences, especially for the decay patterns of these states as we shall see below.

A simple picture of S=-1 baryons emerges once it is realized that when the strange quark mass differs from the non-strange quark mass it is no longer necessary to construct baryon wave functions that are totally antisymmetric in space, spin, flavour, and colour. The above solutions to the unequal mass harmonic oscillator problem have already indicated this: the heavier strange quark tends to lie closer to the baryon centre-of-mass. It is not too difficult to convince yourself--- depending on the extent to which the generalized antisymmetrization principle has become second nature to you--- that in this situation we are free to single out the strange quark as quark 3 and that it is only the $1 \leftrightarrow 2$ symmetry of the states which remains relevant. We have accordingly in Appendix A also introduced the isospin wave functions $\phi_{\Sigma} = \frac{1}{\sqrt{2}}(ud + du)s$ and $\phi_{\Lambda} = \frac{1}{\sqrt{2}}(ud - du)s$ appropriate to the description of the Λ and Σ^{o} states in this framework (by isospin invariance this is sufficient for the whole S=-1 sector). Since the states (IV.23) to (IV.32) are either symmetric or antisymmetric under 1 $\leftrightarrow$ 2, one can with the aid of the usual spin and colour wave functions construct states that are antisymmetric under interchange of quarks 1 and 2. The resulting states --- which we call the uds basis states --- are discussed in Appendix A where they are labelled by $|Y\,^{2S+1}L_{\sigma}J^{P}>$ where Y = Λ or Σ^{0}, S, L, and J^P are as in the non-strange sector, and σ = ,ρ,λ,$\rho\rho$,$\rho\lambda$, etc. One can of course also discuss these states in terms of the SU(6) basis $|Y\,^{2S+1}L_{\pi}J^{P}>$ and we have accordingly also discussed these states in Appendix A along with the relation between the two descriptions. The SU(6) basis is often useful for calculational purposes and may more directly be compared with the non-strange sector and with previous SU(6)-type analyses of baryons, but we shall continually emphasize the uds basis since we shall see that it sheds a great deal of light on the physics of these states.

C. Turning on the Non-harmonic Potential U

With our zeroth-order wave functions and energies established it remains to do perturbation theory in the perturbation U and H_{hyp}. We first consider U to find the complete Hamiltonian matrix of the confining potential.

We begin by discussing the S=0 sector where in the harmonic limit we have only three equally spaced energy levels. Since V_{conf} is SU(6) symmetric even when U $\neq$ 0, the degeneracy of $\psi_{11}^{M_\rho}$ and $\psi_{11}^{M_\lambda}$ remains unbroken; this degeneracy is a consequence only of the complete permuational symmetry of the S=0 sector. For the same reason the N=2 levels $\psi^{M\rho}_{oo}$ and $\psi^{M\lambda}_{oo}$ and the levels $\psi^{M\rho}_{22}$ and $\psi^{M\lambda}_{22}$ will remain degenerate, but the degeneracy of the non-strange sectors of the five N=2 SU(6) supermultiplets $[56',0^+]$, $[70,0^+]$, $[56,2^+]$, $[70,2^+]$, and $[20,1^+]$, which was a consequence of the U=0 limit, will in general be broken. Thus in the general case although there are ten S=0 wave functions, there will only be seven energies which we may denote by $E[S_S] = E[56,0^+]$, $E[P_M] = E[70,1^-]$, $E[S'_S] = E[56',0^+]$, $E[S_M] = E[70,0^+]$, $E[D_S] = E[56,2^+]$, $E[D_M] = E[70,2^+]$, and $E[P_A] = E[20,1^+]$. Without specifying U--- which we really don't know --- it would seem that it would be necessary to introduce seven unknown parameters, one to describe the energy of each supermultiplet in the full confining potential V_{conf}. There are so many baryon states known that the introduction of seven parameters --- which are a description of the unknown confining potential plus short range spin-independent effects --- would not be that terrible. Alternatively, one could presumably parameterize U in some more economical manner and actually calculate its matrix elements. While this is a very interesting program to undertake, in practice it is unnecessary for our present purposes as it turns out that these seven energies are constrained by a remarkable rule[19]: in first order perturbatation theory <u>any potential U will split the harmonic oscillator spectrum into exactly the same pattern</u>. This pattern, which is shown in Figure IV.1, has the property that the energies of the seven supermultiplets are always given by

$$E[S_S] = E_o \qquad \text{(IV.35a)}$$

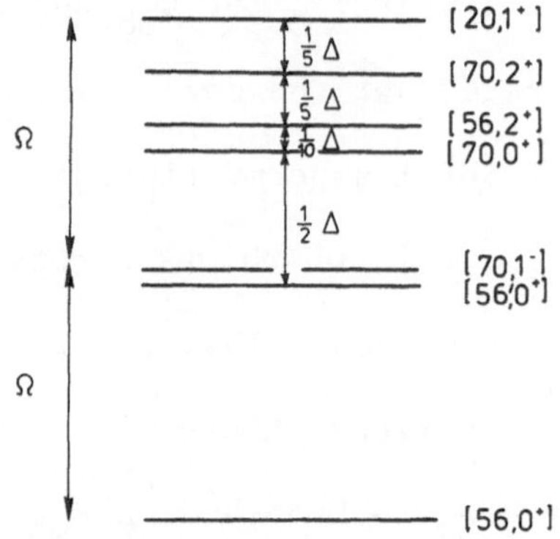

Figure IV.1: The pattern of low-lying supermultiplets from first order perturbation theory.

$$E[P_M] = E_o + \Omega \qquad \text{(IV.35b)}$$

$$E[S'_S] = E_o + 2\Omega - \Delta \qquad \text{(IV.35c)}$$

$$E[S_M] = E_o + 2\Omega - \tfrac{1}{2}\Delta \qquad \text{(IV.35d)}$$

$$E[D_S] = E_o + 2\Omega - \frac{2}{5}\Delta \qquad \text{(IV.35e)}$$

$$E[D_M] = E_o + 2\Omega - \frac{1}{5}\Delta \qquad \text{(IV.35f)}$$

$$E[P_A] = E_o + 2\Omega \qquad \text{(IV.35g)}$$

This rule is proved in Appendix B, which also includes a discussion of the values for E_o, Ω and Δ appropriate to the description of these states. As intimated previously, for an attractive potential U, $\Omega > \omega$ and $\Delta > 0$ in accordance with the observed energies of these states. Note in particular that the gap between the $[56,0^+]$ and the $[70,1^-]$ states, Ω, is automatically larger than the gap between the $[70,1^-]$ and the $[56,2^+]$ as observed and that of all the N=2 levels the $[56',0^+]$ automatically comes down farthest from its harmonic-oscillator-like position at 2Ω above the ground state in accordance with the observation of the very low-massed Roper resonance. These observations will become more quantitative in what follows; for now we simply note that the good agreement of this pattern with the one observed empirically indicates that this approach to the description of these states may be sensible.

The case $U \neq 0$ is somewhat more complicated in the S=-1 sector since now the ρ and λ oscillators behave differently. Of course the energies of the states before turning on U are as previously discussed: a unit of λ-excitation has less energy than a unit of ρ-excitation by a factor $(\frac{m}{m_\lambda})^{\frac{1}{2}} = (\frac{2x+1}{3})^{\frac{1}{2}}$. In the presence of a perturbation U this relationship between ρ and λ excitation energies will change, but (especially considering that most of the energy is still harmonic) presumably this relationship remains a reasonable approximation to the actual situation.

The generality of the important qualitative feature of this harmonic oscillator relationship --- that λ excitations have a lower frequency than ρ excitations --- can most easily be seen in operation in the case $m' \gg m$. In this limit baryons become like a helium atom with inverted statistics.[20] In the absence of interaction between quarks 1 and 2 this system will have two degenerate negative parity excited spatial wavefunctions above the ground state, namely,

$$|\tilde{\lambda}> \equiv \frac{1}{\sqrt{2}} \{|sp> + |ps>\} \quad \text{(IV.36)}$$

$$|\tilde{\rho}> \equiv \frac{1}{\sqrt{2}} \{|sp> - |ps>\} \quad \text{(IV.37)}$$

where s and p are the lowest-lying single particle s and p orbitals about quark 3. If one now turns on the residual attractive potential energy of the two light quarks (from any confinement potential), then $|\tilde{\lambda}>$ will end up lying lower than $|\tilde{\rho}>$ since in $|\tilde{\rho}>$ the two quarks are unlikely to be near to each other. Thus though our numerical results will depend on the use of the harmonic oscillator approximations sketched above, we expect our principal conclusions are more general. They amount to saying that just as the strange quarks "spin more slowly about their axis" resulting in a reduced magnetic moment, they also move more slowly in their orbits.

D. Turning on the Hyperfine Interaction

With the approximate eigenfunctions and eigenvalues of the spin-independent potential established as outlined in the previous section, we now turn to a calculation of the matrix elements of the hyperfine interaction. The hyperfine interaction will both shift the energies of states and cause mixings between them. The former effect may of course be directly seen in the spectroscopy of baryons; the latter effects may be related to the production and decay characteristics of baryons as shall be discussed shortly.

There are a few features of the hyperfine matrix elements that it is essential to understand. First of all, as mentioned previously, since the tensor force is an operator of rank two in both space and spin the selection rules $L=0 \not\leftrightarrow L=0$, $L=0 \not\leftrightarrow L=1$, and $S=\frac{1}{2} \not\leftrightarrow S=\frac{1}{2}$ apply. The contact term is on the other hand a scalar in both space and spin so it allows only $L \leftrightarrow L$ and $S \leftrightarrow S$ transitions and is independent of how L and S are coupled to a final total angular momentum J. In fact at the quark level the contact term only operates when two quarks are in a relative S-wave . Thus in the N=0 states the tensor force is absent*and only the contact force is being tested. In the N=1 states both tensor and contact interactions operate side by side since one of the two oscillators is in an S-wave while the other is in a P-wave. Mixings of states with S=1/2 and S = 3/2 can (and do) occur in this sector, and the physical states are predicted (and observed) to be mixtures of the pure SU(6) eigenstates. At N=2 one reaches a further level of complexity not only in that there are a very large number of states to consider, but also that states belonging to different SU(6) multiplets associated with N=2 are predicted (and observed) to be prominently mixed.

---

*more precisely, diagonal tensor terms are absent; in principle the tensor term could mix D-waves in the ground states. It may be construed to be a success of this model that in fact such admixtures are small in accordance with the known absence of a large E2 moment in $\Delta \to N\gamma$.

We shall see shortly that when all of these effects --- colour confining potentials with a short range attractive part, quark mass differences, and hyperfine interactions --- are put together they provide a fairly successful picture of both baryon spectroscopy and decays. In order to appreciate the results, however, we must first briefly consider the calculation of decays in the quark model.

E. Baryon Decays in the Quark Model

A dynamical model for hadron structure is being tested in a quite limited way when it is compared only with spectroscopic data; once a state is predicted at a given mass it is next necessary to decide whether the internal structure attributed to the state by the model will give rise to its observed width and branching ratios. It is not very convincing, for example, to correctly predict the masses of the $N^*1/2^-(1515)$ and $N^*1/2^-(1690)$ but to fail

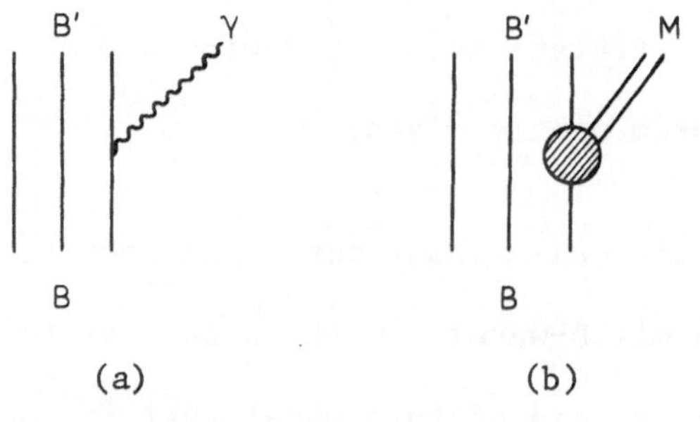

Figures IV.2(a): The Decay $B \rightarrow B'\gamma$ in the quark model.
IV.2(b): The Decay $B \rightarrow B'M$ in the quark model.

to predict that, despite phase space factors and comparable widths, the $N\eta$ decay mode is dominant in the first and negligible in the second.

In the quark model, the decay of a baryon[21)] is presumed to occur through a single quark transition as depicted in Figures IV.2(a) and IV.2(b) for photonic and mesonic transitions.

The T-matrix element for photodecay, obtained by making a non-relativistic reduction of a γ^μ quark-photon interaction is of the familiar form

$$<B'(p's')\gamma(q\lambda)|T|B(ps)> = -\frac{3ie}{(2\pi)^{3/2}} <B'(p's')| \frac{e_3}{e} \times\{\frac{\vec{\sigma}.\vec{q}x\vec{\varepsilon}^*}{2m} + i\frac{\vec{\varepsilon}^*.\vec{p}_3'}{m}\} e^{i\sqrt{\frac{2}{3}}\vec{q}.\vec{\lambda}}|B(ps)> \qquad (IV.38)$$

where the photon polarization vector is $\vec{\varepsilon}(q\lambda)$, and where e_3 and p_3' are the charge and momentum of the third quark in B'. The matrix element for pion emission, on the other hand, must be of the general form

$$<B'(p's')\pi^i(\vec{K})|T|B(ps)> = \frac{-3i}{(2\pi)^{3/2}} <B'(p's')| \times\{ g\vec{K}.\vec{\sigma}_3 + h\,\vec{\sigma}_3.\vec{p}_3'\} e^{i\sqrt{\frac{2}{3}}\vec{K}.\vec{\lambda}}\tau^i|B(ps)> \qquad (IV.39)$$

Pion emission from a non-strange quark in the simplest case comes from the process depicted in Figure IV.3(a) from which it can be seen that strange quarks cannot emit pions. Conversely, kaon emission from a strange quark proceeds via the same diagram where once again a non-strange quark-antiquark pair is created. We may therefore

(a) (b)

Figure IV.3(a): An OZI-allowed meson emission diagram.
IV.3(b): An OZI-forbidden meson emission diagram.

expect SU(3) breaking effects to be rather small in relating kaon to pion emission and this allows us to write down an expression similar to (IV.39) above for kaon emission involving the same constants g and h. Eta emission may not be dealt with so trivially; however, if we consider eta emission from non-strange states only (there is practically no data on eta emission by strange states anyway) then in addition to the diagrams IV.3(a) one could also get contributions from the diagrams IV.3(b) which are OZI-violating. With the assumption that the OZI-violating amplitudes are small, eta emission from a non-strange quark may also be expressed in terms of g and h. There are at least two reasons why diagrams like IV.3(b) might be unimportant: 1) they are OZI-violating (though the mechanism of OZI-violation is not well-understood so one must be cautious) and 2) at least part of these diagrams is automatically taken into account in the η-η' mixing angle (see Section V.B for details).

There is an example of the way in which decays give information about the internal structure of states which is not only of great relevance to baryons, but is also so simple that it may be obtained in a way that bypasses even the simple machinery outlined above. As mentioned in the construction of the strange baryon states, in the SU(6) limit most states are in general a linear superposition of degenerate ρ- and λ-type oscillations. On the other hand we have seen that in the uds basis, at least before turning on hyperfine interactions, pure ρ or pure λ oscillations are eigenstates of the Hamiltonian. That is, the existence of a strange quark-nonstrange quark mass difference tends to mix the SU(6) eigenstates in such a way as to give the observed states an internal structure corresponding to a pure ρ or pure λ oscillation. This possible segregation of strange states into ρ and λ type oscillations is in fact very easy to test by looking at decays because <u>pure ρ-type states will decouple from the $\bar{K}N$ channel</u>. The reason for this is easily understood. For $\bar{K}$ emission the s quark in the excited hyperon must be ejected; the two non-strange quarks are spectators. However, if the state in question had any degree of ρ-excitation (i.e. excitation localized in the two non-strange quarks) then the ρ-variable will have zero overlap with a nucleon in which each non-strange quark pair is in its ground state!

There are of course many features of the decays of baryons predicted by our QCD-inspired model which can only be extracted by detailed calculations with the interactions outlined in this section, and we shall freely refer to such details in the comparison with observations which follow in the next few sections. It is on the other hand fortunate that this latter result on the decoupling of certain states from $\bar{K}N$, which plays a prominent role in the comparison in the S=-1 sector, can be understood very simply.

F. Results in the Ground State Baryons

We begin the comparison of the simple QCD-based model for baryons outlined above in the beginning: with the familiar octet and decouplet associated with the $[56,0^+]$ ground state of SU(6) and the N=0 level of the harmonic oscillator.

The machinery for extracting results in this (and all other sectors) having been described above, it is only necessary to diagonalize the relevant sectors of the Hamiltonian to obtain predictions. In this case the relevant sectors include $J = 1/2^+$ and $J = 3/2^+$ states which are associated with the N=2 levels of the harmonic oscillator, mixing with which leads to a number of interesting effects which will be discussed below.

The spectroscopic results for these states are displayed in Figure IV.4, though certain aspects of these results deserve separate mention. First, as we have said before, the qualitative result $\Delta > N$ is independent of the choice of parameters. Equally impressive is the fact that the sign and magnitude of the Σ-Λ mass difference emerges in a natural and simple way[13]. The main

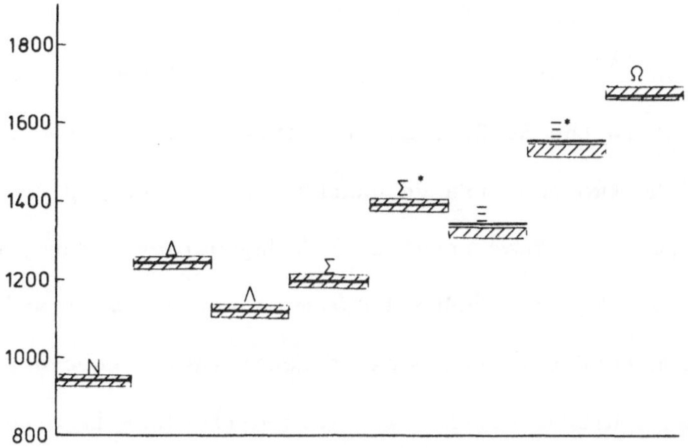

Figure IV.4: The spectrum of ground state baryons (hatched regions) compared to the model (bars).

features of this effect are in fact independent of the complication of the N=2 interband mixing and are easily understood in the uds basis. In that basis the Σ and Λ eigenfunctions are

$$|\Sigma^2S1/2^+> = \phi_\Sigma \chi^\lambda\psi_{oo} \qquad \text{(IV.40)}$$

$$|\Lambda^2S1/2^+> = \phi_\Lambda \chi^\rho\psi_{oo} \qquad \text{(IV.41)}$$

where the presence of χ^λ and χ^ρ are dictated by the requirement that the states be overall antisymmetric in the two equal mass non-strange quarks 1 and 2. The spin wave function χ^ρ (see Appendix A) has the property that quarks 1 and 2 are in S=0, so that quark 3 sees each of the other two quarks with an uncorrelated spin.

The hyperfine stabilization in the $\Lambda 1/2^+$ is therefore accomplished entirely by the two non-strange quarks and so is equal to the stabilization in the nucleon ($\sim$ 150 MeV). On the other hand, in the spin wave function χ^λ quarks 1 and 2 are in S=1 and tend to destabilize the Σ so that in the SU(3) limit the interaction of the strange quark with the two non-strange quarks would be providing the stabilization to give the then required Σ-Λ degeneracy. Here, however, the strange quark has a reduced chromomagnetic moment and the stabilization energy it produces is accordingly reduced giving $\Sigma > \Lambda$. The magnitude of the resulting difference is also well reproduced in terms of $x \equiv \frac{m_d}{m_s}$ and the Δ-N mass difference, providing a rather impressive first confirmation of the presence of at least the contact part of the hyperfine interaction (III.5). In the interest of brevity we forego further discussion of the spectroscopy of this sector, which is obviously satisfactory, and turn to other effects.

There is little to say in this sector with regard to decays since the admixture of N=2 components into these states does not drastically modify their decays so the old successful SU(6) results continue more or less to apply. The admixture of N=2 levels into these states is not, however, without important effects on the decays of excited states into them as we shall see shortly.

In addition, the resulting large admixture of $|N^2S_M 1/2^+\rangle$ into the nucleon (with an amplitude of $\simeq -1/4$) has one very directly observable effect: it gives the neutron an internal charge distribution which explains in sign and magnitude its observed electric form factor.[22)] This may be simply understood. The two d-quarks in the neutron have parallel spins and so repel each other relative to the ud interaction which is on the whole attractive as required to give $\Delta > N$. The three quarks are thus not symmetric --- there is some $[70, 0^+]$ mixed into $[56, 0^+]$ --- and the asymmetry corresponds to the two d quarks being further from the centre of mass of the neutron than the u quark. The neutron will therefore have a positive core, as observed.

G. Results in the Low-lying Negative Parity Baryons

When we turn to the negative parity baryons the model faces several very severe tests. First of all, if hyperfine interactions are really in operation their tensor part should appear for the first time* in the P-wave baryons with exactly the correct strength and, moreover, the short range character of the contact term should become apparent. The P-wave baryons are in a sense a perfect testing ground for hyperfine interactions since they simultaneously have one relative coordinate in an S-wave and one in a P-wave. When two quarks are in a relative S-wave, the contact force can operate but the tensor force vanishes, while when two quarks are in a relative P-wave the contact force vanishes (since there is zero amplitude to find the quarks "in contact") while the tensor force

*apart from the small admixture of D-waves into the ground states.

operates. Thus in P-wave baryons the two pieces of the hyperfine interaction operate side by side and their relative strength ($\frac{8\pi}{3}$?), as well as the short range character of the contact force, is testable.

Another feature of the P-wave baryons is that there are many states with the same IJ^P quantum numbers which are degenerate in the SU(6) limit; such states will in general be very strongly mixed by the SU(6)-violating terms in our Hamiltonian. Apart from the anomalously large matrix element responsible for the interband mixing of the N=2 level $[70,0^+]$ state into the nucleon, the resulting intraband mixing angles, which can be "measured" via an analysis of baryon decays, provide the first real test of whether the model is correctly reproducing the internal structure as well as the spectroscopy of baryons.

There is so much information available on the negative parity baryons that we separate our discussion into two parts, beginning with the non-strange negative parity states. The result of diagonalizing the appropriate sectors of the Hamiltonian are given in Figure IV.5. The solid bars and the compositions written just above them are the predictions of the model. The shaded regions and the associated compositions are the experimentally determined positions and internal structures of these states. We stress that these predictions are completely determined by parameters which come from the ground state baryons. For example, in the $N1/2^-$ sector

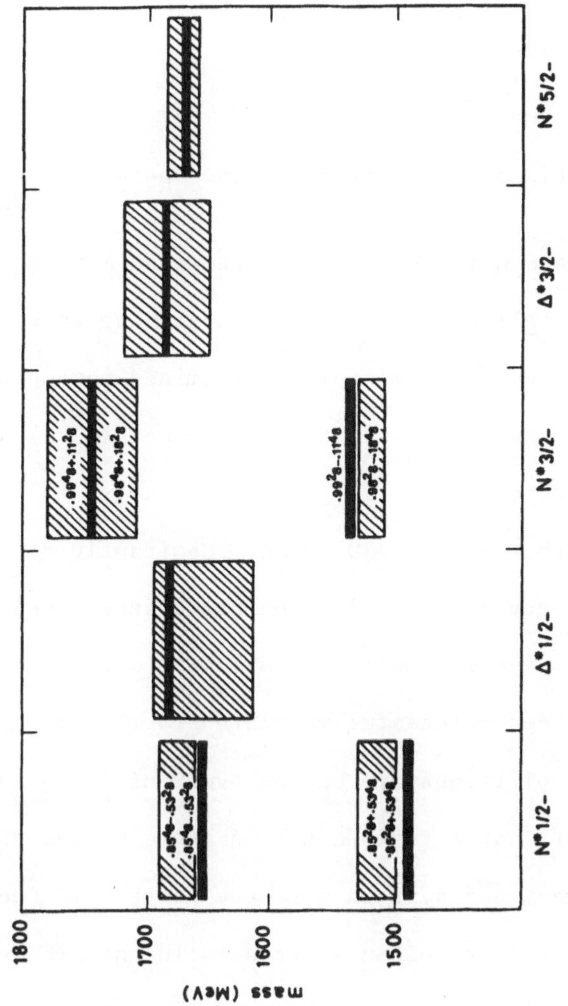

Figure IV.5: The spectrum and composition of S=0 P-wave baryons compared to the model predictions.

one simply has

$$<N^2P_M 1/2^- | H_{contact} | N^2P_M 1/2^- > = -\frac{1}{4}\delta \tag{IV.42a}$$

$$<N^4P_M 1/2^- | H_{contact} | N^4P_M 1/2^- > = +\frac{1}{4}\delta \tag{IV.42b}$$

$$<N^4P_M 1/2^- | H_{tensor} | N^4P_M 1/2^- > = -\frac{1}{4}\delta \tag{IV.42c}$$

$$<N^4P_M 1/2^- | H_{tensor} | N^2P_M 1/2^- > = +\frac{1}{4}\delta \tag{IV.42d}$$

where all other matrix elements are zero and where $\delta = \frac{4\alpha_s\alpha^3}{3\sqrt{2\pi}m_d^2}$ is equal to the Δ-N mass difference. The predicted mixing angles are thus <u>independent of all parameters</u> and the good agreement in sign and magnitude with the experimentally determined compositions indicate the presence of the tensor term with the correct ($\frac{8\pi}{3}$!) strength relative to the contact term. It is, incidentally, this mixing which explains the fact that $N1/2^-(1520)$ decays dominantly to Nη while $N1/2^-(1700)$ has a very small branch to that channel even though for unmixed states the reverse would be true. The size of the splittings has, in addition, been successfully related to the Δ-N splitting δ. The fact that the splittings are on the order of $\frac{1}{2}\delta \simeq 150$ MeV is related to the fact that a given quark in these states spends half its time in an S-wave and half in a P-wave so that it feels the contact force only half as often as it does in the Δ-N system. The correct prediction of the size of the splittings here is therefore a test of the short-range character of the contact term. A more sensitive test involves the two negative parity Δ's. Note that for the nucleonic states the dominantly S=3/2 states lie above the dominantly S=1/2 states as one naively expects from the Δ-N system. On the other hand the two S=1/2 Δ's lie up with the S=3/2 nucleonic states. For a long-range $\vec{S}_i . \vec{S}_j$ force the Δ's would indeed lie with the S=1/2 nucleonic states; that they lie higher is due to the short range of the contact interaction coupled with a correlation between the spin and spatial wave functions for these states. Their observed position is therefore a strong indication for the short range character of the $\vec{S}_i . \vec{S}_j$ forces as expected from hyperfine contact forces.

We now consider the strange negative parity states. The new feature which comes in here is most easily illustrated by considering "another" Σ-Λ mass difference: $\Sigma 5/2^- - \Lambda 5/2^-$. As these are unique states they avoid the complications of mixing. In the uds basis these states are simply

$$|\Sigma\, {}^4P_M 5/2^-\rangle = \phi_\Sigma \chi^s_{3/2} \psi^\lambda_{11} \qquad \text{(IV.43)}$$

$$|\Lambda\, {}^4P_M 5/2^-\rangle = \phi_\Lambda \chi^s_{3/2} \psi^\rho_{11} \qquad \text{(IV.44)}$$

Note the appearance of the spatial wave functions ψ^λ and ψ^ρ correlated with the uds isospin wave functions ϕ_Σ and ϕ_Λ. As in the $\Sigma 1/2^+ - \Lambda 1/2^+$ sector the hyperfine interactions try to make $\Sigma 5/2^-$ heavier than $\Lambda 5/2^-$, though here by only 25 MeV. This effect is, however, almost completely overwhelmed by the previously discussed non-degeneracy of the ρ and λ orbital modes which make a λ-excitation 75 MeV less massive than a ρ-excitation. The net result is that the $\Lambda 5/2^-$ is predicted to be about 50 MeV heavier than the $\Sigma 5/2^-$, a reversal of the familiar case and in good agreement with experiment. This effect plays an equally important role throughout the rest of the strange sector and is responsible in large part for the pattern of mixing angles observed in these states. Since, as discussed in Section E, the ρ oscillations decouple from the $\bar{K}N$ channel this tends to lead to a pattern of decouplings in these states which had been previously noted empirically, and called "ideal mixing for baryons"[23) in analogy with the decoupling of the ideally mixed ϕ from non-strange states. The success of the orbital splitting mechanism in explaining $\Sigma 5/2^-$-$\Lambda 5/2^-$ and the pattern of observed mixed angles may be considered as support for the flavour independence of the confinement potential as this is the main assumption on which it depends.

The results of diagonalizing the relevant matrices for these states --- matrices which are once again determined by our previous results --- are shown in Figure IV.6 where they are compared with the data from multichannel $\bar{K}N$ partial wave analyses[24) and with mixing angle determinations based on decay analyses.[25) The agreement of both the spectroscopy and the mixing angles (in magnitude and sign) is once again very encouraging. The worst feature of the predictions is the discrepancy with the observed mass of the $\Lambda 1/2^-(1405)$ (though the composition is well-predicted). There are several possible interpretations of this discrepancy: 1) this state is predicted to be very strongly coupled to the $\bar{K}N$ channel and its proximity to threshold would tend to suppress its mass, 2) if the absence of

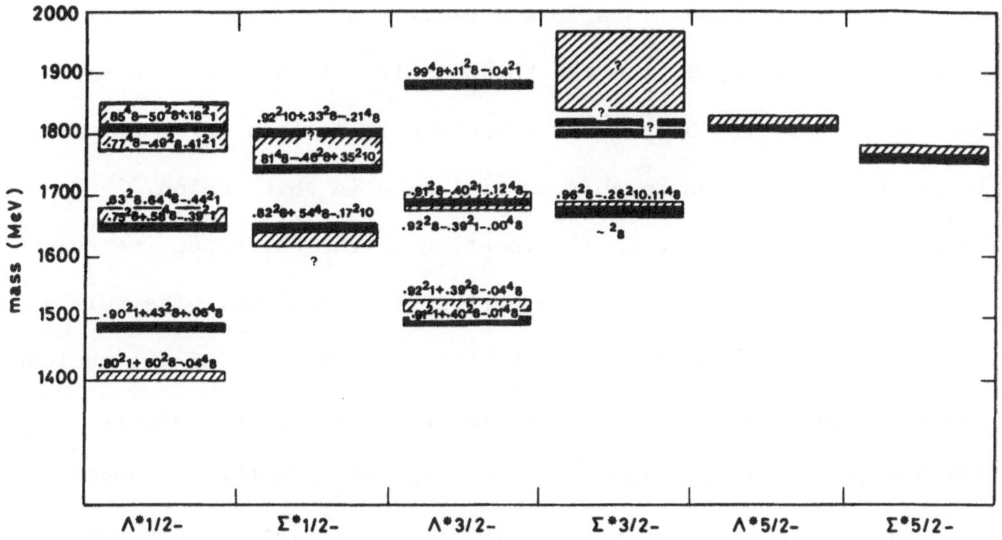

Figure IV.6: The spectrum and composition of S=-1 P-wave baryons compared to the model predictions.

spin-orbit interactions is due to a cancellation between one-gluon effects and Thomas precession in the confinement potential, the "goodness" of the cancellation will depend on the quark masses; the splitting between $\Lambda 1/2^-(1405)$ and $\Lambda 3/2^-(1520)$ may result. In fact, in model calculations which neglect three-body spin-orbit forces, this "uncancellation" significantly affects only these two states and produces in them a sizeable splitting in the right direction.

Another feature of the spectrum which might be thought to be cause for concern is the absence of the predicted $\Lambda 3/2^-$ state at around 1900 MeV. This is not a difficulty, however, but a success since this state is predicted to be an almost pure ρ-excitation and so to decouple from the phase-shift analyses!

This state is in fact the L-S coupling cousin to the $\Lambda 5/2^-$

which (see eqn. IV.44) is also a ρ-excitation and therefore also predicted to decouple from the $\bar{K}N$ channel. Its $\bar{K}N$ width is in fact very small ($\Gamma[\Lambda 5/2^- \to K\bar{N}] \simeq 4$ MeV) and it is seen mainly because it is the only action in this partial wave in this region. Incidentally, even this small violation of the $\bar{K}N$ decoupling rule can be understood in terms of hyperfine interactions. The rule only applies to transitions to the $[56,0^+]$ component of the nucleon but is allowed for transitions to the $[70,0^+]$ admixed into it. The violations of both the $\bar{K}N$ decoupling rule and the Moorhouse selection rule in photoproduction[26)] (which violations are substantial) can be understood in both magnitude and sign as arising from transitions to the $[70,0^+]$ part of the nucleon[27)].

H. Results for the Positive Parity Excited Baryons

There are no new ideas required to extend the model to include the positive parity baryons associated with the N=2 level of the harmonic oscillator model. There is, however, a new effect which begins to play a role and should be mentioned: we can expect in this band to find in general large mixings between the once degenerate SU(6) multiplets. These mixings have two sources: hyperfine interactions and $m_s \neq m_d$. Thus the states $N^2D_S5/2^+$, $N^2D_{\mathcal{M}}5/2^+$, and $N^4D_{\mathcal{M}}5/2^+$ can be mixed by the hyperfine interaction and in fact the low-lying $N5/2^+$ should be a strong mixture of ^2D_S (i.e. $[56,2^+]$) and ^2D_M (i.e., $[70,2^+]$) as a result of mixing via the contact interaction. The non-degeneracy of ρ and λ oscillators in the S=-1 sector also results in strong mixing between SU(6) multiplets since, for example,

$$\psi_{22}^{\rho\rho} = \frac{1}{\sqrt{2}} [\psi_{22}^{S} + \psi_{22}^{M_\lambda}] \qquad \text{(I V.45)}$$

$$\psi_{22}^{\lambda\lambda} = \frac{1}{\sqrt{2}} [\psi_{22}^{S} - \psi_{22}^{M_\lambda}] \qquad \text{(IV.46)}$$

so that we may once again expect to find evidence of "ideal mixing for baryons" in this band. We shall in fact find that a great many states, especially in the I=0 sector, are predicted to practically decouple from the $\bar{K}N$ channel; this conclusion is essential for understanding the observed spectroscopy of strange positive parity states.

The results of diagonalizing the Hamiltonian in this sector are shown in Figures IV.7 and IV.8 for S=0 and S=-1 respectively. We

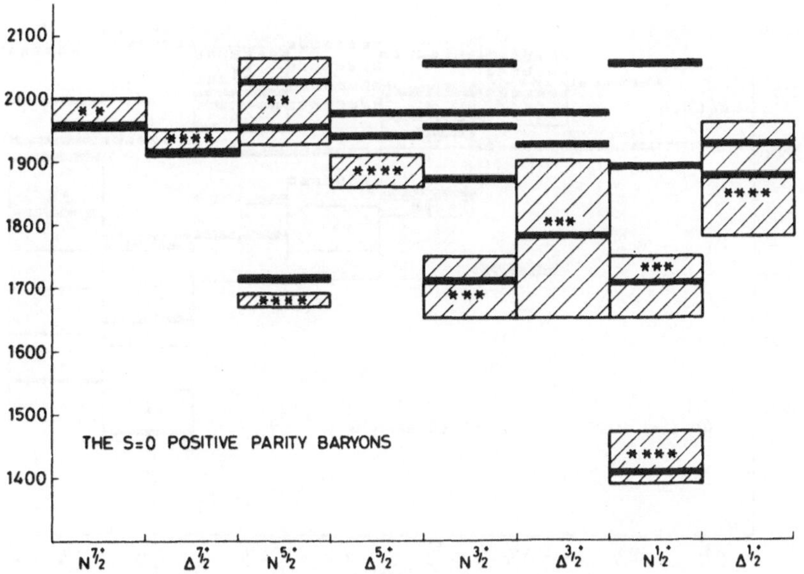

Figure IV.7: The spectrum of S=0 positive parity excited baryons compared to the model predictions.

discuss first the non-strange states. It should to begin with be noted that the position of the first one or two resonances in every partial wave have been adequately predicted. The absence of some of the resonances predicted at higher masses will presumably be understood once their predicted compositions have been used in a decay analysis. For example, the highest $N3/2^+$ and $N1/2^+$ states are dominantly $[20,1^+]$ states and will certainly decouple from πN scattering. Where information on the mixing angles of these states has been checked, it agrees with the predicted mixing angles. The best example is the $\Delta 5/2^+$ resonance at around 1900 MeV.[28] For a pure SU(6) (either $[56,2^+]$ or $[70,2^+]$) state one predicts

$$\frac{\Gamma(\Delta 5/2^+ \rightarrow (\Delta\pi)_F)}{\Gamma(\Delta 5/2^+ \rightarrow (\Delta\pi)_P)} = 1 \qquad \text{(IV.47)}$$

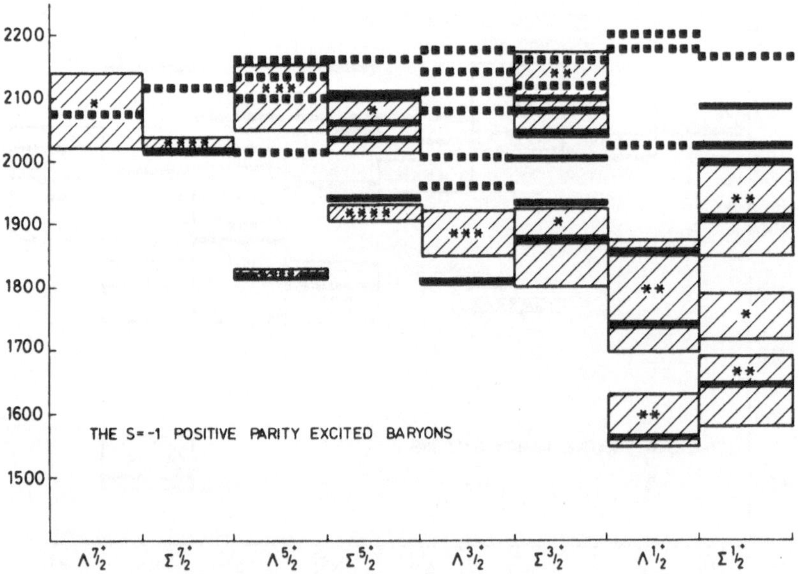

Figure IV.8: The spectrum of S=-1 positive parity excited baryons compared to the model predictions.

Experimentally this channel is known to be completely dominated by the F-wave decay. The model predicts that the tensor force will have caused strong $[56,2^+]$-$[70,2^+]$ mixing of the correct magnitude and sign to make this state dominantly decay to $(\Delta\pi)_F$ by a factor of about 10. Moreover, for this mixing angle the orthogonal partner of this state is predicted to nearly decouple from πN scattering.[28)] The model thus predicts that only one $\Delta 5/2^+$ resonance will be seen in πN scattering, that it will be seen at around 1900 MeV, and that its dominant decay mode will be to $\Delta\pi$ in an F-wave; this is just what is observed experimentally.

The S=-1 sector has many more states predicted than are known experimentally. However, once one takes into account the decoupling of ρ-excitations (the dominantly ρ-excited states are denoted in Figure IV.8 by broken bars) a rather satisfactory picture emerges. The most spectacular example is to be seen in the $\Lambda 3/2^+$ sector where the model predicts (after diagonalization of the appropriate 7x7 matrix!) that six of the seven states expected in this channel will be decoupled from $\bar{K}N$ scattering and that the one coupled state should lie near to the position of the one state that is observed in this channel. To go into more detail would require a complete analysis of the decays of these states which is not yet available, but it can be noted in the meantime that in each partial wave the lowest one or two resonances are associated by the model with a state that will couple to $\bar{K}N$ scattering.

I. Final Comments on Baryons

In view of the detailed agreement of the calculations discussed in this section with experiment, it seems possible that the simple model of approximately non-relativistic and point-like quarks moving in a flavour independent confining potential perturbed by hyperfine interactions will lead to an understanding of the properties of all the low-lying baryons.

This is not to say that this model could possibly be the whole story; certainly other effects will play some role. For example, there will certainly be mass shifts and mixings due to interactions of these resonances with their decay channels, and there may well be residual spin orbit effects (see below). What is possible is that the model accounts for the dominant characteristics of these states.

It remains unclear, however, how to derive such a model from QCD. The major difficulty at the present very crude level of understanding of the connection to QCD is the apparent absence of the spin-orbit forces. Separate analyses of the N=1 and N=2 levels certainly show that such effects are reduced to less than 1/5 of the strength one would expect from one gluon exchange. While the idea that the one gluon exchange effects are being cancelled by Thomas precession in the confining potential is an attractive one, it has not yet been possible to demonstrate that such a mechanism could succeed in cancelling spin-orbit effects so universally as seems to be the case. Here the three body kinematics become very troublesome and in this case it may be that the study of mesons will provide the clearest answers [29]; we shall address this question briefly in the next section.

Of course in some sense it is of little consequence whether the success of this model for baryons is evidence for QCD or not: to the extent that the model is "correct" it should certainly provide important clues in the search for an understanding of the strong interaction. Despite the obvious difficulties, however, it is very difficult not to associate the success of these calculations with QCD, which inspired them. The model may not yet be a consequence of QCD in the mathematical sense, but it is certainly a consequence in the psychological sense.

V. A QCD-based Model for Mesons

As previously mentioned, experimental meson spectroscopy is in a considerably less satisfactory state than is the case for baryons; since the sort of model building we are doing relies at least initially on data, this at least partially accounts for the fact that the theoretical situation is also somewhat more unsettled. Nevertheless there are unquestionably many features of mesons that are rather neatly explained in terms of soft QCD ideas. My initial plan for these lectures was to make a patchwork discussion of mesons dealing with specific aspects in terms of QCD ideas, explaining the difficulties that arise in other areas. Instead I have decided that it would be more fun, and at least equally instructive, to try something quite different: let's exploit fully the soft QCD relations between the $q\bar{q}$ and qq forces and try to predict[11] absolutely what mesons ought to look like on the basis of what we now know about baryons.

Given that there are some uncertainties --- for example the treatment of spin-orbit terms --- I prefer to call what follows a sketch rather than a calculation; this also indicates that one should not take the exact numerical results of this section very seriously, but should rather pay more attention to the general features. Still, within the present framework there is little freedom and given the experimental situation this seems a reasonable approach.

A. A Sketch of Mesons from a Baryonic Perspective

On the basis of our studies of baryons we can take as our Hamiltonian for I=1 and I= 1/2 mesons (I=0 mesons will be discussed separately in Section V.B)

$$H = \sum_i m_i + H'_0 + H'_{hyp} + H'_{so} \tag{V.1}$$

where

$$H'_0 = \sum_{i=1}^{2} \frac{p_i^2}{2m_i} + V'_{conf} \tag{V.2}$$

with

$$V'_{conf} = 2V^{12}_{conf} \tag{V.3}$$

$$H'_{hyp} = 2H^{12}_{hyp} \tag{V.4}$$

and where $H'_{so} \equiv f(r)\vec{L}.\vec{S}$ is the residual spin orbit interaction that remains after subtracting Thomas precession in the confinement potential from the one-gluon exchange effects. We explicitly include this term since, unlike the case of baryons, here it should be at least roughly calculable: there are certainly no three body terms in the mesons!

Apart from the spin-orbit term --- with which we will exercise due caution --- the connection to the baryon Hamiltonian is completely trivial: it is achieved simply by the insertion of some factors of 2.

We can now proceed to solve this system in the same way as we did for baryons. Writing

$$V'_{conf}(r) = \tfrac{1}{2}K'r^2 + V'(r) = Kr^2 + 2U^{12}(r) \qquad (V.5)$$

we begin by solving the reduced mass harmonic oscillator problem

$$\tilde{H}'_o = \frac{p^2}{2\mu} + Kr^2 \qquad (V.6)$$

where

$$r \equiv |\vec{r}_1 - \vec{r}_2| \qquad (V.7)$$

$$\vec{p} \equiv \mu \frac{d\vec{r}}{dt} \qquad (V.8)$$

and

$$\mu = \frac{m_1 m_2}{m_1 + m_2} \qquad (V.9)$$

The low-lying eigenfunctions ψ_{nlm} we need are simply

$$\psi_{000} = \frac{\beta^{3/2}}{\pi^{3/4}} e^{-\frac{1}{2}\beta^2 r^2} \qquad (V.10)$$

$$\psi_{111} = \frac{\beta^{5/2}}{\pi^{3/4}} r_+ e^{-\frac{1}{2}\beta^2 r^2} \qquad (V.11)$$

$$\psi_{200} = \sqrt{\frac{2}{3}} \frac{\beta^{7/2}}{\pi^{3/4}} (r^2 - \tfrac{3}{2}\beta^{-2}) e^{-\frac{1}{2}\beta^2 r^2} \qquad (V.12)$$

where $\beta = (2K\mu)^{1/4} = (Km)^{1/4}$, and where of course

$$E = (\frac{3}{2} + n)\omega' \qquad (V.13)$$

where

$$\omega' = (\frac{2K}{\mu})^{1/2} = 2\,(\frac{K}{m})^{1/2} \qquad (V.14)$$

We note the important relations to the baryonic parameters $\alpha = (3Km)^{1/4}$ and $\omega = (\frac{3K}{m})^{1/2}$:

$$\beta = 3^{-1/4}\alpha \qquad (V.15)$$

$$\omega' = (\frac{4}{3})^{1/2}\omega \qquad (V.16)$$

We can immediately check two aspects of these wave-functions against experiment. If we calculate the charge radii of the proton and the pion we find

$$\frac{<\sum_i e_i r_i^2>^{1/2}_{\pi^+}}{<\sum_i e_i r_i^2>^{1/2}_{p}} = \frac{(\frac{3}{8})^{1/2}\beta^{-1}}{\alpha^{-1}} = (\frac{27}{64})^{1/4} \simeq 0.81 \qquad (V.17)$$

consistent with the measured ratio. Moreover, the wave function at the origin for the ρ meson is

$$|\psi_\rho(0)| = \frac{\beta^{3/2}}{\pi^{3/4}} = \frac{1}{\left[\frac{8\pi}{3} <\sum_i e_i r_i^2>_{\pi^+}\right]^{3/4}} \qquad (V.18)$$

which upon inserting the measured value for the pion charge radius gives

$$|\psi_\rho(0)| \simeq (.05 \pm .02)\ \mathrm{GeV}^{3/2} \qquad (V.19)$$

while from the decay $\rho \to \ell^+\ell^-$ one can conclude (after including a $\sqrt{3}$ that comes from colour!) that $|\psi_\rho(0)| = (.05 \pm .01)\ \mathrm{GeV}^{3/2}$, in surprisingly good agreement. Incidentally, the values of $|\psi(0)|$ deduced from the leptonic widths of ρ,ω,ϕ,ψ, (and even Υ if $e_q = -1/3$) provide further evidence for the presence of a short range attractive force in addition to the confinement potential since they are roughly proportional to the $q\bar{q}$ reduced mass μ, a behaviour intermediate between that expected for $1/r$ and r or r^2.

With this indication that our results are not totally absurd, we are encouraged to continue. As with the baryons, the next step is to find the eigenvalues of the confining potential by taking into account the perturbation U' which includes the short range attractive (Coulomb-like) effects and distortions from the harmonic potential. Unfortunately, this brings us up against our first uncertainty in making the transition from baryons to mesons: in the baryons we simply took certain expectation values of U^{12} as parameters and we cannot directly relate these integrals to the ones that arise here. Fortunately, this uncertainty is not very significant since although we don't know $U^{12}(r)$, our baryon parameters were essentially the first three <u>moments</u> of $U^{12}\ e^{-\frac{1}{2}\alpha^2 r^2}$ and these are sufficient to satisfactorily determine the expectation values we need here. A straightforward numerical exercise then gives in the I=1 sector the results

$$E_o \simeq 650 \text{ MeV} \tag{V.20}$$

$$E_1 \simeq 1275 \text{ MeV} \tag{V.21}$$

$$E'_o \simeq 1125 \text{ MeV} \tag{V.22}$$

where there is an uncertainty of the order of 25 MeV in these estimates. With the eigensolutions of the confining potential now in hand, we can proceed as in the baryons to calculate the effects of the hyperfine interactions.

We begin with the S-waves where only contact interactions play a role.* The contact interactions depend on the wave function at the origin which as can be seen from equations (V.10) and (V.12) increase with radial excitation for a harmonic oscillator, while from charmonium we know that

$$\frac{|\psi_{200}(0)|}{|\psi_{000}(0)|} \simeq 0.8 \tag{V.23}$$

This effect could presumably be obtained by correctly taking into account the effect of the short-range attraction U' on the wave function. This corresponds to an interband mixing by U' which we have been neglecting and will certainly correct the harmonic oscillator value in the direction of the charmonium result. We choose instead, in the spirit of the sketch that this is intended to be, to short circuit matters by simply taking over the <u>measured</u> charmonium values; we then get

$$\langle\pi|H'_{hyp}|\pi\rangle = -3\langle\rho|H'_{hyp}|\rho\rangle = -\frac{2\sqrt{2}}{3^{3/4}}\,\delta \tag{V.24}$$

$$\langle\pi'|H'_{hyp}|\pi'\rangle = -3\langle\rho'|H'_{hyp}|\rho'\rangle = -\frac{2\sqrt{2}}{3^{3/4}}\left|\frac{\psi_{200}(0)}{\psi_{000}(0)}\right|^2\delta \tag{V.25}$$

$$\langle\pi'|H'_{hyp}|\pi\rangle = -3\langle\rho'|H'_{hyp}|\rho\rangle = -\frac{2\sqrt{2}}{3^{3/4}}\left|\frac{\psi_{200}(0)}{\psi_{000}(0)}\right|\delta \tag{V.26}$$

---

*we neglect mixing of S and D-wave 1^{--} states for simplicity.

where $\delta = \Delta - N$ is a baryonic parameter. If we now diagonalize, we obtain

$$M_\pi = 160 \text{ MeV} \qquad M_{\pi'} = 1010 \tag{V.27}$$

$$M_\rho = 755 \text{ MeV} \qquad M_{\rho'} = 1225 \tag{V.28}$$

The large ρ-π splitting and the low pion mass are independent of our fiddling with $\psi_{200}(0)$; the main thing accomplished by this (justifiable, but nevertheless reprehensible) cheating was to obtain a pion mass which is also unreasonably close to its actual value. We note that according to this sketch the first radial excitation of the ρ should be at about 1225 MeV so that we may wish to believe the evidence for the existence of the $\rho'(1250)$. An important qualitative point which emerges from this calculation is that a sizeable fraction of the large ρ-π splitting arises from the fact that π-π' mixing is much stronger than ρ-ρ' mixing[30].

We next turn to the I=1 P-wave mesons. The contact force vanishes in these states and the hyperfine interaction becomes represented by the previously absent tensor force alone. One easily obtains the results that

$$\langle {}^1P_1 | H'_{tens} | {}^1P_1 \rangle = 0 \tag{V.29}$$

$$\langle {}^3P_2 | H'_{tens} | {}^3P_2 \rangle = -\frac{1}{5}\, t \tag{V.30}$$

$$\langle {}^3P_1 | H'_{tens} | {}^3P_1 \rangle = t \tag{V.31}$$

$$\langle {}^3P_0 | H'_{tens} | {}^3P_0 \rangle = -2t \tag{V.32}$$

$$\text{where } t = \frac{2\sqrt{2}}{3\cdot 3^{3/4}}\,\delta \tag{V.33}$$

while

$$<^1P_1|H'_{so}|^1P_1> = 0 \quad (V.34)$$

$$<^3P_2|H'_{so}|^3P_2> = s \quad (V.35)$$

$$<^3P_1|H'_{so}|^3P_1> = -s \quad (V.36)$$

$$<^3P_0|H'_{so}|^3P_0> = -2s \quad (V.37)$$

where

$$s = \frac{2\beta^5}{3\pi^{3/2}} \int d^3r\, r^2\, f(r)\, e^{-\beta^2 r^2} \quad (V.38)$$

A reliable calculation of s remains unattainable, but we can easily illustrate the kinds of cancellations that can occur between one-gluon exchange and the confinement potential. It is not hard to show that the spin-orbit interaction from one-gluon exchange (including $\vec{\mu}\cdot\vec{B}_{Lorentz}$ and Thomas precession in the 1/r potential) is

$$H_{SO(1G)} = \frac{2\alpha_s}{m^2 r^3} \vec{L}\cdot\vec{S} \quad (V.39)$$

while from the harmonic potential in (V.6) one gets (from Thomas precession alone)

$$H_{SO(HO)} = -\frac{2K}{m^2} \vec{L}\cdot\vec{S}. \quad (V.40)$$

If we were to assume that U is comprised of only the 1/r potential (which is unlikely) then the above would be the only spin orbit effects and we would have

$$s = \frac{4\sqrt{2}}{3^{3/4}} \delta \left[1 - \frac{1}{2} - \frac{3^{3/4} K}{2\sqrt{2} m_d^2 \delta}\right] \quad (V.41)$$

The term in brackets, which is numerically ≃ 1/4, represents the reduction in the $\vec{\mu}\cdot\vec{B}'_{Lorentz}$ spin-orbit interaction by Thomas precession in the 1/r and harmonic potentials. If at this point we admit that we can't calculate the term in the bracket, but instead arbitrarily choose it to be 1/10, the limit we obtained from baryon spectroscopy, we get the results

$$M(^1P_1) = 1275 \text{ MeV} \tag{V.42}$$

$$M(^3P_2) = 1325 \text{ MeV} \tag{V.43}$$

$$M(^3P_1) = 1325 \text{ MeV} \tag{V.44}$$

$$M(^3P_0) = 875 \text{ MeV} \tag{V.45}$$

which in view of the uncertainty in the mass of the A_1 and the confusion surrounding the scalar mesons doesn't seem all that bad.*

---

*The uncertainty associated with the A_1 is well known, but I should mention the present 0^{++} situation: The decay pattern of the states normally associated with the 0^{++} nonet ($\delta, S^*, \varepsilon, \kappa$) are very difficult to reconcile with a $q\bar{q}$ picture for these states. On the other hand they seem quite reasonable candidates for cryptoexotic mesons of the type $qq\bar{q}\bar{q}$ mentioned in Section III.A.[10)] At the same time there have been some experimental developments which indicate that the scalars are complicated: there certainly at this point seem to be more than nine of them, and this must be considered as support for the existence of cryptoexotics. To sort this situation out will probably require an analysis that simultaneously considers both kinds of states and possible mixings between them.

Let us next consider the I=1/2 mesons. Since once again strange quarks "spin more slowly", we get a shift in the eigenvalues in the confinement potential which we can deal with as in the baryons. We must consider in addition, of course, the related change in the wave function at the origin and the reduced chromomagnetic moment of the strange quark. The results are

$$m_K = 510 \text{ MeV} \qquad m_{K^*} = 965 \text{ MeV} \qquad (V.46)$$

$$m_{K'} = 1230 \text{ MeV} \qquad m_{K^{*\prime}} = 1405 \text{ MeV} \qquad (V.47)$$

which once again seems a quite reasonable sketch of the observed situation. As a result of the uncertainties in the spin-orbit term, we don't try to extend these results to the I=1/2 P-waves.

Finally, of course, one may attempt to extend this picture to charmed mesons and charmonium. In the case of charmed mesons, the reduced mass of the meson is sufficiently similar to old mesons that the extension is straightforward. For example, the naive result

$$\frac{D^*-D}{\rho-\pi} \simeq \frac{m_d}{m_c} \qquad (V.49)$$

works very well. The extension to charmonium is not without difficulty however; not only is the reduced mass very different, which may bring with it a rather different effective value for α_s [31], but also the charmonium system has I=0 and is therefore subject to annihilation interactions (see the next section). Nonetheless, since with $\psi(0) = 0$ the P-wave charmonia are unlikely to be subject to large annihilation contributions they probably provide the cleanest test available of the idea of a cancellation between $\vec{\mu}\cdot\vec{B}'_{\mathrm{Lorentz}}$ and Thomas precession. It is therefore gratifying and a good omen for the eventual successful resolution of the spin-orbit problem in baryons and old mesons that in P-wave charmonia such a cancellation scheme seems to work very well.[29)]

B. Some Comments on Isoscalar Mesons

In a potential model the isoscalar mesons

$$|M_{ns}\rangle = \frac{1}{\sqrt{2}}\left[|u\bar{u}\rangle + |d\bar{d}\rangle\right] \qquad (V.50)$$

$$|M_s\rangle = |s\bar{s}\rangle \qquad (V.51)$$

$$|M_c\rangle = |c\bar{c}\rangle \qquad (V.52)$$

$$|M_b\rangle = |b\bar{b}\rangle \quad ,\mathrm{etc.} \qquad (V.53)$$

are necessarily eigenstates of the Hamiltonian with $|M_{ns}\rangle$ degenerate with its associated I=1 meson

$$|M_{I=1}\rangle = \frac{1}{\sqrt{2}}\left[|u\bar{u}\rangle - |d\bar{d}\rangle\right] \qquad (V.54)$$

This state of affairs is actually a reasonably good approximation to the observed states

ρ(770), ω(780), ϕ(1020), ψ(3095), Υ(9500), ...

A_2(1310), f(1270), f'(1515), χ(3555), ? , ...

g(1680), ω(1670), ? , ? , ...

so that at the SU(3) level these mesons seem very close to having the "ideal" mixing angle of around 35° corresponding to the complete segregation of $\frac{1}{\sqrt{2}}[|u\bar{u}\rangle + |d\bar{d}\rangle]$ and $|s\bar{s}\rangle$ into separate mesons.

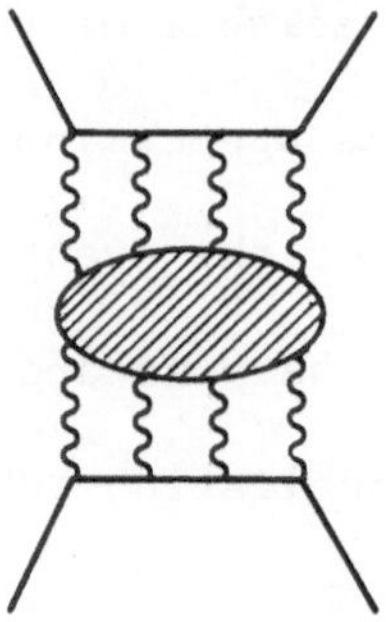

Fig. V.1: Meson mixing via gluon annihilation channels

On the other hand $m_\eta \gg m_\pi$ and the pseudoscalar mixing angle of -10° is far from the ideal value. While these aspects of isoscalar mesons are not well understood, it is nevertheless possible to give a somewhat comforting phenomenological description of this behaviour.[32)] In addition to "t-channel" gluon exchanges which presumably give rise to our potentials, a $q\bar{q}$ pair (carrying no net flavour) may annihilate into another (or the same) quark-antiquark pair $q'\bar{q}'$ as indicated schematically in Figure V.1. Let us now specialize to the SU(3) sector. If we neglect SU(3) breaking in the annihilation amplitude (so that in this sector the amplitude for $q\bar{q} \to q'\bar{q}'$ is A for all q, q') then this interaction will lead to a meson mass matrix in a given J^{PC} sector of the form

$$\begin{bmatrix} m_{u\bar{d}}+A & A & A \\ A & m_{u\bar{d}}+A & A \\ A & A & m_{s\bar{s}}+A \end{bmatrix} \begin{bmatrix} |u\bar{u}\rangle \\ |d\bar{d}\rangle \\ |s\bar{s}\rangle \end{bmatrix}$$

where $M_{u\bar{d}}$ and $M_{s\bar{s}}$ are the eigenvalues obtained in a potential model for $u\bar{u}$ ($d\bar{d}$) and $s\bar{s}$ bound states (so that $M_{s\bar{s}} - M_{u\bar{d}}$ will be of order $2(m_s - m_d)$, and where A will depend on J^{PC}. It is instructive to diagonalize this matrix in two stages: in terms of the basis vectors $|M_{I=1}>$, $|M_{ns}>$, and $|M_s>$ we have

$$\begin{bmatrix} M_{u\bar{d}} & 0 & 0 \\ 0 & M_{u\bar{d}} + 2A & \sqrt{2}A \\ 0 & \sqrt{2}A & M_{s\bar{s}} + A \end{bmatrix} \begin{bmatrix} |M_{I=1}> \\ |M_{ns}> \\ |M_s> \end{bmatrix}$$

so that $|M_{I=1, I_3=0}>$ has the same mass as $|M_{I=1, I_3=\pm 1}>$ as it must. In terms of the remaining two by two matrix we can distinguish two interesting situations that can arise for reasonable values of A:

1) if A is very small relative to $2(m_s - m_d)$, then the physical mesons will be very nearly ideally mixed; if A>0 then the mostly-nonstrange isoscalar will be heavier than $|M_{I=1}>$ and the mixing angle will be greater than 35°, while if A<0 then the mostly-nonstrange isoscalar will be lighter than $|M_{I=1}>$ and the mixing angle will be less than 35°.

2) if A is positive and substantial compared to $2(m_s - m_d)$, it can make the two diagonal entries nearly degenerate relative to $\sqrt{2}A$; in this case the observed isoscalar mesons will be a complete (45°) mixture of $|M_{ns}>$ and $|M_s>$, both will be heavier than $|M_{I=1}>$ with the mostly singlet state highest, and they will have a mixing angle of $35^o - 45^o = -10^o$.

We may thus interpret (ρ,ω,ϕ) as an example of A small and positive $[\omega>\rho,\ \theta_V>35^o]$, $(A_2,\ f,\ f')$ as an example of A small and negative $[f<A_2,\ \theta_T<35^o]$, and $(\pi,\ \eta,\ \eta')$ as an example of complete flavour mixing corresponding to a substantial positive value of A giving

$$\eta\ (\eta') \simeq \frac{1}{\sqrt{2}}\Big[\ |M_{ns}> \mp |M_s>\Big] \tag{V.55}$$

$$\simeq \frac{1}{\sqrt{2}}\Big[\frac{1}{\sqrt{2}}(|u\bar{u}> + |d\bar{d}>) \mp |s\bar{s}>\Big] \tag{V.56}$$

i.e., $\theta_P \simeq -10^o$.

Finally, we comment that it is possible that some problems associated with charmonium spectroscopy may have these annihilation interactions as their cause. In particular, a possible resolution of the η_c puzzle involving such interactions has recently been proposed[33).].

Acknowledgements

It is a pleasure for me to have the opportunity to acknowledge the great degree to which I have benefited from conversations with Chris Llewellyn Smith during my stay at Oxford.

I would also like thank Professor R. H. Dalitz for many valuable conversations and to thank both him and the other members of the Departments of Theoretical Physics and Nuclear Physics for the very friendly reception I have received as a visitor.

Finally, of course, I would like to express my debt to Gabriel Karl without whom references (14) would have been much shorter on both physics and humour.

References

1. See, for example, O.W. Greenberg, to appear in Ann. Rev. of Nucl. and Particle Phys. 28, 1978; A.W. Hendry and D.B. Lichtenberg, to appear in Rep. on Prog. in Phys., 1978; H. Fritzsch, lectures at the XVIII Internationale Universitätswochen für Kernphysik, Schladming, 1978.

2. M.Y. Han and Y. Nambu, Phys. Rev. 139B, 1006(1965); H. Fritzsch, M. Gell-Mann, and H. Leutwyler, Phys. Lett. 47B, 365(1971).

3. O.W. Greenberg, Phys. Rev. Lett. 13, 598(1964).

4. D. Gross and F. Wilczek, Phys. Rev. Lett. 30, 1343(1973); H.D. Politzer, Phys. Rev. Lett. 30, 1346(1973).

5. P.C. Bosetti et al., Oxford Nuclear Physics preprint 16/78 (1978).

6. C.H. Llewellyn Smith in lectures at the XVIII Internationale Universitätswochen für Kernphysik, Schladming, 1978 and references therein; Yu.L. Dokshitser, D.I. D'Yakanov, and S.I. Troyan in lectures at the 13th Leningrad Winter School, 1978 (SLAC Translation 183).

7. This model is very similar to some preconfinement models. See H.J. Lipkin, Phys. Lett. 45B, 267(1973); Y. Nambu, in Preludes in Theoretical Physics, ed. by A. de Shalit, H. Feshbach, and L. van Hove (North Holland, Amsterdam, 1966).

8. See reference 13).

9. Chan Hong-Mo and H. Høgaasen, Phys. Lett. 72B, 121(1977) and references therein.

10. R.L. Jaffe, Phys. Rev. D15, 267,281(1977) and references therein.

11. See especially H.J. Lipkin, Phys. Lett. 74B, 399(1978) for a recent application of such relations between qq and $q\bar{q}$ forces.

12. W. Heisenberg, Z. Phys. 39, 499(1926); G. Breit, Phys. Rev. 36, 383(1930); E. Fermi, Z. Phys. 60, 320(1930).

13. A. De Rújula, H. Georgi, and S.L. Glashow, Phys. Rev. D12, 147(1975).

14. The model discussed here was developed in: Nathan Isgur and Gabriel Karl, Phys. Lett. 72B, 109(1977); Phys. Lett. 74B, 353(1978); "P-wave Baryons in the Quark Model", to be published in the Physical Review; "Positive Parity Excited Baryons in a Quark Model with Hyperfine Interactions", submitted to the Physical Review. For related recent work on baryons, see references 15). Of course, all recent work on baryons is built on the foundations created in the classic papers on baryons by Dalitz[16], Greenberg[17], and their collaborators.

15. D. Gromes and I.O. Stamatescu, Nucl. Phys. B112, 213(1976); W. Celmaster, Phys. Rev. D15, 1391(1977); D. Gromes, Nucl. Phys. B130, 18(1977); L.J. Reinders, Univ. College London preprint (1978); U. Elwanger, Heidelberg preprint HD-THEP-78-1 (1978).

16. R.H. Dalitz in High Energy Physics, Ecole d'Eté de Physique

Theorique, eds. C. DeWitt and M. Jacob (Les Houches, 1965), (Gordon and Breach, New York, 1966).

17. O.W. Greenberg, Phys. Rev. Lett. 13, 598(1964); O.W. Greenberg and M. Resnikoff, Phys. Rev. 163, 1844(1967); D.R. Digvi and O.W. Greenberg, Phys. Rev. 175, 2024(1968).

18. G. Karl and E. Obryk, Nucl. Phys. B8, 609(1968).

19. This rule is a special case of an SU(6) relation. See R.H. Dalizt in the Proceedings of the Triangle Meeting (VEDA Publishing House, Bratislava, 1975) and R. Horgan and R.H. Dalitz, Nucl. Phys. 66B, 135(1973). I am grateful to R.H. Dalitz and R.J. Reinders for clarification of the relation of the rule (IV-35) to the SU(6) relations. This rule has also been noted in the case where V^{ij}_{conf} is a power law potential by D. Gromes and I.O. Stomatescu in reference 15.

20. I am grateful to H.J. Lipkin for discussions of this limit.

21. For a discussion of particle decays in the quark model, see: H.J. Lipkin, Phys. Rep. 8C, 173(1973); J.L. Rosner, Phys. Rep. 11C, 189(1974); R. Horgan in Proceedings of the Topical Conference on Baryons, Oxford, 1976, eds. R.T. Ross and D.H. Saxon (Rutherford Laboratory, 1976).

22. R. Carlitz, S.D. Ellis, and R. Savit, Phys. Lett. 64B, 85(1976); Nathan Isgur, Acta Physica Polonica B8, 1081(1977).

23. W.P. Petersen and J.L. Rosner, Phys. Rev. D6, 820(1972); D. Faiman, Phys. Rev. D15, 854(1977).

24. We show in Figure IV.6 states seen in at least two of the four multichannel $\overline{K}N$ partial wave analyses: G.P. Gopal et al., Nucl. Phys. B119, 362(1977); J.K. Kim, Phys. Rev. Lett. 27, 356(1971); A. Lea et al., Nucl. Phys. B56, 77(1973); W. Langbein and F. Wagner, Nucl. Phys. B47, 477(1972).

25. The compositions quoted in Figures IV.5 and IV.6 are taken from either D. Faiman and D.E. Plane, Nucl. Phys. B50, 379(1972) or A.J.G. Hey, P.J. Litchfield, and R.J. Cashmore, Nucl. Phys. B95, 516(1975).

26. See, for example, references 25) for discussion of the $\overline{K}N$ selection rule. For the photoproduction selection rule see R.G. Moorhouse, Phys. Rev. Lett. 16, 771(1966).

27. Nathan Isgur, Gabriel Karl, and Roman Koniuk, "Quark Hyperfine Interactions and Violations of SU(6) Selection Rules", University of Toronto preprint.

28. That the properties of the Δ5/2+ resonance could be understood in terms of $[56,2^+] - [70,2^+]$ mixing was proposed by D. Faiman, J.L. Rosner, and J. Weyers, Nucl. Phys. B57, 45(1973). Quark hyperfine interactions provide a dynamical mechanism that supplies the sign and magnitude of mixing that was suggested by these authors on empirical grounds.

29. The suppression of spin-orbit interactions has also been noted in charmonium; A.B. Henriques, B.H. Kellet, and R.G. Moorhouse in fact suggested in Phys. lett. 64B, 85(1976) that this reduction could be understood if the confining potential were a world scalar. In the non-relativistic limit this is just the Thomas precession mechanism for spin-orbit suppression. See also H. Schnitzer, Phys. Lett. 65B, 239(1976); 69B, 477(1977); Brandeis preprint, March 1978; Lai-Him Chan, Phys. Lett. 71B, 422(1977).

30. For a fuller discussion of this and related issues see P.J. O'Donnell and R.H. Graham, U. of Toronto preprint, April 1978; Seiji Ono, U.C. Davis preprint, August 1977; and I. Cohen and H.J. Lipkin (private communication from H.J. Lipkin).

31. Howard J. Schnitzer, Phys. Lett. 69B, 477(1977).

32. Nathan Isgur, Phys. Rev. D13, 122(1976). See also I.Cohen and H.J. Lipkin (reference 30) for a more recent discussion.

33. H.J. Lipkin, H.R. Rubinstein, and N. Isgur, "Where is the η_c?", to be published in Physics Letters.

Appendix A: Zeroth Order States in the S=0 and -1 Sectors.

Here we briefly discuss the construction of baryon states. Aside from the spatial wave functions of Sections IV.A and IV.B and the colour wave-functions(III.3) we need isospin wave functions which may be chosen to be

M_ρ -type (octet)

$$\begin{cases} \phi^\rho_p = \frac{1}{\sqrt{2}}(udu - duu) \\ \phi^\rho_\Lambda = -\frac{1}{\sqrt{12}}(2uds - 2du + usd - dsu - sud + sdu) \\ \phi^\rho_{\Sigma^0} = \frac{1}{2}(usd - sud + dsu - sdu), \text{ etc.} \end{cases}$$

M_λ -type (octet)

$$\begin{cases} \phi^\lambda_p = -\frac{1}{\sqrt{6}}(udu + duu - 2uud) \\ \phi^\lambda_\Lambda = -\frac{1}{2}(usd - dsu + sud - sdu) \\ \phi^\lambda_{\Sigma^0} = -\frac{1}{\sqrt{12}}(usd + sud + dsu + sdu - 2uds - 2dus), \text{ etc.} \end{cases}$$

S-type (decouplet)

$$\phi^S_{\Delta^{++}} = uuu, \text{ etc.}$$

A-type (singlet)

$$\phi^A_\Lambda = \frac{1}{\sqrt{6}}(uds - dus - usd + dsu + sud - sdu)$$

uds-type

$$\begin{cases} \phi_\Lambda = \frac{1}{\sqrt{2}}(ud - du)s \\ \phi_\Sigma = \frac{1}{\sqrt{2}}(ud + du)s \end{cases}$$

and spin wave functions which may be chosen to be

M_ρ-type (doublet)
$$\begin{cases} \chi_+^\rho = \frac{1}{\sqrt{2}}(\uparrow\downarrow\uparrow - \downarrow\uparrow\uparrow) \\ \chi_-^\rho = \frac{1}{\sqrt{2}}(\uparrow\downarrow\downarrow - \downarrow\uparrow\downarrow) \end{cases}$$

M_λ-type (doublet)
$$\begin{cases} \chi_+^\lambda = -\frac{1}{\sqrt{6}}(\uparrow\downarrow\uparrow + \downarrow\uparrow\uparrow - 2\uparrow\uparrow\downarrow) \\ \chi_-^\lambda = \frac{1}{\sqrt{6}}(\uparrow\downarrow\downarrow + \downarrow\uparrow\downarrow - 2\downarrow\downarrow\uparrow) \end{cases}$$

S-type (quartet)
$$\chi_{3/2}^S = \uparrow\uparrow\uparrow, \text{ etc.}$$

As examples, we now explicitly construct some states, referring to references (14) for a more complete listing. In the S=0 sector we wish to construct states that are totally symmetric in space, spin, and isospin. Thus at N=0 where we have ψ_{oo}^S we can make only

$$|N^2S_S 1/2^+\rangle = \psi_{oo}^S \frac{1}{\sqrt{2}} [\chi_+^\rho\phi_N^\rho + \chi_+^\lambda\phi_N^\lambda]$$

$$|\Delta^4S_S 3/2^+\rangle = \psi_{oo}^S \chi_{3/2}^S\phi_\Delta^S \;;$$

at N=1 there are more possibilities like

$$|N^4P_M 5/2^-\rangle = \chi_{3/2}^S \frac{1}{\sqrt{2}} [\phi_N^\rho\psi_{11}^\rho + \phi_N^\lambda\psi_{11}^\lambda]$$

$$|N^2P_M 3/2^-\rangle = \frac{1}{2}[\chi_+^\rho\phi_N^\rho\psi_{11}^\lambda + \chi_+^\rho\phi_N^\lambda\psi_{11}^\rho + \chi_+^\lambda\phi_N^\rho\psi_{11}^\rho - \chi_+^\lambda\phi_N^\lambda\psi_{11}^\lambda]$$

$$|\Delta^2P_M 3/2^-\rangle = \phi_\Delta^S \frac{1}{\sqrt{2}} [\chi_+^\rho\psi_{11}^\rho + \chi_+^\lambda\psi_{11}^\lambda]$$

and even more possibilities at N 2 like

$$|\Delta^4 D_s 7/2^+> = \psi^S_{22}\chi^S_{3/2}\phi^S_\Delta$$

$$|\Delta^2 D_M 5/2^+> = \phi^S_\Delta \frac{1}{\sqrt{2}}[\chi^\rho_+\psi^{M_\rho}_{22} + \chi^\lambda_+\psi^{M_\lambda}_{22}]$$

(The notation used here is more carefully defined in Section IV.A.) Note that we have in our examples shown only the top state possible from combining a given L and S; states of lower J may be constructed in the usual way.

One may also construct S=-1 states which are totally symmetric in space, spin, and isospin if the difference $m_s - m_d$ is neglected. Such states are often useful. However, as stressed in the text, when $m_s \neq m_d$ it is not necessary to (anti-) symmetrize with respect to the strange quark and it is often very convenient to use instead of an SU(6)-basis the uds-basis in which one (anti-) symmetrizes only with respect to non-strange quarks. Thus, labelling the strange quark as quark 3 we have at N=0

$$|\Lambda^2 S 1/2^+> = \phi_\Lambda \psi_{oo}\chi^\rho_+$$
$$|\Sigma^2 S 1/2^+> = \phi_\Sigma \psi_{oo}\chi^\lambda_+$$
$$|\Sigma^{*4} S 3/2^+> = \phi_\Sigma \psi_{oo}\chi^S_{3/2}$$

At N=1 we have, for example

$$|\Lambda^4 P_\rho 5/2^-> = \phi_\Lambda \psi^\rho_{11}\chi^S_{3/2}$$
$$|\Sigma^4 P_\lambda 5/2^-> = \phi_\Sigma \psi^\lambda_{11}\chi^S_{3/2}$$
$$|\Lambda^2 P_\rho 3/2^-> = \phi_\Lambda \psi^\rho_{11}\chi^\lambda_+$$
$$|\Lambda^2 P_\lambda 3/2^-> = \phi_\Lambda \psi^\lambda_{11}\chi^\rho_+ \qquad \text{etc.}$$

and at N=2 a tremendous variety of states like

$$|\Lambda^4 D_{\rho\lambda} 7/2^+> = \phi_\Lambda \psi_{22}^{\rho\lambda} \chi_{3/2}^S$$
$$|\Lambda^2 D_{\lambda\lambda} 5/2^+> = \phi_\Lambda \psi_{22}^{\lambda\lambda} \chi_+^\rho$$
$$|\Lambda^2 D_{\rho\lambda} 5/2^+> = \phi_\Lambda \psi_{22}^{\rho\lambda} \chi_+^\lambda$$
$$|\Lambda^2 D_{\rho\rho} 5/2^+> = \phi_\Lambda \psi_{22}^{\rho\rho} \chi_+^\rho$$
$$|\Lambda^4 P_{\rho\lambda} 5/2^+> = \phi_\Lambda \psi_{11}^{\rho\lambda} \chi_{3/2}^S$$

For an example of the relation between these uds basis states and the usual SU(6) basis, see equations (IV.45) and (IV.46), from which it can be seen that the uds basis in some sense maximally violates SU(6) symmetry.

Appendix B: Turning on Non-harmonic Forces.

Here we prove the formulae IV.35. We know by symmetry, first of all, that U will leave the pairs of wave functions of mixed symmetry degenerate. This may also be verified explicitly. In any event one easily shows by taking the wave functions (IV.8) to (IV.14) that in lowest order

$$E[S_S] = 3m + 3\omega + a$$
$$E[P_M] = 3m + 4\omega + \frac{1}{2}a + \frac{1}{3}b$$
$$E[S'_S] = 3m + 5\omega + \frac{5}{4}a - b + \frac{1}{3}c$$
$$E[S_M] = 3m + 5\omega + \frac{5}{8}a - \frac{1}{6}b + \frac{1}{6}c$$
$$E[D_S] = 3m + 5\omega + \frac{1}{2}a + \frac{2}{15}c$$
$$E[D_M] = 3m + 5\omega + \frac{1}{4}a + \frac{1}{3}b + \frac{1}{15}c$$
$$E[P_A] = 3m + 5\omega + \frac{2}{3}b$$

where

$$a = \frac{3\alpha^3}{\pi^{3/2}} \int d^3\rho U(\sqrt{2}\rho)\ e^{-\alpha^2\rho^2}$$

$$b = \frac{3\alpha^5}{\pi^{3/2}} \int d^3\rho U(\sqrt{2}\rho)\rho^2\ e^{-\alpha^2\rho^2}$$

$$c = \frac{3\alpha^7}{\pi^{3/2}} \int d^3\rho U(\sqrt{2}\rho)\rho^4\ e^{-\alpha^2\rho^2}$$

If we define

$$E_o = 3m + 3\omega + a$$

$$\Omega = \omega - a/2 + \frac{1}{3}b$$

$$\Delta \equiv -\frac{5}{4}a + \frac{5}{3}b - \frac{1}{3}c$$

then we get the results (IV.35). Empirically one finds a good description of the hyperfine-unperturbed multiplet levels with

$$E_o \simeq 1150 \text{ MeV}$$

$$\Omega \simeq 440 \text{ MeV}$$

$$\Delta \simeq 440 \text{ MeV}$$

Note that

1) $E_o > \frac{1}{2}(M_N + M_\Delta)$ as a result of second order effects.

2) Ω, the "effective" oscillator spacing, is considerably greater than ω which would be roughly 250 MeV.

The facts that Ω is much greater than ω and that Δ is comparable to Ω raise serious questions of how badly the harmonic oscillator wave functions will be distorted by U. Such questions need to be settled by solving the three-body problem by numerical methods with realistic potentials.

DISCUSSION SESSIONS

CHAIRMAN: Nathan Isgur

SCIENTIFIC SECRETARIES: G. Battistoni, H. Jacob-Hoffman, L. Karsten

DISCUSSION No. 1

- ZICHICHI:

How do we understand scale breaking in QCD?

- ISGUR:

First of all, we can understand approximate scaling in QCD since $\alpha_s(q^2) \to 0$ as $q^2 \to \infty$ so that in this regime the theory behaves like a free field theory. On the other hand at any finite q^2, $\alpha_s(q^2)$ is non-zero and will lead to gluon bremsstrahlung from the quarks, $q\bar{q}$ pair creation, etc. These effects, which change logarithmically with increasing q^2, give deviations from free quark behaviour. Naturally, when a quark bremsstrahlungs it moves toward x=0 and this sort of behaviour is to be expected from almost any theory on the basis of very general arguments of the type made, for example, by Kogut and Susskind. It is the prediction of logarithmic scaling violations that is specific to QCD. A physical picture of scaling violations along these lines has been given by, for example, Altarelli and Parisi.

- THIRRING:

Antishielding (c.f. the (-) sign in the expression for α_s) seems to contradict the Lehmann-Kallen representation. Do you know perhaps the reason why this representation is not valid in non-Abelian gauge theories?

- ISGUR:

I would have to look at the assumptions that go into the derivation, but a guess would be that it has to do with the presence of the massless vector gluons and the $-g^{\mu\nu}$ that appears in their propagator.

- THIRRING:

Yes, these loops perhaps invalidate the Lehmann argument. It is possible that for vector particles the indefinite metric causes the sign change.

- HOFFMAN:

In what way, if any, is this potential model superior to the MIT bag model that, after all, can also give a good fit to the baryon spectrum (perhaps because it contains the same hyperfine interaction?). Is there anything one can calculate with one model that cannot be done with the other?

- ISGUR:

There are indeed many similarities between the two kinds of models and many predictions in common. I believe they should be thought of as complementing each other rather than competing, but since you have given me the opportunity I won't resist pointing out what I think are some of the advantages of the potential model approach. One advantage seems to me to be that in such a model one knows how to separate out the centre-of-mass motion while in the bag all three quark coordinates are independent and can move against the bag (of course this is not a disadvantage if such modes are seen, but I believe that at the moment the bag suffers from the presence of such states). A related issue is the difficult bag boundary conditions that must be faced in dealing with excited hadrons (for example P-wave baryons); in potential models there is of course no problem.

- BHANOT:

Why do you assume two-body confining potentials in baryons and not three-body ones?

- ISGUR:

The two-body potentials have so far proved to be adequate for dealing with these systems, but there are good reasons to investigate the possibility of three body forces (e.g. QCD on a lattice). I believe several people are investigating such questions at the moment; we certainly are. On the other hand you will see that the strongest conclusions we draw are independent of the detailed nature of the confining force and have to do with the way multiplets split up under the action of hyperfine interactions and quark mass differences. Nevertheless, there is some weak evidence for the dominance of two-body forces and I'll point this evidence out in my third lecture.

- BERTINI:

Could you show an example of how you would construct baryonium bound states?

- ISGUR:

Briefly, if you have a qq $\overline{q}\overline{q}$ system, since 3x3 = $\overline{3}$+6 and $\overline{3}$x$\overline{3}$ = 3+$\overline{6}$, one can couple the diquark to the antiquark to get an overall allowed colour singlet either as $\overline{3}$+3 or 6+$\overline{6}$. There are then many possibilities. First of all one can show that $(qq)_{\overline{3}}$ binds and so does $(\overline{q}\overline{q})_3$ and of course these composites bind to each other. On the other hand $(qq)_6$ is an antibound state, but as a system it will be confined to $(\overline{q}\overline{q})_{\overline{6}}$ so with a sufficiently high angular momentum barrier the 6+$\overline{6}$ configuration is possible. These two configurations will have different Regge slopes and different decay characteristics and may be associated with the various types of baryonia which have been seen.[9,10)]

- SURSOCK:

What is the mechanism suppressing the decay of baryonium into mesonic states in QCD?

- ISGUR:

A $qq\overline{q}\overline{q}$ state could of course just fall apart into $q\overline{q}$ + $q\overline{q}$ and there are candidates for such states among the scalar mesons.[10)] However, if qq and $\overline{q}\overline{q}$ are separated by a high angular momentum barrier, the decay may preferably go by $q\overline{q}$ pair creation into $B\overline{B}$. Actually, with a 6+$\overline{6}$ state even this may not be enough since 1 $\not\subset$ 3x6 so of the $B\overline{B}$ states some may be broad ($\overline{3}$+3) and some narrow $(6+\bar{6})$; there even seems to be some evidence for such a situation.[9,10)]

- HOFFMAN:

Do you have any idea as to why potential models that are after all a non-relativistic reduction --- the validity of which depends on heavy quark masses --- work so well even though the quarks are light as in this case?

- ISGUR:

There are two points to be made here. First of all, the relativistic character of the quarks may not be as great as is often supposed. It is not difficult to show that 350 MeV quarks in a proton with its observed charge radius can have less than 100 MeV of kinetic energy. In addition, however, even if the motion were relativistic I am not prevented from adding (in the sense of Foldy-Wouthuysen) relativistic corrections to my Hamiltonian (so long as they converge!) and then solving for the wave function of the system. As I mentioned earlier, the strongest conclusions we draw are based on the way states split up and not on the calculation of absolute energy levels which is mainly what is at stake if the relativistic effects are neglected.

- THIRRING:

The hyperfine interaction leaves the Hamiltonian unbounded from below and needs a cut off mechanism. What is this mechanism in QCD?

- ISGUR:

That's right. The Schrödinger equation only allows attractive potentials less singular than r^{-2} and in an S=0 state we have an attractive δ-function . If uncontrolled the entire wave function would collapse into the origin. Of course the same objection may be raised in QED for the hyperfine structure of the ground state of hydrogen, but there we know that first order perturbation theory correctly gives the 21-cm line. This problem has its resolution in realizing that the hyperfine interaction is only valid in lowest order in α_s; if used alone in higher order by computing mixing matrix elements to higher states it will indeed fail unless one simultaneously considers other effects of higher order in α_s like two-gluon exchange. Physically, such processes have the effect of smearing out the δ-function over a region of dimension m_q^{-1}. For the ground state of hydrogen this smeared δ-function is still much smaller than the radius of the state and the effect is weak so first order perturbation theory works fine. In hadrons the smearing size is not negligible and the effect may be strong so one must explicitly consider higher order processes to know how the δ-function is tamed.

DISCUSSION NO. 2

- ZICHICHI:

Considering the problems mentioned by Professor Wolf concerning the η_c in the potential model, how can you say that this model works so very well?

- ISGUR:

You are right. The state at 2.83 GeV is an outstanding problem. It's peculiar characteristics may, however, be related to the question of strong annihilation channels in pseudoscalar mesons which I will mention in my fourth lecture. Lipkin, Rubinstein, and I have recently suggested on the basis of the η-π system that these annihilation reactions make the true η_c a broad state lying quite close to the ψ where it would be difficult to see; we further suggest on the basis of the charm-strange analogy that the state at 2.83 GeV may be a 0^{++} $q\bar{q}c\bar{c}$ cryptoexotic like the possible 0^{++} $q\bar{q}s\bar{s}$ cryptoexotics (S*,δ, etc.) lying below the ϕ. This is a possible resolution of the puzzles associated with the state.

- THIRRING:

How big is the spin-orbit force for the linear potential?

- ISGUR:

The clearest test of whether the spin-orbit contributions from Thomas precession in the confinement potential are sufficient to account for the observed suppression of spin-orbit couplings relative to the level expected from one-gluon exchange alone comes from the P-wave levels in charmonium. I will discuss these in my fourth lecture, but for now I can just mention the idea. There are three levels here: 2^{++}, 1^{++}, 0^{++}. For a given value of α_s, you will need a particular linear potential to get the ψ'-ψ splitting, etc. correct. One can then ask whether there is any value of α_s

that, in conjunction with Thomas precession, gives the observed P-wave levels. The answer is that this emerges quite naturally with about a factor of 3 reduction of spin orbit forces due to cancellation between one-gluon exchange and Thomas precession in the confinement potential.

- JACOB:

Gluons take about one-half of the proton momentum. Why do they not contribute to, for example, the angular momentum of the baryons, i.e., why do the baryons behave like a three quark system?

- ISGUR:

From deep inelastic scattering we indeed find that about half of the proton's momentum is carried by gluons and about 10% by quark-antiquark pairs. To reconcile this with constituent models we must remember that the proton's constitution appears to be a function of q^2: we have already discussed how when we probe with increasing q^2, gluon bremstrahlung and $q\bar{q}$ pair creation become more and more important so that the valence quarks appear to carry ever less of the proton's momentum, i.e., scale breaking effects push the quark distribution functions into x=0. To connect the parton picture to the constituent models we must ask the reverse question: if at deep inelastic q^2 we see a certain distribution of quarks, antiquarks, and gluons, from what distribution did they evolve at $q^2=0$? This question has been considered by Parisi et al. who find that the pure three quark proton of the constituent models naturally evolves in QCD via gluon bremsstrahlung, etc., into the proton of the parton model by $q^2 \sim 2\ GeV^2$. So the two pictures are complementary and not at all in contradiction.

- MUSSET:

I would like to make a comment. There are two indications in neutrino reactions which point to a u and d "effective" quark mass $\sim 0.3\ GeV/c^2$. One comes from the total inelastic cross section on point-like quarks. It gives from $\sigma_T \sim G^2 mE_\nu$ a value $m = 0.3\ GeV/c^2$. The second is the ratio, averaged, $< \frac{Q^2}{2\nu} >$ since in elastic reactions on quarks with mass m one has $Q^2 = 2m\nu$. Here again one obtains $m \sim 0.3\ GeV/c^2$.

- KLEINERT:

Several students showed surprise at the high values of your quark masses as compared with the small numbers m_u, $m_d \simeq$ 5-15 MeV, $m_s \simeq$ 200-400 MeV found in determinations via hard QCD calculations. It should be pointed out that probably there is complete consistency with your values 350, 580 MeV. The reason is that the hard QCD determinations give what one may call bare mass parameters $M = \text{diag}\ (m_u, m_d, m_s)$ in the QCD Lagrangian. For $M = 0$ the Lagrangian is (neglecting c quarks) chiral SU(3)xSU(3) invariant. The vacuum is supposed to break this symmetry giving rise to massless π, K mesons. Dynamically this is caused, as in superconductors, by a non-zero solution of the quark self-energy equation (called the gap equation) which to lowest order looks like the figure below, i.e., one gives the quark a trial mass $M_0 \neq 0$ and sees whether this remains stable under self-energy corrections.

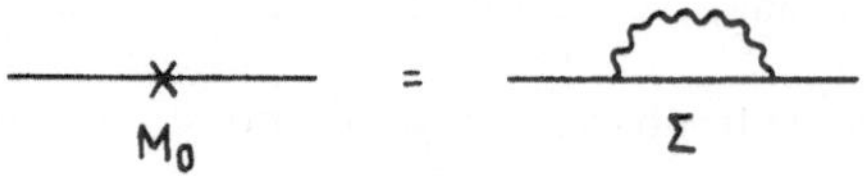

The mass M_o arises from a self-consistent representation of short distance diagrams. Now if there is long distance confinement, the massive quarks will never appear in any asymptotic state. But within the calculations of the potential approach exactly these masses should be used. In a simple model of spontaneous symmetry breakdown proposed a long time ago by Nambu and Jona-Lasinio one can trace this mechanism quantitatively and one finds that the spontaneously generated quark mass M_o for zero bare quark masses in the Lagrangian shows up in the vector and scalar mesons as

$$6M_o^2 = M_{A_1}^2 - M_\rho^2$$

$$4M_o^2 = M_\sigma^2$$

Numerically this amounts to M_o = 310 MeV. Now nature has massive π and K mesons. This can be accounted for only by choosing non-zero bare mass values M. It turns out that m_π^2 and m_K^2 are then proportional to $2m_u$ and m_u+m_s. In the model one needs (neglecting electromagnetic breakings)

$m_u \simeq m_d \simeq 10\text{MeV}$

$m_s \simeq 430\text{MeV}$

to explain the very small π and the moderate K masses. Notice the strong SU(3) violation. If one takes these values into account in the gap equation one finds that also the spontaneously generated quark masses split with M_o receiving an additional matrix piece. In this way one easily obtains Dr. Isgur's masses. For details see Phys. Lett. B62, 429 (1976) and my 1976 Erice Lectures.

- HANSL:

Can you calculate G_A/G_V in your model?

- ISGUR:

The SU(6) quark model gives the result $G_A/G_V = 5/3$ to be compared to the experimental value which is 1.25, so the discrepancy is a factor of 3/4. Since our model has substantial SU(6)-breaking effects in it we certainly hoped to understand this factor. In particular, we have a large $[70,0^+]$ amplitude in our proton and a $[70,0^+]$ nucleon would have $G_A/G_V = 1/3$. We have only made a preliminary analysis of this issue, but it doesn't seem that the effect is strong enough: though the amplitude of $[70,0^+]$ is large the probability (which is what comes in here) is small and gives only about a 6% reduction instead of a 25% reduction. We have also looked at the possible presence of D-wave configurations like $[70,2^+]$ in the proton (they give $G_A/G_V = -1/3$!) but once again their effect is small. I would tentatively conclude that the remaining discrepancy must be attributed to neglected relativistic effects as conjectured by Bogoliubov et al. and Le Yaouanc et al. This is quite consistent, as mentioned earlier, with the accuracy we can expect from the non-relativistic model.

- KLEINERT:

You claim good agreement for the baryonic masses and decay rates. However, old quark models also gave reasonable agreement with the baryonic spectrum without having used the colour idea. What is the basic ingredient in your model that makes it superior to those other models?

- ISGUR:

The original models of Dalitz, Greenberg, and others were more limited in their ambitions. They laid the foundations for this work by showing that baryons(and mesons)could be described in a very general broken SU(6) framework as non-relativistic constituent states. Their approach was to parameterize all effects in terms of the allowed SU(6) and space-time invariants, and as a result they had many parameters and no ability to relate their parameters in going from one multiplet to another or in going from baryons to mesons. The principle difference with the approach I am presenting here is that, for better or for worse, our parameters are all prescribed in terms of some simple dynamics (e.g. hyperfine interactions). Thus all our parameters are highly constrained. To give just one example: all our SU(3) breaking is given in terms of the quark masses m_d and m_s whereas in old models each kind of operator had to have its SU(3) breaking parameterized independently.

Finally, it can be mentioned that because of the complexity of being so general, from our point of view these old models dropped some essential terms, namely tensor terms. As a result, though their spectroscopy was okay, they in fact often failed to obtain the observed decay rates.

DISCUSSION NO. 3

- ROTH:

Why do you think perturbation theory is justified in your calculation when the variation of the potential from harmonic is large in some regions?

- ISGUR:

This is in fact our main worry at present, and we have a program under way to actually solve for the eigenfunctions and eigenvalues variationally. As a partial answer I can say that we have found that for commonly discussed potentials (e.g., linear plus Coulomb) the eigenfunctions with given quantum numbers are of harmonic form to a reasonable approximation. So I am fairly confident that apart from an overall scale factor our results, for example for the P-wave baryons, are valid. The main remaining question is whether the relations we have found between multiplets will survive.

- THIRRING:

Have you tested your calculational procedure in the three nucleon problem $H^3 - He^3$? Also I wondered if you have looked at electromagnetic splittings.

- ISGUR:

Both things you have mentioned are part of programs that are under way, but we don't have any results yet.

- KLEINERT:

Why do you have only two parameters (Δ and Ω) in your splitting formulas?

- ISGUR:

I could explain the result mechanically, but why it is possible to express these six splittings in terms of two numbers, I confess, remains a bit of a mystery to me. That one is taking expectation values of U in polynomials of order ≤ 2 accounts for most, but not all, of the result.

- KLEINERT:

Perhaps the symmetry group U(3) of the harmonic oscillator provides the organizer --- as a Clebsh-Gordon supplier.

- LACKNER:

How is the strength of the harmonic oscillator determined? Can one choose this in such a way as to make the perturbation U as small as possible?

- ISGUR:

That is basically how our variational program works, but for our present purposes the oscillator constant gets absorbed into the parameters E_o and Ω and we never have to discuss its value.

- LACKNER:

Is it true that you get a considerable mixing of SU(6)

multiplets and is that why you found another basis more useful?

-- ISGUR:

Bascially yes, but we must distinguish between SU(6) breaking from hyperfine interactions and from quark mass differences. Both contribute some very large effects. The advantage of the uds basis is that it automatically diagonalizes the large effects due to quark mass differences.

- JACOB:

Is there a connection between your $\bar{K}N$ selection rule for ρ-excitations and planar duality?

- ISGUR:

Yes, there is, both physically and sociologically. We first obtained our mixing angles mechanically without realizing what they meant, but were led to recognize the connection after a comment by Johnathan Rosner. The ρ-λ splitting essentially provides a dynamical basis for the empirical suggestion of "ideal mixing for baryons" of Rosner et al. and Faiman.

- WOLF:

The earlier SU(6) approach to baryon spectroscopy also included hyperfine splitting. Is the QCD hyperfine splitting in better agreement with the data than the old SU(6) one?

- ISGUR:

There are several differences. The early SU(6) calculations certainly had spin-spin interactions, but they didn't interpret them as hyperfine interactions so they didn't include the tensor supported by the data indicates the presence of massless vector exchange coupled to the colour λ-matrices.

DISCUSSION No. 4.

- SURSOCK:

Have you compared your model's prediction with respect to Regge trajectories (baryonic and mesonic)?

- ISGUR:

Not in a systematic way, but since our spectroscopy is good we would agree for low-lying states; on the other hand, for very high excitations our model is unreliable.

- FELTESSE:

Can you predict the branching ratios to different channels for the baryons?

- ISGUR:

Yes, and where the calculations are completed (mainly the P-waves) there is good agreement. A new feature which makes the calculations more difficult, but also perhaps more interesting, is the relatively large admixture of S_M (i.e., $[70,0^+]$) in the proton.

- BHANOT:

How do you explain the fact that the ratio of the wave functions squared at the origin for 2S and 1S is 0.6 in charmonium and 1.5 for the harmonic oscillator?

term. As a result they failed to describe the composition of baryons correctly. They also didn't know the radial dependence of their spin-spin interaction so they had to describe each multiplet independently. Of course they also had the freedom to make Δ lighter than N and/or π heavier than ρ while in QCD, as I mentioned, $\Delta>N$ and $\rho>\pi$ are automatic. Finally, I'll mention that in this approach the SU(3) breaking in the spin-spin interaction is prescribed by the Dirac-like character of the chromomagnetic moments; in the older models these effects were simply parameterized.

- BLASI:

Would you say that your model, so far, does not really include very much of QCD and that the same results could just as well be obtained by group theoretical methods?

- ISGUR:

No I wouldn't agree. There is a great deal of dynamics in our conclusions, and our results certainly depend on the assumptions of soft QCD. For an example of dynamics I would pick out the ρ-λ splitting which is simply the dynamical fact that heavy quarks are slower; for an example of the importance of QCD I would certainly pick the many distinctive features of colour hyperfine interactions which seem essential for understanding the data.

- ZICHICHI:

Can the hyperfine interaction be disentangled from QCD?

- ISGUR:

Not very easily. The hyperfine interaction we have found

- ISGUR:

This result is evidence that the confining potential (be it linear or harmonic) must be perturbed by a short range attractive part as expected in QCD. Since for a Coulombic potential this ratio is $\frac{1}{8}$, a suitable mixture of $-\frac{4}{3}\alpha_s/r$ and harmonic (or linear) will produce this result.

- BHANOT:

You could perhaps also argue that strong renormalization effects change the interaction at the origin in some way, so that the wave function at the origin is not a reliable criterion.

- MONTGOMERY:

Some time ago the motivation for electroproduction experiments was that the form factors tested the dynamics of various quark models. Do you have any predictions for these form factors? In particular the $S_{11}(1530)$ has a strange behaviour which it would be nice to see come out.

- ISGUR:

Low $|\vec{q}|^2$ behaviour can be dealt with in non-relativistic models like these, and we are studying such questions as part of our general program to predict the couplings of the baryons.

- WOLF:

It would be very valuable to test the model even in a small range of $-q^2$, say $-q^2 \lesssim 0.2$ GeV^2, because there the fall-off is very minute.

- MONTGOMERY:

In fact the form factor rises at first; from a recent experiment at DESY we know that this cannot be attributed to the longitudinal part of the cross-section.

NEUTRINO PHYSICS

K. Winter

CERN

Geneva, Switzerland

1. INTRODUCTION

The past of neutrino physics at high-energy accelerators has been glorious. It started with the discovery, in 1962, of the muon-neutrino as a separate particle. The linear rise of the total cross-section was discovered at CERN in 1964 and not immediately understood; however, in 1967 it was the main supporter of the scaling and parton interpretation of deep inelastic electron scattering experiments at SLAC. A new era started with the discovery of the weak neutral currents at CERN in 1973. The following years, evidence for the quark-parton model was emerging from deep inelastic neutrino scattering. In 1975, dilepton events were discovered, both at CERN and at Fermilab, and taken as evidence for particles with new quantum numbers *(Charm)*.

In these lectures I am reviewing recent advances in neutrino physics in three main subjects.

The weak neutral current coupling of the leptons is now emerging from experiments on neutrino-electron scattering. I am also discussing other leptonic channels, e.g. $e^+e^- \rightarrow \mu^+\mu^-$, and the status of evidence for the τ-lepton and its neutrino.

The weak neutral current coupling of hadrons has been extensively studied and is now well understood. I am following the beautiful analysis of Sehgal and Sakurai and include the recent result on parity violation in electron-deuterium scattering. For the first time a new energy scale is defined by these results, the mass of the charged and neutral bosons of the weak interaction. Do bosons with this mass exist? I am giving an outlook on the experimental possibilities of finding them in the near future.

With the increased sensitivity and accuracy of experiments on deep inelastic neutrino scattering, the picture of the quark structure of the nucleon is evolving. New results show the effects of the quark-quark interaction, by scaling violation, and are on the verge of discovering the field boson of this interaction, the gluon.

2. THE STANDARD MODEL OF THE UNIFIED GAUGE THEORY

The unified theory of weak and electromagnetic interactions is based on the group

$$G = SU(2) \otimes U(1) \ . \tag{1}$$

The weak neutral current is given by the mixing of the weak isospin current, which it has in common with the charged current and the electromagnetic current. The strength of mixing is not predicted by the theory; it is denoted by a mixing angle:

$$J_\mu^{NC} = J_\mu^3 - \sin^2 \theta J^{em} \ . \tag{2}$$

At present we know of three families of fermions; they are composed of quarks and of leptons. From their couplings, as determined by the experiments to be discussed in the next sections, we know that they occur in isospin doublets of left-handed fermions and in singlets of right-handed fermions.

We assign weak isospin $1/2$ to the "up" member of a doublet and $-1/2$ to the "down" member:

$$\begin{pmatrix} u \\ d \end{pmatrix}_L \quad \begin{pmatrix} \nu_e \\ e^- \end{pmatrix}_L$$
$$\begin{pmatrix} c \\ s \end{pmatrix}_L \quad \begin{pmatrix} \nu_\mu \\ \mu \end{pmatrix}_L \tag{3}$$
$$\begin{pmatrix} t \\ b \end{pmatrix}_L \quad \begin{pmatrix} \nu_\tau \\ \tau \end{pmatrix}_L$$

The members of the first two families have all been discovered. The third family is not yet complete. The b quark is identified with the hidden quarks of the Υ state discovered at FNAL in 1977. Its assignment to weak isospin $I^3 = -\frac{1}{2}$ (down) is supported by the width $\Gamma_{ee} \sim 1.5$ keV as measured at DESY this year. The t (from top) quark has not yet been found; a search for it is one of the main aims of the experimental programme of the 10-30 GeV e^+e^- storage rings PETRA and PEP. The charged τ-lepton has been discovered; however, its neutral member with $I^3 = \frac{1}{2}$ has not yet been directly detected.

Each factor of the group G has its own gauge coupling constant g and g′, which are analogous to the electric charge. Four vector bosons are mediating the weak and the electromagnetic interactions:

2 charged bosons: $W^\pm$
1 neutral heavy boson: Z^0
1 neutral massless boson: the photon .

The gauge coupling constants are related by the electric charge,

$$e^{-2} = g^{-2} + g'^{-2} \ . \tag{4}$$

If more heavy bosons were found, more terms would be added to the right-hand side of Eq. (4). The mixing of the two neutral vector bosons γ and Z^0 is described by a mixing angle θ:

$$\tan\theta = g'/g \ . \tag{5}$$

The charged current beta-decay interaction is mediated by $W^{\pm}$ exchange with a coupling strength given by

$$\frac{G_F}{\sqrt{2}} = \frac{g^2}{8m_W^2} , \quad \text{with} \quad \frac{e}{g} = \sin\theta . \tag{6}$$

The mass of the charged vector boson follows directly from Eq. (6):

$$m_W = 37.3 \text{ GeV}/\sin\theta . \tag{7}$$

If the Higgs mechanism of spontaneous symmetry breaking is assumed, the expression

$$\frac{m_W}{m_Z \cos\theta} \tag{8}$$

equals unity and the mass of the neutral vector boson is determined as well:

$$m_Z = \frac{37.3 \text{ GeV}}{\sin\theta \cdot \cos\theta} . \tag{9}$$

The value of expression (8) can be determined experimentally.

The fermions and the intermediate bosons are coupled according to two kinds of charges -- the three-colour charges and the flavours -- as shown schematically in Table 1.

Table 1

Bosons	Quarks	Leptons
$W^{\pm}$, Z^0, γ gluons	flavour colour	flavour

The neutral current axial-vector and vector coupling of the fermions and the Z^0 according to Eq. (2) is given by their weak isospin and by their charge

$$a_f = (I_L^3 - I_R^3) \tag{10a}$$

$$v_f = (I_L^3 + I_R^3) - 2Q_f \sin^2\theta . \tag{10b}$$

The experimental results which I shall discuss in the following sections favour the assignment of fermions to left-handed doublets and right-handed singlets, and hence $I^3_R = 0$. The Glashow-Weinberg-Salam model predicts a universal coupling to the Z^0 of all leptons and of all quarks, according to their third component of weak isospin I^3_L and their charge Q_f. Hence there are four groups of fermions which are predicted to couple universally to the Z^0:

1) The neutrinos

$$a_\nu = \frac{1}{2} \qquad v_\nu = \frac{1}{2} \tag{11}$$

2) The leptons with charge -1:

$$a_\ell = -\frac{1}{2} \qquad v_\ell = -\frac{1}{2} + 2\sin^2\theta \tag{12}$$

3) The "up" quarks with charge 2/3,

$$a_u = \frac{1}{2} \qquad v_u = \frac{1}{2} - \frac{4}{3}\sin^2\theta \tag{13}$$

4) The "down" quarks with charge -1/3,

$$a_d = -\frac{1}{2} \qquad v_d = -\frac{1}{2} + \frac{2}{3}\sin^2\theta \ . \tag{14}$$

These predictions can be studied in neutrino experiments for the first family of fermions in Eq. (3). Some other processes involving these couplings will be shown in Fig. 1. The universality of the coupling of all u-quarks and of all d-quarks can only be investigated when using future large e^+e^- storage rings[1]. In the neutrino experiments done so far, only the coupling of the ordinary valence u and d quarks has been studied.

3. MIXING OF FLAVOURS AND THE UNIVERSAL STRENGTH OF THE WEAK HADRONIC CURRENT

The notion of universality of the charged current coupling of leptons and hadrons is based on the Cabibbo mixing of the d and s quarks:

$$\begin{pmatrix} u \\ d \end{pmatrix}_L \rightarrow \begin{pmatrix} u \\ d \cos \theta_C + s \sin \theta_C \end{pmatrix}_L . \tag{15}$$

However, this mixing induces a strangeness-changing neutral current coupling, e.g. $K_L^0 \rightarrow \mu^+\mu^-$, which is not observed experimentally with the universal strength. The most precise measurement gives

$$\frac{\Gamma(K_L^0 \rightarrow \mu^+\mu^-)}{\Gamma(K^+ \rightarrow \mu^+\nu)} \sim (4 \pm 1) \times 10^{-9} . \tag{16}$$

This rate is compatible with a charged-current electromagnetic second-order transition $[\mathcal{O}(\alpha^4)]$. Glashow, Iliopoulos and Maiani (GIM) invented a second left-handed doublet of quarks with a charm-up quark c and a Cabibbo-rotated down-quark

$$\begin{pmatrix} c \\ s \end{pmatrix}_L \rightarrow \begin{pmatrix} c \\ s \cos \theta_C - d \sin \theta_C \end{pmatrix}_L \tag{17}$$

which cancels the strangeness-changing $s \rightarrow d$ neutral current transition.

The weak interaction Lagrangian can be written in a simple way as a product of currents. The charged lepton currents are given by

$$\begin{aligned} j_\alpha^- &= \bar{e}\gamma_\alpha(1 + \gamma_5)\nu_e + \text{other leptons} \\ j_\alpha^+ &= \bar{\nu}_e\gamma_\alpha(1 + \gamma_5)e + \text{other leptons} , \end{aligned} \tag{18}$$

and the neutral lepton current by [see Eqs. (11) and (12)] by

$$\begin{aligned} j_\alpha^N = \bar{\nu}\gamma_\alpha(1 + \gamma_5)\nu &+ v_e[\bar{e}\gamma_\alpha e + \bar{\mu}\gamma_\alpha\mu + \bar{\tau}\gamma_\alpha\tau] \\ &+ a_e[\bar{e}\gamma_\alpha\gamma_5 e + \bar{\mu}\gamma_\alpha\gamma_5\mu + \bar{\tau}\gamma_\alpha\gamma_5\tau] . \end{aligned} \tag{19}$$

Similarly, we can write the isospin raising and lowering charged-hadron currents of universal strength, following Eq. (15):

$$J_\alpha^- = \cos\theta_C \bar{d}\gamma_\alpha(1+\gamma_5)u + \sin\theta_C \bar{s}\gamma_\alpha(1+\gamma_5)u$$
$$J_\alpha^+ = \cos\theta_C \bar{u}\gamma_\alpha(1+\gamma_5)d + \sin\theta_C \bar{u}\gamma_\alpha(1+\gamma_5)\bar{s} \;. \tag{20a}$$

The GIM mechanism introduces the additional charged currents

$$J_\alpha^- = -\sin\theta_C \bar{d}\gamma_\alpha(1+\gamma_5)u + \cos\theta_C \bar{s}\gamma_\alpha(1+\gamma_5)c \;. \tag{20b}$$

This term cancels the strangeness-changing neutral current, leaving only the flavour-conserving currents:

$$J_\alpha^{NC} = \bar{u}\gamma_\alpha(v_u + a_u\gamma_5)u + \bar{d}\gamma_\alpha(v_d + a_d\gamma_5)d \;, \tag{21}$$

where u and d are symbolic for all quark flavours of charge $^2/_3$ and $-^1/_3$, respectively.

4. MIXING OF A SIX-QUARK SYSTEM

There are reasons to believe that we are dealing with a system of three families of fermions consisting of six quarks. If this were the case, the notion of universality of the coupling of leptons and quarks would have to be extended. We would then need three mixing angles. The transition matrix element between "up" and "down" quarks,

$$(\bar{t},\ \bar{c},\ \bar{u})\ (A)\ \gamma_\alpha(1+\gamma_5)\begin{pmatrix} b \\ s \\ d \end{pmatrix} , \tag{22}$$

contains a mixing matrix A [2)]

$$A = \begin{pmatrix} 1 & -\delta & \varepsilon \\ \delta & 1 & -\theta \\ -\varepsilon & \theta & 1 \end{pmatrix}. \tag{23}$$

For clarity I have dropped all cos θ terms, assuming that all mixing angles are small, and omitted the sine. The mixed quark doublets are given by

$$\begin{pmatrix} u \\ d \end{pmatrix}_L \rightarrow \begin{pmatrix} u \\ d\cos\theta + s\sin\theta - b\sin\varepsilon \end{pmatrix}_L$$

$$\begin{pmatrix} c \\ s \end{pmatrix}_L \rightarrow \begin{pmatrix} c \\ s\cos\theta - d\sin\theta + b\sin\delta \end{pmatrix}_L \qquad (24)$$

$$\begin{pmatrix} t \\ b \end{pmatrix}_L \rightarrow \begin{pmatrix} t \\ b\cos\theta + d\sin\varepsilon - s\sin\delta \end{pmatrix}_L .$$

The GIM mechanism is thereby extended and allows the cancelling of all flavour-changing neutral currents. Experimental errors on the decay rates of strange particles and on the K_S^0-K_L^0 mass difference already give upper limits on the two new mixing angles[2)]. Denoting the old Cabibbo angle by θ_C (d-s mixing), we find ε (b-d mixing) < < 0.05 and δ (s-b mixing) < 0.5.

On general grounds the new mixing angles are expected to be smaller than θ_C, and may be of the order of m_c/m_b and m_u/m_b, respectively. They could also be zero. However, a search for stable mesons, composed of the new b quarks ($m_b \sim 4.5$ GeV) has led to an upper limit[3)] on their lifetime

$$\tau(\text{b meson}) < 5 \times 10^{-8} \text{ sec } (90\% \text{ c.l.})$$

and hence

$$|\varepsilon|^2 + |\delta|^2 > 10^{-6} .$$

5. THE HYBRID MODELS OF UNIFIED GAUGE THEORY

Many variations of the standard model have been invented to explain different scenarios. They can now all be eliminated by the present picture of experimental results. I will therefore restrict the discussion to some typical models.

The hybrid character of all these models is reflected by additional right-handed doublets of leptons or of quarks, whereas the other fermions remain in left-handed doublets only.

Here the leptons of all families (e, μ, τ) are supposed to form the following multiplets of weak isospin:

$$\begin{pmatrix}\text{Left-handed}\\ \text{doublet}\end{pmatrix} \quad \begin{pmatrix}\text{right-handed}\\ \text{doublet}\end{pmatrix} \quad \begin{pmatrix}\text{left-handed}\\ \text{neutral singlet}\end{pmatrix}$$

for instance,

$$\begin{pmatrix}\nu_e\\ e^-\end{pmatrix}_L \qquad \begin{pmatrix}E^0\\ e^-\end{pmatrix}_R \qquad (E^0)_L$$

whereas the quarks remain in left-handed doublets. This model has been invented to cancel parity violation in atoms.

Similarly, right-handed $\binom{u}{b}_R$ doublets have been proposed in order to explain the so-called high-y anomaly.

I shall confront these models with the present experimental situation in Sections 6.1 and 7.1.

6. NEUTRAL CURRENT WEAK INTERACTIONS OF LEPTONS

The discussion will be restricted to neutral weak currents built from products of vector and axial-vector currents. The processes to which they may give rise can be depicted by a tetragon (Fig. 1), in analogy to the famous Puppi triangle of charged-current reactions.

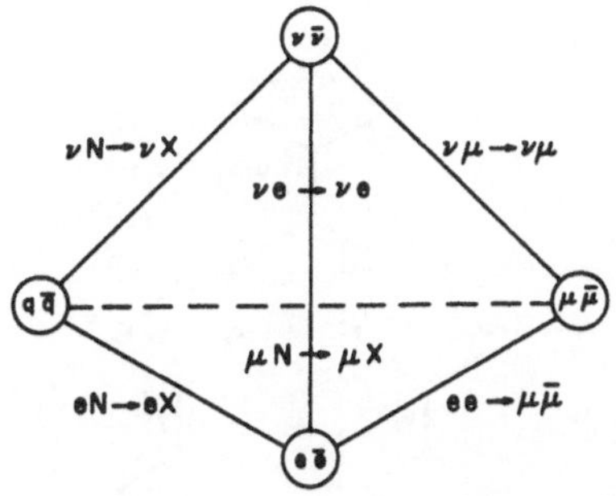

Fig. 1 The tetragon of neutral currents

The neutral current reactions

$$\nu N \to \nu X$$
$$\nu e \to \nu e$$
$$eN \to eX$$

have been observed, and have been shown to be parity violating. The other reactions

$$e\bar{e} \to \mu\bar{\mu}$$
$$\nu\mu \to \nu\mu$$
$$\mu N \to \mu X$$

remain to be discovered and, of course, the analogous reactions involving the τ-lepton and its neutrino.

In these lectures I shall discuss the neutrino-electron reactions (Section 6.1), the neutrino-hadron sector (Section 7), and electron-hadron scattering (Section 7.5).

6.1 Neutrino-electron scattering

The electronic current ($e\bar{e}$) can be studied in three processes,

$$\nu e \to \nu e \quad (25a)$$
$$eN \to eX \quad (25b)$$
$$e^+e^- \to \mu^+\mu^- \, . \quad (25c)$$

According to Fig. 1 they correspond to the coupling with the neutrino current, with the hadronic current, and with the muonic current. There are four different reactions of neutrino-electron scattering:

$$\nu_\mu e^- \to \nu_\mu e^-$$
$$\bar{\nu}_\mu e^- \to \bar{\nu}_\mu e^-$$

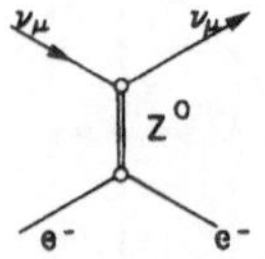

(26)

$$\nu_e e^- \to \nu_e e^-$$
$$\bar{\nu}_e e^- \to \bar{\nu}_e e^-$$

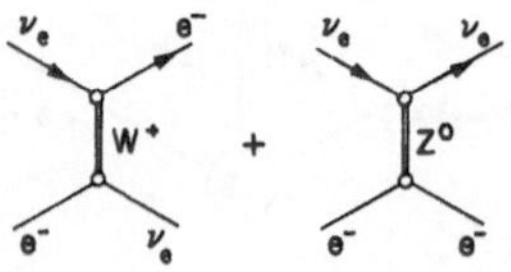

(27)

The scattering of ν_μ can proceed only by the weak neutral current; the scattering of ν_e can proceed also by the weak charged current and we expect a term due to the interference of both. For this reason an experimental study of $\nu_e e^-$ scattering is very interesting. The first experimental observation of a neutral current phenomenon was, however, due to the reaction $\nu_\mu e^-$ [Eq. (26)], which can proceed through the neutral current only.

The unified theory of weak and electromagnetic interactions can be compared to experimental results of reactions (26) and (27) without the uncertainties introduced by using hadronic targets.

The main goal of studying these reactions is to determine the coupling constants of the electronic neutral current, the axial-vector coupling a_e, and the vector coupling v_e, as defined in Eq. (10). The predictions of the Weinberg-Salam model [see Eqs. (11) and (12)] are summarized in Table 2.

Table 2

Weinberg-Salam model predictions for the weak neutral coupling of leptons

Lepton	a	v
ν	$1/2$	$1/2$
e	$-1/2$	$-1/2 + 2 \sin^2 \theta$

The corresponding weak interaction Lagrangian has been defined by Eq. (19).

For comparison with the experiments, we are, however, attempting a phenomenological, model-independent analysis. We are assuming that leptons are point-like and that the outgoing neutrino is identical to the incoming neutrino, and we are restricting the analysis to vector and axial-vector currents.

We obtain for the energy distribution ($y = E_e/E_\nu$) of reaction (26),

$$\frac{d\sigma^\nu}{dy} = \frac{\sigma_0}{2} E_\nu \left[\left(v_e + a_e\right)^2 + \left(v_e - a_e\right)^2 (1 - y)^2 + \frac{m_e}{E_\nu}\left(v_e^2 - a_e^2\right)\right] \tag{28}$$

$$\frac{d\sigma^{\bar\nu}}{dy} = \frac{\sigma_0}{2} E_\nu \left[\left(v_e + a_e\right)^2 (1 - y)^2 + \left(v_e - a_e\right)^2 - \frac{m_e}{E_\nu}\left(v_e^2 - a_e^2\right)\right] \tag{29}$$

The nominal cross-section σ_0 is

$$\sigma_0 = \frac{G^2 m_e}{\pi} = 8.6 \times 10^{-42}\ \mathrm{cm}^2 . \tag{30}$$

Note that a completely model-independent analysis is difficult. The ambiguity between $v_e \leftrightarrow a_e$ in Eqs. (28) and (29) can only be resolved by the terms of order m_e/E_ν. Effects of this term may be detected in experiments using low-energy neutrinos, e.g. from decays of stopped pions or from nuclear reactors.

Assuming μ-e universality of neutral current coupling, we can describe the reactions (27) by the same expressions, by replacing a_e and v_e by

$$\begin{aligned} a'_e &= 1 + a_e \\ v'_e &= 1 + v_e \end{aligned} \tag{31}$$

to account for both charged and neutral currents.

Other measurable quantities can be derived from Eqs. (28) and (29), which again apply also for ν_ee scattering with the substitution of Eq. (31); the total cross-sections

$$\begin{aligned} \sigma^\nu &= \frac{\sigma_0}{2} E_\nu \left[\left(v_e + a_e\right)^2 + \frac{1}{3}\left(v_e - a_e\right)^2\right] \\ \sigma^{\bar\nu} &= \frac{\sigma_0}{2} E_\nu \left[\frac{1}{3}\left(v_e + a_e\right)^2 + \left(v_e - a_e\right)^2\right] \end{aligned} \tag{32}$$

and the mean inelasticities, defined as

$$\langle y \rangle = \frac{\int_0^1 y(d\sigma/dy)dy}{\int_0^1 (d\sigma/dy)dy} \; . \qquad (33)$$

It is convenient to introduce the chiral coupling constants

$$c_L = (v_e + a_e) \; ,$$
$$c_R = (v_e - a_e) \; , \qquad (34)$$

to express the mean inelasticity

$$\langle y \rangle^{\nu} = \frac{6c_L^2 + c_R^2}{12c_L^2 + 4c_R^2} \; , \qquad (35a)$$

$$\langle y \rangle^{\bar{\nu}} = \frac{6c_R^2 + c_L^2}{12c_R^2 + 4c_L^2} \; , \qquad (35b)$$

and the total and differential cross-sections.

The Weinberg-Salam model has predicted a cross-section for these elusive processes which is yet another order of magnitude smaller than the nominal value for $v = a = 1$. These predictions are summarized in Table 3.

Table 3

Predictions for $\nu_\mu e$ scattering

Assumptions	c_L	c_R	σ_{tot}/E_ν (cm^2)	$\langle y \rangle$
nominal, ν	2	0	17.2×10^{-42}	0.5
nominal, $\bar{\nu}$			5.7×10^{-42}	0.25
WS [a)] ν	-0.6	0.4	1.77×10^{-42}	0.468
WS [a)] $\bar{\nu}$			1.20×10^{-42}	0.390

a) (for $\sin^2 \theta = 0.20$).

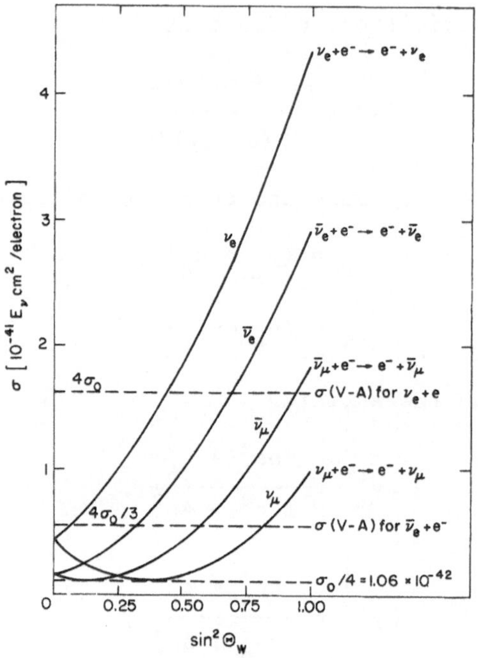

Fig. 2 Cross-section of leptonic reactions as a function of the mixing angle

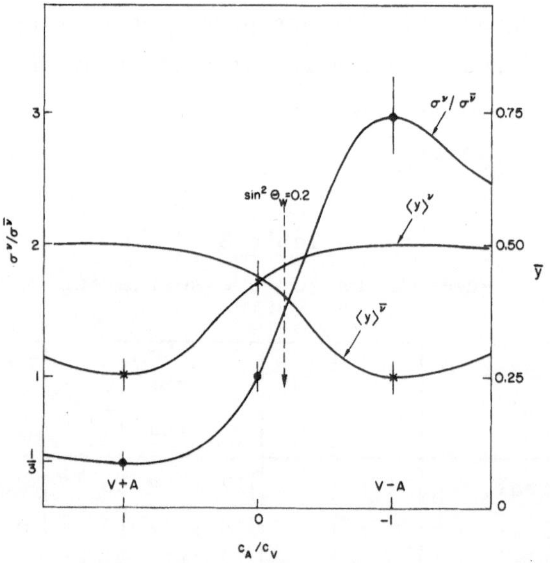

Fig. 3 Mean inelasticities and cross-sections of neutrino-electron scattering as a function of the coupling constant. The error bars represent the results of a simulated experiment with 200 events in each channel.

The dependence of the mean inelasticities and of the total cross-sections on the ratio a_e/v_e and on $\sin^2 \theta$ are shown in Figs. 2 and 3, respectively. Figure 3 also shows error bars as expected for a hypothetical experiment with 200 events for both reactions (26).

The present experimental situation concerning $\nu_\mu e$ scattering[4-7] is summarized in Table 4; the results of several experiments are contradictory. A weighted average is nevertheless formed.

Table 4

Summary of experiments on $\nu_\mu e$ scattering

Experiment	Sample of $\nu_\mu N \to \mu^- X$	Number of $\nu_\mu e$ candidates	Background	$\sigma_{tot}(\nu_\mu e)$ $\times 10^{-42}\ E_\nu\ (cm^2)$
GGM [a)] CERN-PS		≤ 1	0.3 ± 0.1	< 3 (90% c.l.)
Aachen-Padova [b)] CERN-PS		32	21	1.1 ± 0.6
GGM [c)] CERN-SPS	41,000	9	0.4 ± 0.4	$4.3^{+2.0}_{-1.5}$ to $4.9^{+2.2}_{-1.7}$
Columbia-BNL [d)] FNAL 15' BCh	106,000	11	0.7 ± 0.7	1.8 ± 0.8
Weighted average				1.7 ± 0.5

a) J. Blietschau et al., Nuclear Phys. B114 (1976) 189.
b) H. Faissner et al., Phys. Rev. Letters 41 (1978) 213.
c) P. Alibran et al., Phys. Letters 74B (1978) 422. P. Musset, private communication.
d) A.M. Cnops et al., Phys. Rev. Letters 41 (1978) 357.

It should not, however, be ignored that this average value has no clear meaning as it is formed from contradictory results.

The results concerning $\bar{\nu}_\mu e$ scattering are summarized in Table 5. A weighted average value is given here as well and again requires precautions.

A discussion of the procedures used to extract these results can give a better appreciation of the discrepancies. The main

Table 5

Summary of experiments on $\bar{\nu}_\mu e$ scattering

Experiment	Sample of $\bar{\nu}_\mu N \to \mu^+ X$	Number of $\bar{\nu}_\mu e$ candidates	Background	$\sigma_{tot}(\bar{\nu}_\mu e)$ $\times 10^{-42}$ E_ν (cm^2)
GGM [a)] CERN-PS		3	0.4 ± 0.1	$1.0^{+2.1}_{-0.9}$
Aachen-Padova [b)] CERN-PS		17	7.4 ± 1.0	2.2 ± 1.0
BEBC [c)] CERN-SPS	7,500	≤ 1	0.4 ± 0.2	≤ 3.5 (90% c.l.)
FNAL-Mich-IHEP-ITEP [d)] FNAL 15' BCh	6,300	0		≤ 2.9 (90% c.l.)
GGM [e)] CERN-SPS	4,000	0		≤ 3.3
Weighted average				(1.8 ± 0.9)

a) J. Blietschau et al., Nuclear Phys. B114 (1976) 189.
b) H. Faissner et al., Phys. Rev. Letters 41 (1978) 213.
c) B. Armenise et al., Upper limit to the cross-section for $\bar{\nu}_\mu e \to \bar{\nu}_\mu$ at high energy, Contributed paper to the 19th Internat. Conf. on High-Energy Physics, Tokyo, 1978.
d) B. Roe et al., An upper limit to the cross-section for $\bar{\nu}_\mu e \to \nu_\mu e$, Contributed paper to the Oxford Neutrino Conference, 1978.
e) P. Alibran et al., Phys. Letters 74B (1978) 422. P. Musset, private communication.

selection criterion is based on the kinematical consequences of the small mass of the electron target. The most appropriate quantity to express this is

$$E_e \theta^2 = 2m_e(1 - y) \le 2m_e \ , \qquad (36)$$

as it is independent of E_ν.

The distributions of this quantity observed in the Columbia-BNL experiment[7)] are shown in Fig. 4. There is a clear distinction between $\nu_\mu e$ scattering and the two main background processes: quasi-elastic $\nu_e/\bar{\nu}_e$ scattering, e.g.

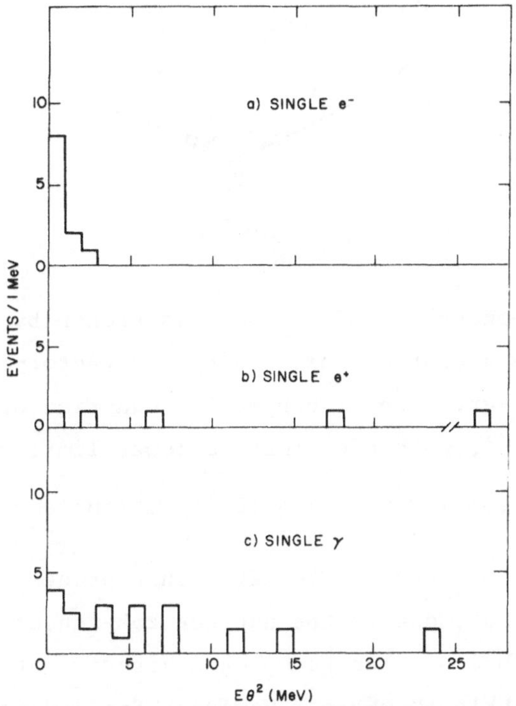

Fig. 4 Evidence for small target mass in the neutrino-electron scattering experiment of the BNL-Columbia group (Ref. 7)

$$\nu_e \mu \rightarrow e^- p \tag{37.}$$

due to the ν_e contamination of the ν_μ beam of typically 1%, and production of single photons. This latter process can either proceed through the neutral current reaction

$$\nu_\mu N \rightarrow \nu_\mu \pi^0 N \; , \tag{38}$$

yielding low-energy photons, or through the reaction

$$\nu_\mu N \rightarrow \nu_\mu \gamma N \; , \tag{39}$$

yielding high-energy photons. Process (39) is the most dangerous and unexpected source of background.

Reaction (39) can proceed via vector meson coupling, according to

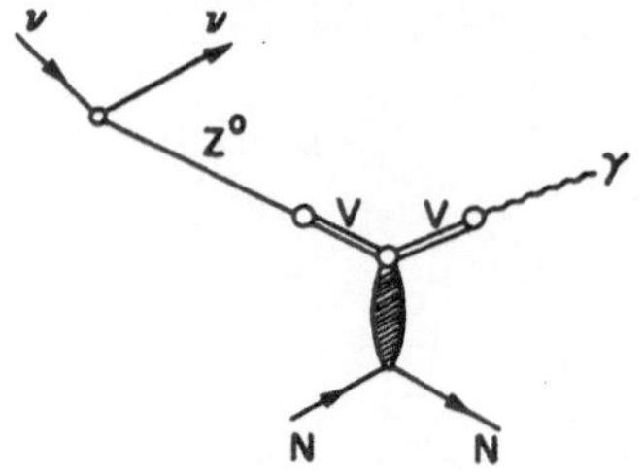

The expected orthogonality of Z^0 and γ is broken by the nuclear scattering of the vector mesons. Using the vector-dominance model and the recent observation of coherent production of ρ^0 in charged current reactions[8], we can estimate a lower limit of

$$\sigma(\nu N \to \nu\gamma N) \leq 7 \times 10^{-41} \text{ cm}^2/\text{nucleon} \qquad (40)$$

for a mean neutrino energy of 30 GeV. This process gives background at low values of $E\theta^2$, due to the nuclear coherence; to be rejected it requires a method of electron-photon discrimination. This can be achieved in bubble chambers. However, for counter and calorimeter experiments, these methods have still to be developed.

The observed value of the cross-section and its relation to the mixing angle in the Weinberg-Salam model is shown in Fig. 5.

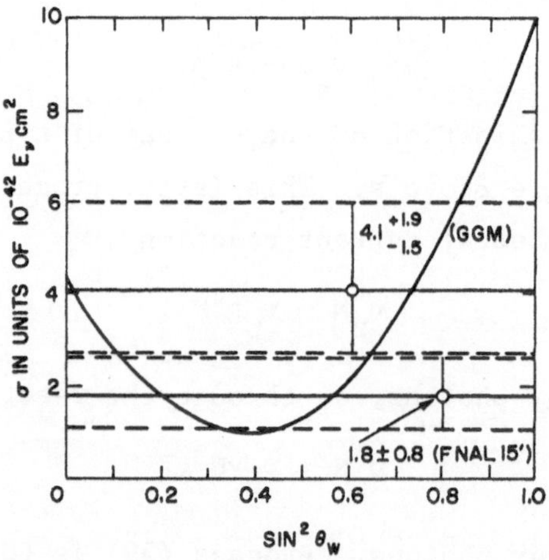

Fig. 5 Cross-section of neutrino-electron scattering as a function of $\sin^2\theta$

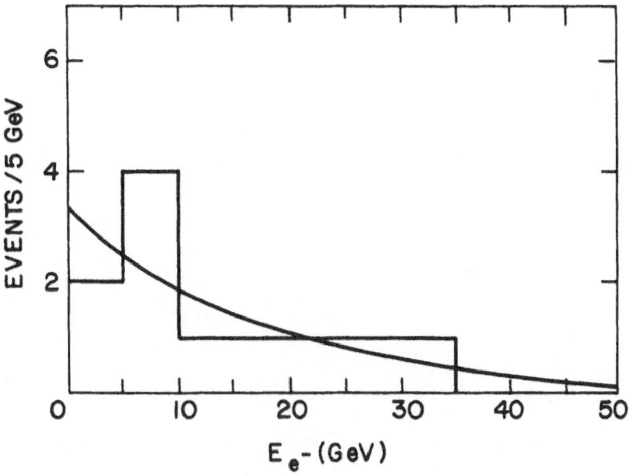

Fig. 6 Electron energies observed in the BNL-Columbia experiment (Ref. 7). The curve is the expectation for $\sin^2 \theta = 0.2$.

The weighted average cross-section of the contradictory experiments on $\nu_\mu e$ scattering would agree with a mixing angle of

$$\sin^2 \theta = 0.21 \begin{array}{l} + 0.09 \\ - 0.06 \end{array}$$

The energy distribution of the 11 events observed in the Columbia-BNL experiment[7] agrees quite well with this prediction, as shown in Fig. 6. The weighted average total cross-section of the $\bar{\nu}_\mu e$ experiments implies

$$\sin^2 \theta = 0.30 \begin{array}{l} + 0.10 \\ - 0.30 \end{array}$$

and has less restrictive consequences.

The most significant confrontation with the theory is achieved by combining the results on $\nu_\mu e$ and $\bar{\nu}_\mu e$ scattering with data from an experiment on $\bar{\nu}_e e$ scattering at a nuclear reactor by Reines et al.[9]. Each of these experiments defines an allowed area of values of a_e and v_e, as shown in Fig. 7 where I have used results of the Aachen-Padova experiment[5] which is the only one having measured both cross-sections of $\nu_\mu e$ and of $\bar{\nu}_\mu e$ scattering. It can be

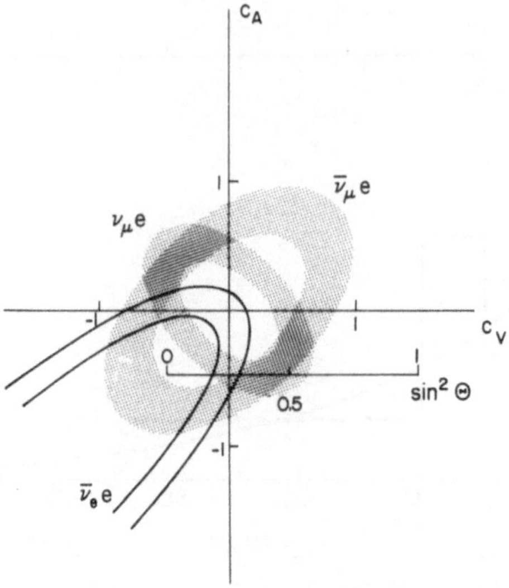

Fig. 7 Confidence area of C_V and C_A as determined from $\nu_\mu e$ scattering (Ref. 7), $\bar{\nu}_\mu e$ scattering (Ref. 5), and $\bar{\nu}_e e$ scattering (Ref. 9)

seen that there is indeed a region of overlap between all three experiments, allowing for two solutions (due to the ambiguity of v and a coupling):

$$
\begin{aligned}
&\text{solution (a)} \qquad v_e = 0.0 \pm 0.1 \ , \qquad a_e = -0.5 \pm 0.1 \ , \\
&\text{solution (b)} \qquad v_e = -0.5 \pm 0.1 \ , \qquad a_e = 0.0 \pm 0.1 \ .
\end{aligned} \tag{41}
$$

Clearly this overlap requires confirmation by new experiments. However, it must be noted that both solutions give a negative value for the chiral coupling constant $c_L = (a_e + v_e)/2$, in agreement with the predictions based on left-handed doublets of weak isospin in the SU(2) ⊗ U(1) gauge theory. Solution (a) agrees with the standard model for $\sin^2\theta = 0.25 \pm 0.05$, whereas solution (b) agrees, for the same value of $\sin^2\theta$, with a hybrid model in which a right-handed doublet

$$\begin{pmatrix} E^0 \\ e^- \end{pmatrix}_R$$

is added to the left-handed doublet.

6.2 Other leptonic neutral current reactions induced by neutrinos

The possibility of studying $\nu_\mu \mu^-$ scattering using the nuclear Coulomb field as a target has been discussed in the literature[10] before the discovery of neutral currents. As in $\nu_e e^-$ scattering, we expect contributions from two processes:

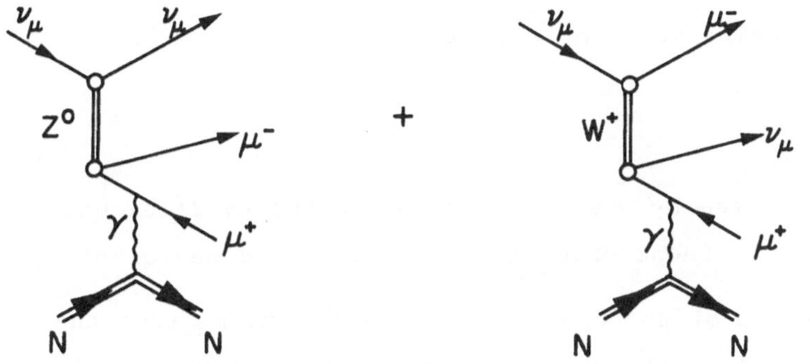

and interference between them. The differential cross-section with respect to the momenta k_1 and k_2 of the two muons is given by two terms:

$$\frac{d\sigma}{dk_1 dk_2} = \frac{c_L^2 + c_R^2}{G_F^2}\left[d\sigma_0(k_1,k_2) + d\sigma_0(k_2,k_1)\right]$$
$$+ \frac{c_L^2 - c_R^2}{G_F^2}\left[d\sigma_0(k_1,k_2) - d\sigma_0(k_2,k_1)\right] , \tag{42}$$

where $d\sigma_0$ is the cross-section for V-A interaction. The value of the chiral coupling constants c_L and c_R is predicted by the standard model,

$$c_L^2 + c_R^2 = \left[\sin^4\theta + (\tfrac{1}{2} + \sin^2\theta)^2\right] G_F^2 , \tag{43}$$

and can be compared to the total cross-section of this reaction. The ratio of c_L/c_R can in principle be determined by studying distributions which are antisymmetric in the exchange of k_1 and k_2, for instance $E(\mu^-) - E(\mu^+)$ or $p_T(\mu^-) - p_T(\mu^+)$.

Although this reaction can give valuable results and is, in principle, accessible for study, using the intense ν_μ beams at high-energy accelerators, it must be pointed out that a strong background process exists in the production of $\mu^-\mu^+$ pairs from the decay of charmed particles.

6.3 Study of neutral current coupling of fermions using e^+e^- annihilation

The annihilation of e^+e^- into a pair of fermions

$$e^+e^- \rightarrow \bar{f}f \tag{44}$$

can be mediated by one-photon exchange and by Z^0 exchange. The interference of these processes produces three measurable effects.

1) The shape of the cross-section is deviating from the 1/s behaviour due to the one-photon exchange (σ_γ) process which is dominating the reaction at low energies, $\sqrt{s} \ll m_{Z^0}$. At the Z^0 pole the cross-section is rising to a value which is R_f times larger than σ_γ; R_f is given by the neutral current coupling constants[11])

$$R_f = \left(g \frac{m_Z^3}{\Gamma_Z}\right)^2 (a_f^2 + v_f^2)\ (a_e^2 + v_e^2)\ . \tag{45}$$

Some values, as expected in the standard model, are given in Table 6. The shape of the Z^0 pole is shown in Fig. 8.

Table 6

Production of various fermion pairs at the Z^0 pole

Fermion pair	R_f
$\nu\bar{\nu}$ (per flavour)	332
$\mu^+\mu^-$, $\tau^+\tau^-$	166
$u\bar{u}$ (per flavour)	550
$d\bar{d}$ (per flavour)	720

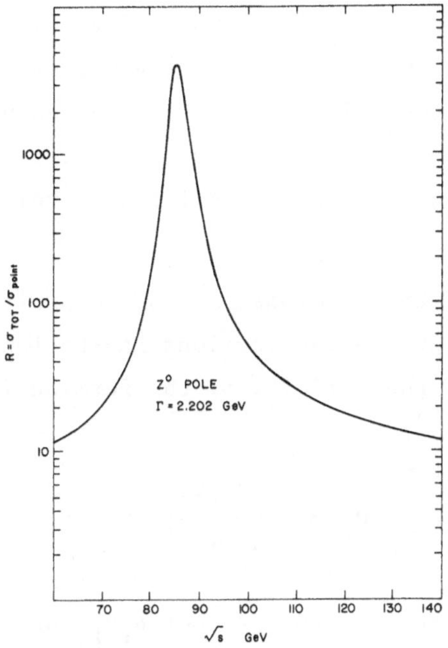

Fig. 8 Ratio of total cross-section of e^+e^- to charged particles and the point-like cross-section to $\mu^+\mu^-$ at the Z^0 pole, as predicted by the standard model

The value of R is predicted to be universal for all u quarks of charge $^2/_3$ and for all d quarks of charge $-^1/_3$, irrespective of their flavour. Hence, a measurement of the cross-sections of reaction (43) at the Z^0 pole gives $(a_f^2 + v_f^2)$ for all fermions with mass less than $m_{Z^0}/2$.

2) The angular distribution of the fermion f with respect to the incident e^- is forward-backward asymmetric. The fractional asymmetry

$$A_f = \frac{F - B}{F + B}$$

is again determined by the neutral current coupling constants. At the Z^0 pole the model predicts[11])

$$A_f = \frac{3v_e a_e v_f a_f}{(a_e^2 + v_e^2)\ (a_f^2 + v_f^2)} \ . \tag{46}$$

The energy dependence of A_f is shown in Fig. 9. A measurement of these large and characteristic effects gives $|a_f|/|v_f|$ for all fermions. For the leptonic final states these measurements are easy to perform. For the quark final states there is the difficulty of determining the quark charge and flavour from the fragmentation products[1,12].

3) The helicity of the fermions. Owing to parity violating neutral current interactions the fermions may be longitudinally polarized. The value of the helicity of the fermion is predicted by the standard model[11]:

$$H_f = - \frac{2a_f v_f}{a_f^2 + v_f^2} . \tag{47}$$

Its measurement determines the product $a_f v_f$ and its sign.

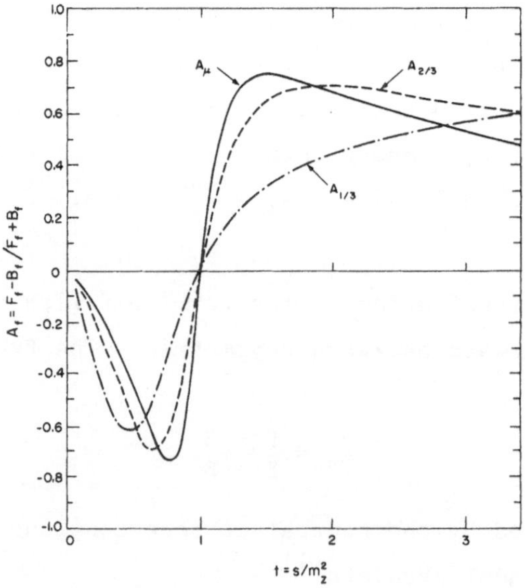

Fig. 9 Angular asymmetry of μ pairs, quark (Q = 2/3) pairs and quark (Q = 1/3) pairs in e^+e^- annihilation near the Z^0 pole at m_{Z^0}

In the case of $e^+e^- \to \mu^+\mu^-$ the helicity can be measured using the parity violation in the decay process

$$\mu^+ \to e^+\nu_e\bar{\nu}_\mu ,$$

which favours emission of e^+ in the direction of the μ^+ spin. The helicity of τ leptons can be determined using the two-body decay mode

$$\tau^+ \to \nu_\tau\pi^+ ,$$

which favours emission of ν_τ in the direction of the τ helicity owing to helicity conservation. A quark polarimeter has not yet been invented.

These experiments may one day be performed at a large (e^+e^-) colliding beam facility. CERN has a project called LEP, to be built starting in 1982. A first attempt at these experiments may be achieved using the new (e^+e^-) facilities PETRA and PEP.

7. THE NEUTRAL CURRENT COUPLING OF HADRONS

Here I shall follow the attempts to analyse the results obtained on neutrino-hadron scattering within the most general phenomenological framework[13]. My starting point is a Lagrangian based on invariance principles only, and incorporating only vector and axial-vector currents of the quarks:

$$L^{NC} = \rho \frac{G}{\sqrt{2}} \bar{\nu}\gamma_\alpha(1 + \gamma_5)\nu \Big\{\bar{u}\gamma_\alpha\left[u_L(1 + \gamma_5) + u_R(1 - \gamma_5)\right]u$$

$$+ \bar{d}\gamma_\alpha\left[d_L(1 + \gamma_5) + d_R(1 - \gamma_5)\right]d\Big\} . \quad (48)$$

$\bar{u}$, u, and $\bar{d}$, d denote the quark fields and u_L, u_R, d_L, and d_R their chiral coupling constants:

$$u_L = v_u + a_u , \quad u_R = v_u - a_u$$

$$d_L = v_d + a_d , \quad d_R = v_d - a_d . \quad (49)$$

Which experimental data are required to determine these coupling constants? And what do the values of these coupling constants tell us about the structure of the weak neutral current interaction? Neutrino interactions with the nucleons can give information on the elementary quark reactions with the valence quarks u and d:

$$\nu u \to \nu u \tag{50a}$$

$$\nu d \to \nu d \tag{50b}$$

$$\bar{\nu} u \to \bar{\nu} u \tag{50c}$$

$$\bar{\nu} d \to \bar{\nu} d \ . \tag{50d}$$

For instance, a study of the y-dependence of the cross-section for these reactions would determine the four chiral coupling constants as defined in Eq. (48). Up to now we have, however, only data of experiments performed using isoscalar targets composed of equal numbers of neutrons and protons and therefore of equal numbers of u and d quarks. These experiments therefore determine only the sums of the chiral coupling constants squared, e.g.

$$\begin{aligned} \frac{d^2\sigma^{\nu}}{dxdy} &= \frac{G^2M}{2\pi} E_{\nu} \left[A_L + A_R(1-y)^2\right] , \\ \frac{d^2\sigma^{\bar{\nu}}}{dxdy} &= \frac{G^2M}{2\pi} E_{\nu} \left[A_L(1-y)^2 + A_R\right] . \end{aligned} \tag{51}$$

The functions A_L and A_R represent the probabilities to observe neutrino or antineutrino scattering on left-handed and on right-handed quarks carrying a fraction x of the nucleon's momentum:

$$\begin{aligned} A_L &= (u_L^2 + d_L^2)q(x) , \\ A_R &= (u_R^2 + d_R^2)q(x) , \end{aligned} \tag{52}$$

with $x = Q^2/2M\nu$, in conventional notation. In analysing the data we must, however, take into account the scattering of neutrinos on antiquarks from the sea, and therefore modify Eqs. (52):

$$A_L = (u_L^2 + d_L^2)q(x) + (u_R^2 + d_R^2)\bar{q}(x) ,$$
$$A_R = (u_R^2 + d_R^2)q(x) + (u_L^2 + d_L^2)\bar{q}(x) . \tag{53}$$

Another free parameter is the factor ρ in front of the Lagrangian in Eq. (48). In the Weinberg-Salam model,

$$\rho = \frac{m_W^2}{m_Z^2 \cos^2 \theta} = 1 . \tag{54}$$

We shall determine the value of ρ from experimental data.

The results of this analysis can then be confronted with the predictions of the standard model and of its variations incorporating additional right-handed doublets. These predictions are summarized in Table 7 for further reference[13].

Table 7

Predicted values of neutral current coupling constants ($x = \sin^2 \theta$)

Model	u_L	d_L	u_R	d_R	v_e	a_e
Weinberg-Salam	$\frac{1}{2} - \frac{2}{3}x$	$-\frac{1}{2} + \frac{1}{3}x$	$-\frac{2}{3}x$	$\frac{1}{3}x$	$-\frac{1}{2} + 2x$	$-\frac{1}{2}$
$(u,b)_R$	"	"	$\frac{1}{2} - \frac{2}{3}x$	$\frac{1}{3}x$		
$(t,b)_R$	"	"	$-\frac{2}{3}x$	$-\frac{1}{2} + \frac{1}{3}x$		
$(u,b)_R$, $(t,b)_R$	"	"	$\frac{1}{2} - 2x$	$-\frac{1}{2} + \frac{1}{3}x$		
$(E^0,e^-)_R$					$-1 + 2x$	0

7.1 Analysis of inclusive data on isoscalar targets

The neutrino facilities at the 400 GeV proton synchrotrons at FNAL and at CERN have given results with high statistics from the study of inclusive charged and neutral current reactions on isoscalar targets. Neutrino scattering is now approaching a level of precision comparable to electron and muon scattering.

The simplest quantities measured are the ratios of the cross-sections

$$R = \frac{\sigma(\nu N \to \nu N)}{\sigma(\nu N \to \mu^- N)} , \quad \bar{R} = \frac{\sigma(\bar{\nu} N \to \bar{\nu} N)}{\sigma(\bar{\nu} N \to \mu^+ N)} \tag{55}$$

on isoscalar targets. It has been shown[14] that if the target is isoscalar, certain conclusions can be drawn without invoking a model for the scattering process. In particular, the relation

$$(u_L^2 + d_L^2) - (u_R^2 + d_R^2) = \frac{\sigma(\nu \to \nu) - \sigma(\bar{\nu} \to \bar{\nu})}{\sigma(\nu \to \mu^-) - \sigma(\bar{\nu} \to \mu^+)} \tag{56}$$

is valid for any kinematical domain provided it is the same for all four reactions. It is therefore not necessary to correct the experimental data for event losses due to selection criteria which are imposed by the necessity to reject the background, e.g. the requirement that the hadron energy produced is larger than 10 GeV [15]. In terms of the ratios of inclusive cross-sections defined by Eq. (55) and using also $r = \sigma(\bar{\nu} \to \mu^+)/\sigma(\nu \to \mu^-)$,

$$(u_L^2 + d_L^2) - (u_R^2 + d_R^2) = \frac{R - r\bar{R}}{1 - r} . \tag{57}$$

To evaluate relation (57) we use the data compiled in Table 8 [16], and for $r = \sigma(\bar{\nu} \to \mu^+)/\sigma(\nu \to \mu^-)$ the weighted mean of results obtained by the CITF[17], BEBC[18], and CDHS[19] experiments,

$$r = 0.48 \pm 0.01$$

and obtain, after correcting for the neutron excess of the target,

$$(u_L^2 + d_L^2) - (u_R^2 + d_R^2) = 0.269 \pm 0.035 . \tag{58}$$

In the standard model

$$(u_L^2 + d_L^2) - (u_R^2 + d_R^2) = \tfrac{1}{2} - \sin^2\theta .$$

Table 8

Ratios of inclusive neutral current reactions on isoscalar targets[16)]

Experiment	E_ν (GeV)	E_μ cut (GeV)	Raw ratios		Corrected ratios	
			R	$\bar{R}$	R	$\bar{R}$
GGM, CERN-PS	1-10	1	0.25 ± 0.04	0.56 ± 0.03	0.26 ± 0.04	0.39 ± 0.06
HPWF, FNAL	30-200	4	0.28 ± 0.03	0.39 ± 0.10	0.30 ± 0.04	0.33 ± 0.09
CITF, FNAL	30-200	12	0.27 ± 0.03	0.35 ± 0.11	0.27 ± 0.02	0.40 ± 0.08
CDHS, CERN-SPS	30-200	12	0.293 ± 0.01	0.35 ± 0.11	0.295 ± 0.01	0.34 ± 0.03
BEBC, CERN-SPS	30-200	15	0.33 ± 0.03	0.39 ± 0.07		

Comparison with Eq. (58) gives

$$\sin^2 \theta = 0.231 \pm 0.035 \,. \tag{59}$$

Let me pause here to discuss the experimental and theoretical problems encountered in obtaining this result.

The neutral current data sample is subject to a number of corrections; typical values are summarized in Table 9. The largest correction is due to the failure to observe low-energy muons owing to the limited acceptance of the external muon identifier of BEBC, or to the required track length in a target calorimeter of high density, e.g. in the CDHS detector.

Among the theoretical uncertainties in deriving relation (57), the corrections of the simple quark-parton picture due to effects

Table 9

Summary of corrections to the inclusive neutrino neutral current data sample (NC)

Cause	Mixing of samples	Correction to NC
Loss of muons p_μ < 5 GeV/c	CC → NC	-20%
Accidental muon association	NC → CC	+5%
Neutron/K^0 interaction E_h > 15 GeV	→ NC	-5%
ν_e interaction	→ NC	-5%
Wide-band background	→ NC	-1%

predicted by quantum chromodynamics (QCD) could be large. However, they do not affect the results if all cross-sections are measured at the same energy.

To obtain further information, it is necessary to use a model for the process of neutrino-nucleon scattering. For clarity we restrict the discussion to the naive quark model without scaling violation and neglect the effects of charm quarks.

We introduce, in conventional notation, the moments of the quark distributions in x

$$\begin{aligned} q &= \int x(u + d)\,dx \\ \bar{q} &= \int x(\bar{u} + \bar{d})\,dx \\ q_S &= \int x(s + \bar{s})\,dx \end{aligned} \tag{60}$$

and derive the y-distributions of the four inclusive reactions. In units of G^2ME_ν/π:

$$\frac{d\sigma}{dy}(\nu \to \mu^-) = (q + q_S) + \bar{q}(1 - y)^2 ,$$

$$\frac{d\sigma}{dy}(\bar{\nu} \to \mu^+) = (\bar{q} + q_S) + q(1 - y)^2 ,$$

(61)

$$\frac{d\sigma}{dy}(\nu \to \nu) = (u_L^2 + d_L^2)[q + \bar{q}(1 - y)^2] + (u_R^2 + d_R^2)[\bar{q} + q(1 - y)^2] + (s_L^2 + s_R^2)q_S[1 + (1 - y)^2] ,$$

$$\frac{d\sigma}{dy}(\bar{\nu} \to \bar{\nu}) = (u_R^2 + d_R^2)[q + \bar{q}(1 - y)^2] + (u_L^2 + d_L^2)[\bar{q} + q(1 - y)^2] + (s_L^2 + s_R^2)q_S[1 + (1 - y)^2] .$$

Comparing the expressions for charged and neutral currents in Eq. (61) we note that we can substitute the charged current cross-sections for the quark distributions in the expressions for the neutral current cross-sections:

$$\begin{aligned}\sigma(\nu \to \nu) &= (u_L^2 + d_L^2)\sigma(\nu \to \mu^-) + (u_R^2 + d_R^2)\sigma(\bar{\nu} \to \mu^+) + \mathcal{O}(s,\bar{s}) \\ \sigma(\bar{\nu} \to \bar{\nu}) &= (u_R^2 + d_R^2)\sigma(\nu \to \mu^-) + (u_L^2 + d_L^2)\sigma(\bar{\nu} \to \mu^+) + \mathcal{O}(s,\bar{s}) \ .\end{aligned} \tag{62}$$

We thus find separate relations for left-handed and right-handed coupling constants and express them in terms of the inclusive cross-section ratios defined in Eq. (55)

$$\begin{aligned}u_L^2 + d_L^2 &= \frac{R - r^2\bar{R}}{1 - r^2} + \mathcal{O}(s,\bar{s}) \ , \\ u_R^2 + d_R^2 &= \frac{\bar{R} - R}{r^{-1} - r} + \mathcal{O}(s,\bar{s}) \ .\end{aligned} \tag{63}$$

Sehgal[20] has evaluated relations (63) using the data of Table 8 and assuming $\bar{q}/q = 0.15$, $q_S/q = 0.05$, with the result

$$\begin{aligned}u_L^2 + d_L^2 &= 0.29 \ \pm 0.02 \\ u_R^2 + d_R^2 &= 0.026 \pm 0.0125 \ .\end{aligned} \tag{64}$$

Thus right-handed coupling of the u and d quarks is ten times smaller than the left-handed one.

The deviation of the ratio

$$\frac{d\sigma}{dy}(\nu \to \nu) \Big/ \frac{d\sigma}{dy}(\nu \to \mu^-)$$

from a constant value is a direct measurement of the relative strength of right-handed and of left-handed coupling. Analysis of the inclusive CDHS data gives[21] the result

$$\frac{u_R^2 + d_R^2}{u_L^2 + d_L^2} = 0.10 \pm 0.03 \ , \tag{65}$$

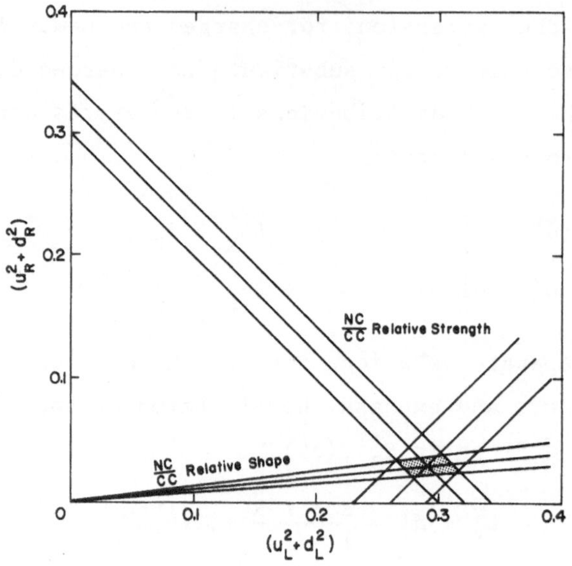

Fig. 10 Left-handed and right-handed neutral current coupling derived from the relative strength and shape of NC and CC inclusive reactions

which establishes right-handed coupling with 3 standard deviations. Figure 10 illustrates the relative weight of the ratios of cross-sections and of y-distributions in the determination of the coup-

Table 10

Neutral current coupling constants

Experiment	$u_L^2 + d_L^2$	$u_R^2 + d_R^2$	Ref.
GGM, CERN PS	0.24 ± 0.05	0.06 ± 0.03	22
CITF, FNAL	0.23 ± 0.04	0.08 ± 0.05	23
HPWF, FNAL	0.29 ± 0.06	0.02 ± 0.06	24
BEBC, CERN SPS	0.33 ± 0.05	$0.02 \pm {0.04 \atop 0.02}$	25
CDHS, CERN SPS	0.29 ± 0.02	0.03 ± 0.01	21

ling constants[21]. Results of various experiments[21-25] are summarized in Table 10; we note satisfactory agreement.

In view of the high precision of this determination of the coupling constants we must pause here and examine to what extent the results may be affected by corrections to the quark model, e.g. due to the observed scaling violations and due to possible violations of the Callan-Gross relation. Inasmuch as a violation of the Callan-Gross relation is the same in all channels, its effects cancel in deriving Eq. (63). The effects of scaling violations have been evaluated in a very preliminary way[20]; they may cause a shift in the results of Eq. (64) by about one standard deviation. Future analysis should incorporate the scaling violations observed in charged current reactions.

Comparing these results with the predictions of models as summarized in Table 7, we note agreement with the standard WS model for $\sin^2\theta = 0.24 \pm 0.02$. The small value of $u_R^2 + d_R^2$ *rules out* all models with right-handed u or d quark doublets, in particular all b-type quark models. Comparison with the Glashow model[26] gives

$$\begin{aligned} \rho &= 0.98 \pm 0.05 \\ \sin^2\theta &= 0.23 \pm 0.03\ , \end{aligned} \qquad (66)$$

in remarkable agreement with the prediction of the standard model, $\rho = m_W^2/m_Z^2 \cos^2\theta = 1$.

The relation of the neutral current coupling constants determined from the inclusive reactions on isoscalar targets with the Lorentz structure of the current is illustrated in Fig. 11. We conclude that pure V+A and pure V or pure A interactions are excluded. Pure V-A interaction is unlikely to describe the data. Hence the neutral current is neither pure in chirality nor in parity.

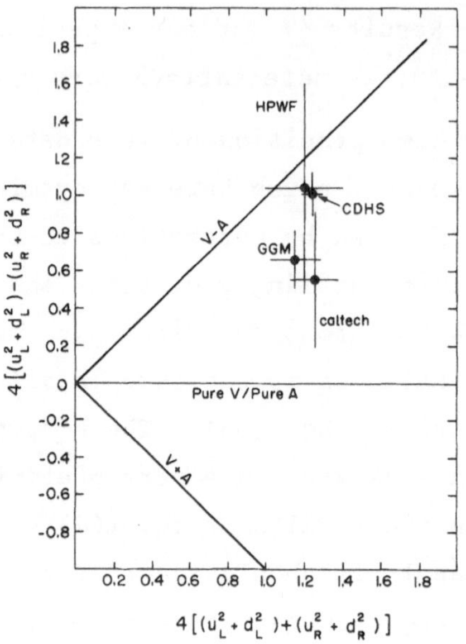

Fig. 11 Coupling constants and the Lorentz structure of the neutral current

7.2 Analysis of inclusive data on protons

Measurements using isoscalar targets cannot determine differences between the u and the d couplings; data on proton targets are needed to obtain this information. Denoting these differences by

$$\Delta_L = u_L^2 - d_L^2 \quad \text{and} \quad \Delta_R = u_R^2 - d_R^2 , \tag{67}$$

we expect to obtain for proton targets

$$\begin{aligned}
\sigma(\nu p \to \nu X) &= \sigma^{I=0}(\nu \to \nu) + \left(\Delta_L + \frac{1}{3}\Delta_R\right)(u - d) + \mathcal{O}(s,\bar{s}) \\
\sigma(\bar{\nu} p \to \bar{\nu} X) &= \sigma^{I=0}(\bar{\nu} \to \bar{\nu}) + \left(\Delta_R + \frac{1}{3}\Delta_L\right)(u - d) + \mathcal{O}(s,\bar{s}) \\
\sigma(\nu p \to \mu^- X) &= \sigma^{I=0}(\nu \to \mu^-) - (u - d) \\
\sigma(\bar{\nu} p \to \mu^+ X) &= \sigma^{I=0}(\bar{\nu} \to \mu^+) + \frac{1}{3}(u - d) ,
\end{aligned} \tag{68}$$

where $u = \int x\,u\,dx$ and $d = \int x\,d\,dx$ are the first moments of the u-

and d-quark x-distributions. The quantity u-d can in principle also be determined by electron scattering on protons and on neutrons. Recent results[27)] suggest that

$$\frac{u - d}{u + d} \approx \frac{1}{3} \,. \tag{69}$$

Assuming $\bar{q}/q = 0.15$ and $q_S/q = 0.05$, Sehgal has derived the relation

$$\left(\frac{\sigma^p - \sigma^n}{\sigma^p + \sigma^n}\right)_{\nu\to\nu} = \Delta_L + \frac{1}{3}\Delta_R \,, \quad \left(\frac{\sigma^p - \sigma^n}{\sigma^p + \sigma^n}\right)_{\bar{\nu}\to\bar{\nu}} = 1.9\left(\Delta_R + \frac{1}{3}\Delta_L\right) . \tag{70}$$

The last quantity (for $\bar{\nu} \to \bar{\nu}$) is zero for $\sin^2\theta = 0.23$, and larger or smaller than zero for $\sin^2\theta$ larger or smaller than 0.23. The data are still rather scarce; first results

$$\begin{aligned} R^p &= 0.48 \pm 0.17 \quad \text{(ref. 28)} \\ \bar{R}^p &= 0.42 \pm 0.13 \quad \text{(ref. 29)} \\ \left(\frac{\sigma^n}{\sigma^p}\right)_{\nu\to\nu} &= 1.31 \pm 0.38 \quad \text{(ref. 30)} \end{aligned} \tag{71}$$

give some constraints but do not allow the u and d couplings to be determined:

$$\begin{aligned} &\Delta_L < 0.025 \quad \text{or} \quad \frac{\Delta_L}{u_L^2 + d_L^2} < 0.1 \\ &\Delta_R < 0.030 \,. \end{aligned} \tag{72}$$

We expect future experiments to measure cross-sections on hydrogen with improved statistical accuracy, maybe to ±10%.

7.3 Analysis of semi-inclusive data

The primary neutrino-quark interaction produces a quark beam with a composition which depends on the structure of the neutral current. For the four reactions on isoscalar targets we expect:

$$\begin{aligned}
&(\nu \to \mu^-) \quad && u : d = 1 : 0 \\
&(\bar{\nu} \to \mu^+) \quad && u : d = 0 : 1 \\
&(\nu \to \nu) \quad && u : d = u_L^2 + \frac{1}{3} u_R^2 : d_L^2 + \frac{1}{3} d_R^2 \\
&(\bar{\nu} \to \bar{\nu}) \quad && u : d = u_R^2 + \frac{1}{3} u_L^2 : d_R^2 + \frac{1}{3} d_L^2 \;.
\end{aligned} \tag{73}$$

The fragmentation of the quarks gives the observed final state of hadrons. The ratio π^+/π^- is, in principle, sensitive to the primary quark composition. In practice, pions with fractional momentum $z > 0.2$ have to be selected to exclude the region to which the fragmentation of the spectator quarks may contribute. For the fragmentation of, for example, the u quark, we have to assume factorization of the fragmentation function $d_u^{\pi^+}(z)$ and of the x-distribution of u quarks, u(x). This assumption does not seem to be supported by recent experiments[31]. Data obtained at low energy by the Gargamelle Collaboration have been analysed by Kluttig et al.[32], selecting events with $E_\mu > 1$ GeV and $0.3 < z < 0.7$ with the following result:

$$\begin{aligned}
(\pi^+/\pi^-)_{\nu \to \nu} &= 0.77 \pm 0.14 \\
(\pi^+/\pi^-)_{\bar{\nu} \to \nu} &= 1.64 \pm 0.36 \;.
\end{aligned} \tag{74}$$

A quantitative analysis based on Eq. (73) and the assumption of factorization has been performed by Sehgal[33]. Using the results on inclusive reactions as constraints, he has determined separate values of the squares of the chiral coupling constants of u and d quarks, as summarized in Table 11.

Table 11

Chiral coupling constants of u and d quarks determined from inclusive and semi-inclusive data

	u_L^2	d_L^2	u_R^2	d_R^2
Data	0.10 ± 0.03	0.19 ± 0.03	0.03 ± 0.015	0 ± 0.015
Weinberg-Salam for $\sin^2\theta = 0.25$	0.11	0.17	0.028	0.007

We note agreement with the standard model. However, in a critical analysis as attempted in these lectures, the model dependence of this result must be stressed. The analysis of the π^+/π^- ratio must be repeated at higher energies, and the underlying assumption of factorization must be demonstrated to be valid.

The relative signs of the chiral coupling constants cannot be determined by these experiments. Choosing one sign, e.g. $u_L > 0$, and supposing $d_R = 0$, there remain two sign ambiguities:

$$\begin{aligned} &\text{sign } (u_L d_L) \\ &\text{sign } (u_L u_R) \;. \end{aligned} \tag{75}$$

Hence there are four possible solutions, which are shown in Fig. 12. They correspond to ambiguities between V and A couplings and between I = 0 (isoscalar) and I = 1 (isovector) coupling, as given in Table 12.

Table 12

Ambiguities of the neutral current coupling constants

Solution	Sign $(u_L d_L)$	Sign $(u_L u_R)$	Isospin dominance	v, a dominance
A	−	−	1	v
B	−	+	1	a
C	+	+	0	v
D	+	−	0	a

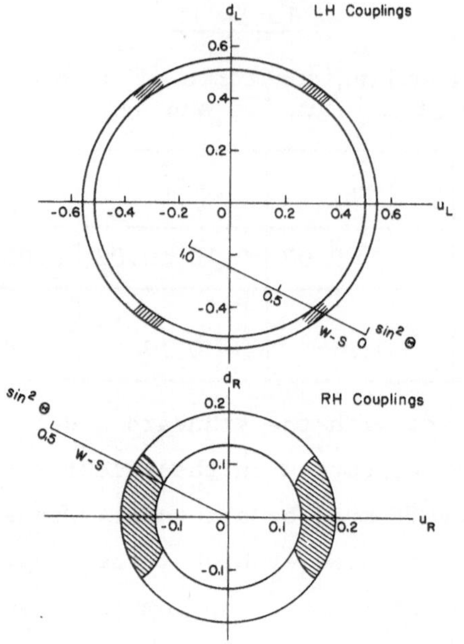

Fig. 12 Solutions for left-handed and right-handed neutral current coupling

For completeness we note here some relations between the chiral coupling constants and the isospin couplings:

$$u_L + d_L = \tfrac{1}{2}(v^0 + a^0)$$
$$u_L - d_L = \tfrac{1}{2}(v^1 + a^1)$$
$$u_R + d_R = \tfrac{1}{2}(v^0 - a^0)$$
$$u_R - d_R = \tfrac{1}{2}(v^1 - a^1)\ .$$

Solution A corresponds to the prediction of the standard model; for $\sin^2\theta = \tfrac{1}{4}$ this model predicts dominance of left-handed isovector coupling.

7.4 Analysis of exclusive data

Analysis of data[34] from the elastic reactions

$$\begin{array}{l} \nu p \to \nu p \\ \bar{\nu} p \to \bar{\nu} p \end{array} \tag{76}$$

by Hung and Sakurai[35] and analysis of data[36] from single-pion production by Abbot and Barnett[37] favour solution A. The

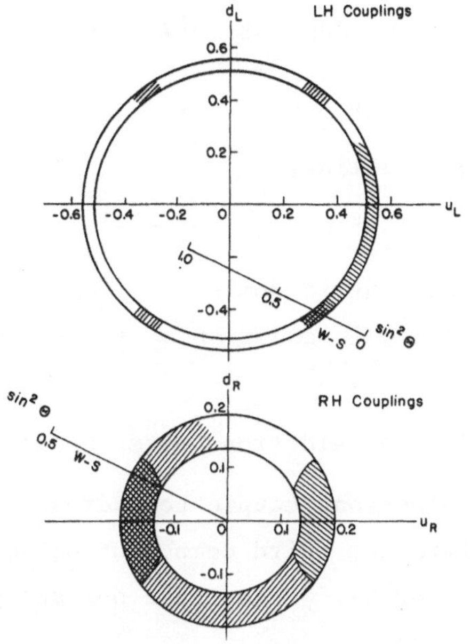

Fig. 13 Favoured solutions of quark coupling to the neutral current, due to constraints from exclusive reactions

constraints in favour of this solution are shown in Fig. 13. The coupling constants corresponding to this solution are given in Table 13, together with the predictions of the standard model.

Table 13

Summary of quark chiral coupling constants

	u_L	d_L	u_R	d_R
Data	0.32 ± 0.03	-0.43 ± 0.04	-0.17 ± 0.05	0 ± 0.12
WS for $\sin^2\theta = \frac{1}{4}$	$\frac{1}{3}$	$-\frac{5}{12}$	$-\frac{1}{6}$	$\frac{1}{12}$

I would again like to stress the preliminary character of these conclusions.

In terms of isospin coupling this solution gives a dominant isovector coupling,

$$v^1 = 1 \ , \quad a^1 = \tfrac{1}{2} \ , \tag{77}$$

and a weak isoscalar coupling,

$$v^0 = 0 \ , \quad a^0 = -\tfrac{1}{6} \ . \tag{78}$$

Hence, the neutral current is dominantly V-A and isovector, as the charged current, and has in addition a weak isoscalar axial-vector term.

7.5 Study of electron-deuteron scattering

Assuming that electrons couple to hadrons like neutrinos, apart from their different third component of weak isospin ($-\tfrac{1}{2}$ for e^- and $+\tfrac{1}{2}$ for ν), we can base our analysis on the Lagrangian derived from Eq. (48):

$$L^{NC} = \frac{G}{\sqrt{2}} \bar{e}\gamma_\alpha(v_e + a_e\gamma_5)e\{\bar{u}\gamma_\alpha[u_L(1+\gamma_5) + u_R(1-\gamma_5)]u$$

$$+ \bar{d}\gamma_\alpha[d_L(1+\gamma_5) + d_R(1-\gamma_5)]d\} \tag{79}$$

in which we denote the quark fields by $\bar{u}$, u and by $\bar{d}$, d.

Studying the reaction

$$e^-d \rightarrow e^-X \tag{80}$$

using left-handed (e_L^-) and right-handed (e_R^-) electron beams, a SLAC group[38] has measured the asymmetry

$$A = \frac{\sigma(e_L^-) - \sigma(e_R^-)}{\sigma(e_L^-) + \sigma(e_R^-)}$$

for q^2 in the range 1-1.9 GeV^2 and y = 0.2. If reaction (80) were induced by one-photon exchange, the asymmetry would be zero. However, interference with the weak neutral current can give parity violating asymmetries, if $a_e \cdot v_q \neq 0$. The experimental result of

$$A_{exp} = (-9.5 \pm 1.6) \times 10^{-5} \ (Q^2/\text{GeV}^2) \tag{81}$$

confirms this expectation. Figure 14 shows that the asymmetry is

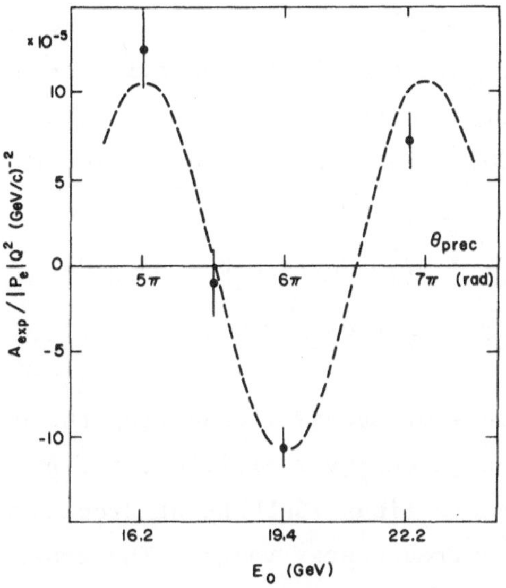

Fig. 14 Asymmetry of polarized electron scattering (Ref. 38) on deuterons as a function of spin orientation. Because of (g - 2) precession, the spin is longitudinal at odd multiples of π and transversal at even multiples of π.

indeed due to longitudinal polarization of the beam. The result can be interpreted by a value of the mixing angle of

$$\sin^2 \theta = 0.20 \pm 0.03 \ . \tag{82}$$

The hybrid model discussed in Section 5 is restricted to $\sin^2 \theta < 0.1$ (90% c.l.). Further measurements at smaller values of y may rule out this version of the model completely.

7.6 Conclusions

In trying to summarize our present knowledge of the weak neutral current, I am a little hesitant. On the one hand there is now rather impressive support for a unified description of the weak and electromagnetic interactions. Clearly this needs experimental confirmation, in particular in the neutrino-electron sector and in the determination of the isospin structure of the quark coupling. On the other hand, the most essential phenomena in support of a unified description by a spontaneously broken gauge theory have still to be observed.

One such phenomenon is the discovery of the charged and neutral vector bosons. All sectors of neutral current interactions seem to be adequately described by the same value of the mixing angle, $\sin^2\theta \simeq 0.20$. The standard model predicts the masses of the bosons,

$$\begin{aligned} m_{W^\pm} &= 37.3\ \text{GeV}/\sin\theta = 83.4\ \text{GeV}\ , \\ m_{Z^0} &= m_{W^\pm}/\cos\theta = 93.3\ \text{GeV}\ . \end{aligned} \tag{83}$$

What are the prospects of actually observing these bosons? The next generation of high-energy facilities, the $\bar{p}p$ collider at CERN and the pp storage rings ISABELLE at Brookhaven, have been designed for the required energy range. The cross-sections are related, by the conserved vector current hypothesis, to lepton pair production:

$$\begin{aligned} &\bar{q}q \to \gamma^* \to e^+e^- \\ &\bar{q}q \to W^- \text{ or } Z^0\ . \end{aligned} \tag{84}$$

Our present knowledge of lepton pair production, when extrapolated to this new energy range, is predicting reasonable production rates. The main difficulties seem to be connected with the detection of a clear signal above background[39]. The prospects seem to be best for the leptonic decay modes

$$\begin{aligned} &W^\pm \to \mu^\pm\nu \\ &Z^0 \to \mu^+\mu^- \end{aligned} \tag{85}$$

with branching ratios of 8% and 3%, respectively. Ultimately the bosons will be found at the new very high energy e^+e^- facility LEP which is now being designed at CERN.

None of these phenomena would demonstrate, however, that gauge theories are required to describe the unification. One has to look for phenomena which can only be described by spontaneously broken gauge theories in order to obtain a proof. One of these phenomena is the self-coupling of vector bosons, e.g. in the reaction

$$e^+e^- \to W^+W^- \tag{86}$$

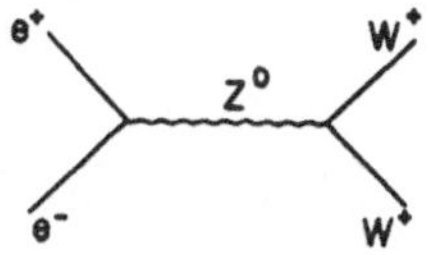

which will be accessible for study[40] at the highest LEP energies. The other is the Higgs particle, a scalar boson which is used to describe the mass spectrum produced by spontaneous symmetry breaking. Its mass is not predicted by the theory and may be anywhere between 5 GeV and 300 GeV. The most sensitive search[41] for its effects can again be conducted at LEP.

Hence we will have to wait for many more years before we can expect to have proofs of a unified gauge theory.

8. THE QUARK STRUCTURE OF THE NUCLEON

There is now a large amount of experimental evidence for quark structure of the nucleon. For a recent review, the reader is referred to reference 42. In these lectures I shall discuss the radiative corrections due to gluon emission and absorption as predicted by quantum chromodynamics (QCD).

The dominant effects in deep inelastic neutrino scattering are shown diagramatically in Figs. 15a and 15b.

In Fig. 15a the quark, which is struck by the weak current W, is radiating a gluon, either before or after the scattering process. The fractional momentum x of the quark is thereby reduced. The coupling of the gluon to the quark depends on the momentum transfer Q^2 in the scattering process. QCD is predicting that the strong coupling constant α_s (for four flavours),

$$\frac{\alpha_s}{\pi} = \frac{12}{25 \ln (Q^2/\Lambda^2)} \tag{87}$$

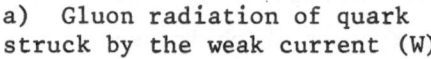

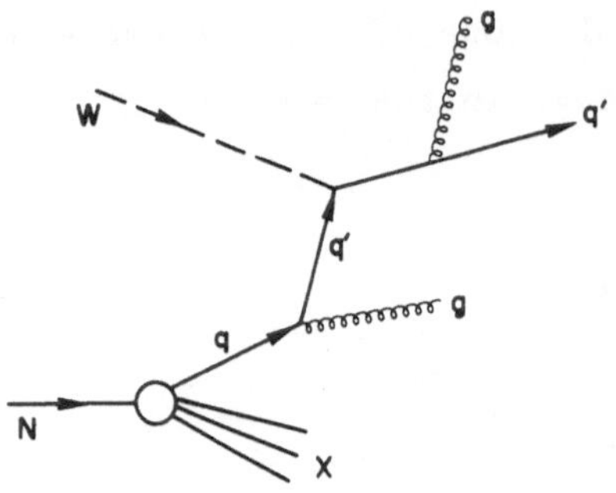

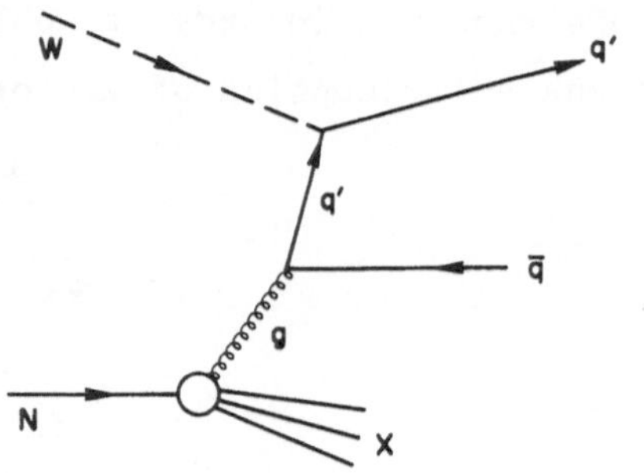

a) Gluon radiation of quark struck by the weak current (W)

b) A virtual gluon is dissociating into quark-antiquark

Fig. 15

decreases with increasing Q^2, leading to quasi-free quarks (asymptotic freedom) in the limit $Q^2 \to \infty$ and to confined quarks (infrared slavery) in the limit $Q^2 \to 0$.

In Fig. 15b a virtual gluon has been emitted, and has dissociated into a quark-antiquark pair. Either the quark or the antiquark may interact with the weak current. Nucleon structure is therefore not only given by valence quarks u and d but also by all quark pairs which can be created from the vacuum, and which are called the "sea" in the literature. The momentum x of sea quarks is expected to grow with Q^2. The scale parameter Λ in Eq. (87) is not predicted by QCD; its value determines the range of Q^2 in which the radiative effects will vary most rapidly.

Hence, gluon emission will lead to scaling violations. As QCD is claimed to be a precise theory, it pays to perform precision experiments. Gluon radiation may manifest itself also directly in the structure of the hadronic final state. It is expected to affect the azimuthal distribution of hadrons, to broaden the transverse momentum distribution, and to give rise to additional gluon jets.

In the following two sections, I shall discuss the experimental evidence as it was available in July 1978.

8.1 Radiative gluon effects on nucleon structure functions

In deep inelastic neutrino scattering on nucleons, the cross-section is conventionally described by structure functions in terms of the variables Q^2 and ν. At large Q^2 and ν the structure functions are expected to scale and to depend on $x = Q^2/2M\nu$ alone. However, as discussed in the preceding introduction, gluon radiation is expected to lead to additional Q^2 dependence of the structure functions and thereby to scaling violations. For isoscalar targets and neglecting flavour-changing currents by assuming $\theta_C = 0$, the cross-section is given by

$$\left.\frac{d^2\sigma}{dxdy}\right|_{\nu,\bar{\nu}} = \frac{G^2ME}{\pi}\left[\left(1 - y - \frac{Mxy}{2E}\right)F_2(x,Q^2) + \frac{y^2}{2} \cdot 2xF_1(x,Q^2) \right.$$

$$\left. \pm\, y\left(1 - \frac{y}{2}\right)xF_3(x,Q^2)\right] . \qquad (88)$$

Integrating Eq. (88) over y and E we obtain expressions for $d\sigma/dx$ and, comparing with the observed distributions, values for the structure functions.

Conventionally it is assumed that according to a theorem by Callan and Gross[43] for the limit $Q^2 \to \infty$,

$$A = \frac{2xF_1(x,Q^2)}{F_2(x,Q^2)} = 1 , \qquad (89)$$

owing to the fermion character (spin ½) of quarks. This assumption has been tested[44] over a wide range of Q^2 values, as shown in Fig. 16. The SLAC-MIT data on ed scattering and the FNAL data on muon scattering are also indicated. Averaged over Q^2 and x they find

$$\begin{array}{llll} Q^2 > 1\ \text{GeV}^2 , & \bar{A} = 0.89 \pm 0.12 & (\text{BEBC}) & \\ Q^2 < 1\ \text{GeV}^2 , & \bar{A} = 0.68 \pm 0.10 & (\text{GGM}) & (90) \\ Q^2 > 10\ \text{GeV}^2 , & |\bar{A} - 1| \lesssim 0.05 , & (\text{CDHS}) & \end{array}$$

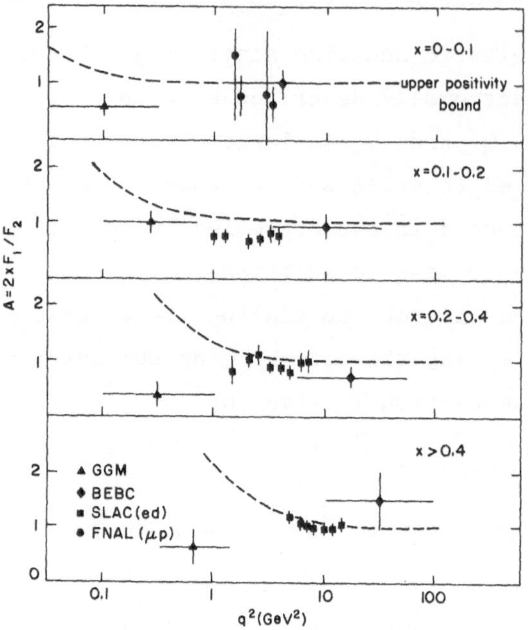

Fig. 16 Test of the Callan-Gross relation (Ref. 44) as a function of Q^2. The dotted line is a positivity bound.

showing evidence for small deviations from the asymptotic relation (89) of Callan and Gross. In the following we shall be mainly concerned with data at $Q^2 > 1$ GeV^2 and therefore assume for simplicity A = 1.

The domain available for a study of the structure functions is shown in Fig. 17. It is limited at high Q^2 and low x by the maximum energy of presently available neutrino beams, typically 200 GeV; at low Q^2 and high x it is limited by the availability of low-energy neutrino beams of known flux and by the resolution in measuring low values of $\nu \simeq E_{hadron}$.

Combining data from neutrino and antineutrino scattering, we can form their sum and difference and obtain, according to Eq. (88), in units of G^2M/π:

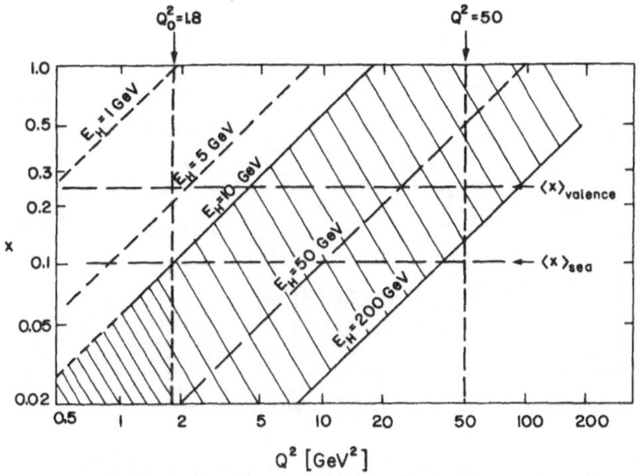

Fig. 17 Kinematical domain of x and Q^2 accessible for studying structure functions (Ref. 45)

$$\frac{d\sigma^{\nu}}{dx}(Q^2) + \frac{d\sigma^{\bar{\nu}}}{dx}(Q^2) = F_2(x,Q^2) \; ,$$
$$\frac{d\sigma^{\nu}}{dx}(Q^2) - \frac{d\sigma^{\bar{\nu}}}{dx}(Q^2) = xF_3(x,Q^2) \; , \tag{91}$$

which describe the distribution of all quarks (F_2) and of the valence quarks (xF_3). Integrating over all Q^2, we find the expected difference between valence and sea quarks. Figure 18 shows the dependence of F_2 on x and Q^2. There is good agreement between

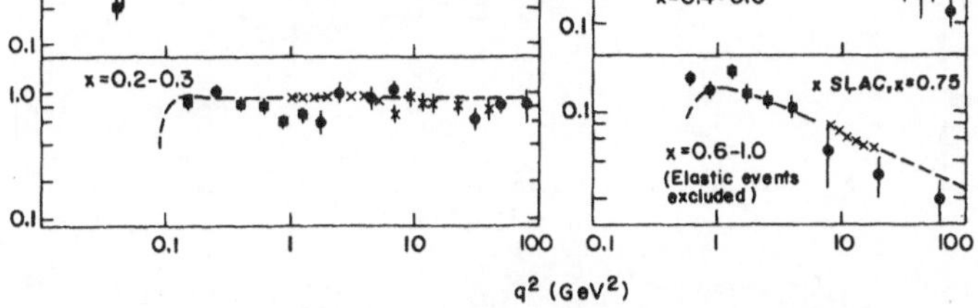

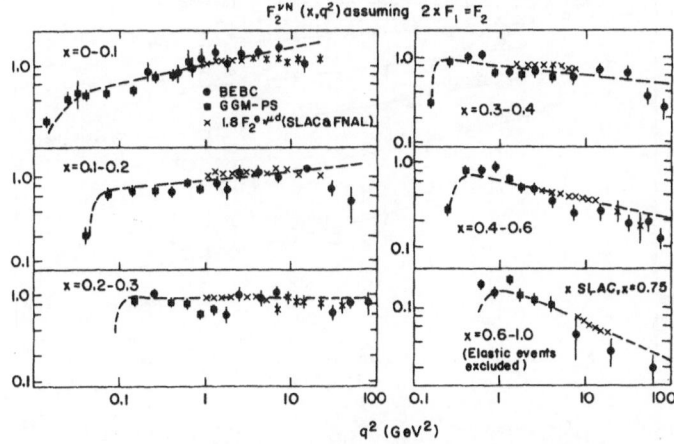

Fig. 18 The structure function $F_2^{\nu N}(x,Q^2)$ and $(9/5)F_2^{ed}(x,Q^2)$ showing scaling violations (Ref. 44)

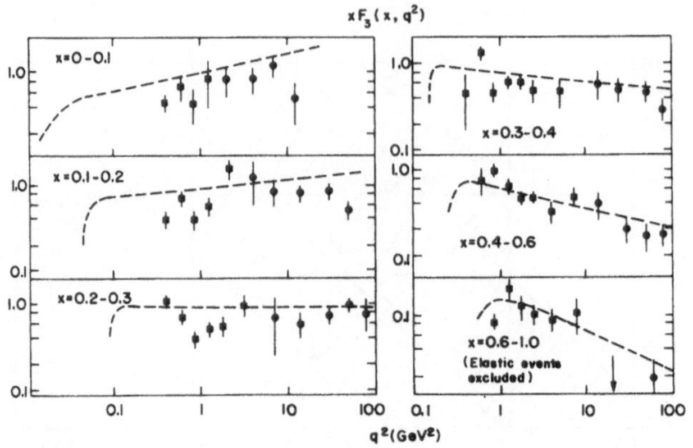

Fig. 19 The structure function $xF_3^{\nu N}(x,Q^2)$

neutrino scattering and ed and μd results and the quark model prediction $F_2^{\nu N}(x,Q^2) = (9/5)F_2^{ed}(x,Q^2)$. A rapid decrease of F_2 at large x with increasing Q^2 is seen.

Figure 19 shows the dependence of xF_3 on x and Q^2. The errors are much larger owing to the smaller statistics of $\bar{\nu}N$ data. The trend of the Q^2 dependence is also visible here. Figure 20

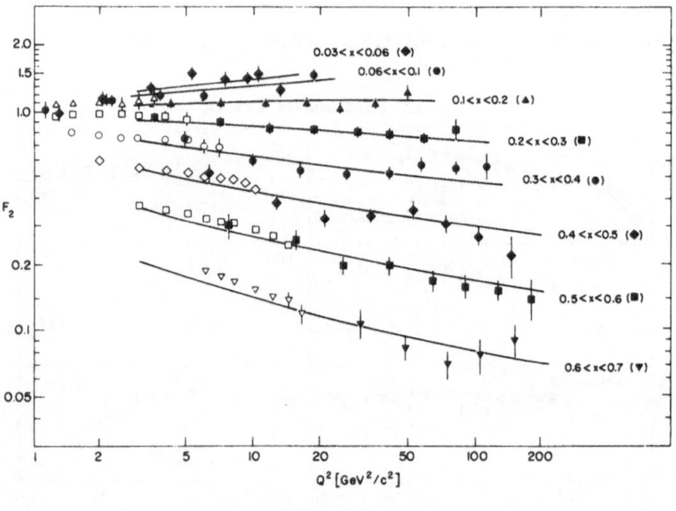

Fig. 20 The structure function $F_2^{\nu N}(x,Q^2)$ (full symbols) as determined by the CDHS collaboration (Ref. 45) and $(9/5)F_2^{ed}(x,Q^2)$ (open symbols)

shows unpublished data of the CDHS collaboration[45] which is extending the evidence to larger values of Q^2.

These deviations from Bjorken scaling may be due to a number of different causes. The rise of F_2 with Q^2 at small x may be accounted for by threshold behaviour, e.g. due to production of strange and charmed particles. These effects are being studied. However, it is clear that these threshold effects cannot account for the deviations observed at larger x. We shall confront these effects with the predictions of QCD.

The integrals of the structure functions can be compared most directly with the predictions of QCD in the limit of $Q^2 \to \infty$,

$$\begin{aligned} \int F_2^{\nu N}(x)\,dx &\to 0.24 \quad \text{for} \quad m = 3 \\ &\to 0.43 \quad \text{for} \quad m = 4 \\ \int xF_3(x)\,dx &\to 0 \ , \end{aligned} \tag{92}$$

where m is the number of quark flavours. Figure 21 shows[44] the integral of the structure function F_2 as a function of Q^2. In

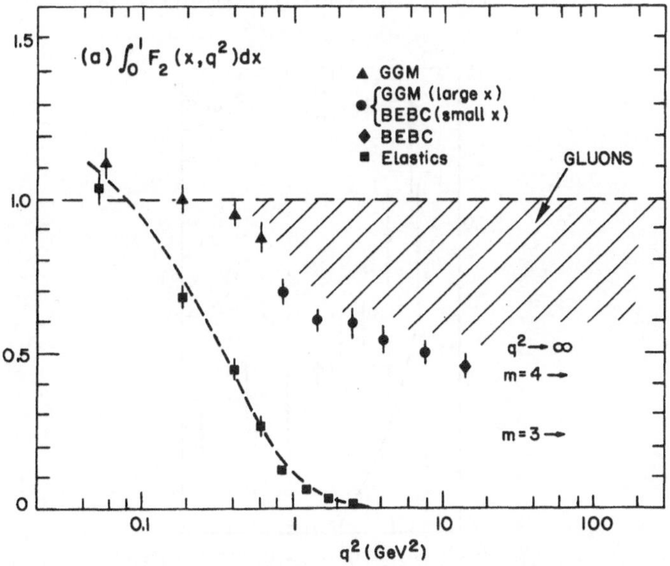

Fig. 21 Integral of the structure function (Ref. 44) F_2 as a function of Q^2. The elastic channel is shown separately. m = 3,4 are the predictions of QCD for 3 and 4 flavours at $Q^2 \to \infty$.

the quark model, $\int F_2(x,Q^2)\,dx$ represents the fraction of the momentum of the nucleon carried by the constituent quarks and antiquarks. Elastic scattering has been included in evaluating the integral; its contribution is dominating at $Q^2 < 1\ GeV^2$, as shown separately in the figure, owing to the rapid variation with Q^2 of the elastic form factors. At low Q^2 we note a tendency to exceed the value of 1. At large Q^2 the data are in qualitative agreement with QCD, which predicts a decrease of the fraction of the nucleon momentum carried by quarks and an increase of the fraction carried by gluons. Experiments with larger statistics may discriminate between the predictions for three and four flavours. The approach of this value in a region of Q^2 in which no single channel such as elastic scattering dominates the integral of F_2 has to be confronted with the prediction of QCD.

The integral of xF_3 represents the fraction of momentum carried by the valence quarks. The data[44] of the BEBC group are shown in Fig. 22. We note qualitative agreement with the predictions of QCD; as Q^2 increases, the fraction of momentum carried by the

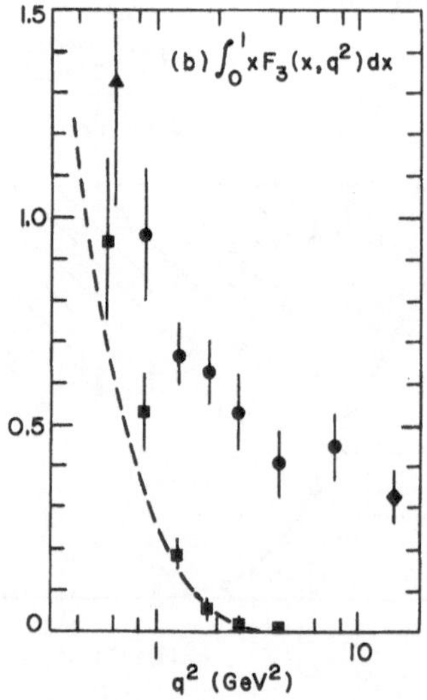

Fig. 22 Integral of the structure function (Ref. 44) xF_3. The elastic contribution is shown separately.

valence quarks decreases asymptotically to zero, and is transferred to gluons and to $q\bar{q}$ pairs. However, most of the observed Q^2 dependence is again due to the elastic contribution, as shown by the dotted line in Fig. 22.

The moments of the structure functions are expected to be Q^2-dependent in a characteristic way. The structure function xF_3 describes the valence-quark distribution, and the predicted[46] Q^2 dependence of its Nth moment is given by

$$M_3(N,Q^2) = \text{const}/(\ln Q^2/\Lambda^2)^d , \tag{93}$$

with

$$d = \frac{4}{(33-2m)}\left[1 - \frac{2}{N(N+1)} + 4\sum_{2}^{N}\frac{1}{j}\right] . \tag{94}$$

The constant in Eq. (93) is not predicted by the theory; m is the number of flavours expected to contribute. The authors of Ref. 44 have used $m = 3$ and argue that c-quarks contribute little to the cross-sections in the energy range of their data. Equation (93) assumes $Q^2 \gg \Lambda^2$ and also large compared to quark masses or transverse momenta.

Equation (93) predicts simple power-law relations between the different moments of xF_3. Some pairs of moments for $Q^2 > 1$ GeV^2 are shown in Fig. 23. The data indeed seem to favour constant slopes indicating power laws. The observed slopes are consistent with the predictions of QCD and disagree, for example, with the values calculated for scalar gluons. This test of QCD is independent of the number of flavours and colours in the gauge model and independent of the value of Λ, provided $Q^2 \gg \Lambda^2$ [47]. This result is the most significant evidence in support of a non-Abelian, asymptotically free gauge theory describing the strong interactions of quarks.

Let us pause here to examine to what extent these results may be affected by experimental uncertainties. It is important to note

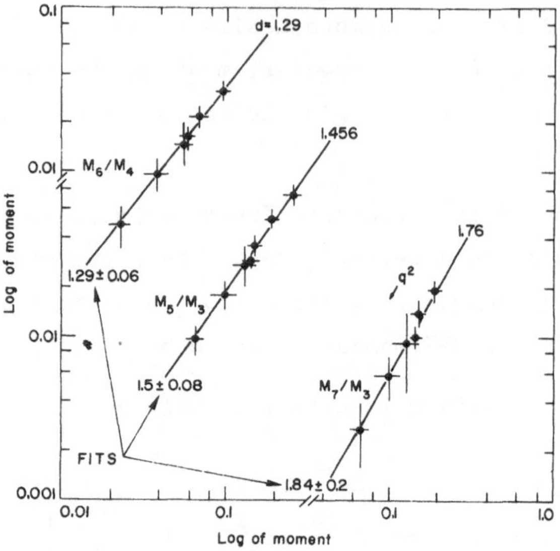

Fig. 23 Relations between moments of the structure function (Ref. 44) xF_3 are well described by a power law as predicted by QCD

that the statistical significance of the data[44] is good for $Q^2 < 10$ GeV^2 only where the validity of Eq. (93) may be marginal. At larger values of Q^2 data for $x < 0.1$ are not available because of the energy limit of present neutrino beams; the lower moments can therefore not be evaluated. The higher moments are strongly dependent on data with $x > 0.5$ and are therefore subject to experimental smearing effects which require important corrections. Other effects, such as the Fermi motion of the nucleons in the target nuclei and radiative corrections, are large as well. Since, however, the main concern is the Q^2 dependence of the structure functions and of their moments, these effects are negligible in comparison with the other errors. Experimental uncertainties due to the normalization of neutrino and antineutrino data and of data at low and high neutrino energies are important, mainly because of the present lack of data at $E_\nu \sim 30$ GeV.

In conclusion we note that scaling violations have been observed which are in qualitative agreement with the effects of gluon radiation. New data at low and at high Q^2 are required in order to confront the predictions of QCD. To obtain these data the present neutrino detectors have to be improved to reduce the minimum value of E_h and the smearing effects at large x.

8.2 Gluon effects in the hadronic final state

Since gluon radiation seems to play an important part in deep inelastic neutrino scattering, it is interesting to investigate whether the large effects observed in the Q^2 dependence of the structure functions also manifest themselves directly by gluons in the final hadronic state.

The investigation of these effects is more model-dependent because the process of quark fragmentation has not yet been described by the theory. Neutrino-induced charged current reactions have the advantage of known quark flavour and direction. However, owing to the large uncertainty in the neutrino energy this direction is not well known for each event.

In order to investigate the transverse momentum distribution of hadrons, the BEBC group[48)] is using the direction of the *charged* hadron system, which they can measure well. The missing neutral secondaries are expected to be distributed symmetrically about this direction. The observed rise of $\langle p_T^2 \rangle$ with Q^2, as shown in Fig. 24, is in qualitative agreement with the prediction of QCD that the struck quark will acquire an additional term in p_T^2 due to gluon radiation, which is approximately[48)] given by $\alpha_s^2 \cdot W^2 \cdot x$. Here α_s is the coupling of gluons to colour charge as defined in Eq. (87) and W^2 is the invariant mass squared of the hadron final state.

The scaling violations observed in the structure functions are also expected to give a Q^2 dependence of the quark fragmentation functions, and hence violations of Feynman scaling.

The increase of the mean transverse momentum due to gluon radiation is expected to limit the shrinkage of jets. Various

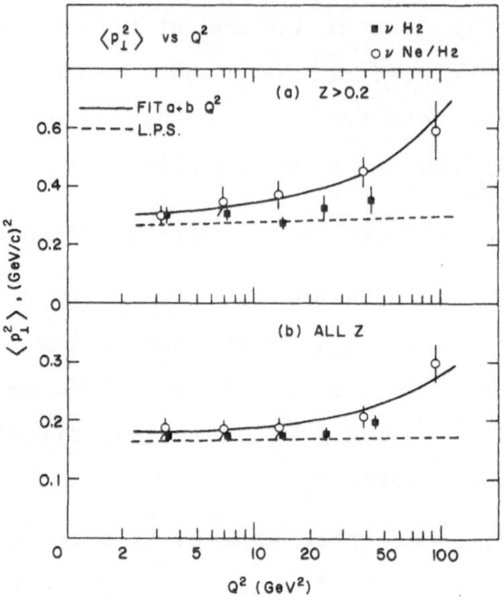

Fig. 24 Observed (Ref. 50) increase of mean transverse momentum with Q^2

quantities such as sphericity, spherocity, and thrust[49] have been invented to describe the energy flux distribution, $dE/d\Omega$, expected in quark jets. If the transverse momentum distribution were limited as expected in quark jets, the opening angle of the jets would shrink with increasing jet energy, without limits. However, as the energy is increasing we are expecting an increasing contribution due to gluon radiation, and hence asymptotically a constant opening angle of jets. This prediction is at present based on qualitative arguments, and a rigorous confrontation with data is not yet possible. However, these qualitative arguments are reflected in the data, e.g. in the thrust distribution observed[50] in νN interaction and shown in Fig. 25. The quantity called thrust is defined as

$$T = 2 \max \left(\frac{\Sigma p_i}{\Sigma |p|} \right). \tag{95}$$

For quark jets with limited transverse momentum this quantity will approach a δ function at $T = 1$. The data show an excess of events

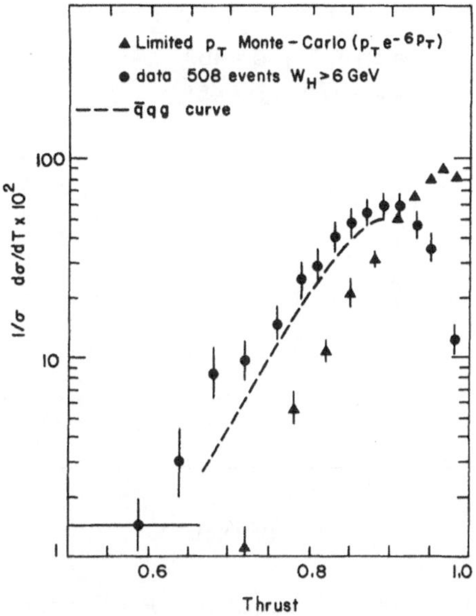

Fig. 25 Distribution of thrust (Ref. 50) and predictions for quark jets (triangles) and for quark plus gluon jets (dashed curve)

with small T, in qualitative agreement with the broadening of jets owing to gluon emission[49]. However, this thrust distribution cannot prove the theory because of the unknown gluon fragmentation function and the unknown primordial p_T distribution of quarks.

Another attempt to predict the asymptotically constant opening angle from QCD, using a rigorous proof, has been made by Sterman and Weinberg[51]. They have calculated the fraction F of events with a fraction $(1-\varepsilon)$ of the jet energy-flow in a cone of opening angle δ, and find

$$F = \frac{1 - \frac{4\alpha_s}{3\pi}\left(3 \ln \delta + 4 \ln \delta \ln \varepsilon + \frac{\pi^2}{3} - \frac{5}{2}\right)}{1 + \alpha_s/\pi} . \qquad (96)$$

The BEBC Collaboration has made a preliminary analysis[50] based on 500 events with W > 6 GeV. They find it more convenient to

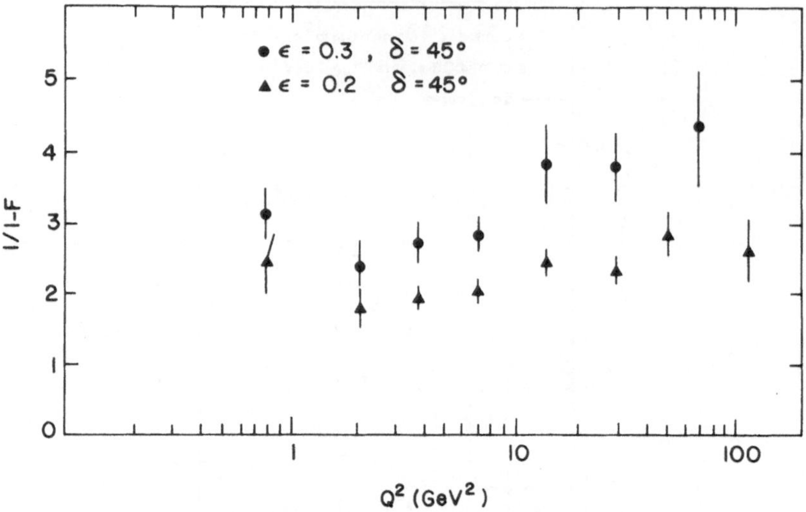

Fig. 26 Q^2 dependence of the fraction F of events (Ref. 50) with $(1 - \varepsilon)$ of the hadron energy flow inside a cone of opening angle σ

investigate the dependence of $1/1-F$ as a function of $\ln Q^2$, as shown in Fig. 26. This quantity can be parametrized in the following way:

$$\frac{1}{1-F} = \frac{25}{12} \ln (Q^2/\Lambda^2) G(\varepsilon,\delta) + G(\varepsilon,\delta) , \tag{97}$$

where $G(\varepsilon,\delta)$ is a function of the energy fraction ε and of the opening angle δ. There is qualitative agreement with a $\ln Q^2$ dependence for $Q^2 > 4$ GeV^2.

9. CONCLUSIONS

The Weinberg-Salam model of a unified gauge theory of weak and electromagnetic interactions seems to be compatible with all experimental data. There is also evidence that the strong interaction of quarks is mediated by spin-one gluons, coupled to the colour charges. The present data are, however, not sufficient to conclude that no other theory can describe them. Also the theory itself, despite the claims for precise results, has not yet given enough hard predictions to be confronted with experiment.

The most essential phenomena in support of a unified description, the field bosons in the mass range of 80-100 GeV and their self-coupling, as well as the Higgs boson, remain to be observed.

Acknowledgements

A survey of a rapidly evolving field, as attempted in these lectures, necessarily remains incomplete and preliminary. It would, however, not have been possible without the generous communication of recent and often unpublished data from the leading experimental groups. I wish to thank many of my colleagues for their help and contributions, in particular H. Faissner, L.M. Sehgal, T. Hansl, F. Dydak, J. Steinberger, W. Scott, D. Cundy and P. Musset.

Nino Zichichi and the good spirits of Erice have helped and inspired me. Thanks to all of them.

REFERENCES

1) See, for example, K. Winter, LEP Summer Study 1978/1-4.

2) See, for example, H. Harari, Physics Reports 42C (1978) 235.

3) R. Vidal et al., Phys. Letters 77B (1978) 344.

4) J. Blietschau et al., Nuclear Phys. B114 (1976) 189.

5) H. Faissner et al., Phys. Rev. Letters 41 (1978) 213.

6) P. Alibran et al., Phys. Letters 74B (1978) 422.
P. Musset, private communication.

7) A.M. Cnops et al., Phys. Rev. Letters 41 (1978) 357.

8) V. Khovansky and K. Winter, private communication.

9) F. Reines, H.S. Gurr and H.W. Sobel, Phys. Rev. Letters 37 (1976) 315.

10) See, for example, Ch. Llewellyn Smith, Physics Reports 3C (1972) 263.

11) See, for example, J. Ellis and M.K. Gaillard *in* Physics with very high energy e^+e^- colliding beams, CERN 76-18.

12) H. Grote and K. Winter, LEP Summer Study ECFA/LEP 20 and ECFA/LEP 31.
M. Davier and F. Richard, ECFA/LEP 34.

13) See, for example, J.J. Sakurai, UCLA/78/TEP/9, and L.M. Sehgal, PITHA-102 (1978).

14) E.A. Paschos and L. Wolfenstein, Phys. Rev. D 7 (1973) 91.
A. Pais and S. Treiman, Phys. Rev. D 6 (1972) 2700.

15) M. Holder et al., Phys. Letters 72B (1977) 254.

16) GGM, CERN-PS: J. Blietschau et al., Nuclear Phys. B118 (1977) 218.
HPWF, FNAL : T.Y. Ling, Proc. Internat. Neutrino Conf., Aachen, 1976 (ed. H. Faissner) (Vieweg, Braunschweig, 1977), p. 296.
CITF, FNAL : see Ref. 17; CDHS, CERN-SPS: see Ref. 19; BEBC, CERN-SPS: see Ref. 18.

17) B.C. Barish et al. (CalTech-Fermilab-Rockefeller Collaboration), CALT 68-605 (1977).

18) P. Bossetti et al. (BEBC Collaboration), Phys. Letters 70B (1977) 773.

19) M. Holder et al. (CDHS Collaboration), New results on possible scaling violations and total cross-sections from CDHS, Contribution to the "Neutrino-78" Internat. Conf. on Neutrino Physics, Purdue Univ., 1978.

20) L.M. Sehgal, Status of neutral currents in neutrino interactions, Invited talk given at the "Neutrino-78" Internat. Conf. on Neutrino Physics, Purdue Univ., 1978; and PITHA-102 (1978).

21) M. Holder et al., Phys. Letters 72B, 254 (1977).

22) J. Blietschau et al., Nuclear Phys. B118, 218 (1977).

23) F. Merritt et al., CALT 68-601 (1977).

24) T.Y. Ling et al., Proc. Internat. Neutrino Conference, Aachen, 1976 (ed. H. Faissner) p. 296.

25) J. Morfin et al., Analysis of Lorentz structure of the neutral current, Contribution to "Neutrino-78" Internat. Conf. on Neutrino Physics, Purdue Univ., 1978.

26) S. Glashow, Nuclear Phys. 22 (1961) 579.

27) A.J. Buras and K.J.F. Gaemers, Nuclear Phys. B132 (1978) 249.

28) F. Harris et al., Phys. Rev. Letters 39 (1977) 437.

29) M. Derrick et al., ANL-HEP-PR-77/69 (1977).

30) J. Mariner, Ph.D. thesis, LBL-6438 (1977).

31) Aachen-Bonn-CERN-London-Oxford-Saclay Collaboration, private communication from D. Cundy.

32) H. Kluttig, J.G. Morfin and W. van Doninck, Phys. Letters 71B (1977) 446.

33) L.M. Sehgal, Phys. Letters 71B (1977) 99.

34) D. Cline et al., Phys. Rev. Letters 25 (1976) 648.

35) P.Q. Hung and J. Sakurai, Phys. Letters 72B (1977) 208.

36) W. Kreuz et al. (GGM), Nuclear Phys. B135 (1978) 45.
D. Erriques et al. (GGM), Phys. Letters 73B (1978) 350.

37) L.F. Abbot and R.M. Barnett, SLAC-PUB-2097 (1978).

38) C.Y. Prescott et al., Phys. Letters 77B (1978) 347.

39) See, for example, F. Halzen and D.M. Scott, Phys. Letters 78B (1978) 318.

40) See, for example, Ch. Llewellyn Smith, e^+e^- physics beyond PETRA energies, LEP Summer Study/1-2 October 1978.

41) J. Ellis, M.K. Gaillard and D.V. Nanopoulos, Nuclear Phys. B106 (1976) 292.

42) D.H. Perkins, Quark structure of the nucleon, Review talk given at the 5th Hawaii Topical Conf. on Particle Physics, 1973; and Oxford Univ. preprint 67-73 (1973).

43) C.G. Callan and D.G. Gross, Phys. Rev. Letters 21 (1968) 211.

44) See, for example, P.C. Bossetti et al., Nuclear Phys. (to be published), Oxford Univ. preprint 16-78 (1978).

45) F. Eisele et al. (CDHS Collaboration), New results on possible scaling violations and total cross-sections, Contribution to the Internat. Conf. on Neutrino Scattering, Oxford Univ., July 1978 (to be published).

46) D.G. Gross and F. Wilczek, Phys. Rev. D 8 (1973) 3633; ibid. D 9 (1974) 980.
H. Georgi and H.D. Politzer, Phys. Rev. D 9 (1974) 416.

47) The favoured value of Λ is 0.5; see, for example, Ref. 44.

48) H.D. Politzer, Harvard preprint HUTP-77/38 (1977).

49) A. De Rújula et al., CERN/TH-2455 (1978).

50) P.C. Bossetti et al., (BEBC Collaboration), Nuclear Phys. (to be published), Oxford Univ. preprint 58-78 (1978).

51) J. Sterman and S. Weinberg, Phys. Rev. Letters 39 (1977) 1436.

D I S C U S S I O N

CHAIRMAN: K. Winter

Scientific Secretaries: L. Barone, G. Hansl, I. Liede, R. Orava

DISCUSSION No. 1

- BLASI:

The Weinberg-Salam model is a spontaneously broken SU(2)xU(1) gauge model. The spontaneous symmetry breaking takes place via scalar boson field acquiring a non-zero vacuum expectation value and thus determining the masses of the vector bosons (Higg's mechanism). What can one say about this scalar boson from the experimental point of view?

- WINTER:

It would couple to the heaviest leptons; so e.g. if it were heavier than two τ masses, it would preferentially decay into $\tau^+\tau^-$. One of the best places to observe it would be in e^+e^-, where, if you sit right on the Z^0 pole, the Z^0 could then decay into a Higgs and presumably a photon. The Higgs you would see in the invariant mass as $\tau^+\tau^-$ pairs. One could probably also make it with a sufficient cross-section in pp-interactions. It is clearly one of those objects one would like to discover, because QCD deals with all kinds of things nobody has ever seen: confined quarks, gluons and these Higgs fields.

- BLASI:

If you have the value of the coupling constants and of the W mass, do you then also know the mass of the scalar boson?

- WINTER:

No, I think the mass of the scalar boson is completely open. It can be 10^{10} GeV, for instance; it must be heavier than the kaon, because otherwise it would already have been seen.

- FARRAR:

Please explain how the "Lederman's limit" $|\delta|^2+|\varepsilon|^2>10^{-6}$ is obtained. If I guess correctly from what you said in your lecture on how it must be done, it seems to me that the limit must be very model-dependent: depending on $B\bar{B}$- production, spectrum and B lifetime.

- WINTER:

Let me refer to the publication of an MIT group (Busza et al.) and a forthcoming paper of the Lederman group and start by saying that the result in both experiments is identical: the cross-section for stable B-mesons is $\lesssim$ 0.1 pb. Now how does this work out? I think the basic thing is that you have to predict the cross-section for a pair of particles of a certain mass, which is notoriously extremely unreliable. As you remember after the discovery of the J/ψ in pBe-collision people looked immediately for naked charm and for many years failed to find it until the recent discovery of charm production at the ISR as well as at Fermilab. The cross-section, however, turned out to be very well determined by the existing model considerations. So the people now estimate the cross-section for the B-meson pair production and as they observe none after the flight path they simply work out a limit on the lifetime, and from that the quoted lower limit on $|\delta|^2+|\varepsilon|^2$.

- ISGUR:

I'd like to comment on a calculation of the production of heavy quark flavors in QCD. The calculation is applicable to this question and is quite stable once you are well above threshold, whereas we found in the calculations done by Steven Wolfram that the cross-section near threshold could vary by five orders of magnitude depending on exactly how we turn on the cross-section. Our conclusion was that there probably is a stable heavy particle at 5 GeV, but it is still quite a model-dependent statement on the basis of those calculations. I think we shouldn't completely rule it out, though I agree it is probably unlikely.

- WINTER:

The numbers I gave are based on the assumption that b-quark production goes in a normal way, although we really know nothing about the nature of that flavor.

- ISGUR:

The only thing I'm really claiming is that in the region where this experiment is done they're on a steeply falling part of threshold behavior and it's very difficult to accurately predict the cross-section.

- FARRAR:

Is any beam dump experiment under way to look for prompt ν_τ's? How will the ν_τ + hadr. $\rightarrow \tau$ + X be identified?

- WINTER:

Under way are maybe big words, but the hope one has is the following. We know the mass of the F meson to be 2.04 GeV and the τ mass is 1.78 GeV, so there is a sizeable mass difference and therefore the F can decay into the τ and ν_τ just as the π can decay into the μ and ν_μ. One can work out the branching ratio and one finds (for that mass difference) 4%. The D gives the bulk of these prompt ν's. It will predominantly decay into $Ke\nu_e/K\mu\nu_\mu$ with a branching ratio of 9%. However, this τ will immediately decay into ν_τ + something else, so in this channel you get two ν_τ's at once and from the point of view of the branching ratios these are equally abundant sources of prompt ν's and so it now comes to be discussed how abundantly you make them in pp-collisions. There one would expect to produce F/D approximately like K/π , because the only difference in quark content is the strange quark. We know K/π to be about 0.1 and thus as many as 10% of the prompt ν's could be ν_τ's. According to the published data they can accommodate up to 30% ν_τ's; in other words the experiment is not precise enough and there are plans of repeating it.

- FARRAR:

Is there a way in the detector to tell that it was a τ that was produced?

- WINTER:

No, but the way the τ best shows up is by its decay; so if you make a τ in $\nu_\tau N \rightarrow \tau^- X$, we know that it decays with a branching ratio of 16% into each leptonic mode. Therefore there are 32% of leptonic modes like $e\nu_\tau\nu_e$ or $\mu\nu_\tau\nu_\mu$. The remaining 68% have no charged lepton in the final state. Those one would classify as neutral currents and you get an apparent neutral to charged current ratio approximately equal to 2 and that is how the thing would sign itself. I think that is the most sensitive signature you would give to ν_τ interactions.

- MOUZOURAKIS:

Can you comment on the status of the three-muon events observed by the HPWF collaboration and attributed to heavy lepton production? Do the upper limits obtained for the M^+ and E^+ masses apply to them?

- WINTER:

What is "super" in these "super events" is that in addition to the μ^- there is a $\mu^-\mu^+$ pair and all of these three muons have high momenta as if they all would come from the lepton sector. The CDHS collaboration has not observed trimuon events of this kind. The "superness" one expects in a heavy lepton or heavy quark decay would be rather a large transverse momentum of two muons in these trimuon events. This is not observed in the HPW events.

- HANSL:

We do not see these so called "super events" in the CDHS experiment at CERN. This could be due to the softer energy spectrum in our experiment as compared to the HPW experiment.

- WHITAKER:

I would like to mention that 76 trimuons are observed in the CDHS experiment at CERN with a horn focused neutrino beam with the energy spectrum peaking at 50 GeV. The HPWF collaboration has observed 14 trimuon events in (mostly) a quad-focused beam which has a higher fraction of its flux above 100 GeV. Comparing the integrated fluxes above 100 GeV, we would expect one "super event" with the total energy greater than 200 GeV whereas HPWF collaboration finds two of these events.

- HANSL:

I would like to remark that there are two graphs contributing to the decay of a heavy lepton with muon number to a muon-pair:

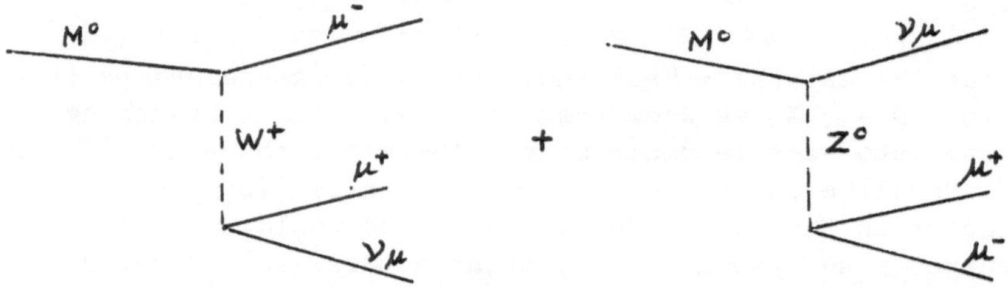

These graphs (figure above) might interfere so that the decay mode $M^0 \to \mu^+\mu^-\nu_\mu$ is suppressed.

In addition I want to mention that the Aachen-Padova groups has reported events of the type: $\nu_\mu N \to \mu e N$ where the electron has an has an energy larger than 2 GeV. The events are not explained by the charm hypothesis or π^0 production or any other conventional process. The signal consists of 13 events, whereas the number of events expected from conventional sources is 5.7±0.8. The rate is about 10^{-4} as compared to the total number of charged current events.

- MONTGOMERY:

Referring to the heavy lepton candidate mentioned by Dr.Winter in his lecture I would like to know what the statistical significance of the possible $\mu^-\pi^+$ signal is.

- ORAVA:

The statistical significance of the possible effect in the $\mu^-\pi^+$ invariant mass spectrum is 5 s.d. in the BEBC NB neutrino experiment in Ne/H_2 and 4 s.d. in the BEBC WB neutrino experiment in hydrogen.

- WOLF:

How are the muons and pions identified in the BEBC experiments? Could this effect be due to the decay $D^o \to \pi^+K^-$, K^- being misidentified with μ^-?

- ORAVA:

The pions are identified in the normal bubble chamber analysis chain where the measured tracks are first reconstructed and then fitted by different mass hypotheses. Part of the pions are uniquely identified by ionization or the decay seen in the chamber and part of them belong to the events giving a 3 constraint fit in kinematical analysis.

The muons are identified by the two plane BEBC External Muon Identifier (EMI), where a cut of 3 GeV/c in the muon momentum is to be applied due to a geometrical acceptance. The charged tracks observed in the chamber are followed through the BEBC-EMI structure taking the measurement errors and multiple scattering into account and the positions of hits in the EMI planes are predicted. Predicted points are then compared with the actual hits recorded on a magnetic tape during the experiment. I would like to add that part of the data comes from the 3C fit events $\nu p \to \mu^-\pi^+p$ where one possibly has evidence for the quasielastic type of production of M_o.

- RÖSSLER:

Do you see anything in the $\mu^+\pi^-$ mass spectrum?

- ORAVA:

Incorporating this type of a hypothetical heavy lepton into the standard SU(2)xU(1) gauge group it should be helicity suppressed in $\bar{\nu}N$ interactions. Unfortunately, our present statistics does not allow any conclusive analysis in $\bar{\nu}N$ reactions.

DISCUSSION No. 2 (Scientific Secretaries: L.Barone, T.Hansl, I.Liede, I.Liede, R.Orava)

- GROSSE:

Could you describe the source of polarized electrons in the SLAC e^-d experiment?

- WINTER:

Linearly polarized light from a laser was used; it was converted into circularly polarized light by a Pockels cell. Longitudinally polarized electrons were produced by optical pumping with these circularly polarized photons between the valence and conduction band in a gallium arsenide crystal.

The plane of polarization of the incident light could be varied by rotating a calcite prism.

A random pulse generator randomly reversed the voltage driving Pockels cell and consequently the helicity of the photons, which in turn reversed the helicity of the electrons. The polarized electrons were then accelerated to a target of liquid deuterium. A spectrometer consisting of a dipole magnet, a single quadrupole and a second dipole, analyzed outgoing electrons, subsequently intercepted by two separate electron detectors (Čerenkov counter and lead glass shower counter).

The yield Y for each pulse was measured and the asymmetry $A = \frac{\langle Y_+\rangle - \langle Y_-\rangle}{\langle Y_+\rangle + \langle Y_-\rangle}$ is the final result, where "+" and "-", referring to opposite electron helicities, were assigned by the random number generator, "+" meaning right-handed electrons.

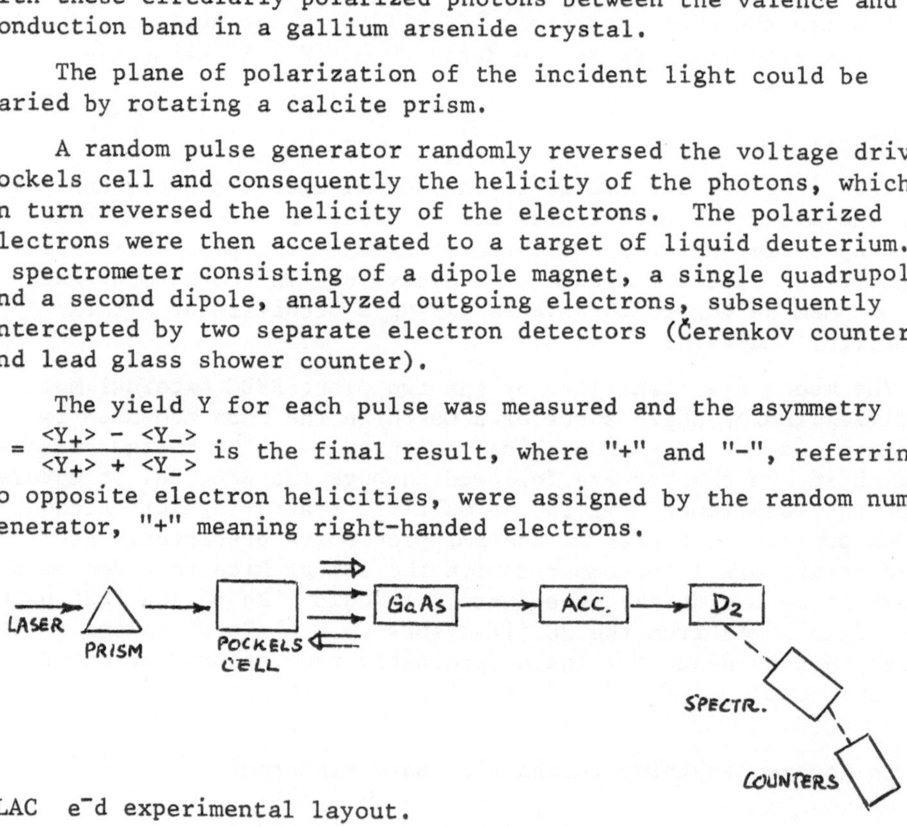

SLAC e^-d experimental layout.

- THIRRING:

Which value for the asymmetry can be expected in an e^- tritium experiment?

- WINTER:

The asymmetry adds up completely incoherently. Tritium is composed of 2 neutrons and 1 proton, i.e. 4 u and 5 d - quarks. Hence the effect of d-quarks is enhanced as compared to deuterium and one would observe a larger asymmetry, because the d- coupling constant is larger.

This argument applies of course only to inelastic electron scattering in which the tritium nucleus is breaking up.

- MONTGOMERY:

High energy muon beams are automatically polarized and one can repeat the SLAC experiment at higher Q^2 and over a wider range of Y, which gives good discrimination against models. Both NA2 and NA4 experiments at CERN have this measurement in their program. In addition there is an elastic ep scattering experiment being planned at the Mainz accelerator.

- WINTER:

Also at Bonn the experiment will be repeated. There is a 2.4 GeV electron synchrotron. Electrons have been accelerated already in the synchrotron and ejected and still have a polarization of ∿ 50%. The SLAC experiment has shown that systematic errors can be cancelled down to a level of 10^{-5}. With a μ- beam one has automatically polarized μ's: in the π^+-decay f.e., the ν_μ is left-handed and therefore the μ going forward in the c.m.s. has also to be left-handed.

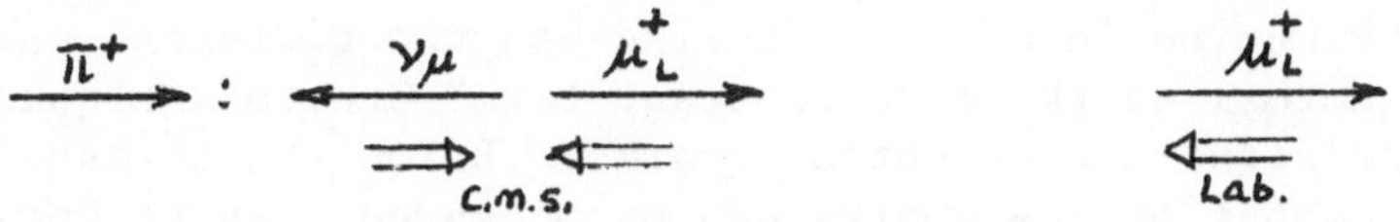

Muons emitted backwards in the c.m.s. are also left-handed, but the Lorentz transformation will boost them forward so that one gets right-handed μ^+'s in the lab-system.

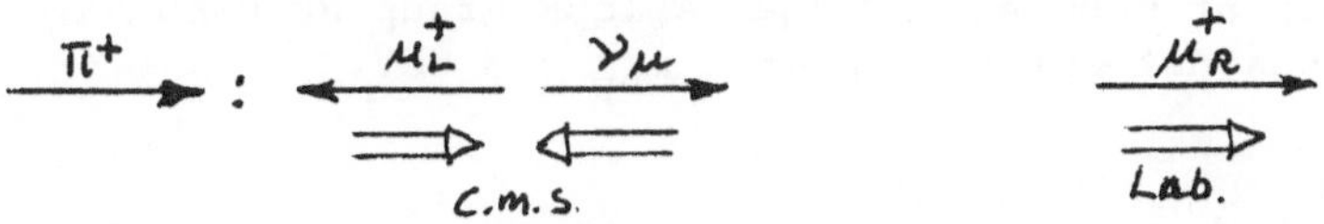

Hence the polarization can be selected by just selecting the momentum of the μ beam.

In fact one can also play another game: in the inverse neutrino reaction $\mu_L^+ N \rightarrow \bar{\nu}_\mu X$

$\mu_R^+ N \rightarrow \bar{\nu}_\mu X$ $\sigma/E \approx 0.3 \times 10^{-38} cm^2/GeV$

one may search for "helicity flipping" weak interactions, or for the existence of right-handed neutrinos.

This will be a very difficult experiment : the vanishing of the μ has to be ascertained and the momentum transfer Q^2 measured via measuring the direction of the hadron final state.

- KENNEDY:

Is there any possibility of observing parity violating effects due to coherent macroscopic effects in He^3?

- KLEINERT:

The proposal was made by A. Legget and there are indeed coherent macroscopic effects, though small. The trouble is that all effects are coherent, also competing directional effects from currents and fields. It is hard to estimate the size and peel out the small dipole moments.

- FELTESSE:

You mentioned that the b-quark is not produced in ν - interactions. But it is produced in the Lederman experiment. Is there agreement between the experimental data?

- WINTER:

I was talking about a right-handed b-quark and I showed its effect on the right-handed coupling constants of the u and d quark and on the y-distribution in $\bar{\nu}N \rightarrow \mu^+X$. A right-handed b-quark behaves like an antiquark.

The y-distribution in CC $\bar{\nu}$ scattering has two different parts: $(1- y^2)$ due to normal quarks + "a constant term" due to antiquarks from the sea and right-handed bottom quarks. Below the threshold for $u \rightarrow b_R$ the amount of sea-quarks can be measured. As E_ν goes up, a sudden rise in the constant term of the y-distribution would indicate the production of the b_R quark. Such a rise had been reported by the HPW group, but since then more precise measurements by the CDHS group at CERN and by the CALTECH group at FNAL have shown that it does not exist with mass $m_{b_R} < 8$ GeV.

- BHANOT:

One can infer the presence of the Z^0 also from the interference between Z^0 exchange and γ exchange in e^+e^- elastic scattering?

- WINTER:

You have in e^+e^- two graphs contributing to the production of an e (μ) pair :

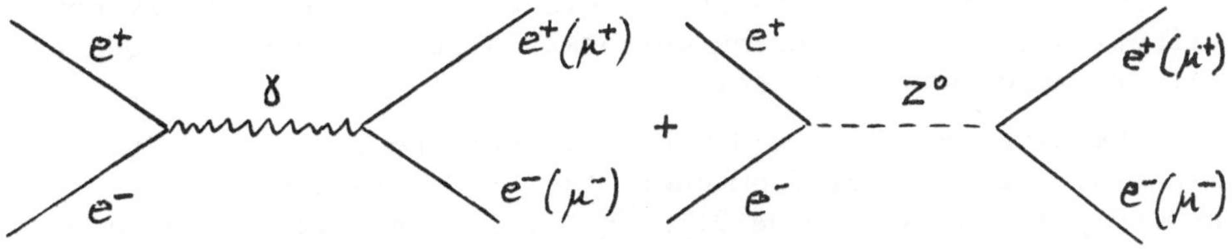

They interfere and lead to a forward - backward charge asymmetry. The effect is of the order of $10^{-4} \times Q^2$ /GeV^2 ($Q^2 \sim s$). At the PETRA storage ring, which will typically operate at energies up to 30-40 GeV, effects up to 10% are expected. If this asymmetry is measured as a function of energy, one could in principle measure the mass of the Z^0.

- MONTGOMERY:

You mentioned an E or Q^2 dependence of the measured values of the Weinberg angle.

- WINTER:

There is some tendency in the data to give higher $\sin^2\theta_W$ values at lower energies. Some experiments with even larger $\sin^2\theta_W$ are not shown in the figure given in my lecture. The reason I mentioned this was more a question to our QCD theorists : could there be any QCD effect creating an E dependence of the Weinberg angle?

- ISGUR:

There is at least one effect. Unless one goes to very high Q^2, the bare V - A structure of the quarks will not be seen. But I would not assume this could be important. For example at $Q^2 = 0$ there is no reason why the axial vector current can't be renormalized.

DISCUSSION No. 3 (Scientific Secretaries: L. Barone, T. Hansl, I. Liede, R. Orava)

- BARONE:

Can you say something about the experimental tests of the Callan-Gross relation?

- WINTER:

Assuming that the constituents on which the neutrino scatters have spin 1/2 or that the (g−2)-factor of quarks is 2, one gets the Callan-Gross relation $F_2=2xF_1$. However, when there is any contribution to the scattering which may also be induced by QCD-effects, this relation would not hold.

Results from BEBC and GGM show that the quantity $A=2xF_1/F_2$ might slightly differ from one: the BEBC data give a value $A=0.89\pm.12$ and GGM $A=0.68\pm0.10$ ($q^2<1.$). The average value of these two experiments is $A=0.80\pm0.12$. This ratio also depends on q^2 as seen from the figure shown in my lecture. The CDHS group quotes an average value $0.95<A<1$. The parameter A is closely related to quantity $r=\sigma_S/\sigma_T=(1+4M^2x^2/q^2-A)/A$ which is determined in the BEBC and GGM experiments, as well. It is a ratio between the scalar and transverse virtual boson absorption cross sections. The experimental value from BEBC and GGM data is $\langle r\rangle=0.15\pm0.10$.

- PETERSEN:

Which value has to be used in the formula for α_s, the value $\Lambda=0.40$ GeV or the "effective" value of 0.79 ± 0.05?

- WINTER:

The lines shown in the plot of the moments of the structure function correspond to $\alpha_s=12\ /\ (33-2n_f)\ln(q^2/\Lambda^2)$ and $\Lambda=0.7$ GeV. In a recent analysis Bardeen et al. take into account higher order corrections of QCD and deduce a value of $\Lambda=0.4$ GeV.

- ISGUR:

The formula for α_s comes from the lowest order perturbation theory in α_s, so it is only valid when α_s is small. QCD predictions relying on this α_s become inaccurate for q^2 near Λ^2. With the naive formula for α_s the two lines in the figure shown by K. Winter in his lecture would intersect for $\alpha_s=\infty$. Taking higher order corrections into account one obtains smaller values for Λ.

- WINTER:

In this connection I want to mention that elastic data are included in the figure in my lecture. Elastic scattering dominates the q^2-dependence at low q^2 values. In elastic data one observes form factor effects rather than gluons. Low q^2 values should consequently be ignored. In addition, there are threshold effects - strange particle production and charm production - which can produce rapid variations at low values of x. We should therefore be cautious in attributing variations at low x-values to scaling violations alone.

The data from larger q^2-values from the ed deep inelastic scattering experiment at SLAC and from the CDHS-experiment contain more information on QCD. There is a striking regularity over a large q^2-range that should be understood within the theory. Experimentally the situation can be improved by measuring the structure functions via the neutral current. In NC-reactions flavor is not changed and hence there are no threshold effects. One can thus see scaling violations only. The WA18-experiment at CERN is designed to measure x-distributions in neutral current reactions.

- ISGUR:

The structure functions you showed as a function of E_ν assumed $\int F_2\, dx = 1$. I would have expected that one could measure this effective q^2-dependence of the structure functions absolutely, so why is this done?

- HANSL:

This plot of the structure functions is mainly meant to demonstrate that there is a change in the shape of the structure function $F_2(x)$ with ν energy, though no energy dependence is seen either in the cross section or in average y (30 GeV < E_ν<200 GeV). The unnormalized $F_2(x)$ distributions are indeed very similar to the ones shown, the reason just being that σ and <y> are about constant within our energy range. Normalizing $\int F_2(x)$ to one eliminates possible flux uncertainties.

- MONTOGOMERY:

I would like to make a comment on the q^2-dependence of the structure functions at a given value of x. Assume that there is a q^2-dependence - which is not a flat one - in the shape of the structure functions. Then the Fermi motion smearing can certainly modify this dependence, especially at large values of x. The tendency would be to underestimate the shrinking of the structure function.

- WINTER:

I agree, but this is not very clearly expressed. What you want to say is that for q_2^2 (see the figure) the derivative d^2F_2/dx^2 is larger than for q_1^2. Therefore we expect larger smearing effects due to Fermi motion at higher q^2.

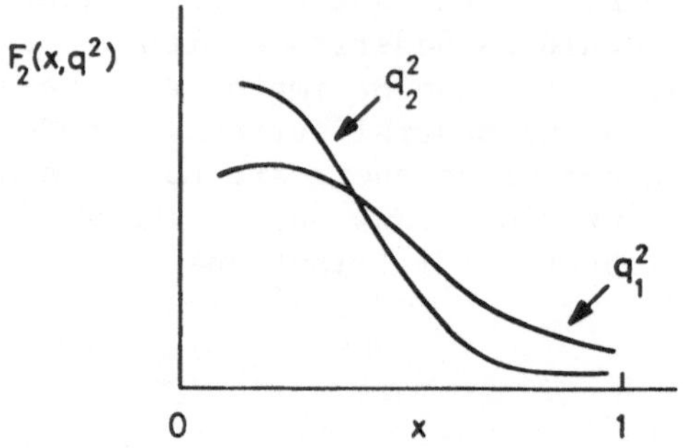

- ZICHICHI:

Does this scaling violation really prove QCD is correct? We need a definitive test of QCD.

- FARRAR:

Some years ago Darrell Jackson and I (G.R. Farrar and D.R. Jackson, Phys. Rev. Lett. 35, 14 (1975)) observed that QCD predicts that the ratio of the d- and u-quark distribution functions in the proton (d/u) approaches 0.2 for $x \to 1$. Predictions of other models are 0.5 and 0.0 as will be seen below. These distributions both vanish as x approaches 1, and on average the d-quark distribution is about half the u-quark distribution, except for the sea (low x). Experimentally u(x) and d(x) can be obtained from ep and en deep inelastic scattering, however practically speaking ed scattering cannot be used to extract en for $x \gtrsim 0.7$, because of distortions at large x due to Fermi motion. Another source of information on d/u are the reactions

$$\nu p \to \mu^- x \quad \text{and} \quad \bar{\nu} p \to \mu^+ x.$$

The data still has large errors (see G. Farrar, P. Schreiner, and W. Scott, Phys. Lett. 69B, 112 (1977)), but it is sufficient to rule out the value d/u = 0.5. At CERN there should soon be sufficiently accurate data to distinguish between the cases d/u = 0.2 or 0.0.

How do these predictions come about? At x = 1 sea contributions are negligible. Most of the time each of the quarks is carrying some finite fraction of the momentum, e.g.,

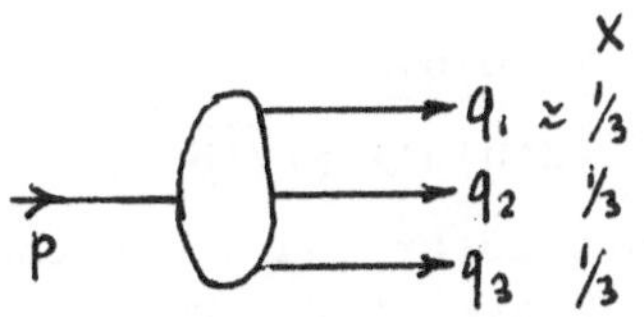

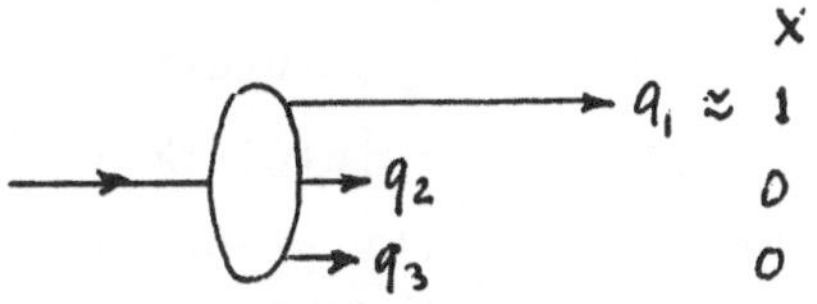

Very exceptionally two quarks give all of their momentum to the third. In the very naive parton days it was thought that since on the average d/u = 1/2, probably d/u ~ 0.5 even for $x \approx 1$. When the ratio was found to be much smaller than 1/2, Feynman argued that the I=0 state of q_2 and q_3 is likely to have lower energy than the I=1 state, so that Q_1 therefore must be a u-quark. However, according to QCD, we are dealing here with a "short distance" phenomenon (when $x \approx 1$). At short distances the flavor should not matter; what differentiates the u and d quark is only their masses. But short distances are connected to very large energies, so mass differences should not matter. Gluons of course couple only to color, which is the same for all quarks, independent of their flavor. As a "short distance" phenomenon we may study it perturbatively.

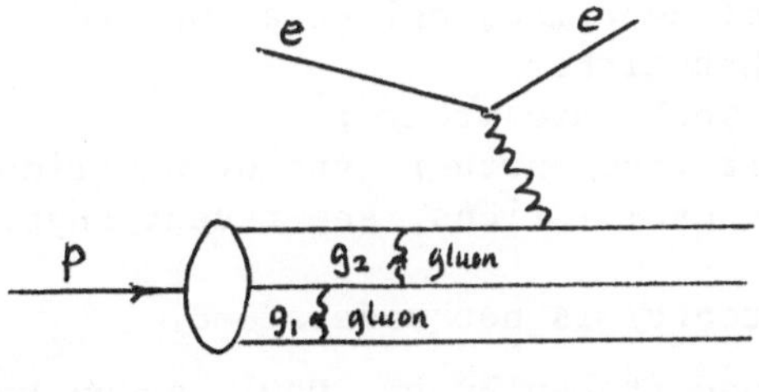

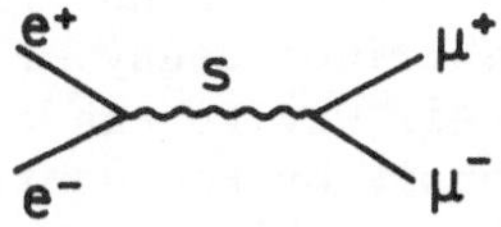

One way q_2 and q_3 could transfer all their momentum to q_1 is indicated by the gluons g_1 and g_2. Gluon g_1 has to be a transverse one and its helicity is therefore ± 1 in the infinite momentum frame. This means that q_2 and q_3 have to have oppositely aligned spins and q_1 has the same spin as the proton. Other graphs which contribute in leading order give the same result. I.e., by this process a part of the wave function is filtered out: the QCD dynamics which generates the $x \approx 1$ wave function from the normal wave function only uses (to leading order in 1-x) that part of the normal wave function in which quark 1 has the same helicity as the proton.

Looking at the SU(6) wave function for the normal proton,

$$|p\uparrow\rangle = \{ 2|u\uparrow u\uparrow d\downarrow\rangle + 2|u\uparrow d\downarrow u\uparrow\rangle + 2|d\downarrow u\uparrow u\uparrow\rangle$$
$$- |u\uparrow u\downarrow d\uparrow\rangle - |u\downarrow d\uparrow u\uparrow\rangle - |d\uparrow u\uparrow u\downarrow\rangle$$
$$- |u\downarrow u\uparrow d\uparrow\rangle - |u\uparrow d\uparrow u\downarrow\rangle - |d\uparrow u\downarrow u\uparrow\rangle \}/\sqrt{18}$$

We see that the probability that a particular quark has the same helicity as the proton is five times greater for a u-quark than for a d-quark. Thus as $x \to 1$ we find from QCD (in particular the spin-1 nature of the gluon in QCD) that $d/u \to 1/5$.

A discussion of the rigor of this prediction including the other graphs, the effect of higher order corrections, the legitimacy of treating the problem perturbatively, and corrections from other components in the SU(6) wave function, can be found in G.R. Farrar, Phys. Lett. 70B, 346 (1977).

- BHANOT:

Could you comment on solar neutrino experiments?

- WINTER:

In the experiment of Davis solar neutrinos are collected in order to check our understanding of the energy producing mechanism in the sun.

For some years the number of neutrinos collected was too small. The possible conclusions were that either
- the ideas on how the sun works are wrong ;
- neutrinos decay or convert into another type of neutrino on their way to the earth (ν oscillations, see recent Phys. Report by Pontecorvo) ;
- the efficiency of the detector is not understood.

Great endeavors have been undertaken by the Davis group to understand their apparatus and the statistics has been improved. By now the deviation is not striking anymore.

There are new ideas for an improved apparatus consisting of large masses of solid state detectors, so that the events can be counted directly, without processing large amounts of material to extract a few atoms that have been transformed.

- FELTESSE:

What is the future program of physics at the neutrino beams of CERN and FNAL?

- WINTER:

At CERN there are 5 experiments aligned along the ν beam :

- BEBC,
- WA1, the iron calorimeter
- WA18, a fine-grained calorimeter, which measures not only the hadron energy but also the direction of the hadron jet
- GGM,
- quark search experiment of Prof. Zichichi.

Three periods of running in a dichromatic beam are foreseen for the second half of 1978, starting with 200 GeV parent momentum in August, and dividing $\bar{\nu}/\nu \approx 2:1$.

In the beginning of 1979 the muon shield will be improved, the proton energy can then be brought up to 450 GeV.

A beam dump run is foreseen for March or April and extensive wideband beam running.

The bubble chambers will mainly analyze the final hadron state, the counter experiments will concentrate on an analysis of structure functions, both in CC and in NC. The polarization of the muon will be measured, for which WA18 has already taken data. WA18 will also get data on νe elastic scattering.

At Fermilab a last run of the HPWF experiment is scheduled for October 1978. At the beginning of the next year, in the new dichromatic beam, the Caltech group will take data with a calorimeter similar to the WA1 detector.

Later on in 1979 a big new device using Conversi flash tubes will compete with the CERN experiment on NC.

e^+e^- INTERACTIONS*

Günter Wolf

Deutsches Elektronen-Synchrotron, DESY, Hamburg

These lectures are dedicated to the memory of the late Professor Martin W. Teucher

The exploration of electron-positron interactions has become a powerful and fascinating tool of particle physics. The analysis of the J/ψ, ψ' family showed that these particles are SU_3 and colour singlets which are best described as quark-antiquark bound states of a new type of quark. The properties of the new quark seemed to agree with those of the charm quark which had been introduced to understand certain phenomena of weak interactions. This picture was confirmed when the first charmed mesons were detected, the nonstrange member D by the SLAC-LBL group and the strange partner F by the DASP group.

Along with searches for D and F mesons a third charged lepton, the τ, was discovered. While it took more than two years to establish its existence the data presented on the τ in the past few months make the τ an almost closed subject.

The recent upgrading of DORIS to 10 GeV total c.m. energy provided access to a new energy regime and allowed to produce the T in e^+e^- collisions. The measurements by PLUTO and DASP2 revealed the T as a narrow state, similar to J/ψ and ψ' and it seems almost certain by now that the T is the onium of a fifth type of quark. In a heroic effort the DORIS machine group pushed the maximum energy beyond 10 GeV which permitted DASP2 and a collaboration from DESY-Hamburg-Heidelberg and Munich to observe the T'. The comparison of the leptonic widths of T and T' showed that the fifth quark has charge 1/3 - and not 2/3.

Perhaps the most fascinating aspect of T studies is the possibility for a thorough test of QCD. In QCD the direct hadronic decays of the T (as well as those of the J/ψ) proceed via a three gluon intermediate state. As a result the spatial configuration of final states from T decay should be markedly different from the two-jet structure observed for nonresonant hadron production (SLAC-LBL). The first data presented by the DORIS experiments show that the final states on and off the T are indeed different in a manner consistent with the QCD predictions.

The lectures cover e^+e^- physics data above 3 GeV and focus on the recent experimental results. They are based in part on a report written in collaboration with B.H. Wiik[1]. Some topics such as the J/ψ family and earlier data on the T which were the objects of previous lectures at this school[2] will be discussed only briefly. Very recent data from SPEAR e.g. on D formation and decay and on inclusive photon spectra will be presented by Prof. L. Barbaro-Galtieri in her lectures. Some of the experimental material on T and T' given below has become available only after the school had finished.

1. INTRODUCTION TO e^+e^- PHYSICS

1.1 Electron-Positron Storage Rings

The diagram 1.1 summarizes the existing or planned e^+e^- storage rings in terms of their c.m. energy range and the luminosity $\mathcal{L}$, defined as

$$\mathcal{L} = \frac{\text{event rate}}{\text{cross section}} .$$

Of the three high energy rings, CESR at Cornell and PEP at SLAC are still under construction while first beams have been stored in the DESY ring PETRA during the passed weeks.

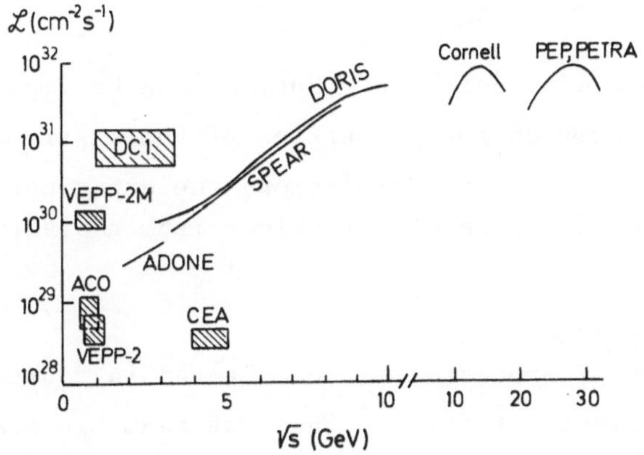

Fig. 1.1 Luminosity versus total c.m. energy for existing or planned e^+e^- storage rings

1.2 Classification of e^+e^- Interactions

1.2.1 Purely electromagnetic processes

We start with a brief discussion of the phenomenology of electromagnetic e^+e^- interactions. The electron will be assumed to have only electromagnetic interactions.

The lowest order processes are of order α^2, such as Bhabha scattering, μ pair production or two photon annihilation.

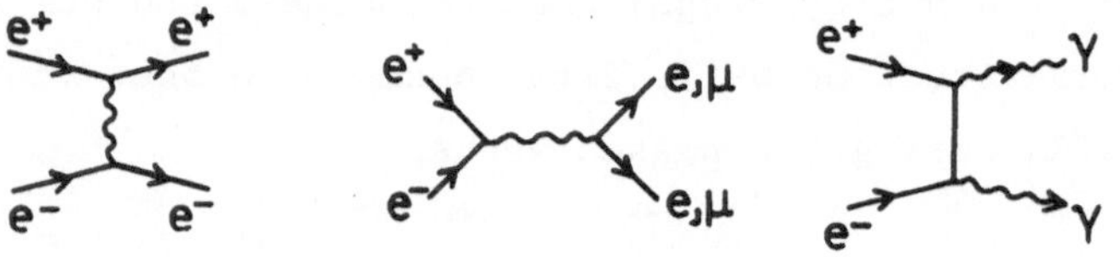

The next higher order processes constitute radiative corrections to the first ones:

Although of order α^3, their contributions can be important. Since they depend strongly on the properties of the experimental setup (such as energy and angular resolution) the experimental results are usually presented with the contributions from radiative corrections removed.

A new class of processes is encountered in fourth order: the virtual photon clouds of the two incident beam interact with each other. After integration over the photon spectra the cross section

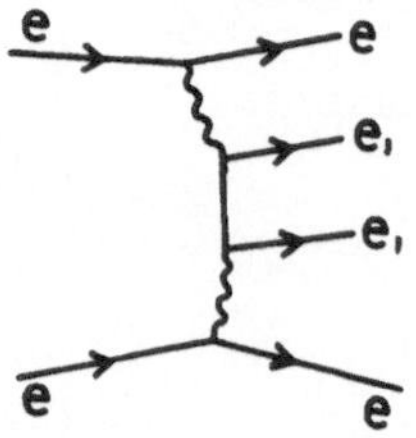

is proportional to $\alpha^4 \ln^2(\frac{E}{m_e})$ where E denotes the beam energy. Since $\ln \frac{E}{m_e} \sim 10$ for energies in the GeV region one power of α is essentially cancelled by the integration.

The simplest of all QED reactions is μ pair production, $e^+e^- \rightarrow \mu^+\mu^-$. It proceeds via time-like photon exchange.

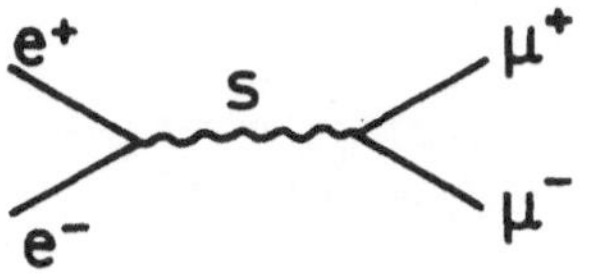

The differential cross section reads

$$\frac{d\sigma}{d\Omega} = \frac{\alpha^2}{4s} \beta_\mu \left\{ (1 + \cos^2\Theta) + (1 - \beta_\mu^2) \sin^2\Theta \right\} \qquad (1.1)$$

where $\alpha = 1/137$, s square of the total c.m. energy, $s = 4E^2$, E = beam energy and $\beta_\mu = p_\mu/E_\mu$. For $p_\mu \approx E_\mu$

$$\frac{d\sigma}{d\Omega} = \frac{\alpha^2}{4s} (1 + \cos^2\Theta) \qquad (1.2)$$

The integrated cross section is given by

$$\sigma_{\mu\mu} \approx \frac{4\pi}{3} \frac{\alpha^2}{s} = \frac{\pi}{3} \frac{\alpha^2}{E^2} = \frac{21.9 \text{ nb}}{E^2} \qquad \text{(E in GeV)} \qquad (1.3)$$

Possible deviations from QED will depend on s and can only be detected by measuring the absolute magnitude of the cross section. This can be done e.g. by comparing μ-pair production to small angle Bhabha scattering.

Fig. 1.2 shows the μ pair cross section between $s = 1$ and 9 GeV^2 as measured by Alles-Borelli et al.[3]. It is seen to agree well with the QED prediction given by the solid curve. Deviations have been observed near the narrow vector states J/ψ, ψ' where resonance excitation contributes significantly to μ-pair production.

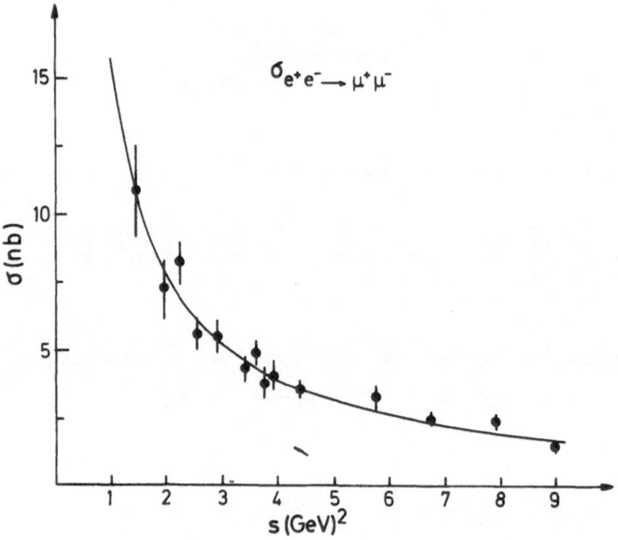

Fig. 1.2 $e^+e^- \to \mu^+\mu^-$: Comparison of the measured cross section with the QED prediction (solid curve); from Ref. 3

1.2.2 Phenomenology of hadron production

The lowest order e^+e^- scattering processes leading to hadron production are:

one-photon annihilation

radiative corrections to one-photon annihilation

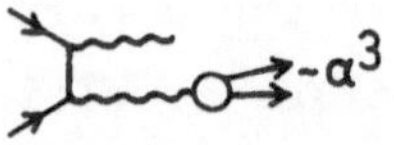

two-photon scattering

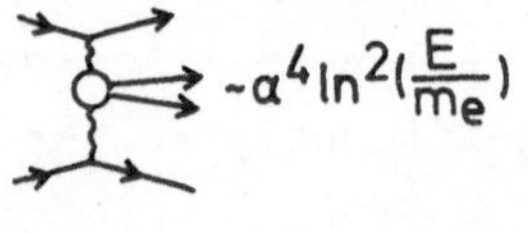

As in the purely electromagnetic case the two-photon scattering contribution is effectively of order α^3 after integration over the photon spectra. Furthermore, while the cross section for one-photon annihilation probably decreases as s^{-1} with energy (see below) the two-photon contribution increases ~ln s and eventually will win over the one-photon contribution. The importance of the two-photon process was first recognized by Low and by Kessler and coworkers[4]. Evidence for this mechanism was found at Frascati, where three events of the type $e^+e^- \to e^+e^-\pi^+\pi^-$, $e^+e^-\pi^+\pi^-\pi^o$ [5] were observed. The kinematics of the two-photon processes favor the emission of hadrons along the beam direction. At presently available energies and because present detectors do not cover angles close to the beams, two-photon contributions can be neglected. From now on we shall only consider the one-photon channel.

1.2.3 Properties of the one-photon channel

The hadron system produced by one-photon annihilation has the quantum numbers of the photon, $J^{PC} = 1^{--}$. For this reason the angular momentum L of the incident e^+ and e^- is limited to 0 and 2. Since $L = R\cdot E$, the radius of interaction will be of order $\frac{1}{E}$ leading to a total cross section for hadron production of

$$\sigma^{tot} \approx \alpha^2\pi R^2 = \frac{\alpha^2\pi}{E^2} \approx \frac{60\text{ nb}}{E^2} \qquad \text{E in GeV} \qquad (1.4)$$

From this simple minded exercise we expect σ^{tot} to decrease with energy as s^{-1}. Because $J^P = 1^-$ the most general angular distribution with respect to the beam direction for a particle h produced via $e^+e^- \to hX$ is of the form

$$\frac{d\sigma}{d\Omega} = a + b\cos^2\Theta \qquad (1.5)$$

which is radically different from typical angular distributions in hadron hadron collisions.

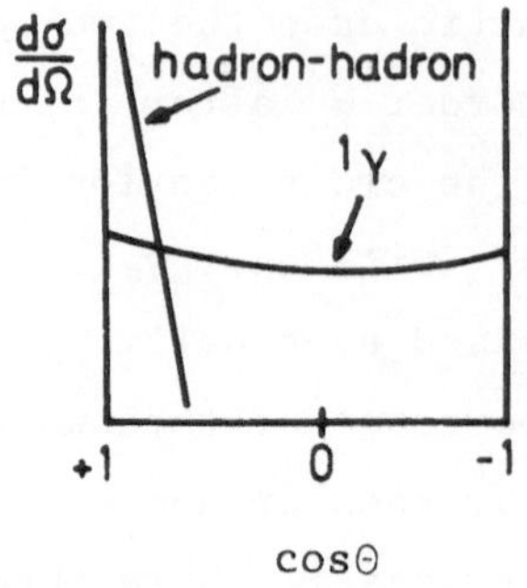

1.2.4 σ^{tot} in the quark-parton model

The experimental results for deep inelastic electron-nucleon scattering led to the hypothesis that the photon-hadron interaction is basically a photon-quark interaction[6].

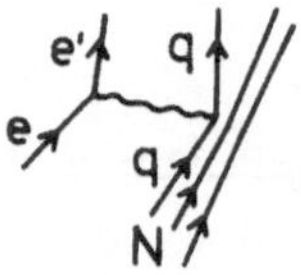

As a consequence we expect $e^+e^- \to$ hadrons to proceed via the formation of a quark-antiquark pair.

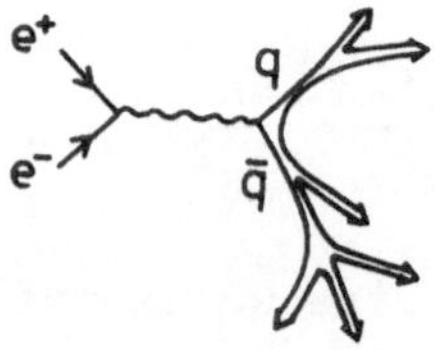

We assume the quarks to have spin 1/2 and are pointlike. Then the cross section for producing a free $q\bar{q}$ pair is the same as for producing a $\mu^+\mu^-$ pair except that the quark charge Q_i replaces the muon charge 1:

$$\sigma(e^+e^- \to q\bar{q}) = Q_i^2 \, \sigma_{\mu\mu} = Q_i^2 \, \frac{4\pi\alpha^2}{3s} \qquad (\beta_q = 1 \text{ is assumed}) \quad (1.6)$$

Assuming further that the produced $q\bar{q}$ pair turns into hadrons with probability one, the total hadron cross section is found by summing over all possible $q\bar{q}$ pairs:

$$\sigma^{tot} = \sum_i Q_i^2 \, \sigma_{\mu\mu} \tag{1.7}$$

As we see from eqn(1.7) the quark model predicts the total hadron cross section to decrease with energy as $\sigma^{tot} \sim s^{-1}$ and its magnitude to be of the order of the μ pair production. The value of Q_i^2 depends on the specific quark scheme:

quark model	Q_i^2
u, d, s	$\frac{4}{9} + \frac{1}{9} + \frac{1}{9} = \frac{2}{3}$
(u, d, s) ⊗ color	$3 \cdot \frac{2}{3} = 2$
(u, d, s, c) ⊗ color	$3 \cdot \frac{10}{9} = \frac{10}{3}$

Because of the expected behaviour (1.7) it is customary to study the ratio $R \equiv \sigma^{tot}(e^+e^- \to \text{hadrons})/\sigma_{\mu\mu}$.

The foregoing discussion was concerned with the high energy behaviour of σ^{tot}. At low energies we expect to produce nonstrange vector mesons which have the same quantum numbers as the photon.

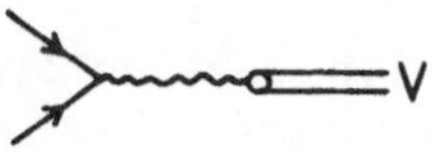

Close to a vector state the cross section can be written in terms of the partial width Γ_{ee} and the total width Γ_V of the vector meson:

$$\sigma_{ee \to V \to f} = \frac{3\pi}{s} \frac{\Gamma_{ee}\,\Gamma_V}{(m_V - \sqrt{s})^2 + \Gamma_V^2/4} \tag{1.8}$$

where m_V is the vector meson mass. Both, the proportionality between σ^{tot} and $\sigma_{\mu\mu}$ and the excitation of vector states is observed in the data. Fig. 1.3 shows a not quite up to date summary of R measurements. Below 1 GeV the dominating feature is ρ, ω and φ excitation. Between 1 and 3 GeV R is roughly equal to two. At 3.09 and 3.68 GeV dis-

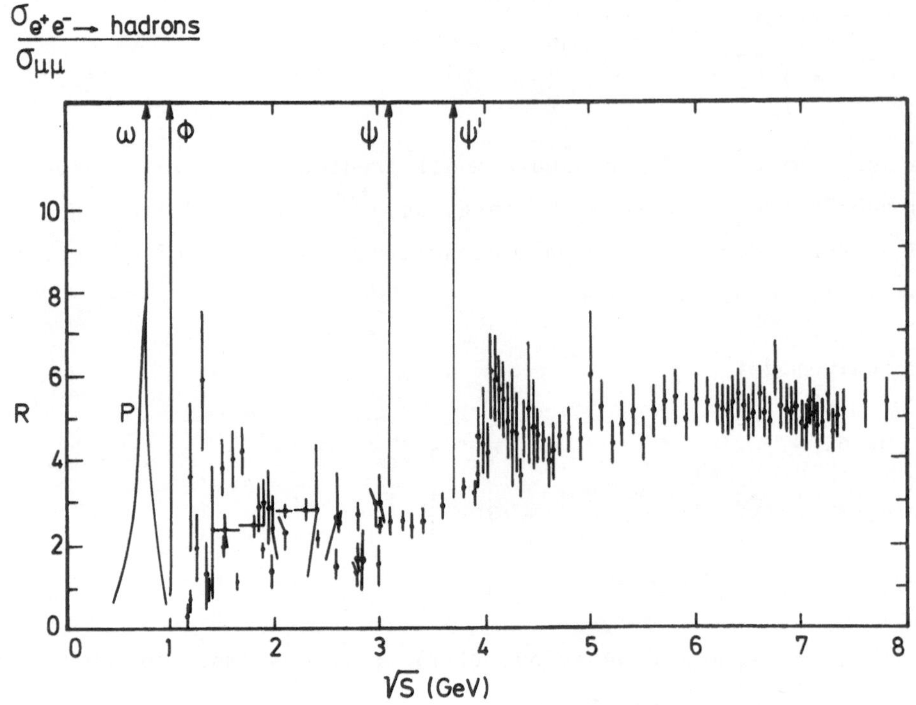

Fig. 1.3 The ratio $R = \sigma_{e^+e^- \to \text{hadrons}}/\sigma_{\mu\mu}$; Summary of data up to 1976 (Ref. 7)

crete lines due to J/ψ and ψ' production are observed. Above 5 GeV the ratio R is rather constant with a value around 5.

2. CHARMONIUM STATES

2.1 The J/ψ and ψ' Particles

The first member of the new family of particles was discovered at Brookhaven as the J-particle in the e^+e^- spectrum of p-Be collisions[8] (see Fig. 2.1),

$$p\ Be \to e^+e^-X$$

and at SPEAR as the ψ particle in e^+e^- annihilation[9] (see Fig. 2.2),

$$e^+e^- \rightarrow \text{hadrons}$$
$$\rightarrow e^+e^-$$
$$\rightarrow \mu^+\mu^-$$

The first evidence for the ψ' observed at SPEAR[9] is shown in in Fig. 2.3.

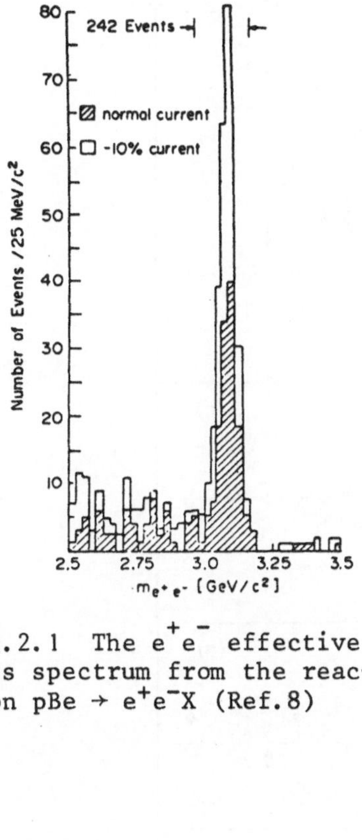

Fig. 2.1 The e^+e^- effective mass spectrum from the reaction pBe → e^+e^-X (Ref. 8)

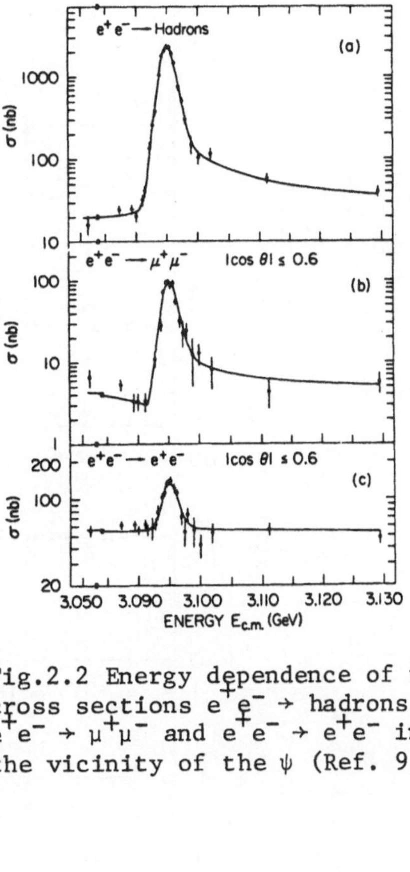

Fig. 2.2 Energy dependence of the cross sections $e^+e^- \rightarrow$ hadrons, $e^+e^- \rightarrow \mu^+\mu^-$ and $e^+e^- \rightarrow e^+e^-$ in the vicinity of the ψ (Ref. 9)

The properties of J/ψ and ψ' were studied at SPEAR and DORIS and a plethora of decay modes was detected and analyzed[1,10]. Both states were found to be nonstrange isoscalar vector mesons, $J^{PC} = 1^{--}$, $I^G = 0^-$. In terms of SU_3 J/ψ and ψ' are singlet states.

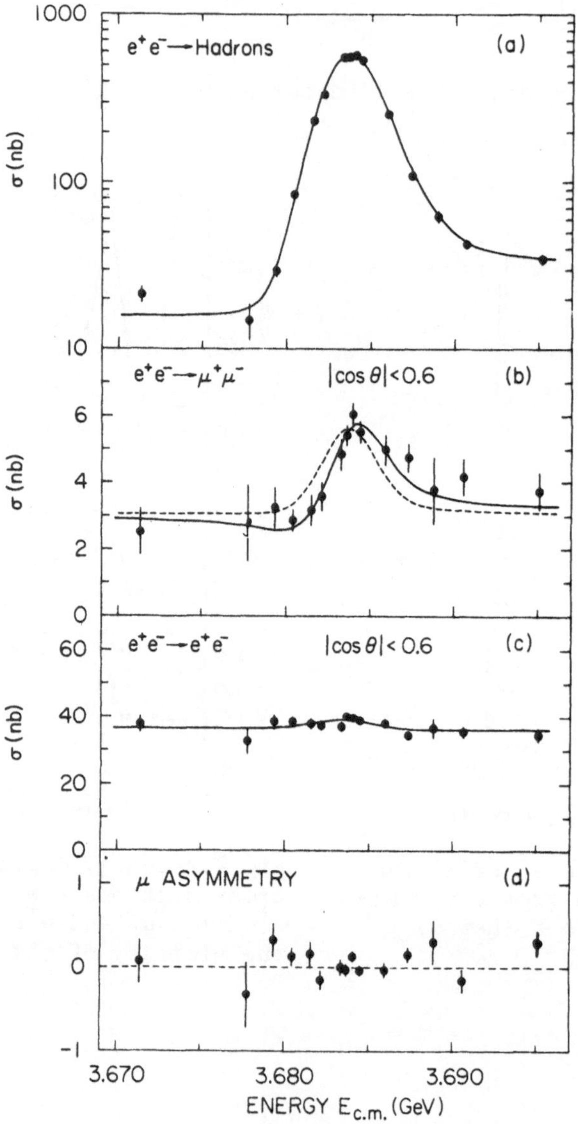

Fig. 2.3 Energy dependence of the cross sections $e^+e^- \to$ hadrons, $e^+e^- \to \mu^+\mu^-$, $e^+e^- \to e^+e^-$ and the forward-backward asymmetry in μ pair production in the vicinity of the ψ' (Ref. 9)

The mass and width parameters of J/ψ and ψ' are listed in table 2.1 and compared to the corresponding values of the ρ meson. Γ_{direct} measures the partial width for the direct decay into hadrons, excluding the cascade decays in the case of the ψ'.

Table 2.1 Mass and widths parameters for ρ, J/ψ and ψ'

	$\rho(770)$	J/ψ	ψ'
m (MeV)	770	3096	3687
Γ (MeV)	150	0.07 ± 0.01	0.23 ± 0.06
Γ_{direct} (MeV)	150	0.05 ± 0.01	~ 0.05
Γ_{ee} (KeV)	6.4 ± 0.8	4.8 ± 0.6	2.2 ± 0.3

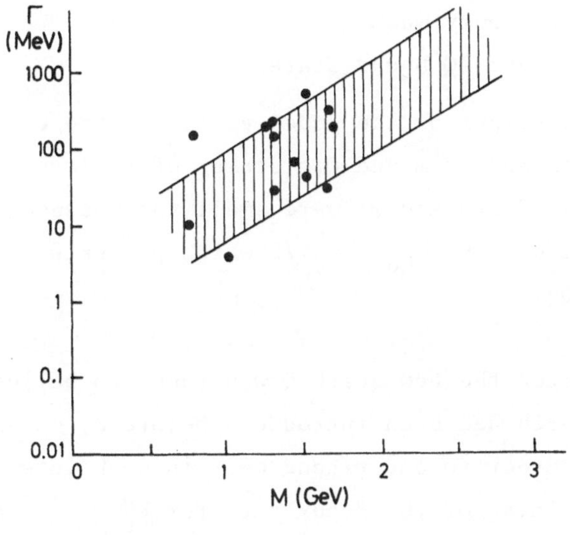

Fig. 2.4 A plot of width versus mass for nonstrange mesons

The most exciting property of J/ψ and ψ' is their small decay widths. The direct decay is suppressed by roughly a factor of 10^4 compared to what one would expect for a conventional meson of 3 to 4 GeV mass. This is illustrated in Fig. 2.4 which gives a plot of the width versus the mass of conventional nonstrange mesons. Despite its small decay width the J/ψ is a strongly interacting particle. This was established by photoproducing the J/ψ off nuclei; the t distribution exhibits a coherent part which leads to a total J/ψ nucleon cross section of ~1 mb near E_γ = 100 GeV.(Ref. 11).

2.2 The Quark Model Interpretation of J/ψ and ψ'

An intuitively simple and by now accepted explanation for J/ψ and ψ' was offered by the quark model. It assumed that

a) besides the u, d and s quarks a fourth quark Q exists which carries a new quantum number such that the decay $Q \to q$, q = u, d or s can proceed only weakly;

b) the J/ψ and ψ' are $Q\bar{Q}$ bound states.

Based on these assumptions the model predicted the existence of further $Q\bar{Q}$ mesons and of mixed Q,q states of the type $(Q\bar{q})$ and $(\bar{Q},q)$. Since the widths of J/ψ and ψ' were small the masses of the $(Q\bar{q}),(\bar{Q}q)$ mesons were assumed to be $M_{Q\bar{q}} > M_{\psi'}/2$ which prevents decays of the type $\psi' \to (Q\bar{q})(\bar{Q}q)$.

In a next step the new quark Q was taken to be identical to the charm quark c which had been introduced before by Glashow, Iliopoulos and Maiani[12] in order to understand certain weak interaction phenomena such as the smallness of the decay rate for $K_L^o \to \mu^+\mu^-$ and the absence of strangeness changing neutral currents. The properties of the hypothesized charm quark c were

charge +2/3
strangeness 0
charm 1, where charm is a new quantum number

c decays only weakly into d or s quarks according to

$c \rightarrow -d \sin\Theta_c + s \cos\Theta_c$

where Θ_c is the Cabibbo angle, $\Theta_c = 0.23$ (GIM mechanism).

2.3 $c\bar{c}$ SPECTROSCOPY

The variety of $c\bar{c}$ states in terms of J^{PC} follows directly from the spin 1/2 nature of the quark and is equivalent to that of positronium. For instance orbital angular momentum L = 0 leads to vector states (both quark spins parallel, also called orthocharmonium), and to pseudo scalars (spins antiparallel, paracharmonium). In order to compute level spacing, transition rates etc. more assumptions about the $c\bar{c}$ system are necessary. In most calculations the c quark is assumed to be sufficiently heavy such that the $c\bar{c}$ system can be described by a nonrelativistic Schrödinger equation.

The forces acting between $c\bar{c}$ quarks in nonrelativistic calculations are approximated by a steeply rising attractive potential[13]. At short distances this force might be represented by gluon exchange. This will give rise to a term $-\frac{4}{3}\alpha_s/r$ in the potential where α_s is the strong interaction coupling constant. A linear term is added to the potential in order to ensure quark confinement:

$$V(r) = -\frac{4\alpha_s}{3\,r} + ar + V_o \qquad (2.1)$$

A potential of this type[13-15] will lead to the level scheme shown in Fig. 2.5. The levels are labeled by J^{PC} with $P = (-1)^{L+1}$ and $C = (-1)^{L+S}$. For each value of L there are two bands of radial excitations with opposite charge conjugation depending on whether S = 0 or 1. The spectroscopic notation $n^{2S+1}L_J$ where n - 1 is the number of radial nodes is used to name the levels.

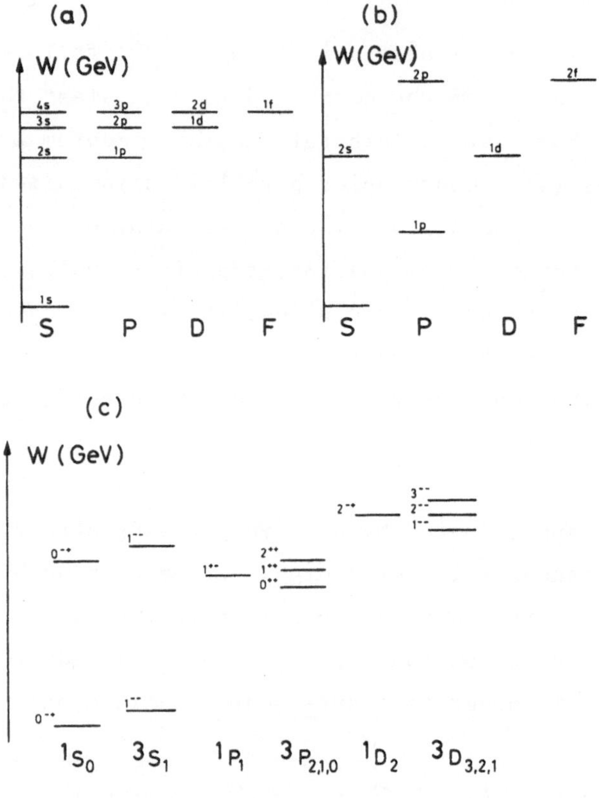

Fig. 2.5 Level scheme predicted for two fermions bound in a rising attractive potential

a) pure Coulomb potential $V = \alpha/r$

b) Harmonic oscillator potential $V = br^2$

c) $V = -\frac{4}{3}\,\alpha_s(m^2)/r + a\,r$

The triplet S states 1^3S_1 and 2^3S_1 are identified with the J/ψ and ψ', respectively. The potential (2.1) will lead to a small spin-spin force and hence to a small mass splitting between triplet ($^3S_1$) and singlet ($^1S_0$) states. The P levels will split into one state ($^1P_2$) with odd and three states ($^3P_{2,1,0}$) with even charge conjugation. In a pure Coulombic potential the first set of P levels would be degenerate in mass with the 2^3S_1 level. The addition of a confining potential pushes the mass of the 1P levels below the mass of the 2^3S_1 level.

The first L = 2 level splits into one state ($^2D_1$) with even and three states ($^3D_{3,2,1}$) with odd charge conjugation. The 1^3D_1 state has the quantum number of the photon. By mixing with the nearby 2^3S_1 state its wave function aquires a finite value at the origin[15]. As a consequence the 1^3D_1 can show up e.g. in the total e^+e^- cross section. The observation of this state was one of the major triumphs of charmonium spectroscopy which had predicted (Eichten et al.[15]) the ψ" two years before its discovery[16,17].

The number and the quantum numbers of the predicted $c\bar{c}$ levels are mainly a consequence of the spin-1/2 nature of the c quark. The level spacing and the level widths on the other hand are strongly model dependent and cannot firmly be predicted.

2.3 THE VECTOR STATES

The leptonic and hadronic widths of the J/ψ and the J/ψ - ψ' spacing can be used to fix the parameters of the potential (2.1). The $^3S_1$ state decays to lowest order into hadrons via a 3 gluon intermediate state. The hadronic width is given by

$$\Gamma(^3S_1 \rightarrow ggg \rightarrow \text{hadrons}) = \frac{160}{81}(\pi^2 - 9)\,\alpha_S^3\,\frac{\left|^3S_1(0)\right|^2}{M^2} \qquad (2.2)$$

where $^3S_1(0)$ describes the wave function at the origin. The width for the decay into a pair of leptons is given by

$$\Gamma(^3S_1 \rightarrow e^+e^-) = 64\pi\alpha^2/9 \frac{|^3S_1(0)|^2}{M^2} \qquad (2.3)$$

where M is the mass of the vector state.

Applying these relations to the J/ψ yields $\alpha_s \approx 0.19$. From the J/ψ − ψ' spacing the slope of the confining term is found to be $a = 0.25\ GeV^2$. These parameter values should be considered as order of magnitude estimates only since higher gluonic corrections could be sizeable. The scale violations observed in neutrino interactions for instance prefer $\alpha_s \approx 0.4$. The R dependence of the potential computed for $\alpha_s = 0.19$, $a = 0.25\ GeV^2$ is shown in Fig. 2.6. It is evident that due to the large DeBroglie wavelength of the charm quark the charmonium spectrum as well as the decay rates will be controlled primarily by the linear part of the potential.

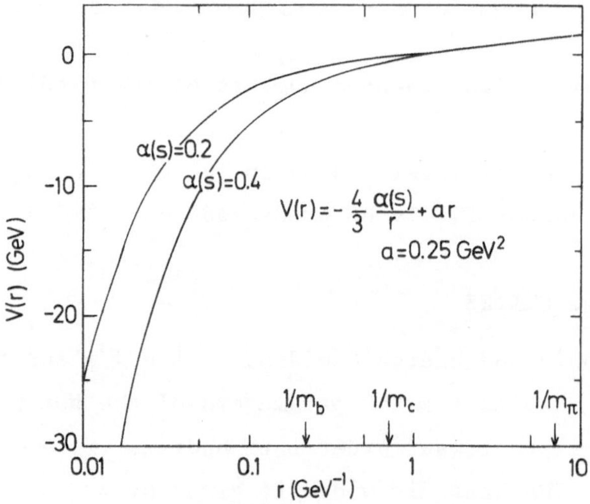

Fig. 2.6 The radius dependence of the charmonium potential.

With the α_s and a parameters determined from J/ψ and ψ' the third vector state $\psi'' = 1^3D$, was predicted to be at 3.755 GeV [15]. Figs. 2.7 and 2.8 show the excitation of the ψ'' in e^+e^- annihilation as measured by LBL-SLAC[16] and the DELCO group[17] at SPEAR.

The properties of the ψ'' are summarized in Table 2.2.

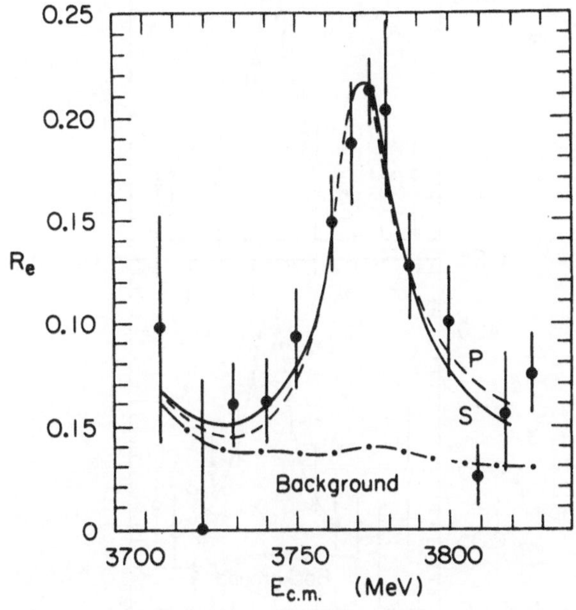

Fig. 2.7 $R = \sigma^{tot}/\sigma_{\mu\mu}$ near the ψ' after radiative corrections (Ref. 16)

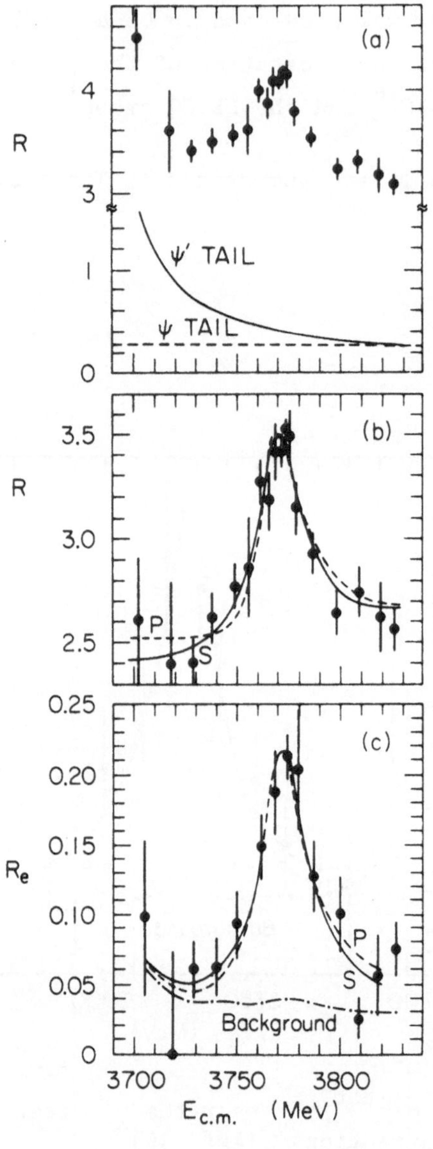

Fig. 2.8 $R = \sigma^{tot}/\sigma_{\mu\mu}$ near the ψ" (Ref. 17)

a) measured R

b) after radiative corrections

c) $R_e = \sigma(e^+e^- \to eX)/\sigma_{\mu\mu}$

Table 2.2 Properties of the ψ"(3771)

Mass (MeV)	Γ(MeV	Γ_{ee} (eV)	
3772 ± 6	28 ± 5	370 ± 90	LBL-SLAC[16]
3770 ± 6	24 ± 5	180 ± 60	DELCO[17]

branching ratios:

$\psi'' \to D^o\overline{D^o}$ 49 ± 25 %

$\to D^+D^-$ 50 ± 38 %

Table 2.3 Resonance parameters of vector states between 3 and 4.5 GeV

	Mass(MeV)	Γ(MeV)	Γ_{ee}(keV)	$B_{ee} = \frac{\Gamma_{ee}}{\Gamma}$	
J/ψ	3095±2	0.069±0.015	4.8 ±0.6	$6.9\pm0.9\cdot10^{-2}$	SLAC-LBL[22]
ψ'	3686±2	0.228±0.056	2.1 ±0.3	$9.3\pm1.6\cdot10^{-3}$	SLAC-LBL[22]
ψ"(3771)		28±5	0.37±0.09	$(1.3\pm0.2)\cdot10^{-5}$	LBL-SLAC[16]
		24±5	0.18±0.06	$0.7\pm0.2\cdot10^{-5}$	DELCO[17]
4.04	4040±10	52±10	0.75±0.10	$1.4\pm0.4\cdot10^{-5}$	DASP[20]
4.16	4159±20	78±20	0.77±0.20	$0.9\pm0.3\cdot10^{-5}$	DASP[20]
4.41	4414±7	33±10	0.44±0.14	$1.3\pm0.3\cdot10^{-5}$	SLAC-LBL[18]
		66±15	0.40±0.10	$0.7\pm0.2\cdot10^{-5}$	DASP[20]

In Fig. 2.10 the observed vector mesons are compared to vector state levels predicted by a charmonium model. Between J/ψ and ~4.6 GeV the number of vector states known at present agrees with the number of states expected by theory. Whether the interpretation of the observed states as suggested by Fig. 2.10 is the correct one, namely $\psi(4.04) = 3^3S_1$, $\psi(4.16) = 2^3D_1$ etc. is an open question.

The ψ'' being ~ 45 MeV above the $D\bar{D}$ threshold is found to decay almost exclusively into a pair of D mesons and is therefore an ideal place to study the D properties (see section 3.1).

Evidence for still heavier vector states has been found in studies of the total e^+e^- cross section above 4 GeV. Fig. 2.9 shows the measurements of SLAC-LBL[18], PLUTO[19] and DASP[20] in the region from 3.6 to 5.2 GeV. Preliminary data in this energy region are also available from DELCO[17]. The error bars shown in Fig. 2.9 do not include systematic uncertainties which are primarily due to the limited solid angle. The SLAC-LBL and PLUTO measurements were done in a solenoidal geometry that permitted to detect charged particles and determine their momenta over 65 % and 86 % of 4π, respectively. Extrapolation to the full solid angle was done by means of a Monte Carlo program that included effects such as information on jets and production and decay of charmed particles. In the DASP case the total cross section was determined in the nonmagnetic inner detector which detected charged particles and photons and measured their directions. The efficiency correction was determined directly by detecting inclusively charged particles in the magnetic spectrometers. The systematic errors are estimated to be ±15 % for SLAC-LBL, ±11 % for PLUTO and ±15 % for DASP. The cross section values shown in Fig. 2.9 include the contribution from the heavy lepton τ. The three data sets agree with each other at the level of one unit in R which is of the order of the systematic errors. The cross section data of PLUTO and DASP exhibit structures at 4.04, 4.16 and 4.4 GeV. The first and the third one were seen first by SLAC-LBL. Since many charmed particle channels open up in this energy region (D^*D^*, $D^{**}D^*$, $F\bar{F}$,...) an interpretation of these structures is not straightforward. If they are assumed to be resonances one finds the mass and width values given in Table 2.3. They should be considered as preliminary since a precise determination will probably require a coupled channel analysis[23]. The result of the resonance fit to the DASP data is shown by the solid curve in Fig. 2.9

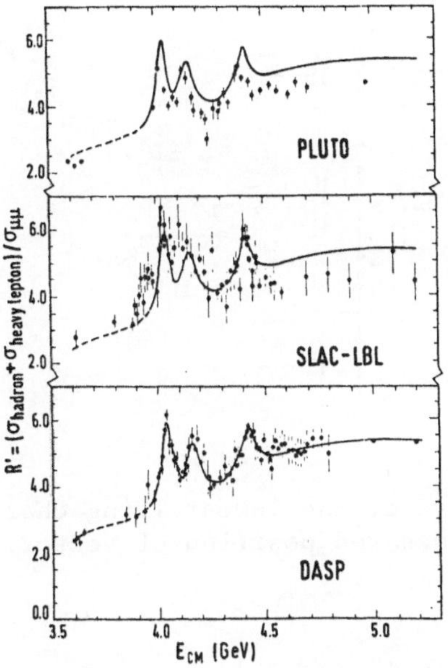

Fig. 2.9 Total hadron production: Plotted is the ratio $R' = (\sigma_{hadon} + \sigma_{heavy\ lepton})/\sigma_{\mu\mu}$ from the Pluto group (Ref. 19), SLAC-LBL (Ref. 18) and DASP (Ref. 20). The curve represents the resonance fit to the DASP data. Fig. taken from Ref. 20.

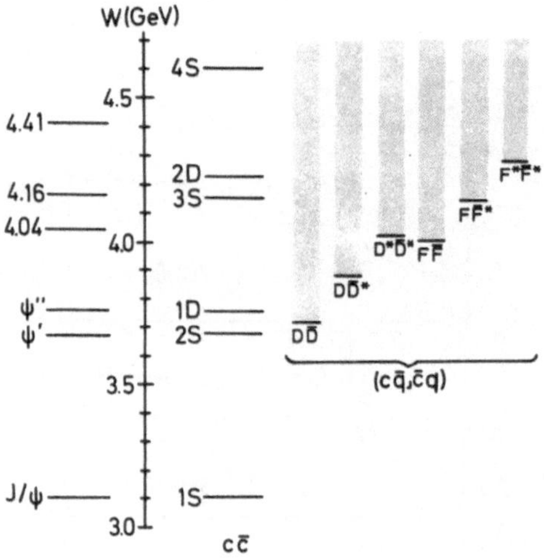

Fig. 2.10 The $c\bar{c}$ levels, the lowest lying charmed meson channels and the measured position of vector mesons (after Ref.23)

2.4 P-WAVE STATES

As mentioned above, three P-wave states $^3P_{2,1,0}$ are predicted between J/ψ and ψ' with even parity and even charge conjugation (Fig. 2.1). If the $\psi' - {}^3P_J$ mass difference is smaller than $2\cdot m_\pi$ these states can be reached from the ψ' by photon emission only,

$$\psi' \to \gamma\, {}^3P_J$$

The ^3P_J decay can either be purely hadronic via gluon annihilation (diagrams a,b) or by radiative decay into the J/ψ (diagram c).

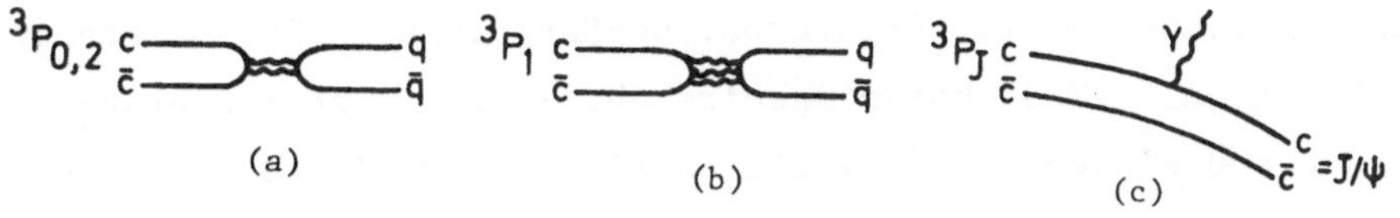

The decay widths for (a,b) are proportional to the wave functions squared at small distances. These widths have been estimated by Barbieri, Gatto and Kögerler[24] as $\Gamma(2^{++})/\Gamma(0^{++}) = 4/15$ and $\Gamma(0^{++}) = 2.4$ MeV. The 1^{++} state is expected to be even narrower.

The decays $2^3S_1 \rightarrow \gamma\, ^3P_J$ can proceed to lowest order by an electric dipole (E1) transition. The rate for an E1 transition is given by

$$\Gamma(2^3S_1 \rightarrow \gamma 1^3P_J) = \left(\frac{16}{243}\right) \alpha(2J+1)\, k^3\, |\langle 1P|r|2S\rangle|^2 \tag{2.4}$$

The rate depends on the overlap between the radial wave functions for the S and P levels. For an E1 transition the angular distribution of the photon with respect to the beam axis is of the form

$$\begin{aligned} 2^3S_1 &\rightarrow \gamma 1^3P_0 && 1 + \cos^2\Theta \\ &\rightarrow \gamma 1^3P_1 && 1 - 1/3 \cos^2\Theta \\ &\rightarrow \gamma 1^3P_2 && 1 + 1/13 \cos^2\Theta \end{aligned} \tag{2.5}$$

For the latter two transitions also higher multipole amplitudes could contribute. For this reason only the prediction for $^3S_1 \rightarrow {}^3P_0$ is unique.

The ^3P_J states were detected in three different ways: as discrete lines in the photon spectrum, through the cascade decay $\psi' \rightarrow \gamma\, ^3P_J$, $^3P_J \rightarrow \gamma J/\psi$ and as peaks in $\pi\pi$, 4π ... mass distributions. The study of the cascade decay led to the first discovery of an intermediate state (P_c(3510)) by DASP[25].

a) Discrete photon lines.

Monochromatic photons from ψ' decay were observed by SLAC-LBL[26] and by the Maryland, Pavia, Princeton, UC-San Diego, SLAC and Stanford Collaboration[27]. The photon spectra obtained by the two experiments from J/ψ and ψ' decay are displayed in Figs. 2.11 and 2.12. The ψ' spectrum shows several maxima superimposed on a smoothly varying background. No structures are observed in the J/ψ spectrum. The photon lines are centered at 121 MeV, 168 MeV, 256 MeV and 383 MeV. The first three transitions correspond to intermediate states with masses of 3561 MeV, 3512 MeV and 3481 MeV now called $\chi(3561)$, $P_c(3512)$ and $\chi(3418)$. The fourth line results from the decay $P_c(3512) \rightarrow \gamma J/\psi$. The branching ratios for the $\psi' \rightarrow \gamma\, ^3P_J$ decays are given in Table 2.4. They account for more than one fourth of all ψ' decays. The theoretically predicted decay widths are consistently lower by a factor of 2 - 3.

Table 2.4 P_c/χ States

Decay	Branching ratio %	Γ (keV)	Reference	Theory (keV)
$\psi' \rightarrow \gamma\ \chi(3.41)$	7.2 ± 2.0	16 ± 5	27	44
	6.5 ± 2.2	15 ± 5	26	
$\psi' \rightarrow \gamma\ \chi(3.45)$	< 2.5 %	< 5.7	27	18
$\psi' \rightarrow \gamma\ P_c(3.51)$	7.1 ± 2.0	16 ± 5	27	38
$\psi' \rightarrow \gamma\ \chi(3.55)$	7.0 ± 2.3	16 ± 5	27	27

b) $\psi' \rightarrow \gamma\, ^3P_J \rightarrow \gamma\gamma J/\psi$

Next we discuss the observation of the P states via the cascade decay. The J/ψ is identified by its $\mu^+\mu^-$ decay. Fig. 2.13 shows the μ-pair mass spectrum from ψ' decay as measured by DASP[25]. Two narrow signals are observed; the higher one corresponds to the $\psi' \rightarrow \mu^+\mu^-$ decay and to μ-pair production by QED, the peak at 3.1 GeV results from

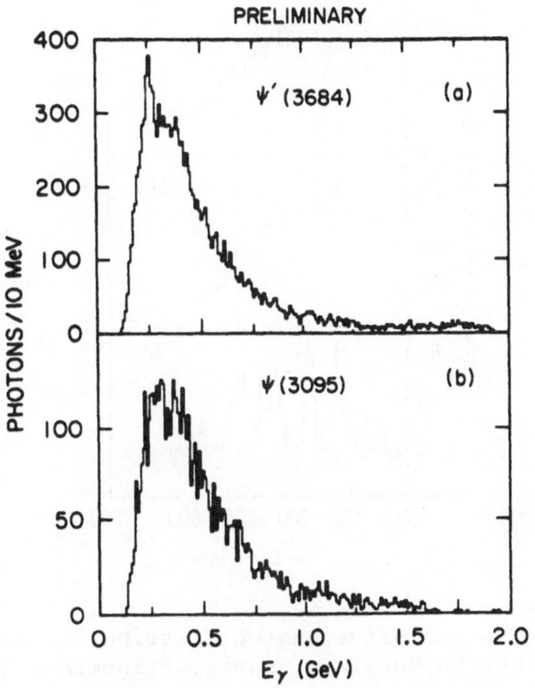

Fig. 2.11 The inclusive photon energy spectra (Ref. 26)

a) from J/ψ decay

b) from ψ' decay

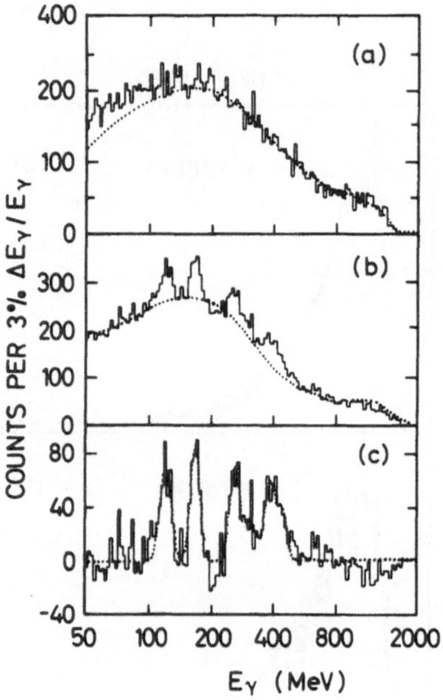

Fig. 2.12 The inclusive photon distribution measured by the Maryland, Pavia, Princeton, UC-San-Diego, SLAC and Stanford Collavoration (Ref. 27) as a function of E_γ for:

a) the J/ψ

b) ψ'

c) shows the difference between the data and the continuum in (b)

$J/\psi \to \mu^+\mu^-$ decay. Events with a $\mu^+\mu^-$ mass in the J/ψ region and two photons are selected and fitted to the hypothesis $\psi' \to \gamma\gamma J/\psi$
$\qquad\qquad\qquad\qquad\qquad\qquad\qquad\qquad\qquad\qquad\qquad\qquad\qquad\qquad\qquad\qquad\qquad\qquad\hookrightarrow \mu^+\mu^-$.

A large fraction of the events are due to the decay $\psi' \to \eta J/\psi$. They are removed by a cut in the $\gamma\gamma$ mass (e.g. by requiring $M_{\gamma\gamma} < 0.52$ GeV in the DASP analysis).

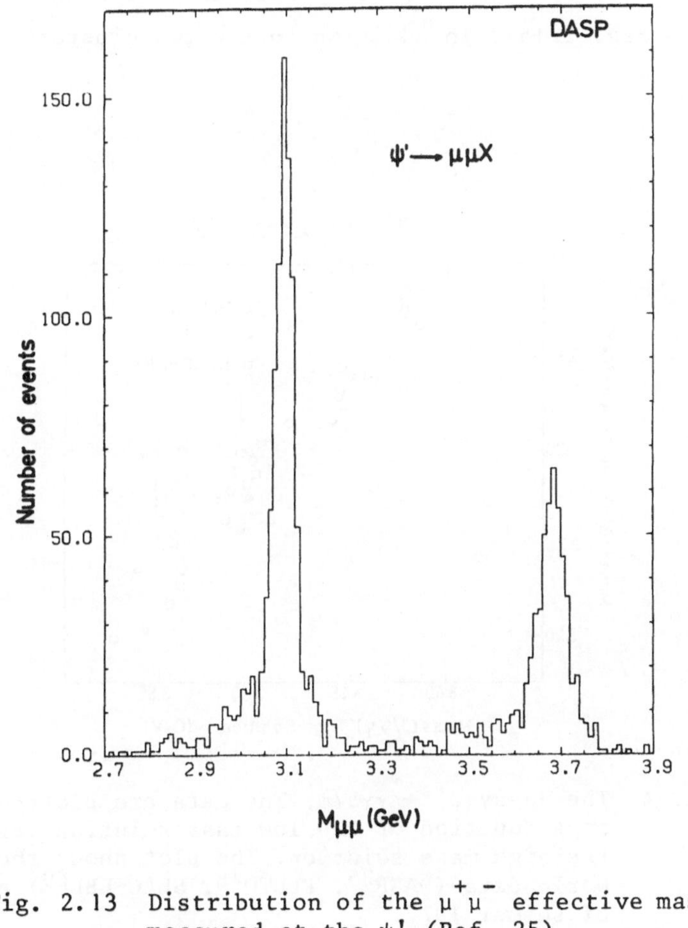

Fig. 2.13 Distribution of the $\mu^+\mu^-$ effective mass measured at the ψ' (Ref. 25)

Fig. 2.14 shows the remaining events obtained in the experiments of DASP[25], PLUTO[28] and SLAC-LBL[29]. The low ($\gamma J/\psi$) mass solution is plotted versus the high ($\gamma J/\psi$) mass solution. A small percentage of these events are due to background from $\psi' \to \pi^0 \pi^0 J/\psi$. Two distinct clusters of events are observed around 3.51 GeV and 3.56 GeV corresponding to the $P_c(3.51)$ and $\chi(3.56)$. Note that the mass spread is smaller for the high mass solution which gives supporting evidence that the high mass solution is the correct one. Clear signals for the $P_c(3.51)$ and $\chi(3.56)$ have also been obtained by the DESY-Heidelberg group[30] (see Fig. 2.15). In addition to the two clusters at 3.51 and

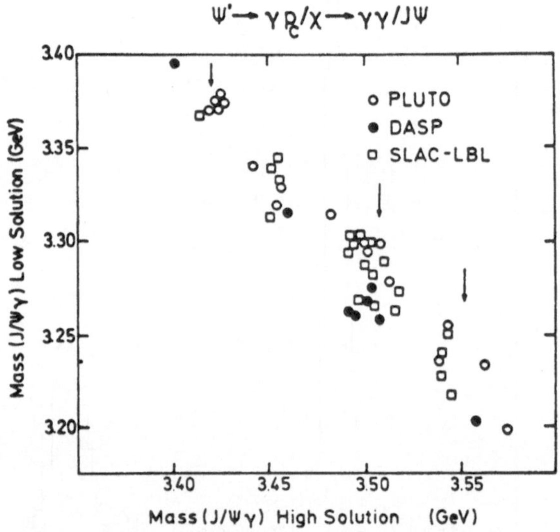

Fig. 2.14 The decay $\psi' \to \gamma\gamma J/\psi$. The data are plotted as a function of the low mass solution versus the high mass solution. The plot shows the world data (DASP[25], PLUTO[28], SLAC-LBL[29]) as of summer 1977.

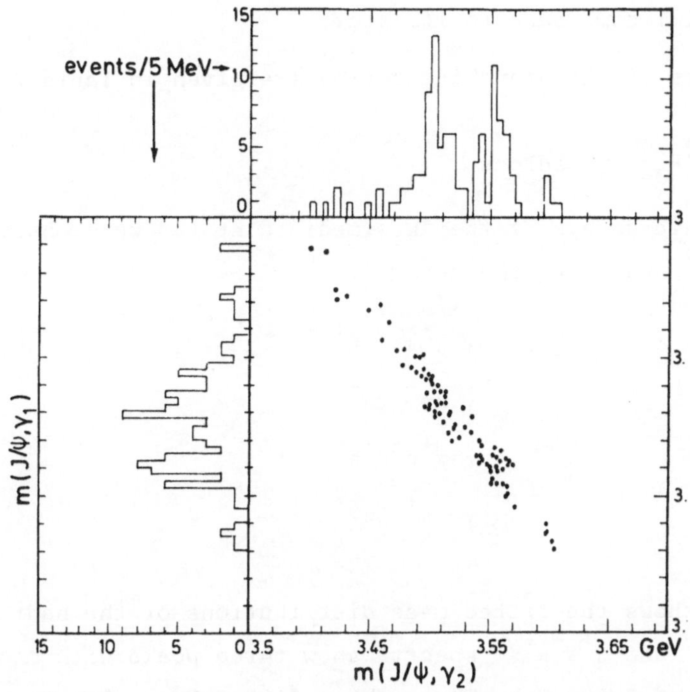

Fig. 2.15 The decay $\psi' \to \gamma\gamma J/\psi$: scatter plot of mass $M(\gamma_1 J/\psi)$ vs. mass $M(\gamma_2 J/\psi)$ as measured by DESY-Heidelberg (Ref. 30).

3.56 GeV Fig. 2.14 shows two groups of events centered at 3.45 GeV and 3.41 GeV. The latter one can be identified with the χ(3418) seen in the photon spectrum and in the hadronic mass spectra.

The DESY-Heidelberg group recently found evidence for a state near 3.60 GeV (see Fig. 2.15)[30]. More data are needed to firmly establish the existence of this particle.

The 3.45 GeV group has not been observed in any other decay mode. We will come back to it later.

The products of the branching ratios are given in Table 2.5.

c) $\psi' \to \gamma\, ^3P_J \to \gamma$ hadrons

Hadronic decays of the intermediate states were observed by SLAC-LBL[26,31,32] analyzing the decays

$$\psi' \to \gamma P_c\, \chi \to \gamma\ \pi^+\pi^-,\ K^+K^-$$
$$\gamma\ 2\pi^+\ 2\pi^-$$
$$\gamma\ \pi^+\pi^-K^+K^-$$
$$\gamma\ 3\pi^+\ 3\pi^-$$

Fig. 2.16 shows the fitted mass distributions of the hadronic systems. The $2\pi^+\ 2\pi^-$ and $\pi^+\pi^-K^+K^-$ spectra show three peaks with masses of 3.415 GeV and 3.550 GeV. The peak at 3.68 GeV results from the direct decay of the ψ'. In the $3\pi^+\ 3\pi^-$ distribution the two upper states are not resolved. There is a clear $\pi^+\pi^-$ (or K^+K^-) signal for the decay of the 3415 MeV state. Eight events are observed with a mass of 3.55 GeV. No signal is found at 3.50 GeV. The branching ratios are listed in Table 2.6.

Note that none of the mass spectra give evidence for a 3.45 GeV state.

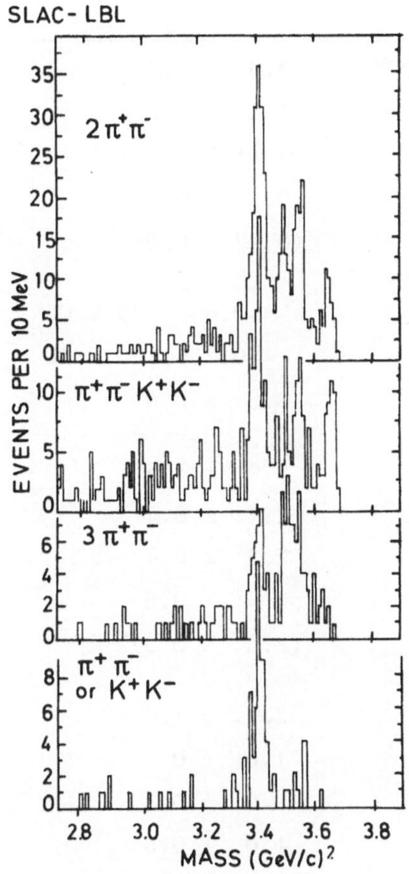

Fig. 2.16 The decay $\psi' \to \gamma$ + charged hadrons. Mass distributions of the hadron system measured by SLAC-LBL[26,31,32]

Table 2.5 Cascade Decays ($\psi' \to \gamma P_c/\psi$) ($P_c/\chi \to \gamma J/\psi$)

Intermediate State	$BR(\psi' \to \gamma P_c/\chi) \cdot BR(P_c/\chi \to \gamma J/\psi)$ %	Reference
$\chi(3.41)$	3.3 ± 1.7	27
	0.3 ± 0.2	25
	0.14 ± 0.09	30
	0.2 ± 0.2	29
	(0.22 ± 0.08)[a]	29
$\chi(3.45)$	< 0.61 90 % CL	25
	< 0.25 90 % CL	30
	0.8 ± 0.4	29
$P_c(3.51)$	5.0 ± 1.5	27
	2.1 ± 0.4	25
	2.5 ± 0.4	30
	2.4 ± 0.8	29
	(2.4 ± 0.3)[a]	
$\chi(3.56$	2.2 ± 1.0	27
	1.6 ± 0.6	25
	1.0 ± 0.2	30
	1.0 ± 0.6	29
	(1.9 ± 0.2)[a]	
$P_c(3.59)$	0.18 ± 0.06	30

a The combined branching ratio obtained by averaging the results of references 25, 26 and 30.

Table 2.6 Properties of the Intermediate States

MASS (MeV)	Decay	$BR(\psi' \to \gamma P_c/\chi)\cdot BR(P_c/\psi$ decay) %	Reference	$BR(P_c/\psi$ decay)*) %
3413 ± 5	$\pi^+\pi^-$	0.07 ± 0.02	31	1.0 ± 0.3
	K^+K^-	0.07 ± 0.02	31	1.0 ± 0.3
	$2\pi^+2\pi^-$	0.32 ± 0.06	31	4.7 ± 0.9
	$\pi^+\pi^-K^+K^-$	0.27 ± 0.07	31	3.9 ± 1.0
	$p\bar{p}\ \pi^+\pi^-$	0.04 ± 0.013	31	0.6 ± 0.2
	$3\pi^+3\pi^-$	0.14 ± 0.05	31	2.0 ± 0.7
	$\gamma J/\psi$	0.22 ± 0.08	a	3.2 ± 1.2
	$\gamma\ \gamma$	< 0.014	25	< 0.2
	$\rho^0\pi^+\pi^-$	0.12 ± 0.04	31	1.8 ± 0.6
	$K^{0*}K^+\pi^-$	0.17 ± 0.06	31	2.5 ± 0.9
3.508 ± 4	$\pi^+\pi^- + K^+K^-$	< 0.015	31	< 0.21
	$2\pi^+2\pi^-$	0.11 ± 0.04	31	1.5 ± 0.6
	$\pi^+\pi^-K^+K^-$	0.06 ± 0.03	31	0.85 ± 0.42
	$p\bar{p}\pi^+\pi^-$	0.01 ± 0.008	31	0.14 ± 0.11
	$3\pi^+3\pi^-$	0.17 ± 0.06	31	2.4 ± 0.8
	$\gamma J/\psi$	2.4 ± 0.3	a	34 ± 4
	$\gamma\ \gamma$	< 0.013	25	< 0.18
	$\rho^0\pi^+\pi^-$	0.026± 0.022	31	0.37 ± 0.31
	$K^{0*}K^+\pi^-$	0.31 ± 0.022	31	0.44 ± 0.31
3.552 ± 6	$\pi^+\pi^- + K^+K^-$	0.02 ± 0.01	31	0.29 ± 0.14
	$2\pi^+2\pi^-$	0.16 ± 0.04	31	2.3 ± 0.6
	$\pi^+\pi^-K^+K^-$	0.14 ± 0.04	31	2.0 ± 0.6
	$p\bar{p}\ \pi^+\pi^-$	0.02 ± 0.01	31	0.29 ± 0.14
	$3\pi^+3\pi^-$	0.08 ± 0.05	31	1.1 ± 0.7
	$\gamma J/\psi$	1.9 ± 0.2	a	27 ± 3
	$\gamma\ \gamma$	< 0.004	25	< 0.06
	$\rho^0\pi^+\pi^-$	0.05 ± 0.03	31	0.71 ± 0.43
	$K^{0*}K^+\pi^-$	0.052± 0.031	31	0.74 ± 0.44

*) The values listed in Table 2.4 for $\psi' \to P_c/\chi$ were used to extract the branching ratios.

a) The average values from Table 2.5 were used.

d) Quantum numbers of the intermediate states

The distributions in Θ, the angle of the photon with respect to the beam axis has been obtained by summing the various decay channels. The result is shown in Fig. 2.17 for the $\chi(3.41)$. The values of a obtained from a fit to the data of the form $1 + a\cos^2\Theta$ are as follows:

State:	$\chi(3.41)$	$P_c(3.51)$	$\chi(3.55)$
a	1.4 ± 0.4	0.1 ± 0.4	0.3 ± 0.4

Remember that for a spin 0 state a = 1. The $\chi(3.41)$ is therefore consistent with J = 0, whereas a J = 0 assignment for the $P_c(3.51)$ and the $\chi(3.55)$ is excluded on the 2σ level.

In the level scheme depicted in Fig. 2.5 there are four levels with even charge conjugation between the ψ' and the J/ψ, one pseudscalar $^1S_0$ with $J^{PC} = 0^{-+}$ and three ^3P states with $J^{PC} = 2^{++}$, 1^{++} and 0^{++}. Are the levels found consistent with these quantum numbers?

All levels have even charge conjugation since they are populated via $\psi' \to \gamma P_c/\chi$.

$\chi(3.413)$ It follows from C = + and the observed decay into $\pi^+\pi^-$ and/or K^+K^- that the state is an isoscalar and has positive parity. The angular distribution is consistent with J = 0. We can therefore safely assign this state to $J^{PC} = 0^{++}$.

$\chi(3.454)$ Needs to be confirmed.

$P_c(3.508)$ The absence of $\pi^+\pi^-$ and K^+K^- decays are consistent with the state belonging to an unnatural spin-parity sequence 0^-, 1^+,... . The angular distribution of the photon suggests $J \neq 0$ and isospin must be even since the resonance decays into an even number of pions.

$\chi(3.552)$ It follows from the observed decay into $\pi^+\pi^-$ and/or K^+K^- that the level is an isoscalar and belongs to a natural spin-parity sequence. The angular distribution of the photon excludes J = 0 with 2σ.

This is all that can be deduced from the data alone. However, comparing the available information with the levels expected in the charmonium model (Fig. 2.5) leads to a unique assignment using the following reasoning. The 3.413 GeV level must be the 0^{++} state. This leaves only one natural spin-parity level - the $^3P_2(2^{++})$ level - which then must be associated with the 3.552 GeV level. The 3.508 GeV level has $J \neq 0$ and must therefore be the $^3P_1(1^{++})$ level. This assignment is also consistent with the fact that $P_c(3.51) \rightarrow \pi^+\pi^-, K^+K^-$ has not been observed.

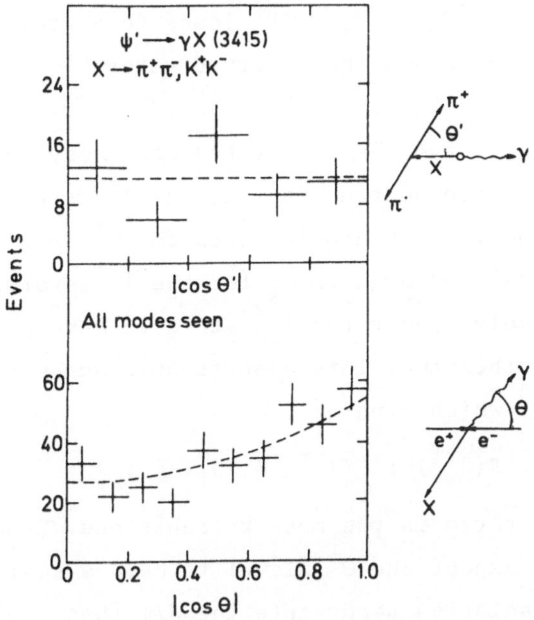

Fig. 2.17 The decay $\psi' \rightarrow \gamma X(3.41)$ measured by SLAC-LBL[26,31,32].

a) Decay $X(3.41) \rightarrow \pi^+\pi^-$: distribution of the π^+ polar angle measured in the X system with respect to the X distribution of flight

b) Photon production angular distribution measured with respect to the incoming e^+ beam

The decay $\psi' \to \gamma\ {}^3P_{2,1,0}$ are to lowest order all electric dipole transitions. The measured widths are listed in Table 2.4 and compared to model calculations assuming E1. Note that the predicted rates are considerably larger than the measured values. If the matrix elements are independent of J then the relative rates for E1 transitions are given by $\Gamma \sim k^3\ (2J + 1)$. In this case we expect

$$\Gamma(2^3S_1 \to \gamma\ {}^3P_0) : \Gamma(2^3S_1 \to \gamma\ {}^3P_1) : \Gamma(2^3S_1 \to \gamma\ {}^3P_2) =$$

$$1 \quad : \quad 0.9 \quad : \quad 0.6$$

compared to the observed ratios 1 : 1.04 : 1.02. Reversing the order of the levels i.e. ${}^3P_0 > {}^3P_1 > {}^3P_2$ leads to a gross disagreement between the predicted and the observed rates.

In QCD the 0^{++} and the 2^{++} states can decay to lowest order via 2 gluon emission into hadrons whereas the 1^{++} state must emit at least three gluons. The total hadronic width for 0^{++} and 2^{++} is therefore proportional to α_s^2 compared to α_s^3 for the 1^{++} state. Since α_s is less than one we naively expect the 1^{++} state to have a smaller hadronic width than the other two. This assertion is supported by detailed calculation[24,33] which find:

$$\Gamma(0^{++}) : \Gamma(2^{++}) : \Gamma(1^{++}) = 15 : 4 : 1$$

where the first ratio is the most reliable one. Experimentally we would therefore expect the 0^{++} level to have a smaller branching ratio for the radiative decay into the J/ψ than the other two and this is indeed supported by experiments.

The ratios of the hadronic widths can be estimated from the radiative transition with the following assumptions:

1) $\Gamma_{tot} = \Gamma({}^3P \to \text{hadrons}) + \Gamma({}^3P \to \gamma\ {}^3S_1) = \Gamma_h + \Gamma_\gamma$

2) $\Gamma({}^3P \to \gamma\ {}^3S_1) \sim k^3$, i.e. independent of the matrix elements.

With these assumptions we can write:

$$\Gamma_h = (\Gamma_{tot} - \Gamma_\gamma) = (\frac{1}{B_\gamma} - 1)\, \Gamma_\gamma \quad \text{where } B_\gamma = \Gamma_\gamma / \Gamma_{tot}.$$

With the radiative branching ratios listed in Table 2.6 we find

$$\Gamma_h(3.41):\Gamma_h(3.55):\Gamma_h(3.51) = (14.2\pm6.0):(3.8\pm1.3):(0.5\pm0.16)$$

consistent with the predictions.

The $^1P_1$ state.

Besides the three ^3P_J states the potential (2.1) predicts a singlet P state with $J^{PC} = 1^{+-}$ (Fig. 2.5). Because of odd C parity this level cannot be reached from the ψ' by a γ transition. The decay $\psi' \to \pi^0\, ^1P_1$ is suppressed by isospin conservation. The strong decay $\psi' \to \pi\pi\, ^1P_1$ is probably forbidden by kinematics. No evidence so far was found for the existence of the $^1P_1$.

2.5 PSEUDOSCALAR STATES

The potential (2.1) predicts the pseudoscalar states ($\eta_c, \eta_c' \ldots$) $1\, ^1S_0$, $2\, ^1S_0$ ... to lie below the corresponding vector states ($n\, ^3S_1$). The pseudoscalar states can decay into ordinary hadrons by two gluon exchange, i.e.

$$\Gamma(\eta_c \to \text{hadrons}) \sim \alpha_s^2$$

Hence one expects the $^1S_0$ states to have a larger width than the vector states for which $\Gamma \sim \alpha_s^2$. More precisely:

$$\Gamma(^1S_0 \to gg \to \text{hadrons}) = \frac{32\pi}{3} \alpha_s^2 \frac{|^1S_0(0)|^2}{M^2} \tag{2.6}$$

The $^1S_0$ states having C = + can also decay into two photons; the corresponding width is given by

$$\Gamma(^1S_0 \to \gamma\gamma) = \frac{256\pi}{27} \alpha^2 \frac{|^1S_0(0)|^2}{M^2} \tag{2.7}$$

A comparison of (2.6) and (2.7) leads to

$$\frac{\Gamma(^1S_o \to \gamma\gamma)}{\Gamma(^1S_o \to \text{hadrons})} = \frac{8}{9}\,\frac{\alpha^2}{\alpha_s^2} \approx 1.3 \cdot 10^{-3} \qquad (2.8)$$

for $\alpha_s = 0.2$. Under the assumption that $^3S_1$ and ^1S_o have the same radial wave functions equ (2.2) and (2.6) predict for the ^1S_o width

$$\Gamma(^1S_o \to \text{hadrons}) = \frac{27}{5}\,\frac{\pi}{(\pi^2 - 9)\alpha_s}\,\Gamma(^3S_1 \to \text{hadrons}) \simeq 5\text{ MeV}$$

If the triplet-singlet level splitting is small, the ^1S_o can be reached from the corresponding $^3S_1$ state by a radiative transition only, $^3S_1 \to \gamma\ ^1S_o$. This is a pure spinflip or M1 transition. The amplitude is proportional to the magnetic moment of the c quark. If this is assumed to be of the Dirac type,

$$\mu_c = \frac{2/3\ e}{2\ m_c} \qquad (2.10)$$

m_c = mass of c quark

the transition rate is given by

$$\Gamma(n_i\ ^3S_1 \to n_f\ ^1S_o) = \frac{16}{27}\,\alpha\,\frac{k^3}{m_c^2}\,|\langle n_f|n_i\rangle|^2 \qquad (2.11)$$

k is the photon energy.

The overlap integral $|\langle n_f|n_i\rangle|^2$ is expected to be unity for $n_f = n_i$ and small for $n_f \neq n_i$. Table 2.7 shows the predicted width for $J/\psi \to \gamma\eta_c$ for various values of the η_c mass.

Table 2.7 Predicted decay width Γ and branching ratio B for the decay $J/\psi \to \gamma\eta_c$ as a function of the η_c mass.

M_{η_c} (GeV)	photon energy (MeV)	Γ(keV)	B (%)
3.04	50	0.2	0.3
2.99	100	1.6	2
2.89	200	14	20
2.79	300	45	70

A search of the photon spectrum from J/ψ decay failed to produce evidence for the $J/\psi \to \gamma\eta_c$ transition [27] (see Figs. 2.11, 2.12).

Evidence for a heavy and narrow resonance below the J/ψ was reported by the DASP group [34] in the decay $J/\psi \to \gamma X \to \gamma\gamma\gamma$.

Events with three photons were selected using the DASP inner detector. For each photon the production angles were measured taking the position of the interaction point from Bhabha events observed concurrently. A 1C fit was made to the hypothesis $J/\psi \to \gamma\gamma\gamma$ and events with a $\chi^2 < 2.7$ were retained. The photon energies for the accepted events were computed. Of the three possible γγ mass combinations only two are independent. The events are plotted in Figs. 2.18 and 2.19 as a function of the lowest and the highest γγ mass in the event. The low mass plot shows a clear η signal and an indication for η' production. The position of the η(547.1 ± 4.2 MeV) agrees with its known mass value; the observed η width (σ = 24 ± 4 MeV) is consistent with the expected resolution of 20 MeV.

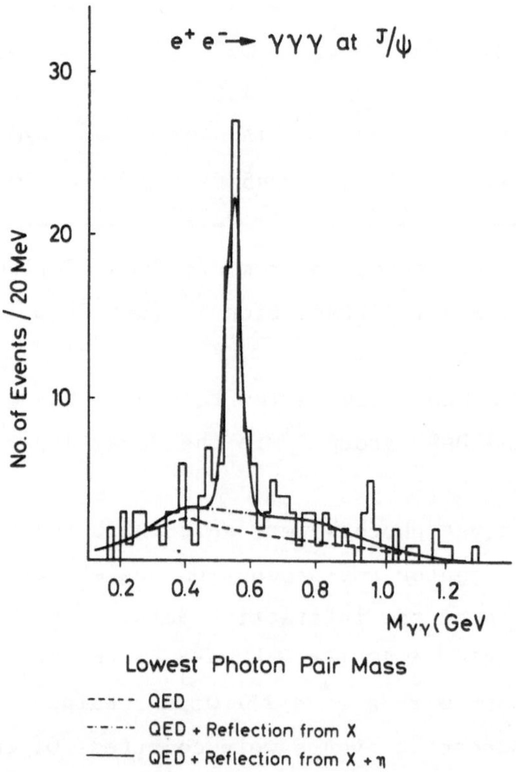

Fig. 2.18 $J/\psi \to \gamma\gamma\gamma$ measured by DASP[34]. Distribution of the lowest photon pair mass

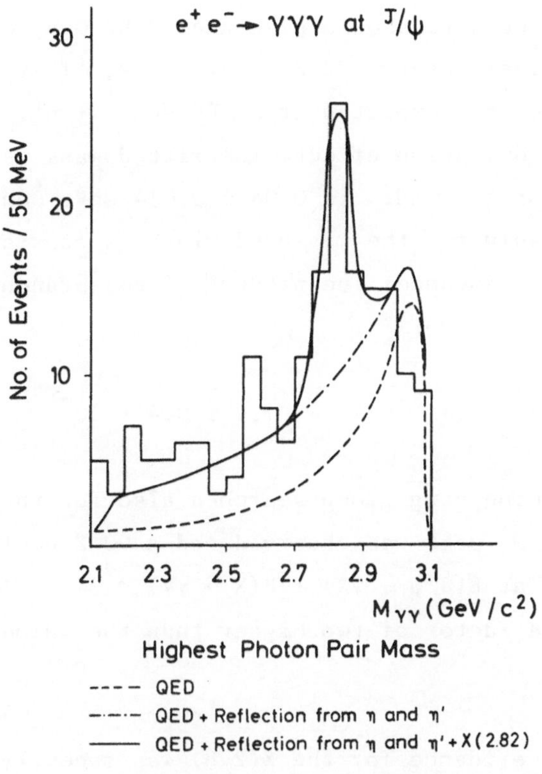

Fig. 2.19 $J/\psi \rightarrow \gamma\gamma\gamma$ measured by DASP[34]. Distribution of the highest photon pair mass

The high mass data are shown in Fig. 2.19 together with the contributions from QED and $\eta + \eta'$. The QED curve was computed from the matrix elements evaluated by Berends and Gastmann [35]. The QED curve is an absolute prediction. The calculation was checked by measuring $e^+e^- \rightarrow \gamma\gamma\gamma$ at a c.m.energy of 3.6 GeV, i.e. away from resonances.

The high mass data show a peak near 2.82 GeV, called the X(2.8). In the mass interval 2.82 ± 0.04 GeV, 41 events are observed, compared to 19 events expected from QED and $\eta + \eta'$. This corresponds to a 5 standard deviation effect. The fitted mass is 2.82 ± 0.014 GeV; the value of the width is 0.04 ± 0.014 GeV. Considering the experimental resolution the measured width is consistent with that of a zero width resonance. The product of the branching ratios was determined to

$$B(J/\psi \rightarrow \gamma X) \cdot B(X \rightarrow \gamma\gamma) = 1.4 \pm 0.4 \cdot 10^{-4}$$

The DESY-Heidelberg group searched also for the decay $J/\psi \rightarrow \gamma X \rightarrow \gamma\gamma\gamma$ [30,36]. No peak was observed and a 90 % confidence upper limit was found at $B(J/\psi \rightarrow \gamma X) \cdot B(X \rightarrow \gamma\gamma) < 3.2 \cdot 10^{-4}$. This limit is about a factor of two higher than the value measured by DASP.

Confirming evidence for the X(2.8) was recently reported by a group working at Serpukov [37]. The experiment using a 40 GeV π^- beam incident on a hydrogen target searched for events of the type

$$\pi^- p \rightarrow \gamma\gamma\ n.$$

The $\gamma\gamma$ mass distribution (Fig. 2.20) shows an enhancement at 2.88 ± 0.06 GeV with a systematic uncertainty of 0.03 GeV. The width of 0.17 GeV is consistent with the experimental resolution. No other enhancement is seen in the mass spectrum and the authors identify the peak with the X(2.82). The photons are emitted iso-

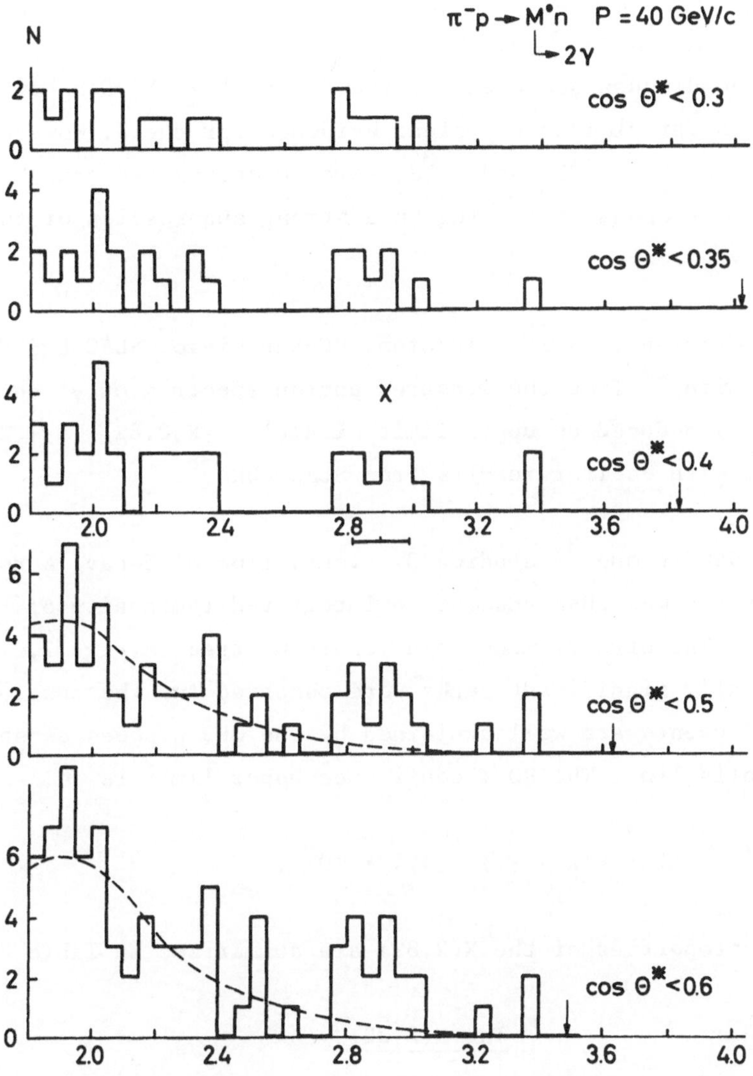

Fig. 2.20 The reaction $\pi^- p \rightarrow \gamma\gamma n$ at 40 GeV/c as measured at Serpukhov[37]. Plotted are the number of events versus the photon pair mass with the cosinus of the c.m. decay angle Θ^* as a parameter. Events are detected with an efficiency of at least 0.8 for photon pair masses below the value shown by the arrow.

tropically in the X rest system as expected for a spin zero particle.

We now discuss searches for the decay $\psi' \rightarrow \gamma X$. None of these searches so far yielded any clear evidence for the X. However, remember that the $1\ ^1S_0$ and $2\ ^3S_1$ wave functions are expected to be nearly orthogonal resulting in a strong suppression of the decay $\psi' \rightarrow \gamma X(2.82)$.

The Maryland, Pavia, Pinceton, UC-San Diego, SLAC and Stanford Collaboration [27] from the measured photon spectrum of ψ' decay (Fig. 2.12) deduced an upper limit of $B(\psi' \rightarrow \gamma X(2.82)) < 1$ % in agreement with earlier results from SLAC-LBL [38].

The DASP group [34] studied 3γ events from ψ' decay. A total of 190 events was observed for an integrated luminosity of 1610 nb^{-1}. The high $\gamma\gamma$ mass solution is plotted in Fig. 2.21. No statistically significant peaks were observed and the absolute number of events are well explained by the QED process as shown by the solid line. The 90 % confidence upper limit is

$$B(\psi' \rightarrow \gamma X) \cdot B(X \rightarrow \gamma\gamma) < 1.4 \cdot 10^{-4}.$$

The properties of the X(2.82) are summarized in Table 2.8.

Table 2.8 The X(2.82)

		Ref.
Mass	2.82 ± 0.014 GeV	34
$B(J/\psi \rightarrow \gamma X) < 1.7$ %		27
$B(J/\psi \rightarrow \gamma X) \cdot B(X \rightarrow \gamma\gamma) = 1.4 \pm 0.4 \cdot 10^{-4}$		34
$B(\psi' \rightarrow \gamma X) \cdot B(X \rightarrow \gamma\gamma) < 1.4 \cdot 10^{-4}$		34

We now have all available data on hand to compare the X(2.82) with the predicted $1\ ^1S_0$ state.

1. $B(J/\psi \rightarrow \gamma\ ^1S_0)$

For a $^1S_0$ of mass 2.82 GeV theory predicts a branching ratio $B(J/\psi \rightarrow \gamma\ ^1S_0) \simeq 50\ \%$ compared to the measured upper limit of 1.7 %. Basic input into the theoretical calculation was the assumption that the singlet and triplet radial wave functions are the same and that the magnetic moment of the c quark is of the Dirac size. Several authors have speculated that the latter may not be true [39].

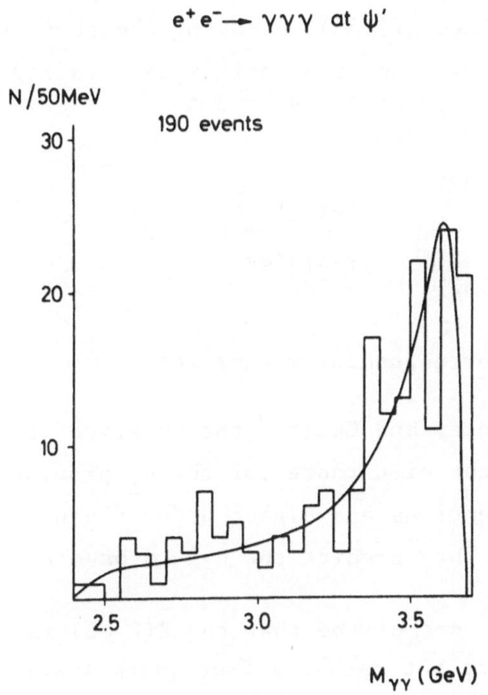

Fig. 2.21 $\psi' \rightarrow \gamma\gamma\gamma$ measured by DASP[34]. Distribution of the highest photon pair mass

A precise measurement of the $D^{*+} \to D^+\gamma$ and $D^{*0} \to D^0\gamma$ decay rates will shed light on this question [39].

2. From the upper limit $B(J/\psi \to \gamma X) < 1.7$ % and from $B(J/\psi \to \gamma X)$. $B(X \to \gamma\gamma) = 1.4 \pm 0.4 \cdot 10^{-4}$ one finds $B(X \to \gamma\gamma) > 8 \cdot 10^{-3}$ which is to be compared with the predicted value of $1.3 \cdot 10^{3}$ (equ. 2.8). The theoretical prediction is based on the value of $\alpha_s = 0.2$ deduced from $J/\psi \to ee$ decay.

3. Theory predicts the overlap integral between $2\ {}^3S_1$ and $1\ {}^1S_0$ to be small. This can be checked by a comparison of the J/ψ and ψ' decay rates into X(2.82). Removing the phase space factor one finds agreement with the theoretical expectation :

$$\frac{\frac{1}{k'^3}\ \Gamma_{\psi' \to \gamma X(2.82)}}{\frac{1}{k^3}\ \Gamma_{J/\psi \to \gamma X(2.82)}} = \frac{|\langle X|\psi'\rangle|^2}{|\langle X|J/\psi\rangle|^2} < 0.11 \qquad (2.12)$$

k, k' are the corresponding γ energies.

4. According to Bradley and Gault[40] the observed features of the X(2.82) are consistent with those for the η_c provided relativistic spin-dependent corrections and higher order gluon corrections are taken into account. They predict the η_c' to have a mass of ~3.6 GeV.

5. Several authors[41] speculated that the X(2.82) is not the lowest $c\bar{c}$ pseudoscalar state but rather a four quark state $c\bar{c}q\bar{q}$. It was further assumed that this state is below the real $1\ {}^1S_0$ level and hence narrow. In this case the decays $\psi' \to \gamma X$ and $J/\psi \to \gamma X$ should both be electric dipole transitions and the widths, assuming equal matrix elements, will be in the ratio $\Gamma(\psi' \to \gamma X)/\Gamma(J/\psi \to \gamma X)$

$= (k'/k)^3 \simeq 25$. This is an order of magnitude larger than the experimental upper limit of 2.3.

2.6 THE 3.45 GeV LEVEL

As discussed above, the 3.45 GeV state is not yet firmly established. It was observed only in the $\gamma J/\psi$ decay mode. Chanowitz and Gilman[42] first suggested to identify it with the $2\ {}^1S_0$ state. However, this causes serious problems. From the branching ratio $B(\psi' \to \gamma\chi(3.45)) \cdot B(\chi \to \gamma J/\psi) = 0.8 \pm 0.4$ % determined by SLAC-LBL and the upper limit $B(\psi' \to \gamma\chi(3.45))$ quoted by Ref. 27 one finds $B(\chi(3.45) \to \gamma J/\psi) > 25$ %, in contradiction to the theoretical predictions that the $2\ {}^1S_0 - 1\ {}^3S_1$ overlap integral should be small (see equ. 2.12).

Recently it was pointed out by Harari[42] that the large singlet-triplet splitting operative if the X(2.82) is identified with the ${}^1S_0$ may push the ${}^1D_2$ ($J^{PC} = 2^{-+}$) level below the ψ' mass [42]. Since this state can decay into $\gamma J/\psi$ the $\chi(3.45)$ could be the ${}^1D_2$ state.

SUMMARY

The observed number of levels and their ordering is that expected for a pair of fermions bound in a steeply rising potential. The fine structure splitting of the P-levels i.e. ${}^3P_2 > {}^3P_1 > {}^3P_0$ is that expected for a vector force. A pure scalar potential would reverse the order and give a level assignment in contradiction to the experiments. The level splittings are larger than predicted but the radiative rates are smaller by a factor of two. Indirect evidence shows that the ratio of the hadronic widths of the P states are in agreement with predictions based on QCD.

3. CHARMED MESONS

The existence of a fourth quark besides the familiar u, d and s quarks implies that the SU(3) nonet of 8 + 1 mesons will be replaced[43] by a hexadecimet of 15 + 1 states as shown in Fig. 3.1. Besides ordinary SU(3) resonances with C = 0 the hexadecimet contains six states with open charm, C = ±1 and one $c\bar{c}$ state with hidden charm. The quark content and the names given to the charmed pseudoscalar mesons are

$$C = +1 : \quad D^+ = c\bar{d} \qquad D^0 = c\bar{u} \qquad F^+ = c\bar{s}$$
$$= 0 : \quad \eta_c = 1^1S_o = c\bar{c}$$
$$= -1 : \quad D^- = \bar{c}d \qquad \overline{D^0} = \bar{c}u \qquad F^- = \bar{c}s$$

Higher mass charmed mesons will cascade by strong and/or electromagnetic decays into these states, which then decay weakly with the following decay modes:

$$D(F) \rightarrow \ell\bar{\nu}_e$$
$$\rightarrow \ell\bar{\nu}_e + \text{hadrons}$$
$$\rightarrow \text{hadrons}$$

The leptonic decay mode is suppressed by kinematics ($J_3^D = 0 \neq J_3^\ell + J_3^{\bar{\nu}} = 1$). Semileptonic decays are precicted with a branching ratio on the order of 20 % and most decays (≈80 %) will therefore yield only hadrons in the final state[43,44]. Any new flavour will produce mixed lepton-hadron final states and show up as narrow resonances above threshold. However, the flavour can be identified by the properties of the final states. The GIM mechanism[12] (Fig. 3.2) leads for charmed mesons to the following Cabibbo favoured decay modes:

$$D^o \to (e^+\nu_e)\ (K^-\ldots)$$
$$\to (\bar{K}n\pi)^o$$
$$D^+ \to (\ell^+\nu_e)\ (\overline{K^o}\ldots)$$
$$\to (\bar{K}n\pi)^+$$
$$F^+ \to (\ell^+\nu_e)\ (\eta,\ \eta',\ K\bar{K},\ldots)$$
$$\to (\eta n\pi)^+,\ (\eta' n\pi)^+,\ (K\bar{K}n\pi)^+$$

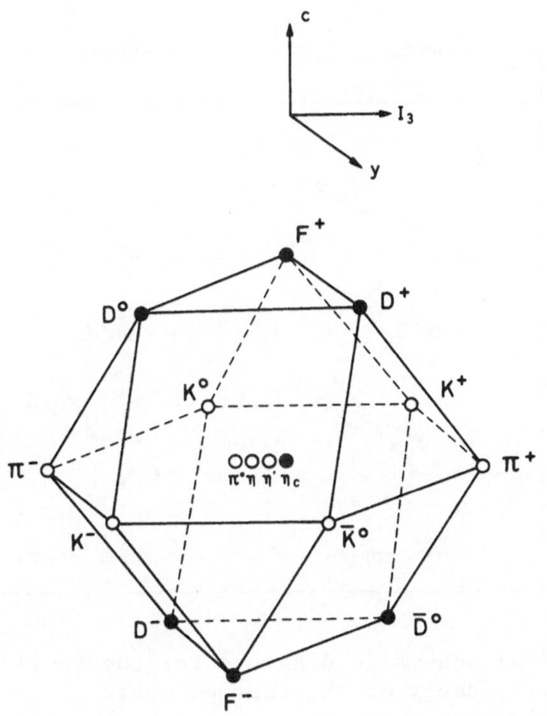

Fig. 3.1 The hexadecimet of the pseudoscalar mesons. Charm is plotted along the z-axis, Y and I_3 along respectively the y-axis and the x-axis. The π^o, η and η' mesons are denoted by the open circles at the origin, η_c by the black circle

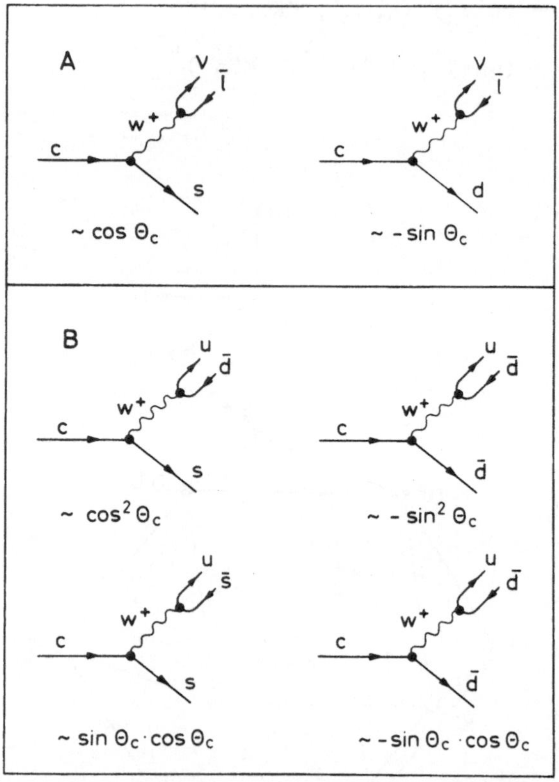

Fig. 3.2 a) Schematic diagrams for the semileptonic decay of the charmed quark

b) Schematic diagrams for the decay of the charmed quark into noncharmed quarks

According to the GIM mechanism associated production of D mesons will show up as an increase in the yield of kaons, a strong correlation between leptons and kaons in multiprong final states and appearant exotic decays like $D^+ \to K^-\pi^+\pi^+$. The production of F mesons will produce an increase in the yield of particle with a large $s\bar{s}$ component like η, η' or ϕ.

A strong increase in the yield of neutral and charged kaons correlated with the step in the total cross section at 4.0 GeV was indeed observed, first by PLUTO[45] and DASP[46] and then confirmed by SLAC-LBL[47]. This is shown in Fig. 3.3 where the ratios $R_{K^\pm} = \sigma(e^+e^- \to K^\pm X)/\sigma(e^+e^- \to \mu^+\mu^-)$ and $R_{K^0} = 2\sigma(e^+e^- \to K^0 X)/\sigma(e^+e^- \to \mu^+\mu^-)$ are plotted as a function of the c.m. energy. The data are in rough agreement. The SLAC-LBL K^0 cross sections are in general larger than those of PLUTO; the PLUTO values have not been corrected for a loss ($\approx$20 %) due to the cut off in K^0 momentum ($P_K > 0.2$ GeV/c). The peak at 4.415 GeV is only reflected in the SLAC-LBL data; no clear evidence is seen for it in the DESY data. This might be due to a lack of statistics at exactly this energy.

A simple kinematical observation can be used to show that the increase in K production is indeed due to charm production. At threshold the charmed mesons are produced at rest, hence kaons resulting from their decays must have energies less than one half of the beam energy. This is demonstrated by Fig. 3.4 where the invariant cross section $(E/4\pi p^2)\, d\sigma/dp$, measured by DASP[46], is plotted as a function of the kaon energy for c.m. energies of 3.6 GeV (below charm threshold) and 4.05 GeV(above charm threshold). The step in the kaon yield is caused by particles with an energy less than one half of the beam energy.

If the decay of each charmed hadron would yield one kaon then $\frac{1}{2}(\Delta R_{K^o,\bar{K}^o} + \Delta R_{K^\pm}) = \Delta R_C$ where $\Delta R_K \cdot \sigma_{\mu\mu}$ is the observed increase in K yield and $\Delta R_C \cdot \sigma_{\mu\mu}$ the contribution of charm production to σ^{tot}. From

Fig. 3.3 one finds $\frac{1}{2}(\Delta R_{K^0,\bar{K}^0} + \Delta R_{K^\pm}) \approx 1.7$ measured as the difference in R_K between 3.6 and 5.0 GeV. The DASP group reported for ΔR_C a value of 2.1 ± 0.3. Thus nearly every charmed hadron decays into final states containing a kaon as predicted by the GIM mechanism.

The SLAC-LBL group has measured the yield of antiprotons and Λ, $\bar{\Lambda}$ as a function of the c.m. energy between 3.7 and 7.6 GeV (Fig. 3.5). The $\bar{p}$ yield is seen to rise above 4.5 GeV which is the threshold for charmed baryon production.

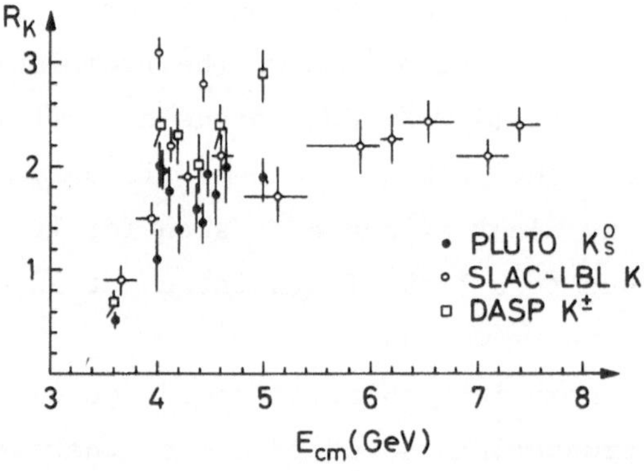

Fig. 3.3 Inclusive K_s^0 and $K^\pm$ production as a function of c.m. energy as measured by PLUTO[45], DASP[46] and SLAC-LBL[47]. Plotted is

$$R_{K^0} = \frac{2\sigma(e^+e^- \to K^0X)}{\sigma(e^+e^- \to \mu^+\mu^-)} \quad \text{and} \quad R_{K^\pm} = \frac{\sigma(e^+e^- \to K^\pm X)}{\sigma(e^+e^- \to \mu^+\mu^-)}$$

as a function of c.m. energy. ($\sigma(e^+e^- \to K^\pm X)$ is the sum of inclusive K^+ and K^- production)

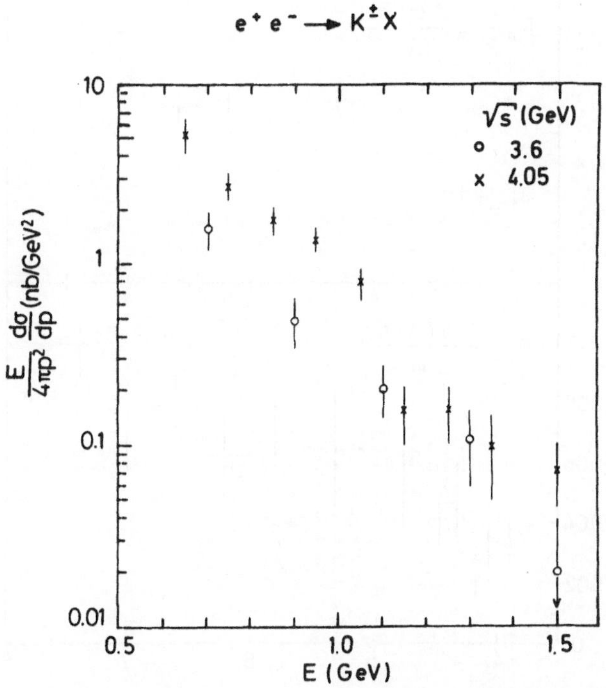

Fig. 3.4 The invariant cross sections $E/(4\pi p^2)\, d\sigma/dp$ for the sum of K^+ and K^- production at c.m. energies of 3.6 GeV and 4.05 GeV (from DASP[46])

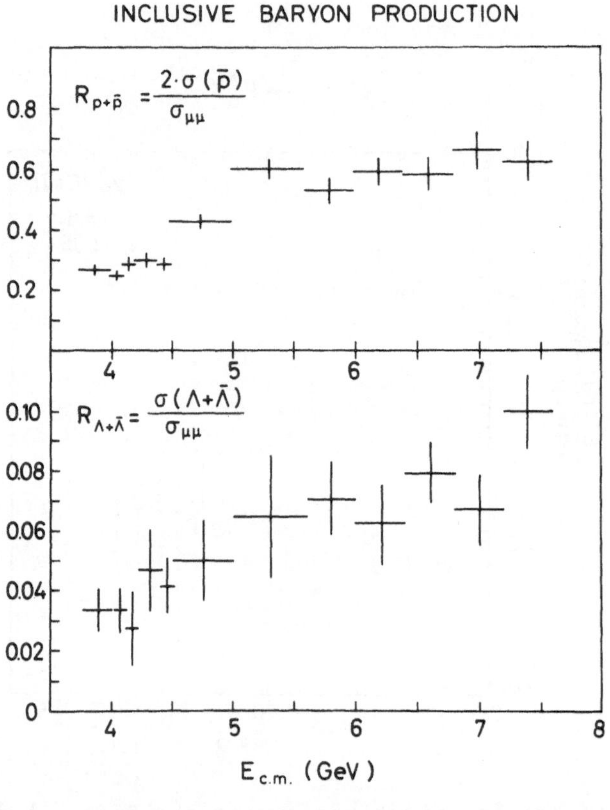

Fig. 3.5 a) $R(p + \bar{p}) = \frac{2\sigma(\bar{p})}{\sigma_{\mu\mu}}$ versus c.m. energy

b) $R(\Lambda + \bar{\Lambda}) = \frac{\sigma(\Lambda + \bar{\Lambda})}{\sigma_{\mu\mu}}$ versus c.m. energy

Data from SLAC-LBL

3.1 The D Mesons

First evidence for charmed mesons was found by SLAC-LBL in a study of hadron events produced between 3.9 and 4.6 GeV and at 4.028 GeV [48]. In the $K^\pm\pi^\mp$, $K^\pm\pi^\mp\pi^\mp$ and $K^\pm\pi^\pm\pi^-\pi^-$ mass distributions narrow peaks were observed at a mass of 1.87 GeV. The small width (Γ < 40 MeV) of this state and the fact that the $K^\pm\pi^\mp\pi^\mp$ decay mode could not come from an ordinary $q\bar{q}'$ system strongly suggested the identification with the D^o, D^+ mesons. Meanwhile the best place to study the D mesons was found to be the $\psi''(3771)$ [16,17] which decays exclusively into $D\bar{D}$ [16]. The ψ'' is only 40 MeV above the $D\bar{D}$ threshold and this has the added advantage that the D's are almost at rest.

Fig. 3.6 shows the sums of charged and neutral $K\pi$, $K\pi\pi$ and $K\pi\pi\pi$ mass spectra measured at the ψ''. Clear signals for the $D^\pm$ and D^o are observed. The favorable kinematics allow a precise determination of the $D^\pm$ and D^o masses. The result is listed in Table 3.1. The charged D is found to be heavier than the neutral D,

$$M_{D^\pm} - M_{D^o} = 5.1 \pm 0.8 \text{ MeV}$$

Table 3.1 D and D^* mass values [16]

	M(MeV)
D^o	1863.3 ± 0.9
D^+	1868.4 ± 0.9
D^{*o}	2006.0 ± 1.5
D^{*+}	2008.6 ± 1.0
$M_{D^+} - M_{D^o}$	5.1 ± 0.8
$M_{D^*} - M_{D^{*o}}$	2.6 ± 1.8

D - branching ratios: Absolute branching ratios for D^o and D^+ decays were determined by LBL-SLAC at the $\psi''(3771)$. Since the ψ'' is only ≈40 MeV above $D\bar{D}$ threshold the only OZI allowed strong decay is into $D\bar{D}$. Noticing that the ψ'' has a strong decay width (Γ ≈ 25 MeV) it is safe to assume that the ψ'' decays almost 100

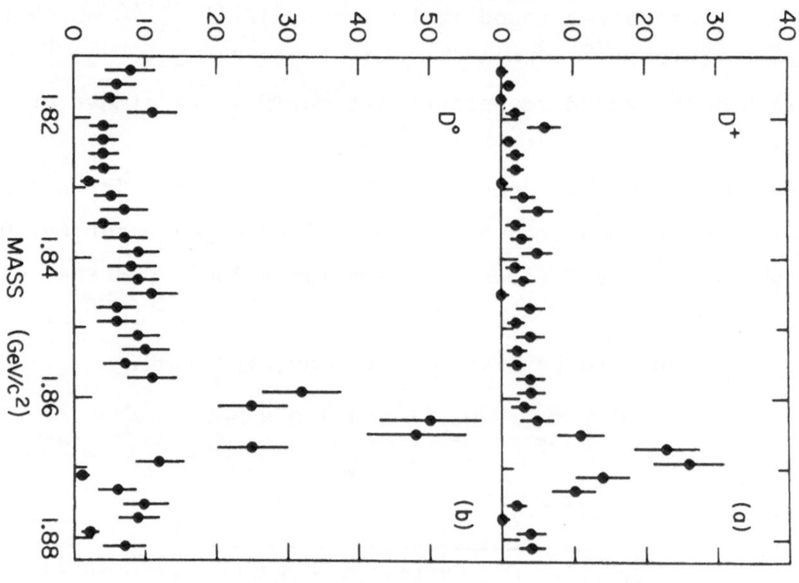

Fig. 3.6 ψ" decay as measured by LBL-SLAC[16]: Invariant mass spectra for the sum of all observed (a) $D^{\pm}$ and (b) D^{0} decay modes

percent into $D\bar{D}$. This assumption was checked by comparing the rates of events with one and two D's identified, respectively, in specific decay channels. The following numbers of events were found:

$$\psi'' \to \underset{\lfloor \to K\pi}{D^o} \; X \qquad 194 \text{ events (one D identified)}$$

$$\to \underset{\lfloor \to K\pi}{D^o} \; \underset{\lfloor \to K\pi}{D^o} \qquad 8 \text{ events (two D's identified)}$$

$$\psi'' \to \underset{\lfloor \to K\pi}{D^\pm} \; X \qquad 82 \text{ events (one D identified)}$$

$$\to \underset{\lfloor \to K\pi}{D^+} \; \underset{\lfloor \to K\pi}{D^-} \qquad 2 \text{ events (both D's identified)}$$

Since the one D event rate is proportional $B(\psi'' \to D\bar{D}) \cdot B(D \to K\pi)$ while the two D rate is $\sim B(\psi'' \to D\bar{D})\ B(D \to K\pi)^2$ both ratios can be computed separately from the data. The result is

$$B(\psi'' \to D^oD^o) = 49 \pm 25\ \%$$
$$B(\psi'' \to D^+D^-) = 50 \pm 38\ \%$$
and
$$B(\psi'' \to D\bar{D}) = 99 \pm 48\ \%$$

which is consistent with $B(\psi'' \to D\bar{D}) = 1$.

Table 3.2 lists the measured D^o and D^+ decay rates.

<u>D^* meson:</u> The data taken by SLAC-LBL at 4.028 GeV provided also evidence for the first excited D states, the D^{*o} and D^{*+} [48,49]. Fig. 3.7 shows the mass spectrum of the system recoiling from $D^\pm$. A narrow signal is found at a mass of 2.008 GeV. Note that no clear evidence is seen for a recoiling D which indicates that the $D\bar{D}$ production cross section at this energy is considerably smaller than for DD^* although the Q value for DD^* is only ~150 MeV.

Table 3.2 D and D^* branching ratios[16]

	B(D → f) (%)	Q(MeV)
$D^o \to K^-\pi^+$	2.2 ± 0.6	
$K^o\pi^+\pi^-$	3.5 ± 1.1	
$K^-\pi^+\pi^+\pi^-$	2.7 ± 0.9	
$D^+ \to K^o\pi^+$	1.5 ± 0.6	
$K^-\pi^+\pi^+$	3.5 ± 0.9	
$D^{*o} \to D^o\pi^o$	55 ± 15	7.7 ± 1.7
$D^o\gamma$	45 ± 15	142.7 ± 1.7
$D^{*+} \to D^o\pi^+$	65 ± 9	5.7 ± 0.5
$D^+\pi^o$	29 ± 8	5.2 ± 0.9
$D^+\gamma$	2 - 10	140.2 ± 0.9

$\Gamma(D^{*o})/\Gamma(D^{*+}) = 0.95 \pm 0.4$

The D^* mass values are given in Table 3.1. The Q values for the $D^* \to D\pi$ decays are only a few MeV or are negative ($D^{*o} \to D^+\pi^-$). This is the reason for the small D^* width. For the same reason radiative D^* decays, $D^* \to D\gamma$, play an important role.Fig. 3.8 illustrates the possible $D^* \to D$ transitions. The observed branching ratios are given in Table 3.2.

From an analysis of the angular distribution for D mesons produced near threshold and of the D^* decay channels, spin and parity of D

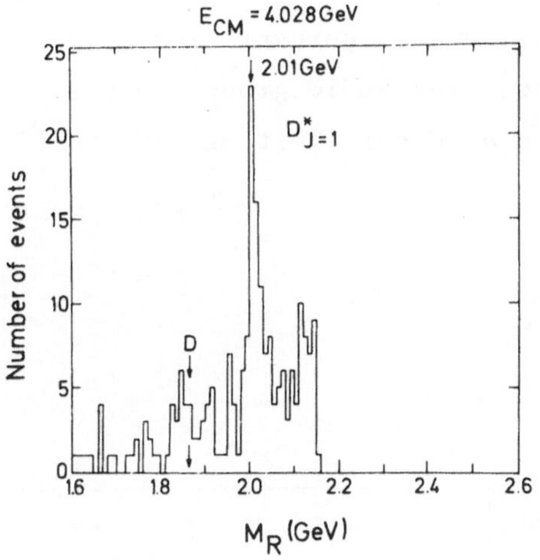

Fig. 3.7 Reaction $e^+e^- \to D^\pm$ R measured by $\hookrightarrow K^\mp \pi^\pm \pi^\pm$ SLAC-LBL[48,49]. Mass spectrum of the recoiling system R

and D^* were found to be $J^P_D = 0^-$, $J^P_{D^*} = 1^-$ where the D was chosen to have negative parity[49].

A detailed study of D production at 4.028 GeV besides DD^* revealed a sizeable fraction of $D^*\bar{D}^*$ production although the Q value is only 16 MeV for the $D^{0*}\overline{D^{0*}}$ and 12 MeV for the $D^{\pm *}D^{\mp *}$ channels[50]. Correcting for phase space effects by a factor p^3 the relative cross section ratios for the D production channels were measured to be

$D^o\overline{D^o}$: $D^o\overline{D}^{*o}$ + $\overline{D}^oD^{*o}$: $D^{*o}\overline{D}^{*o}$ = 0.2 ± 0.1 : 4.0 ± 0.8 : 128 ± 40

Spin counting predicts the ratios 1 : 4 : 7. Hence the 4.04 structure is coupled mainly to the $D^*\overline{D^*}$ channel. This has led to the speculation that the structures seen in σ^{tot} above 4.0 GeV and in particular the 4.04 structure are $c\bar{q}$ $\bar{c}q$ molecule states[51]. As a consequence a large rate (several percent) for J/ψ + X production at 4.04 GeV was expected. The PLUTO group[52] has searched for this process and measured an upper limit of ~10^{-3} σ^{tot} for J/ψX production.

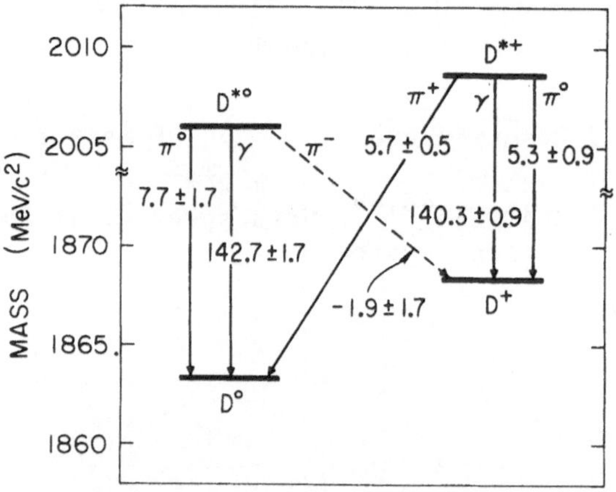

Fig. 3.8 Q values for the D^* → D transitions as measured by LBL-SLAC[16]

3.2 The F Meson

The charm model predicts mesons carrying both charm and strangeness. The ground state is the F^+. The GIM favored decay of the F is into an $s\bar{s}$ system leading to final states containing $K\bar{K}$, Φ, η or η'.

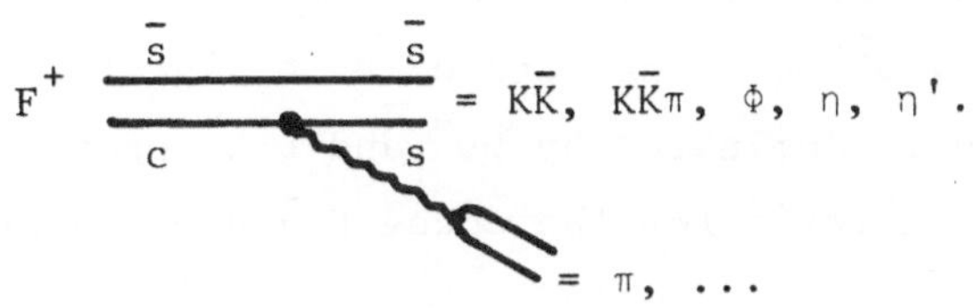

Sizeable η production is therefore a hint for F production. DASP[53] searched for F production by studying events of the type

$$e^+e^- \rightarrow \eta + \geq 2 \text{ charged tracks} + X$$

The η was identified by its decay into two photons. A search of this type is hampered by the $\gamma\gamma$ mass resolution which in the DASP experiment was 80 MeV, and by a large $\gamma_i\gamma_j$ combinatorial background: near 4 GeV on the average two to three π^o's are produced leading to 4-6 photons or 6-15 two-photon mass combinations. The event selection was done as follows. The events accepted were required to have at least two photons and two charged tracks coming from the interaction region. The photons were detected in the inner detector and their angles and energies were measured. Photons with energies between 0.14 and 1.0 GeV were considered candidates in the determination of the two photon mass spectrum $M(\gamma_i\gamma_j)$. The vector sum of the momentum of the two photons (γ_i,γ_j) was required to be between 0.3 and 1.4 GeV. The photon detection efficiency was 50 % at 0.05 GeV rising to 80 % at 0.1 GeV and 95 % above 0.3 GeV.

The data were grouped into five c.m. energy intervals, 3.99 to 4.10 = "4.03" GeV (1178 nb^{-1}), 4.10 to 4.23 = "4.17" GeV (509 nb^{-1}), 4.23 to 4.36 = "4.36" GeV (603 nb^{-1}), 4.36 to 4.49 = "4.42" GeV (2240 nb^{-1}), 4.5 to 4.99 = "4.60" GeV (727 nb^{-1}) and 5.0 GeV (1270 nb^{-1}) . The numbers in quotes are the average energies weighted by the luminosity for a given interval, and those in parentheses are integrated luminosities for each data set.

Figure 3.9 shows the two-photon effective mass distribution for the various center of mass energies; while a clear π^0 peak is observed at all energies, η signals are clearly seen only at 4.17 and 4.42 GeV, and possibly at 4.60 GeV. The lack of an η signal at the 4.03 GeV resonance, which is dominated by DD^* and $D^*\bar{D}^*$ production, indicates that the branching ratio for the decay $D \rightarrow \eta$ + anything is small.

In order to determine the η cross section acceptance and efficiency corrections have to be applied which in this case depend sensitively on the details of the production mechanism. The authors of Ref. 53 considered two different models for η production. Model 1 is the statistical isospin model for F decay by Quigg and Rosner[54]. In the calculation a 16 % semileptonic decay ratio, 38 % hadronic decays with an η and 9 % with an η' in the various final states were assumed. Model 2 is a phase space model which assumes that on the average 4.2 charged particles and 2.8 π^0's are produced together with an η. At 4.17 GeV model 1 gives an η acceptance of 4.2 % whereas model 2 yields 3.5 %. For the π^0 acceptance the discrepancy between the two models is larger: 0.33 % according to model 1 and 0.72 % according to model 2.

The reliability of the analysis was checked by computing from the observed number of π^0's the total π^0 cross section. At all energies the π^0 cross sections calculated with model 1 were consistent within 50 % with $(\sigma(\pi^0) = \sigma(\pi^+) + \sigma(\pi^-))/2$ as measured in the same experiment[50].

Fig. 3.10 shows the total inclusive η cross section $\sigma(\eta)$, as a function of the c.m. energy, calculated using model 1.
While no η signal is seen at 4.03 GeV, a significant production is seen between 4.10 and 4.70 GeV. Note that η production

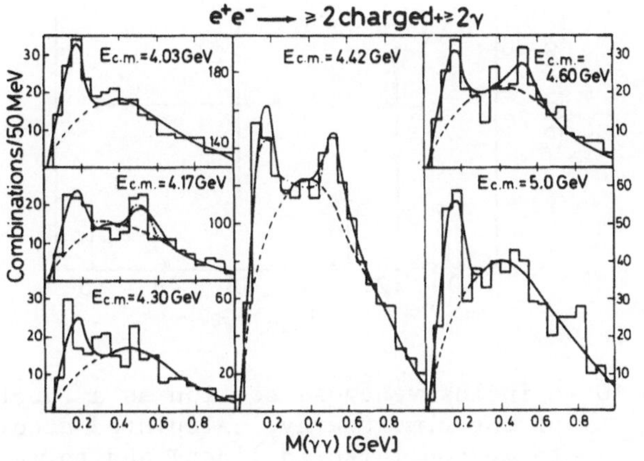

Fig. 3.9 M(γγ) mass distribution at the various c.m. energies. The solid lines are the results of a fit to a sum of background, π^0 and η contributions. The dashed lines correspond to the amount of background required by the fit under the η and π^0 peaks (method I). The dashed dotted lines at 4.17 and 4.42 GeV are the results of a fit corresponding to the sum of F production and the background, described by the M(γγ) mass distribution at $E_{c.m.}$ = 4.03 GeV (method II). (Ref. 53)

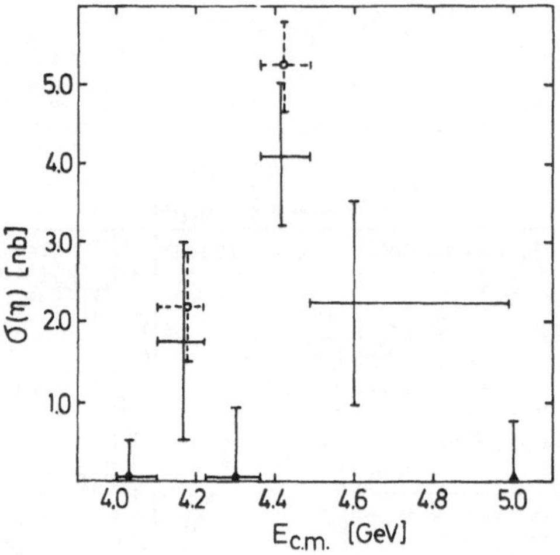

Fig. 3.10 η inclusive cross section as a function of the c.m. energy, calculated according to method I (solid lines) and to method II (dashed lines). (Ref. 53)

below 4.10 GeV is less than 0.5 nb. This is surprisingly small: it is smaller than pion production (i.e. π^+ or π^-) by a factor of 50 - 80 and smaller than kaon production (i.e. K^+ or K^-) by a factor of ten.

The following upper limit was deduced in Ref. 56. B.R. ($D \rightarrow \eta$ + anything) < 2 %. Using the D cross section of Ref.56 at 4.42 GeV, it would imply that $\sigma(\eta$; from D decay) $\leq$ 0.4 nb, which is more than a factor 10 smaller than $\sigma(\eta)$ at 4.42 GeV, implying that η's at 4.42 GeV are produced by a different source. Using $\sigma(\eta)$

at 4.42 GeV, and assuming that all of the 4.42 GeV resonance structure is due to F production, the following lower limit can be obtained: B.R. $(F \rightarrow \eta X) \geq 34$ %.

Fig. 3.11 shows the $M(\gamma\gamma)$ mass distribution at 4.42 GeV for events having an electron in the DASP inner detector. The contamination due to hadrons simulating an electromagnetic shower was estimated by looking at the process $e^+e^- \rightarrow J/\psi \rightarrow \rho\pi \rightarrow \pi^+\pi^-\pi^0$, and it was found to be 1.2 ± 0.5 % per charged track. The contamination due to photons converting in the beam pipe and Dalitz π^0 decay was estimated to be less than 2.5 % of the events. Folding those numbers with the events that contributed to the $M(\gamma\gamma)$ mass distribution in Fig. 3.9 gives rise to the background distribution given in Fig. 3.11 as a $\pm 1\sigma$ band. Clear η and π^0 peaks can be seen above this background, indicating that both are the decay products of weakly decaying states. While the π^0's have their source in τ, D and F production, the η signal cannot be due to the first two sources due to the lack of η signal at 4.03 GeV. Therefore the weakly decaying F meson becomes the most natural source of η production in this energy region.

In an earlier publication the same group found that η's are produced at 4.4 GeV in conjunction with a low energy photon (E_γ < 140 MeV), indicating that the dominant F production mechanism in this region occurs via the FF^* and/or F^*F^* channels. Fig. 3.12 shows the $M(\gamma\gamma)$ distribution for events having a photon of less than 140 MeV momentum at c.m. energies below, in the 4.42 GeV region, and above it. While the first and last distributions do not show any strong η signal, the 4.42 GeV data show a clear η peak. Moreover, this signal lies on a smaller background than the one in Fig. 3.9 for the same c.m. energy. This correlation of the η signal with events having a low energy photon is shown more clearly in Fig. 3.13, where the ratio of events having a low energy photon to those that do not have, is shown as a function of

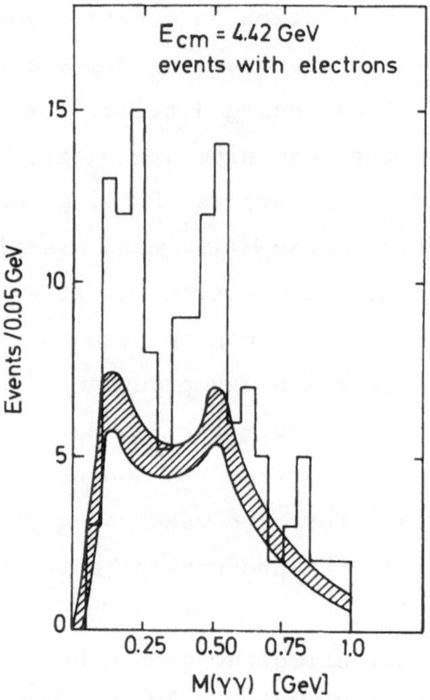

Fig. 3.11 M(γγ) mass distribution at $E_{c.m.}$ = 4.42 GeV, for events having an electron in the DASP inner detector. The shaded band corresponds to a ±1σ uncertainty in the expected background from hadrons simulating an electron and photons converting in the beam pipe (Ref. 53)

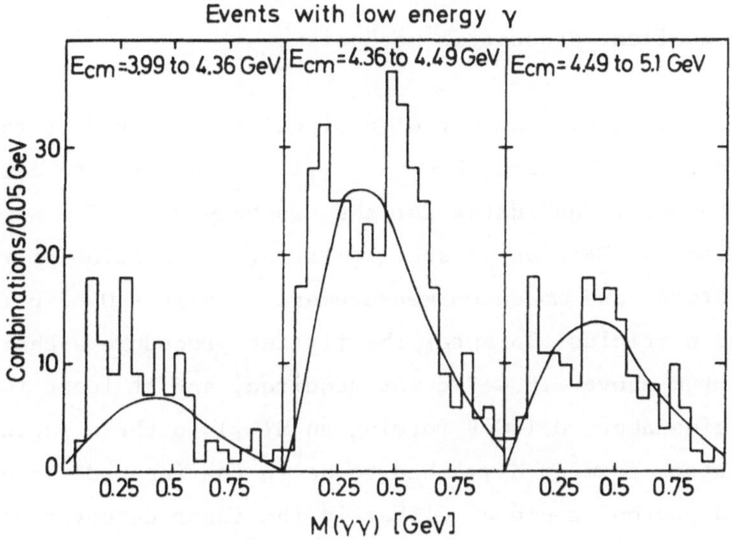

Fig. 3.12 M(γγ) mass distribution for events having a low energy photon (E_γ < 140 MeV), below, at and above the 4.42 GeV $E_{c.m.}$ region. The solid lines are estimates of uncorrelated photon background, normalized for M(γγ) > 0.7 GeV) (Ref. 53)

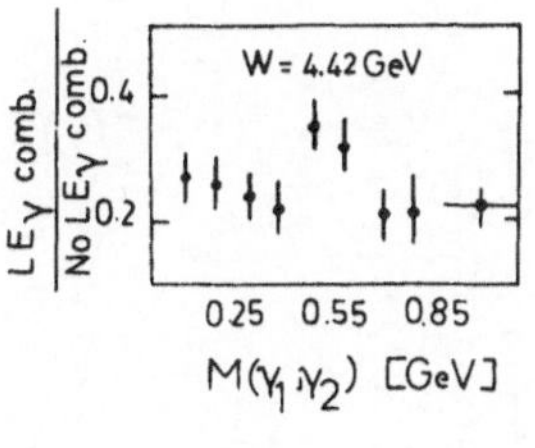

Fig. 3.13 Ratio of the number of combinations for events having a low energy photon to the number of combinations not having a low energy photon as a function of M(γγ), at $E_{c.m.}$ = 4.42 GeV (Ref. 53)

$M(\gamma\gamma)$. A 3.6 σ peak at the η mass is seen, indicating that η's at the 4.42 GeV region are produced in conjunction with a low energy photon. This confirms the previous observation that FF^* or F^*F^* seem to be the dominant production mechanism.

Since the largest amount of η signal is observed at the 4.42 GeV region, the decay $F^{\pm} \to \eta\pi^{\pm} \to \gamma\gamma\pi^{\pm}$ was studied to determine the F meson mass. Candidates for the two body decay $F \to \eta\pi$ were sought using the DASP outer spectrometers, which allow particle identification, and momentum measurement ($\Delta(p)/p = 0.02\ p$ (GeV/c)) of charged particles. To enter the fitting procedure a charged pion with momentum above 0.6 GeV/c was required, and at least two photons with energies above 0.1 GeV forming an $M(\gamma\gamma)$ in the η region. One of the photons forming $M(\gamma\gamma)$ had to be in the inner detector, while the second photon could be either in the inner detector or in the shower counter of the spectrometer arms. Finally there had to be at least one or more photons with an energy below 0.2 GeV (γ_{low}). A total of 43 events satisfied these selection criteria at the 4.42 GeV region, and 79 events at all other energies.
These events were fitted to the reactions:

$$e^+e^- \to (FF^* \text{ or } F^*F) \to F\ F\ \gamma_{low}, \quad F \to \eta\ \pi, \quad \eta \to \gamma\ \gamma \tag{3.1}$$

$$e^+e^- \to F^*F^*, \quad F^* \to \gamma_{low}\ F, \quad F \to \eta\ \pi, \quad \eta \to \gamma\ \gamma \tag{3.2}$$

These are 2C fits because of the mass constraint on $M(\gamma\gamma)$ and the requirement that $\pi\eta$ ($\pi\eta\ \gamma_{low}$) and the missing vector must have the same mass m_F (m_{F*}) in the case of reaction (3.1), ((3.2)).

There were 15 events at the 4.42 GeV region and 11 events at all other energies that gave a fit to reaction (3.1) with $\chi^2 < 8$. By making the additional requirement $|M_{fit}^{(\eta\pi)} - M_{meas}^{(\eta\pi)}| \leq 250$ MeV in order to cut on badly measured events, those numbers reduced furthermore to 12 and 10 events, respectively. Figures 3.14a and c show the fitted $\eta\pi$ mass vs. the fitted recoil mass for the two regions. At 4.42 GeV a strong clustering is seen at $M(\eta\pi) = 2.04$ GeV and $M_{recoil} = 2.15$ GeV (Fig. 3.14a) while no such clustering is seen in Fig. 3.14c for all other energies. All events in Fig. 3.14c lie on a band given by the kinematics of the fit. Figures 3.14b and 3.14d show the projections of Figs. 3.14a and 3.14c along the $M(\gamma\gamma)$ axis. A clear peak containing 6 events is seen at $M(\eta\pi)=2.04\pm0.01$ GeV for the 4.42 GeV data, while no such a peak is seen for all other energies; this implies that the background under the peak at 4.42 GeV is less than 0.2 event. The events at 4.42 GeV also give an acceptable fit to reaction (3.2) with a lower value for the F mass ($M(\eta\pi) = 2.00$ GeV). The spread in the $M(\eta\pi)$ distribution is slightly larger than in the case of hypothesis (3.1), as expected due to the ambiguity in the determination of the $F^* \to F\gamma_{low}$ relation. Allowing for possible systematic uncertainties, the best estimate is $m_F = 2.03 \pm 0.06$ GeV.

The mass difference between F^* and F was directly determined from the energy of the γ_{low} for the six events in the $\eta\pi$ mass peak. The result is $m_{F*} - m_F = 0.11 \pm .046$ GeV. The cross section for those six events is found to be 0.41 ± 0.18 nb, giving

$$\frac{BR(F \to \eta\pi)}{BR(F \to \eta + \text{anything})} = 0.09 \pm 0.05 \qquad (\pm 30\ \%\ \text{syst.})$$

consistent with the assumption made in the Monte Carlo model (0.062) to compute the η acceptance.

Since a strong η signal is seen at the 4.17 GeV region, the authors of Ref. 53 made a search for events that would fit the process

$$e^+e^- \rightarrow F^{\pm}F \quad \lfloor\rightarrow \eta\pi^{\pm} \quad \lfloor\rightarrow \gamma\gamma \tag{3.3}$$

The selection criteria were the same as those imposed for reactions (3.1) and (3.2) except that no requirement was made on the presence of a γ_{low}. After imposing a χ^2 cut at 8, five events survived; only one event was above M = 1.95 GeV. This event had a mass of 2.03 ± 0.02 GeV, which is consistent with the F mass value found at the 4.42 GeV region.

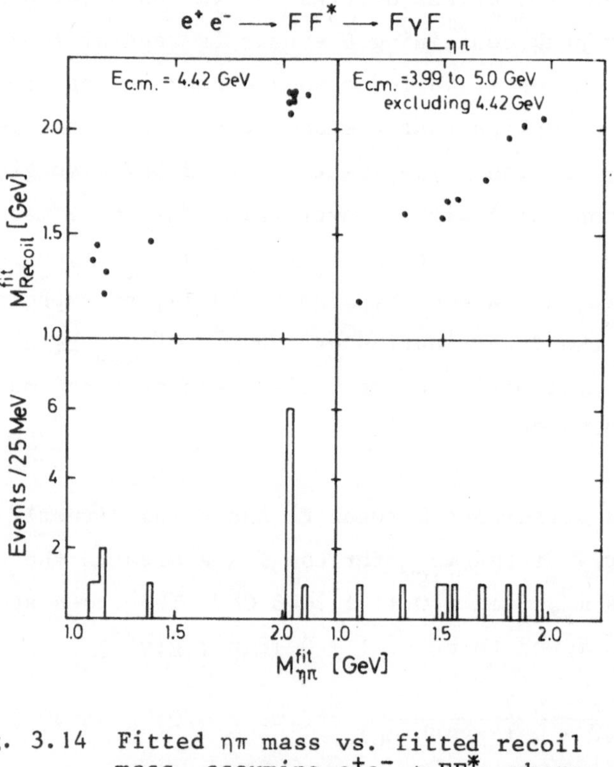

Fig. 3.14 Fitted $\eta\pi$ mass vs. fitted recoil mass, assuming $e^+e^- \rightarrow FF^*$, where $F^* \rightarrow F\gamma$ and $F \rightarrow \eta\pi$ at (a) E_{cm} = 4.42 GeV and (c) at all other energies excluding 4.42 GeV. Histograms (b) and (d) are the projections of Figs. (a) and (c) respectively, along the $M(\eta\pi)$ axis (Ref. 53)

Another group[57] reported a preliminary observation of a peak in the $K^+K^-\pi^\pm$ mass distribution at $E_{c.m.} = 4.16$ GeV. The mass value for this peak is 2039 ± 1.0 MeV, consistent with the mass value for the $\eta\pi$ decay mode given by the DASP group.

A discussion of the F and F^* properties expected by theory can be found e.g. in Refs. 54 and 58.

3.3 Semileptonic Decays of Charmed Particles

3.3.1 Origin of electron inclusive events

The pair production of new particles with large leptonic or semileptonic decay modes will lead to mixed final states containing leptons and hadrons. The observation of such final states at a level above the background expected from higher order electromagnetic interactions and semileptonic pion or kaon decays is direct evidence for the production of new particles. A new lepton or a hadron with a new flavour are but two examples of such particles. As will be discussed later there is now convincing evidence that besides charmed hadrons also a new lepton is pairproduced at c.m. energies above 4 GeV.

The anticipated features, of a new lepton and a new hadron are summarized in Table 3.3. The low multiplicity expected for the decay of a pair of heavy leptons into a final state with an electron (muon) plus hadrons arises as follows. In the decay $L \to \nu_L$ + hadrons, the hadrons come from a low mass current. If the multiplicity is compared to that from a virtual photon of the same mass, it will be small. Specific calculations support this conjecture. The leptonic decay of its partner L contributes in general only one charged track. Inelastic production i.e. $e^+e^- \to L\bar{L}$ + hadrons is negligible at these energies. The high multiplicity expected in the decay of a pair of heavy hadrons will presumably lead to a multiplicity comparable to that observed in the decay of an ordinary hadron of the same mass - i.e. on the average of 2 to 3 charged particles plus a few photons. From the semileptonic decay of its partner we might expect one or two charged particles and a few photons. Thus a cut on the total multiplicity can be used to separate the two classes with $e^+e^- \rightarrow L\bar{L} \rightarrow e(\mu) + X$ predominantly populating the two prong class.

Table 3.3 Properties of Heavy Sequential Leptons and New Hadrons

	L	H
Production	$e^+e^- \rightarrow L^+L^-$ (point cross section)	$e^+e^- \rightarrow H\bar{H}$ (damped by form factors)
	$e^+e^- \rightarrow L^+L^-$ + hadrons (negligibly small, less than α^2 of elastic production near threshold)	$e^+e^- \rightarrow H\bar{H}$ + hadrons or $H^*\bar{H}^*$ (dominant at higher energies, cross section will have structure)
Decay Modes	$L \rightarrow \ell\,\bar{\nu}_\ell\,\nu_L$ $\rightarrow \nu_L \cdot$ hadrons	$H \rightarrow \ell\,\bar{\nu}_\ell$ (suppressed if the lowest flavour state has spin 0) $\rightarrow \ell\,\bar{\nu}_\ell \cdot$ hadrons $\rightarrow$ hadrons
Final States:		
$e\mu$ + neutrinos	important, clear signature (e(μ) from three body decay)	negligible (e(μ) from a multibody decay)
$\ell\,\ell$ + neutrinos + hadrons	negligible, order α^2 at threshold	large (e(μ) from a multibody decay)
e(μ) + neutrino + hadrons	large, lepton spectrum computable and hard, hadrons have low multiplicity	large, lepton spectrum soft, hadrons have high multiplicity

The momentum spectrum of the observed electron (muon) can also be used to classify inclusive lepton events. All particles in the leptonic decay $\tau \to \ell\nu\bar{\nu}$ are pointlike and large values of Q^2, i.e. large electron energies are therefore not suppressed by form factors. Also, at least two of the particles have negligible mass. In a semileptonic decay of a charmed particle the formfactor will disfavour large values of Q^2. The mass of the hadron system will also lead to smaller electron momenta.

We will first discuss the data on semileptonic decay of charmed particles. Information on the τ particle is presented in section 4.

3.3.2 Results on the semileptonic decays of charmed hadrons

Semileptonic decays of charmed hadrons have been widely considered in the literature[43,44,59]. It follows directly from Fig.3.2 that the amplitude for the favoured mode is proportional to $\cos\Theta_c$ with $\Delta S = \Delta Q = \Delta C = +1$ and $\Delta I = 0$. The suppressed mode has an amplitude proportional to $-\sin\Theta_c$ with $\Delta C = \Delta Q$, $\Delta S = 0$ and $\Delta I = 1/2$. A D meson should predominantly decay into final states like $e\bar{\nu}_e K$ or $e\bar{\nu}_e K^*(892)$. Disfavoured modes like $D \to e\bar{\nu}_e\pi$ should be suppressed by $tg^2\Theta_c \simeq 0.05$. The decay into $e\bar{\nu}K^*(1420)$ is suppressed by phase space and soft pion theorems predict that the decay $D \to e\bar{\nu}K(n\pi)$ where the $K(n\pi)$ system does not form a resonance should be small.

The semileptonic branching ratio of charmed hadrons can be determined from a measurement of mixed electron hadron final staes

$$e^+e^- \to C\bar{C} \to (e\ \bar{\nu}\ \ldots)\ (X)$$

The average branching ratio BR($C \to eX$) is obtained from

$$B(C \to eX) = \frac{\text{number of eX-events}}{2\sigma(c\bar{c})}$$

The total cross section for charmed hadron production is defined as $\sigma(e^+e^- \rightarrow c\bar{c}) = \sigma(tot) - R(3.6)\sigma(\mu\bar{\mu})$ where $R(3.6)\sigma(\mu\bar{\mu})$ is the cross section for noncharmed hadron production determined below charm threshold at 3.6 GeV.

In principle mixed muon hadron events can also be used. However, the lepton spectrum must be measured down to low momenta with a rejection factor of 10^3 against hadrons. Such a rejection factor is difficult to obtain for low energy muons.

DASP has measured[60] the single electron spectrum for momenta above 200 MeV/c and c.m. energies between 3.6 GeV and 5.2 GeV for a total integrated luminosity of 6300 nb^{-1}. Electrons were defined as particles which gave a signal in the proper Čerenkov counter and had $\beta = 1$ ($p < 0.35$ GeV/c) or gave a large pulse height ($p > 0.35$ GeV/c) in the shower counter. A pion had a probability of 4×10^{-4} to pass these criteria. Electron pairs from Dalitz decay or pair conversion were rejected by pulse height cuts on the scintillation counters mounted before the magnet.

To reduce the background of electromagnetic origin the event was required to contain at least one nonshowering track. This could be a track in a spectrometer arm identified as a hadron or a muon. A charged track traversing the inner detector could also be called nonshowering if it fired less than 1.5 tubes per layer, averaged over all layers which had at least one tube set. Events containing only two charged particles were particularly sensitive to electromagnetic background. Here it was required that less than 1.25 tubes per plane be activated. Tests of these criteria using well defined pions showed that 95 % of the pions but fewer than 5 % of the electrons satisfied the tight criteria.

The background to this sample from beam gas events, Compton scattering on the material mounted in the front of the Čerenkov counter or from two photon processes was estimated and found to be small. The total background was (11.5 ± 3.5)% for the two prong sample and (15 ± 5)% for the multiprong sample. These values are in agreement with measurements done at the ψ' resonance and at 3.6 GeV.

After all cuts 60 two prongs and 182 multiprong events with an electron momentum above 0.2 GeV/c remained in this sample. The measured charged multiplicity distributions (including the electron) for all inclusive electron events and for those with only charged tracks are plotted in Fig. 3.15. The distribution peaks for $n_{ch} = 2$ but it is rather wide with events up to a charged multiplicity of eleven. As was shown by DASP both charm and τ-pairproduction are needed to explain the observed multiplicity distribution. The observed events are grouped into two prong (predominantly τ-events) and multiprong events (predominantly charm).

The lepton spectrum associated with the multiprong sample is shown in Fig. 3.16. The estimated background due to hadron misidentification or heavy lepton production is also plotted. The background was scaled from measurements below threshold. The heavy lepton contribution was estimated assuming a τ branching ratio of 30 % to decay into final states with three or more charged particles. DASP finds that less than 12 % of the events with $n_{ch} \geq 3$ can be explained as heavy lepton production. The simple cut on hadron multiplicity therefore yields a rather clean sample of charm decays.

The electron spectrum contains information on the semileptonic and the leptonic decay modes of the lowest mass charmed hadrons. Fig. 3.16 demonstrates that semileptonic decays are much more important than leptonic decays because the latter, being two body decays, would produce a peak in the electron spectrum around 1 GeV/c. This is in gross disagreement with the data which peak around an electron momentum of 0.5 GeV with only few events above

0.7 GeV/c. To study the observed momentum spectrum in more detail we consider the spectrum obtained for c.m. energies between 3.99 GeV and 4.08 GeV. The charm cross section in this energy region is dominated by $D\bar{D}^*$ and $D^*\bar{D}^*$ production and is below the threshold for F production. The spectrum, corrected for the background and the heavy lepton contribution, is shown in Fig. 3.17.

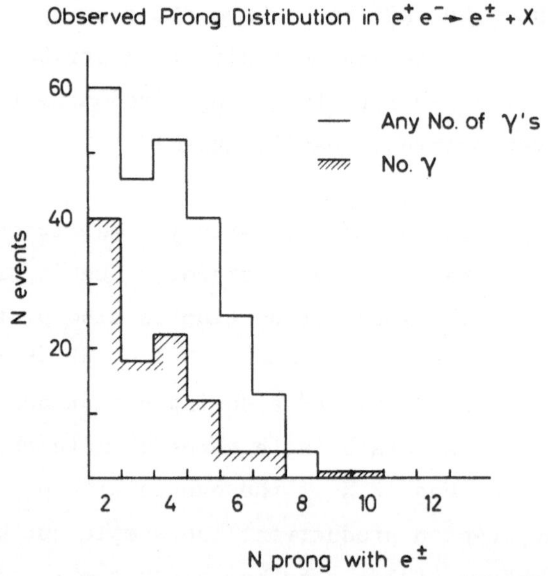

Fig. 3.15 The charged track distribution observed by DASP[60] for inclusive electron events. The electron is included in the prong number. The shaded distribution is for events without photons

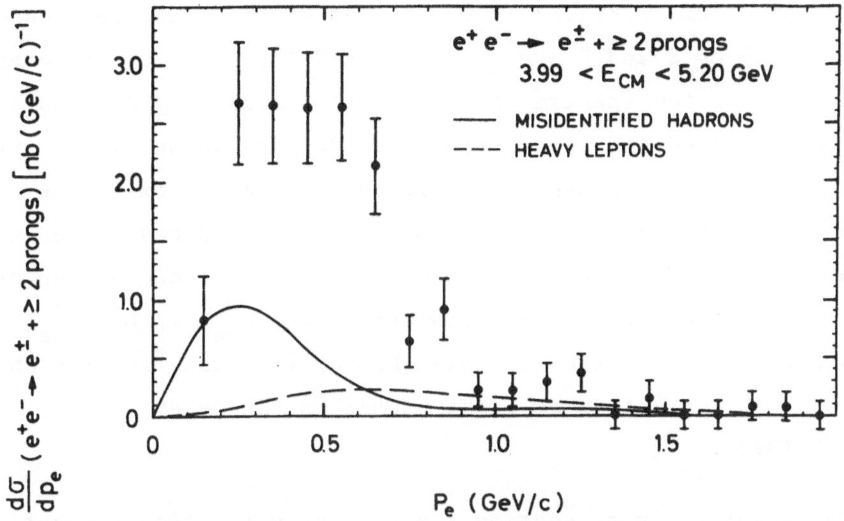

Fig. 3.16 The inclusive electron momentum spectrum measured by DASP[60] between 3.99 GeV and 5.2 GeV for multiprong events

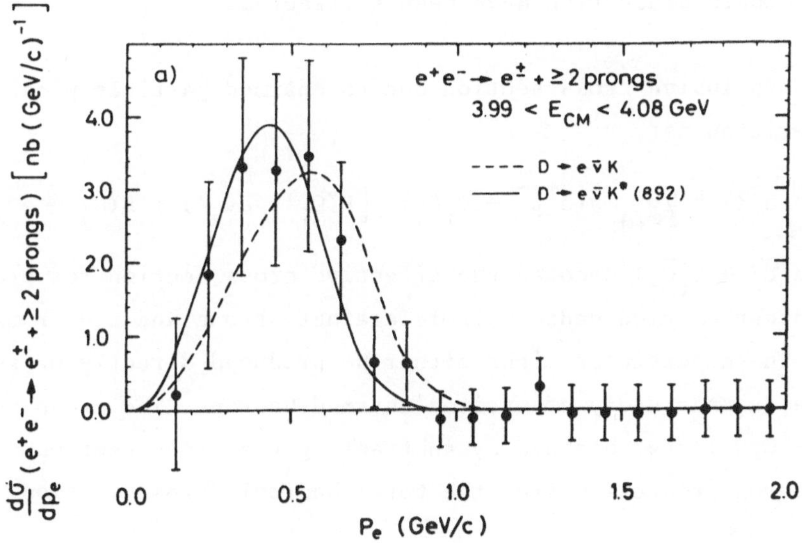

Fig. 3.17 The electron momentum spectrum for D → eνX as measured by DASP[60]. The momentum distributions expected for D → e$\bar{\nu}_e$K and D → e$\bar{\nu}_e$K* (891) are shown by the dashed and the solid curves

The spectrum in Fig. 3.17 was fitted to three possible channels: $D \to e\bar{\nu}_e\pi$, $D \to e\bar{\nu}_e K$ and $D \to e\bar{\nu}_e K^*(892)$. A V-A current was assumed and the form of the spectra was taken from a paper by Ali and Yang[59]. Note that the theoretical spectra are model dependent. These fits gave a χ^2 value for 10 degrees of freedom of: 29.6 for $D \to e\bar{\nu}_e\pi$, 6.3 for $D \to e\bar{\nu}_e K$ and 2.8 for $D \to e\bar{\nu}_e K^*(892)$. The decay $D \to e\bar{\nu}_e\pi$ can therefore be excluded as the sole semileptonic decay mode of the D. The data can be fitted with either $D \to e\bar{\nu}_e K$ or $D \to e\bar{\nu}_e K^*(892)$. A good fit to the spectrum can also be obtained assuming the charm changing weak current to be right handed in the decay $D \to e\bar{\nu}_e K^*(892)$.

The absolute cross section for inclusive electron production $e^+e^- \to e^\pm + X$, where X contains at least two charged tracks and any number of photons, is plotted in Fig. 3.18a as a function of energy. The data have been corrected for radiative effects. The background from hadron misidentification and the contribution from heavy leptonic production have been subtracted.

The inclusive cross section due to charmed particle production can be written as:

$$\sigma(e^+e^- \to e^\pm X) = \sum_{i,j} \sigma(e^+e^- \to C_i\bar{C}_j)\cdot\left\{B(C_i \to e\bar{\nu}_e X) + B(C_j \to e\bar{\nu}_e X)\right\}$$

Here $\sigma(e^+e^- \to C_i\bar{C}_j)$ denotes the effective cross section for producing the lightest charmed hadron stable against strong and electromagnetic decays. These particles might either be produced directly or result from the cascade decay of excited charmed hadrons. The cross section $\sigma(e^+e^- \to C_i\bar{C}_j)$ was obtained by subtracting the cross sections for "old" hadron production from the total hadronic cross section.

Near threshold, where only neutral and charged D production can contribute DASP finds:

$$B(D \to e + X) = 0.08 \pm 0.02$$

This should be compared to the value

$$B(C \rightarrow e + X) = 0.072 \pm 0.02$$

obtained by averaging over all energies between 3.9 GeV and 5.2 GeV (see Fig. 3.18b). These values were extracted using the DASP[20] data for the total cross section and the error quoted is mainly systematic. Evaluating the branching ratio using the SPEAR data[7] on the total cross section as input lead to an average semileptonic branching ratio of 0.08 ± 0.03, compared to 0.11 ± 0.03 obtained using the PLUTO[19] total cross section.
The semileptonic branching ratio can also be determined from the fraction of inclusive electron events containing a second electron. Using this method the DASP group finds $B(C \rightarrow e^- X) = 0.16 \pm 0.06$. Note that this value is independent of the charm cross section.

The DASP results are supported by the results from two recent experiments at SPEAR.

The DELCO group at SPEAR has measured[61] the cross section for $e^+e^- \rightarrow e^{\pm} + \geq 2$ hadrons $+ \geq 0$ photons. Electrons are identified in 65 % of 4π using a Čerenkov counter sensitive only to electrons. (Pions have the threshold at 3.7 GeV/c). The Čerenkov counters are backed by an array of lead scintillator (2 r.l. divided into 3 layers) shower counters which cover 60 % of 4π. The on-line trigger is rather loose requiring only a charged track plus a signal from two shower counters. Electron inclusive multiprong events were selected from this sample demanding that at least two shower counters should fire. The candidate track for an electron was required to give signals in the Čerenkov and the appropriate shower counter. To reduce the background from Dalitz pairs, photon conversions and δ-rays it was required that no other track should be within an angle of 18 mrad/p(GeV/c) (p is the momentum of the softer hadron) with respect to the electron candidate. Hadronic events at the J/ψ resonance passed these selection criteria with an

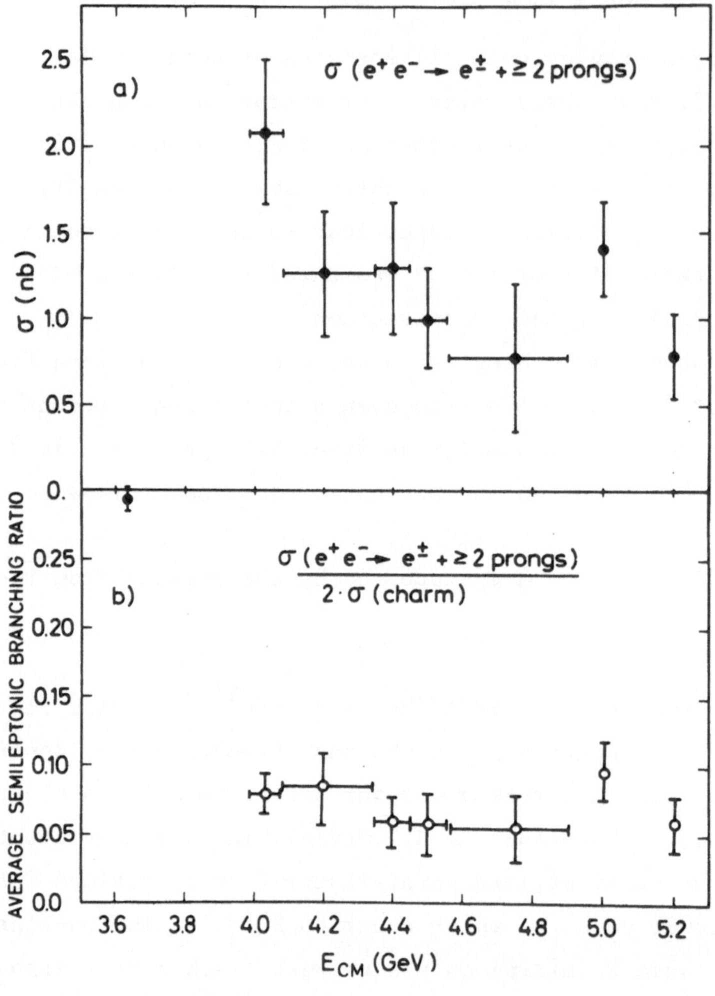

Fig. 3.18 a) The cross section measured by DASP[60] for the inclusive production of electrons plus nonshowering track plus additional charged tracks as a function of c.m. energy

b) The average semilaptonic branching ratio for charmed hadrons as a function of energy. The error bars are statistical only efficiency of $(3.5 \pm 0.3) \cdot 10^{-3}$.

$$R_e = \frac{\sigma(e^+e^- \to e^\pm + \geq 2 \text{ hadrons} + \geq 0 \text{ photons})}{\sigma(e^+e^- \to \mu^+\mu^-)}$$

is plotted in Fig. 3.19 as a function of c.m. energy. The dotted curve represents the estimated background. A clear peak is seen at 3.77 GeV produced by the decays of the ψ''(3772) into pairs of charged

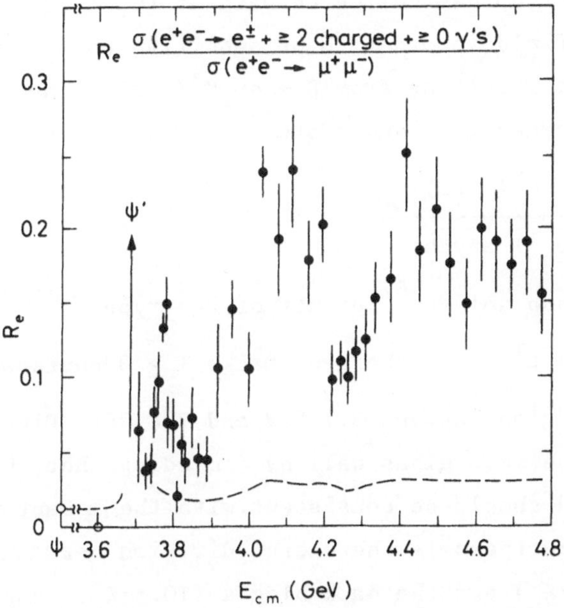

Fig. 3.19 R_e, the cross section for $e^+e^- \to e^\pm + \geq 2$ prongs $+ \geq 0$ photons normalized to the point cross section, plotted versus the c.m. energy. The dotted line indicates the background. The data are from DELCO[61]

and neutral D mesons. The ratio reflects the charm cross section with a rise at 4 GeV, a dip around 4.25 GeV and presumably some structure around 4.4 GeV.

Normalizing to the cross section for $e^+e^- \to \psi''(3772)$ yields $B(D \to eX) = (11 \pm 3)\%$.

The electron momentum spectrum at the $\psi''(3772)$ is plotted in Fig. 3.20. The dashed and the solid curves show the expected shapes for the decay modes $D \to e\bar{\nu}_e K$ and $D \to e\bar{\nu}_e K^*(892)$.

The DELCO group has collected data at the $\psi''(3772)$ with reduced magnetic field in order to be sensitive to electrons with very low momenta resulting from $D \to e\bar{\nu}_e K^*(1420)$. They find no evidence for this decay mode and quote

$$\frac{D \to e\bar{\nu}_e K^*(1420)}{D \to e\bar{\nu}_e + \text{anything}} < 0.1$$

The LBL-SLAC group selects[62] events of the type

$$e^+e^- \to e^\pm + \geq 2 \text{ charged tracks} + \geq 0 \text{ photons}$$

in the energy region between 3.7 GeV and 7.4 GeV. Electrons are identified in the lead glass wall by demanding that the total energy deposited should be consistent with the momentum measured in the magnet. Furthermore the energy division between the active converters ($3.5 \cdot X_o$) and the back blocks ($10.5 \cdot X_o$) should be consistent with that of a showering particle and the time-of-flight should agree within one nanosecond with that expected for an electron.

The probability that these criteria are satisfied by a hadron was measured at the J/ψ resonance. They find this probability to be 1.5 % for a particle of momentum 300 MeV/c decreasing to 0.4 % for a particle of 1.2 GeV/c. The probability averaged over the hadron spectrum is 1.1 % for momenta above 300 MeV/c.

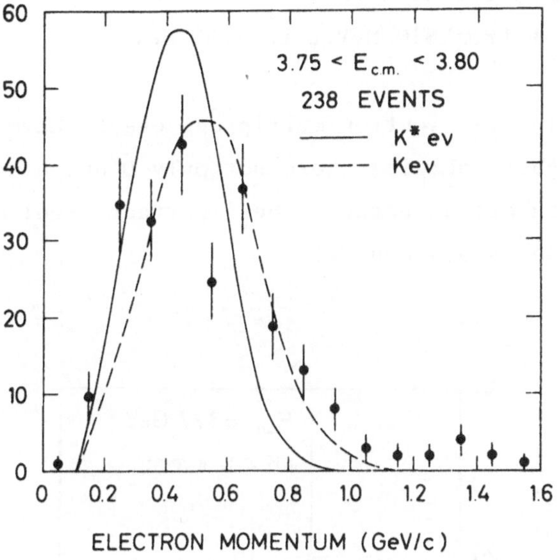

Fig. 3.20 Electron momentum spectrum measured by DELCO[61] for the multiprong events at the ψ''. The dashed and the solid curves show the expected shape for the decay modes $D \to e\bar{\nu}_e K$ and $D \to e\bar{\nu}_e K^*(892)$ (evaluated for V-A couplings).

As discussed above the resonance $\psi''(3772)$ is a clear and well defined source of D mesons. The LBL-SLAC group found a total of 62 candidates for electron inclusive multiprong events at the $\psi''(3772)$. The background, not including τ pairproduction, was estimated to contribute 25 ± 5 events. Assuming that the excess signal results from semileptonic D decays they find:

$$B(D \to eX) = (7.2 \pm 2.8)\%$$

The momentum spectum is plotted in Fig. 3.21 and compared to the spectra predicted for $D \to e\bar{\nu}_e\pi$, $e\bar{\nu}_e K$ and $e\bar{\nu}_e K^*(892)$. The spectrum is not corrected for the contribution from τ pairproduction. The data are consistent with $D \to e\bar{\nu}_e K$ (confidence level 33 %) or $D \to e\bar{\nu}_e K^*$ (C.L. 13 %) but less consistent with $D \to e\bar{\nu}_e \pi$ (C.L. 13%). The data are inconsistent with $D \to e\pi$ which would produce a flat electron spectrum from 810 MeV/c to 1080 MeV/c.

Data on inclusive electron multiprong events have also been collected at higher energies where not only D but also F and charmed baryon production are expected to be important. Preliminary results are listed in Tables 3.4 and 3.5.

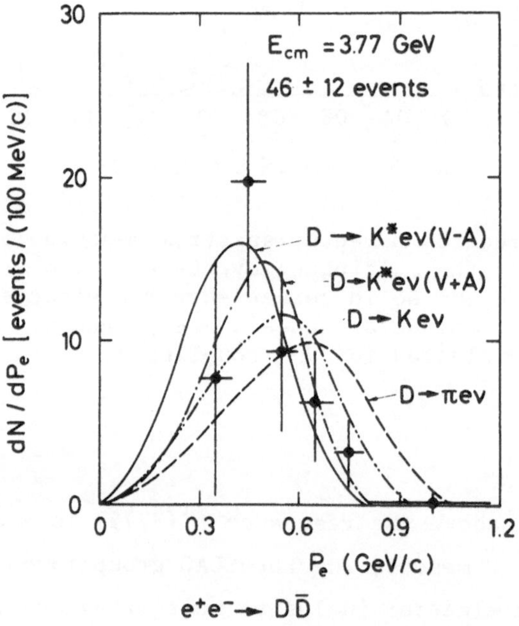

Fig. 3.21 Electron momentum spectrum measured by LBL-SLAC[62] for electron inclusive multiprong events. Predictions for $D \to e\bar{\nu}_e\pi$ $e\bar{\nu}_e K$ and $e\bar{\nu}_e K^*(892)$ are shown.

Table 3.4 Preliminary results on inclusive electron multihadron events (Ref. 61, 63)

E_{CM} (GeV)	R_e	R_{ch}	B(C → eX) %
4.1 - 4.2	0.26 ± 0.08	2.1 ± 0.5	7.7 ± 3.0
4.4 - 5.7	0.23 ± 0.06	1.9 ± 0.5	7.4 ± 2.8
6.4 - 7.4	0.26 ± 0.06	1.9 ± 0.4	8.7 ± 3.2

The data on the semileptonic charm decays are summarized in Table 3.5.

Table 3.5 Semileptonic branching ratio of D mesons

Experiment	B(D → eX)
DASP	8 ± 2 %
DELCO	11 ± 3 %
LBL-SLAC	7.2 ± 2.8 %

The quoted values are valid for a mixture of charged and neutral D mesons. In principle they can be measured separately at the $\psi''(3772)$ resonance by selecting events with one charged or one neutral D meson.
From dimuon production by neutrinos a preliminary value of B(C → μX) ≈ 11 % was obtained[64].

The semileptonic branching ratio is larger than the value of 4 % predicted[43] from the weak decays of strange particles. This indicates that the mechanism responsible for enhancing the non-leptonic channels in strange particle decays are less effective[65] for charmed particle decays. In fact if none of the available channels are selectively enhanced one expects a semileptonic branching ratio of 0.20. This number is obtained by simple counting: the W decay can proceed in five different ways, W → eν, μν and $q\bar{q}'$ times three because of three different colours: Assuming the same coupling strength, each channel has the probability 1 : 5 = 20 %.

One event was found by DASP with 3 electrons plus hadrons. This number is consistent with the expected background leading to an upper limit of

$$\sigma(e^+e^- \to 3e + X) < 0.1 \text{ nb},$$

with 90 % confidence. Events of that type could arise from a charm changing neutral current, which allows a charmed hadron to decay into two electrons plus hadrons[66]. A neutral lepton paired with the electron in a right handed doublet would also yield events with three electrons and hadrons[65].

DASP[60] has determined the number of charged kaons emitted in electron multihadron events. This provides an independent consistency check on the nature of the weak current responsible for charm decay. If it is the GIM current then almost every electron event will have a $K\bar{K}$ pair. The measurement was done with events that had an identified charged hadron (π, K or $\bar{p}$) in the magnetic spectrometer, an electron in the inner detector and possible other charged particles or photons. No $e\bar{p}X$ events were seen. From the observed K to π ratio and the measured charged multiplicity it was found that each multiprong event contained on the average 0.90 ± 0.18 charged kaons per event in agreement with the GIM prediction.

4. THE HEAVY LEPTON τ

In 1975 Perl et al. reported evidence for events of the type

$$e^+e^- \to e^\pm\mu^\mp + \text{nothing} \qquad (4.1)$$

where "nothing" meant that no other particles were registered in the detector[67]. The analysis was hampered by background from purely hadronic processes: hadrons had an 18 % (20 %) probability to fake an electron (a muon). Of the 24 eμ events found 6-8 events were estimated to come from hadronic background. There was also the question whether or not other strongly or electromagnetically interacting particles had been produced together with e and μ. Since the detector covered only two thirds of 4π additionally produced particles had a fair chance to escape detection.

Besides the experimental uncertainties there was the question of interpretation. Electron-muon events can arise, e.g. from the pair production of charmed particles or of a new lepton. The two mechanisms can be distinguished by their different production and decay patterns. Measurements on

1) $e^+e^- \to e\mu$ + nothing
2) $e^+e^- \to e(\mu)$ + minimum ionizing track + any number of photons

provided convincing evidence that above 4 GeV besides charmed particles a new type of weakly decaying particle, τ, is being produced[68]. The proof that this particle indeed exists was given by the DASP collaboration which observed τ-pairproduction below charm threshold at the ψ' [69]. We first discuss the properties expected for a charged heavy lepton[70].

4.1 Expected Properties of a Heavy Lepton

Assuming the lepton τ to be pointlike the e^+e^- production cross section is given by

$$\sigma_{\tau\bar{\tau}} = \frac{4\pi}{3}\frac{\alpha^2}{s}\beta_\tau\left(1 + \frac{1-\beta_\tau^2}{2}\right) \qquad (4.2)$$

where β_τ = P/E is the τ velocity.
If the lepton decays weakly and the decay is mediated by the standard V-A weak current the partial decay widths can be calculated or estimated (see Fig. 4.1).

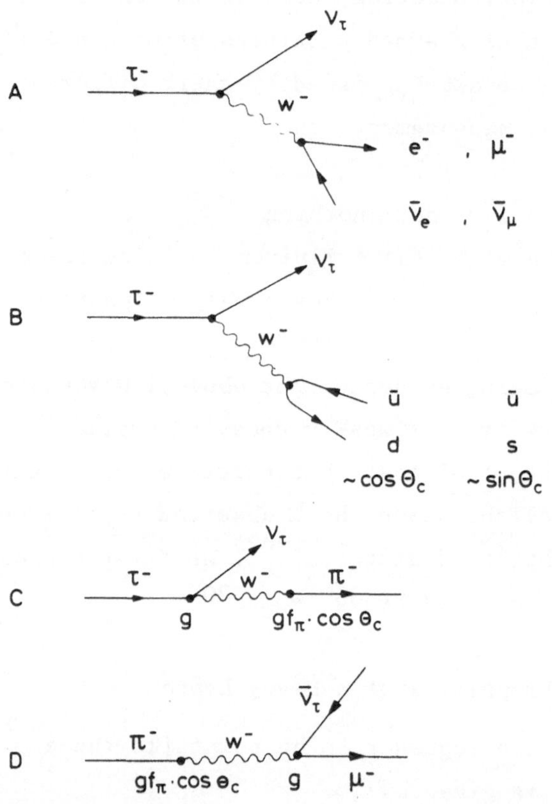

Fig. 4.1 a,b) The graphs for leptonic and semihadronic decays of a heavy sequential lepton

c,d) The relationship between $\pi^- \to \bar{\nu}_\mu \mu$ and $\tau^- \to \nu_\tau \pi^-$

The leptonic decay modes can be computed unambiguously:

$$\Gamma_e \equiv \Gamma(\tau \to \nu_\tau e\bar{\nu}_e) = \frac{G^2 \cdot m_\tau^5}{192\pi^3}$$

$$\Gamma_\mu \equiv \Gamma(\tau \to \nu_\tau \mu\bar{\nu}_\mu) = \Gamma_e \cdot F(y)$$

$$G = \frac{1.02}{m_p^2} \times 10^{-5}(\mathrm{GeV}^{-2}) \qquad \text{and}$$

$$F(y) = (1-8y + 8y^3 - y^4 - 12y^2 \cdot \ln y). \tag{4.3}$$

$F(y = m_\mu^2/m_\tau^2)$ is a small phase space correction.

A lepton decays semihadronically as shown in Fig. 4.1b. Conventional theory predicts that the τ decays into final states of low multiplicity and a small ratio of kaons to pions. This is indeed reproduced in actual calculations[70,71] and an estimate of various branching ratios is listed in Table 4.1.

Some of the decay widths are rather well known:

$\tau \to \nu_\tau\pi$: The pion mode is directly related to the $\pi \to \mu\nu$ decay as shown in Figs. 4.1c,d. Note that the mass squared of the weak current is the same in both cases ($Q^2 = m_\pi^2$). As a consequence the value of the formfactor is the same for both decays. The partial width is given by

$$\frac{\Gamma(\tau^- \to \nu_\tau\pi^-)}{\Gamma_e} = \frac{12\pi^2 f_\pi^2 \cos^2\Theta_c}{m_\tau^2} = \frac{2.1\, m_p^2}{m_\tau^2} \tag{4.4}$$

with $f_\pi = 0.137\, m_p$ (m_p = proton mass)

$\tau \to \nu_\tau K$: This decay is directly related to $K \to \bar{\nu}_\mu\mu$

$\tau \to \nu_\tau\rho$: This decay is related via CVC to $e^+e^- \to \rho$. The width for $\tau \to \nu_\tau\rho$ can be evaluated with an accuracy of 20 % using the measured values of the ρ-coupling constant.

$\tau \to \nu_\tau \cdot (n\pi)$: These decay modes are related via CVC to $e^+e^- \to (n\pi)$.
(n even)

Table 4.1 Branching ratios for a sequential lepton of mass 1.8 GeV. The values are from reference 70.

Decay mode	Branching ratio	$\Gamma(\tau \to \nu_\tau X)/\Gamma(\tau \to \nu_\tau e\ \nu_e)$
$\nu_\tau e \bar{\nu}_e$	0.18	1
$\nu_\tau \mu \nu_\mu$	0.18	0.97
$\nu_\tau \pi$	0.10	0.60
$\nu_\tau \rho$	0.22	1.24
$\nu_\tau A_1$	∿ 0.1	0.41
$\nu_\tau 4\pi$	} 0.2	0.44 ± 0.10
$\nu_\tau 5\pi$		∿ 0.44
$\nu_\tau 6\pi$		∿ 0.11
$\nu_\tau 7\pi$		∿ 0.11
$\nu_\tau K$	< 0.01	0.03
$\nu_\tau K^*(892)$	0.01	0.05
$\nu_\tau Q$	< 0.01	0.02
$\nu_\tau (K \cdot n\pi)$ n > 3	0.01	0.07

Pairproduction of new leptons will lead to mixed electron muon events via:

$$e^+e^- \to \tau\bar{\tau} \to (\nu_\tau e \bar{\nu}_e)(\nu_\tau \mu \bar{\nu}_\mu).$$

The number of eμ-events can be written as

$$N_{e\mu} = 2\sigma_{\tau\bar{\tau}} \cdot L \cdot A_e \cdot A_\mu \cdot B_e \cdot B_\mu$$

The $\tau\bar{\tau}$ cross section (Eq. 4.2) varies rapidly near threshold and a measurement in this energy region can be used to determine the τ mass. Well above threshold the cross section is not very sensitive to the precise value of the τ mass and is given by the point cross section. L is the luminosity, A_e and A_μ the acceptances for electrons and muons. The acceptances depend on the shape of the decay lepton spectrum , i.e. on the form of the weak coupling. The difference in acceptance introduced by a V-A or a V+A coupling is small as long as the leptons are measured down to low momenta.

A measurement of the eμ-yield at high energies can therefore be used to determine $B_e \cdot B_\mu$ without knowledge of the exact mass of the lepton and the form of the weak couplings. If eμ universality is valid as expected both for sequential and ortholeptons[72] (lepton number of the τ^- = lepton number of the e^- or μ^-) then $B_e \equiv B_\mu$ and the B_e value can be directly determined. However if the τ is a paraelectron or a paramuon (lepton number of τ^+ = lepton number of e^- or μ^-) then $B_\mu/B_e = 1/2$ or 2.

The lepton assignment can in principle be determined from a measurement of the final states:

$$e^+e^- \rightarrow \tau\bar{\tau} \rightarrow e^+e^- + \text{missing energy}$$

$$\rightarrow \tau\bar{\tau} \rightarrow \mu^+\mu^- + \text{missing energy.}$$

However, unlike the eμ channel which is rather clean, e^+e^- and $\mu^+\mu^-$ final state events are contaminated with electromagnetic events.

Information on semihadronic decays of the heavy lepton can be obtained from a measurement of inclusive electron (muon) events resulting from:

$$e^+e^- \rightarrow \tau\bar{\tau} \rightarrow (\nu_\tau\ \ell\ \bar{\nu}_e)\ (\bar{\nu}_\tau \cdot \text{hadrons}).$$

The rate for the inclusive events is given by

$$N_{e,h} = 2\sigma_{\tau\bar{\tau}} \cdot L \cdot A_\ell \cdot A_h \cdot B_\ell \cdot B_h$$

with the nomenclature defined above.

As discussed above, τ production can be strongly enhanced relative to charm production either by selecting events with low multiplicity or by demanding the decay lepton to have momenta above 1.0 GeV/c. It was shown in section 3.3 that only a small fraction of the semileptonic charm decays satisfy these conditions. Most experiments reported have therefore measured the lepton two prong cross section:

$$N_{\ell,1p} = 2\sigma_{\tau\bar{\tau}} \cdot L \cdot A_{\ell} \cdot Q_{1p} \cdot B_{\ell} \cdot B_{1p}$$

with $B_{1p} = B(\tau \rightarrow \nu_{\tau} + 1 \text{ charged} + \geq 0 \text{ photons})$.
Electrons are sometimes excluded from the one prong modes.

Table 4.2 Summary of e^+e^- experiments that have searched for heavy leptons

Experiment	cm energy (GeV)	type of final state	events observed	events due to background	comments
Bernadini et al.[73]	1.2 - 3	eμ			no τ with M < 1 GeV
Orito et al.[74]	2.6 - 3	eμ			no τ with M < 1.15 GeV
SLAC-LBL[67,75]	3.8 - 7.8	eμ μX	190 230	46 93	
Maryland-Princeton-Pavia[76,77]	7	μX	13	4	
PLUTO[78,79]	3.6 - 5	eμ μX	23 273	1.9 62	
LBL-SLAC[80]	3.7 - 7.4	eμ eX	32 60	0.4 14	
DASP[69,81]	4.0 - 5.2	eμ eX μX	13 80 25	1.2 14 4	
	3.684	eX	9	1.7	
DELCO[61]	3.77-7.84	eX	230	small	
Iron ball[76]	7	μμX	25	14	

4.2 Experimental Results on the τ

Table 4.2 summarizes the experiments that searched for heavy leptons in e^+e^- annihilation. Two early experiments done at Frascati at c.m. energies up to 3 GeV showed that besides e and μ no leptons exist with a mass below 1.15 GeV [76,77].

4.2.1 eμ events

We first discuss the experiments studying events of the type

$$e^+e^- \rightarrow e^{\mp}\mu^{\pm} + \text{nothing}$$

above c.m. energies of 3.5 GeV. In the experiment of SLAC-LBL the following criteria were used to select the events:

1) Only two charged tracks with opposite charges and no photons.
2) Each track should have a momentum greater than 0.65 GeV/c.
3) One prong is identified as an electron by the pulseheight in the shower counter, the other as a muon by range. The probability of misidentifying a hadron as an electron or as a muon is respectively 18 % and 20 %.
4) The coplanarity angle between the planes defined by the electron and the beam direction and the muon and the beam direction must be greater than 20°.

A total of 190 events[75] with a background of 46 events was observed in the c.m. range from 3.8 GeV to 7.8 GeV. If these events result from charm then the leptons would be accompanied by undetected hadrons. The SLAC-LBL group made a careful evaluation of this possibility and they conclude with 90 % confidence that no more than 39 % of the eμ events can contain additional charged particles or photons. They conclude that the eμ events result from pairproduction of a new lepton:

$$e^+e^- \rightarrow \tau\bar{\tau} \rightarrow (\bar{\nu}_\tau e\, \bar{\nu})(\bar{\nu}_\tau\, \bar{\mu}\, \nu_\mu) \rightarrow e\mu + \text{missing energy.}$$

The observed cross section is plotted in Fig. 4.2 versus the c.m.

energy. The cross section varies smoothly with energy and agrees well with the energy dependence of pairproduction of a heavy lepton plotted as the solid lines in Fig. 4.2. Both curves evaluated for m_τ = 1.8 GeV and 2.0 GeV fit the data well. The momentum distribution is shown in Fig. 4.3. Plotted are the number of electrons and muons versus the normalized momentum variable.

$$r = \frac{P - P_o}{P_{max} - P_o}; \quad P_o = 0.65 \text{ GeV/c}$$

P_{max} is defined as the maximum momentum of the electron (muon) which is consistent with the leptonic decay of a heavy lepton of mass 1.9 GeV with a massless neutrino. The solid curves are distributions predicted for various values of the neutrino mass assuming a V-A leptonic coupling. The dotted curve is for a V+A leptonic coupling and a massless neutrino. For m_τ = 1.9 GeV and a massless τ neutrino they find that the χ^2 probability is about 0.1 % for V+A compared to 50 % for V-A.

The 95 % confidence upper limit on the neutrino mass is $m_{\nu_\tau} < 0.6$ GeV for m_τ = 1.90 GeV and V-A coupling. The mass of the τ was determined to (1.90 ± 0.10) GeV. This is an average over various methods and assumes V-A and a massless neutrino. The leptonic branching ratios were determined from the observed eμ cross section (Fig. 4.2) to

$$B_e \equiv B_\mu \equiv 0.186 \pm 0.01 \pm 0.028.$$

The first error is statistical and the second systematic. This value assumes $M_{\nu_\tau} = 0$, V-A and eμ universality.

The PLUTO Collaboration reported[78] data on $e^+e^- \rightarrow e\mu$ + missing energy with a very small contamination of other events. Electrons above 0.3 GeV/c are identified by showers in two lead converters (0.44 and 1.7 radiation lengths thick) in 55 % of the full solid angle. Muons are required to penetrate the iron yoke and produce

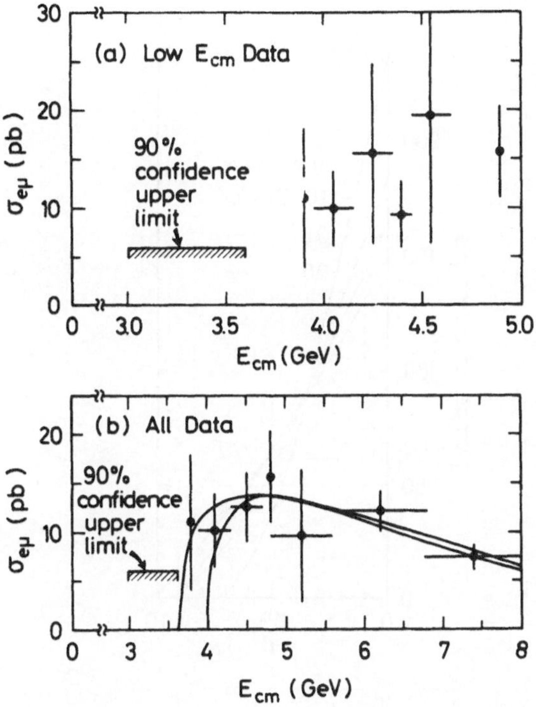

Fig. 4.2 The observed eμ production cross section plotted versus c.m. enery. The data are from SLAC-LBL[75]; only statistical errors are indicated and the horizontal lines show the energy range covered by each point. The 90 % confidence upper limit on the background based on data below 3.6 GeV are shown. The solid lines are predictions for a lepton of mass 1.8 GeV (2.0 GeV) assuming a V-A current

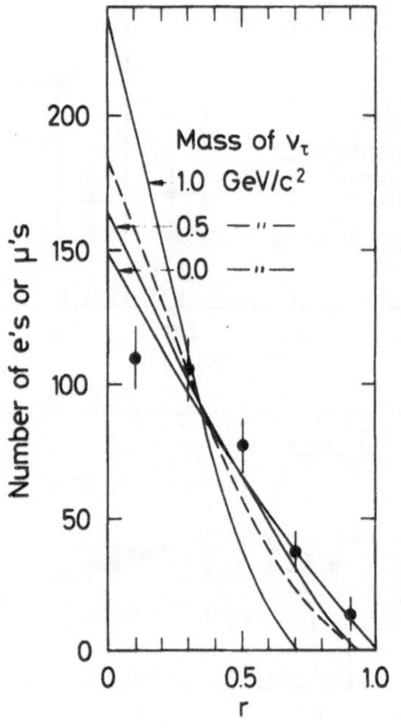

Fig. 4.3 Number of electrons and muons observed in the SLAC-LBL experiment[75] plotted versus

$$r = \frac{p - 0.65\ \mathrm{GeV/c}}{p_{max} - 0.65\ \mathrm{GeV/c}}$$

Solid curves are for $m_\tau = 1.9$ GeV and V-A coupling with the mass of the neutrino as a parameter. The dashed curve is for V + A coupling with $m_\tau = 1.91$ GeV and a massless τ-neutrino a spark in one of the proportional tube chambers. The μ acceptance is 43 % of 4π. The probability for misidentifying a hadron as either electron or muon is small: P_{he} = 3.5 ± 0.7 % and $P_{h\mu}$ = 2.8 ± 0.7 %.

Above 4 GeV 23 $e^{\pm}\pi^{\mp}$ + nothing events were observed. No events with like sign leptons were found. The background due to misidentified hadrons is determined to 2 ± 0.5 events. No such events were found at 3.6 GeV. The number of eμ events accompanied by photons or additional charged particles is small and consistent with coming from background. The scarcity of such events precludes the possibility that the eμ + nothing events come from charmed particle decays: otherwise there should have been three times as many events of the type eμ + γ or eμ + ≥ 1 charged particle as there are eμ + nothing events. The observed number of events is at least an order of magnitude smaller. The total fraction of eμ events that could have originated from charm production is less than 9 %. The energy behaviour of the cross section for eμ events is given in Fig.4.4. The shape and magnitude of the measured cross section is in agreement with the prediction for heavy lepton production (solid curve).

The leptonic branching ratios and the τ mass is determined from the eμ events and the data[79] on the inclusive (μ + 1 prong) cross section discussed below. For V-A couplings they find $B(\tau \to \nu_\tau \mu\bar{\nu}_\mu)$ = 0.14 ± 0.034, $B(\tau \to \nu_\tau e\bar{\nu}_e)$ = 0.16 ± 0.06 and m_τ = 1.93 ± 0.05 GeV. A V+A coupling yields $B(\tau \to \nu_\tau \mu\bar{\nu}_\mu)$ = 0.17 → 0.05, $B(\tau \to \nu_\tau e\bar{\nu}_e)$ = 0.13 ± 0.05 and m_τ = 1.82 ± 0.08 GeV. The V-A fit is slightly favoured over the V+A fit.

The DASP group has measured[69] $e^+e^- \to e\mu$ + missing energy in the c.m. range of 4.0 GeV to 5.2 GeV. Muons of momenta greater than 0.7 GeV/c were identified in the outer detector ($P_{h\mu}$ = 4.2 ± 0.8 %) by range, electrons with momenta above 0.2 GeV/c either in the inner detector (P_{he} = 2 ± 0.9 %) or the outer ($P_{he} = 4 \times 10^{-4}$). A total of 13 eμ events with an estimated background

of 1.2 ± 0.4 events were found. Using the known production cross section and assuming eμ universality yield:
$B_e = B_\mu = 0.182 \pm 0.028 \pm 0.014$ for a V-A current and
$B_e = B_\mu = 0.206 \pm 0.033 \pm 0.015$ for a V+A current. The first error is the statistical, the second the systematic one.

The LBL-SLAC group has recently reported[80] new results on

$$e^+e^- \rightarrow e^\pm + \text{a nonshowering particle} + \geq 0 \text{ photons.}$$

Data are collected for c.m. energies between 4.1 GeV and 7.4 GeV. The electron ($p_e \geq 0.4$ GeV/c) is identified in the lead glass wall. The nonshowering particle must have a momentum larger than 0.65 GeV/c and it is classified as muon or hadron by range. The acoplanarity angle with respect to the beam axis should be at least 20° and the mass squared recoiling against the pair must be greater than (0.8 GeV^2) at lowest c.m. energy and 1.5 GeV^2 at the highest. They find a total of 21 eμ events with an estimated background of 0.4 events. Assuming V-A couplings $m_\tau = 1.9$ GeV, $m_{\nu_\tau} = 0$ and eμ universality they find $B_e = B_\mu = 0.224 \pm 0.055$.

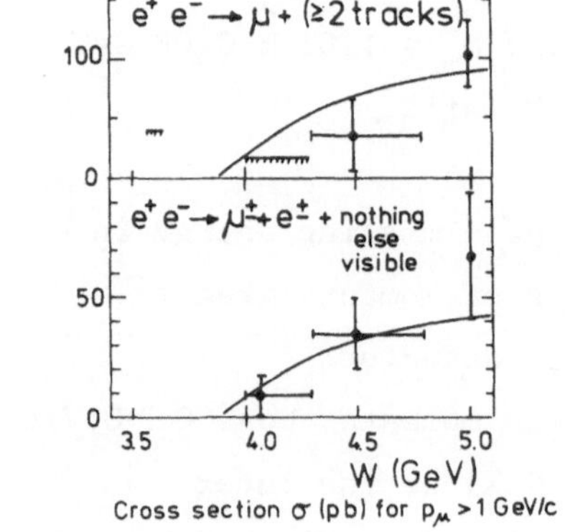

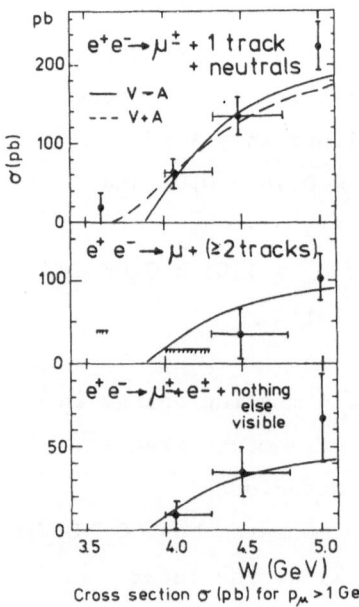

Fig. 4.4 cm energy dependence of muon cross sections measured by PLUTO[78].

a) Inclusive twoprongs (muon + charged track ($T^\pm$) + neutrals)
b) Inclusive multiprongs, and
c) exclusive μe events.

Cross sections are given for muon momenta > 1 GeV/c and have been corrected for trigger and detector acceptance, and for hadron punch-through. a) has also been corrected for QED, c) for electron detection efficiency. The solid curves are calculated for V-A decay with $M(\tau) = 1.91$ GeV, $M(\nu_\tau) = 0$.

4.2.2 ee and μμ events

τ Pairproduction will also lead to events of the type:

$$e^+e^- \to \tau\bar{\tau} \to e^+e^- + \text{missing energy}$$

$$e^+e^- \to \tau\bar{\tau} \to \mu^+\mu^- + \text{missing energy.}$$

The SLAC-LBL group[75] has selected events of this type demanding $P_e > 0.65$ GeV/c, $P_\mu > 0.65$ GeV/c. The acoplanarity angle with respect to the beam axis should be greater than 20°. After large corrections for QED and misidentified hadron events they report:

$$\frac{\text{Number of ee events}}{\text{Number of e}\mu\text{ events}} = 0.56 \pm 0.14 \begin{matrix} +0.16 \\ -0.19 \end{matrix}$$

$$\frac{\text{Number of }\mu\mu\text{ events}}{\text{Number of e}\mu\text{ events}} = 0.70 \pm 0.15 \pm 0.19$$

A sequential lepton or an ortholepton yield 0.5 for both ratios. The authors[75] quote that an electron related paralepton will yield $N_{ee}/N_{e\mu} = 0.86$, $N_{\mu\mu}/N_{e\mu} = 0.29$ for the kinematical cuts used. A muon related paralepton yield the $N_{ee}/N_{e\mu} = 0.29$ and $N_{\mu\mu}/N_{e\mu} = 0.86$. The paralepton assignment is disfavoured by the data.

The Colorado-Pennsylvania-Wisconsin group[76] searched for

$$e^+e^- \to \mu^+\mu^- + \text{missing energy}$$

using the iron ball detector at SPEAR. The detector is a solid magnetized iron spectrometer with an azimuthal field of 15 kGauss and a radius of about 1 m. It identifies muons above 1.3 GeV/c in nearly 90 % of 4π. They select muon pairs with an acoplanarity angle with respect to the beam axis greater than 10° and demand that the system recoiling against the pair should have a mass squared greater than 4 GeV^2. A total of 25 $\mu^+\mu^-$ events satisfy the criteria, the estimated background is 14 events. The remaining 11 events yield a branching ratio $B_\mu = 0.20 \begin{matrix} +0.10 \\ -0.08 \end{matrix}$.

This is evaluated assuming $m_\tau = 1.9$ GeV, V-A coupling and a massless neutrino.

4.2.3 Two prong inclusive electron and muon final states

The first evidence for anomalous muon production in $e^+e^- \to \mu$ + charged track + ≥ 0 photons was reported by the Maryland, Pavia, and Princeton Collaboration[76,77]. They identify muons above 1.0 GeV/c by range in a single arm magnetic spectrometer. The statistics are poor, but the data are consistent with the heavy lepton hypothesis based on the eμ events.

The SLAC-LBL group has measured[82] the cross section for $e^+e^- \to \mu^{\pm}$ + charged particle + ≥ 0 photons for c.m. energies between 3.9 GeV and 7.8 GeV. Muons with momenta above 0.91 GeV/c are defined by range (equivalent of 65 cm of iron) in the "muon tower". A hadron has approximately a 3 % chance to fake a muon. Events should be acoplanar with respect to the beam axis by 20° and the mass squared recoiling against the two prongs should be greater than 1.5 GeV^2. After the background subtraction a significant signal of 103 ± 18 events remained for c.m. energies between 5.8 GeV and 7.8 GeV. The momentum spectrum of the muons is plotted in Fig. 4.5a and it can be compared to the spectrum of muons associated with multiprong events plotted in Fig. 4.5b. The latter spectrum which mainly results from charm, is much softer than the muon spectrum of the two prong sample. A fit to the data assuming V-A couplings, $B(\tau \to \nu_{\tau} + 1 \text{ charged} + \geq 0 \text{ photons}) = 0.85$, $m_{\nu_{\tau}} = 0$ and $m_{\tau} = 1.9$ GeV shown as the solid line in Fig. 4.5a yields $B_{\mu} = 0.175 \pm 0.027 \pm 0.03$. The second error shows the systematic uncertainties. Data from PLUTO[79] and DASP[69] indicate that the one prong branching ratio might be smaller than the value assumed. Decreasing the one prong branching ratio will increase B_{μ}.

PLUTO[79] measured inclusive muon production for μ momenta above 1 GeV/c. The events had to contain at least one extra charged particle with p > 0.2 GeV/c and $|\cos\Theta| < 0.87$. The event sample was divided into two prongs (one extra track + any number of photons)

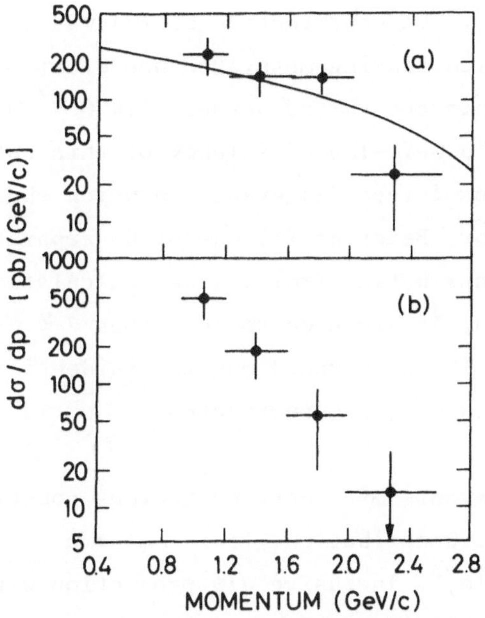

Fig. 4.5 a) The momentum distribution of anomalous muon two prong events. The data were obtained by SLAC-LBL[82] in the c.m.energy range of 5.8 to 7.8 GeV. The solid curve represents the cross section expected from heavy lepton production with the parameters listed in the text.

b) The momentum distribution of inclusive muon multiprong events.

and multiprongs. The two classes contain different contributions from conventional processes.

The main conventional sources of twoprong events are the QED processes (1) $e^+e^- \rightarrow \mu^+\mu^-$, (2) $e^+e^- \rightarrow \mu^+\mu^-\gamma$, and (3) $e^+e^- \rightarrow \mu^+\mu^-\gamma\gamma$. Reactions (1) and part of (2) were removed by requiring an acoplanarity angle of $> 10^o$. The contribution of (2) was further reduced by a cut in the squared missing mass. Because of the changing kinematical resolution this cut varied between 1.4 GeV^2 at $\sqrt{s}$ = 3.6 GeV and 2.7 GeV^2 at $\sqrt{s}$ = 5 GeV. The efficiency of this cut was checked with a 60 % subsample of type (2) events in which the photon converted in the detector. Reaction (3) cannot be separated by kinematical cuts. The contribution from (3) was calculated[83] and subtracted from the data. It amounted to less than 7 % of the remaining muon signal. The background due to misidentified hadrons was typically 15 % and was also subtracted.

For multiprong events misidentifed hadrons constitute the main source of background. Contributions from $e^+e^- \rightarrow e^+e^-\mu^+\mu^-$ were found to be negligible[83]. Inclusive J/ψ production with subsequent $J/\psi \rightarrow \mu^+\mu^-$ decay was eliminated by a cut in the invariant two particle mass. The momentum distributions of the observed events are plotted in Fig. 4.6a-c. The spectra show the triangular upper end characteristic of the 3-body decay of a moving object.

The two prong cross section is plotted in Fig. 4.4a as a function of energy. The data indicate a threshold in the 3.6 GeV to 4.0 GeV region. The data are fitted to the heavy lepton hypothesis for V+A (dotted line) and V-A (solid line) assuming a massless neutrino. The results, plotted in Fig. 6.6 are in agreement with the heavy lepton hypothesis and yield 1.93 ± 0.05 GeV (V-A) and 1.82 ± 0.08 GeV (V+A) for the value of the mass. The size of the cross section near threshold excludes the possibility that the muons result from the pairproduction and weak decays of a spin 0 particle: it leads to unphysical values for the leptonic branching ratios.

The branching ratio for the τ to decay into multiprong and one prong events is determined by assuming that all events in the multiprong sample with muon momenta above 1 GeV/c result from τ production. They find B(1 prong) = 0.70 ± 0.10 and B($\geq$ 3 prong) = 0.30 ± 0.10.

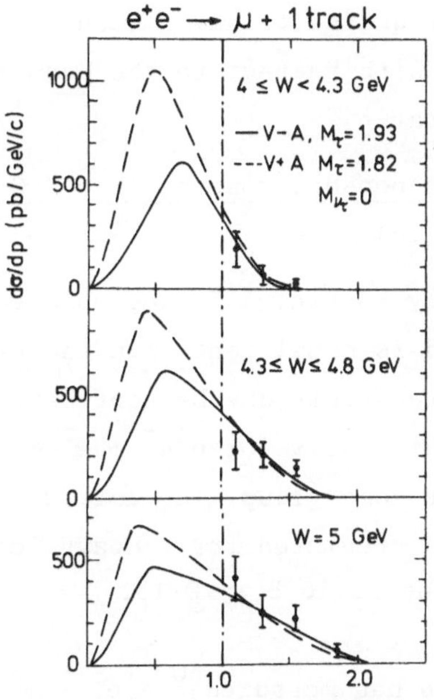

Fig. 4.6 Inclusive muon data for p_μ > 1.0 GeV/c measured by PLUTO[79]. The data are corrected for QED-production and other sources of background. a,b,c) The muon momentum distribution of two prong events for three different c.m. energies. The curves are fits to the data assuming pairproduction of a heavy lepton τ, a massless τ neutrino and V-A (solid line) or V+A (dotted line) leptonic weak couplings.

From an analysis of the vertex distribution the PLUTO group determined an upper limit of $9 \cdot 10^{-12}$ sec for the τ lifetime[79]. This is a factor of 30 larger than the value of $2.9 \cdot 10^{-13}$ sec expected from conventional theory.

The DELCO group measured[61] the cross section

$$e^+e^- \rightarrow e^\pm + \text{nonshowering track} + \geq 0 \text{ photons}$$

for c.m. energies between 3.7 GeV and 7.4 GeV. Electrons were identified by threshold Cerenkov counters and shower counters. In addition both tracks were required to have momenta above 0.3 GeV/c and the acoplanarity angle with respect to the beam axis had to be greater than 20°. The ratio

$$\frac{\sigma(e^+e^- \rightarrow e^\pm + \text{nonshowering track} + \geq 0 \text{ photons}}{\sigma(\mu\bar{\mu})}$$

is plotted versus energy in Fig. 4.7. The observed cross section shows no resonances and is consistent with the cross section expected for a heavy lepton. This can be contrasted with the energy dependence plotted in Fig. 3.19 for $e^+e^- \rightarrow e^\pm + \geq 2$ prongs $+ \geq 0$ photons measured by the same group. The dotted curve in Fig. 4.7 shows the cross section predicted for a heavy lepton of mass 1.85 GeV and a branching ratio $B_e = 0.15$.

The LBL-SLAC group has measured[80] $e^+e^- \rightarrow e^\pm$ + charged $+ \geq 0$ photons for c.m. energies between 4.1 GeV and 7.4 GeV. A total of 31 events were observed with an estimated background of 12.1 events. The data are plotted in Fig. 4.8 versus r. From a fit to these events and the $e\mu$ events discussed earlier they find for the branching ratio B_{1h} for $\tau \rightarrow \nu_\tau$ + 1 charged hadron + any number of photons $B_{1h} = 0.45 \pm 0.19$. The fit was made assuming V-A couplings, $m_{\nu_\tau} = 0$, $m_\tau = 1.9$ GeV. The solid line shows the results of the fit.

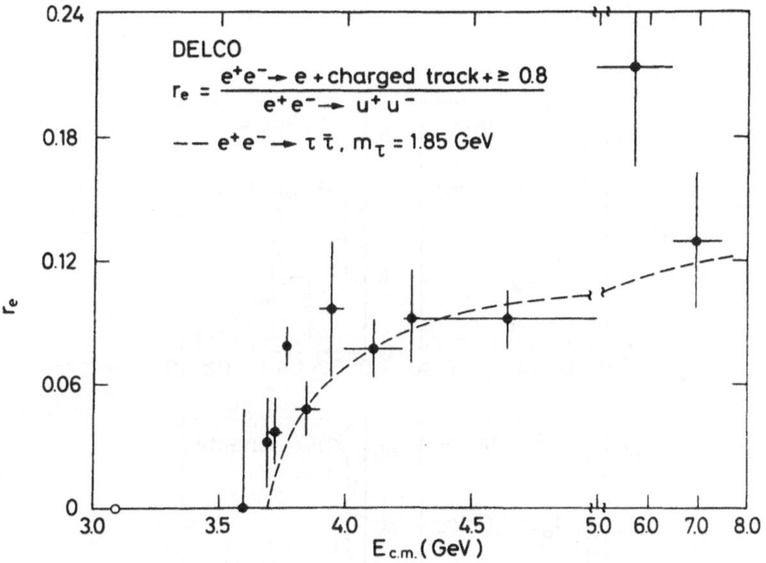

Fig. 4.7 The two prong electron inclusive cross sections measured by DELCO[61] and normalized to the point cross section as a function of c.m. energy. The dotted curve shows the yield expected from τ pairproduction. The absolute normalization is uncertain by ±50 %.

The DASP Collaboration has measured[69]

1) $e^+e^- \to e^\pm$ + nonshowering track + $\geq$ 0 photons

2) $e^+e^- \to \mu^\pm$ + nonshowering track + $\geq$ 0 photons

for c.m. energies between 3.6 GeV and 5.2 GeV.

The selection criteria for the inclusive electron sample have been discussed earlier. It is important to note that electrons are measured down to 0.2 GeV/c momenta and hence the cross sections extracted from these data are nearly independent whether V−A or V+A couplings are assumed.

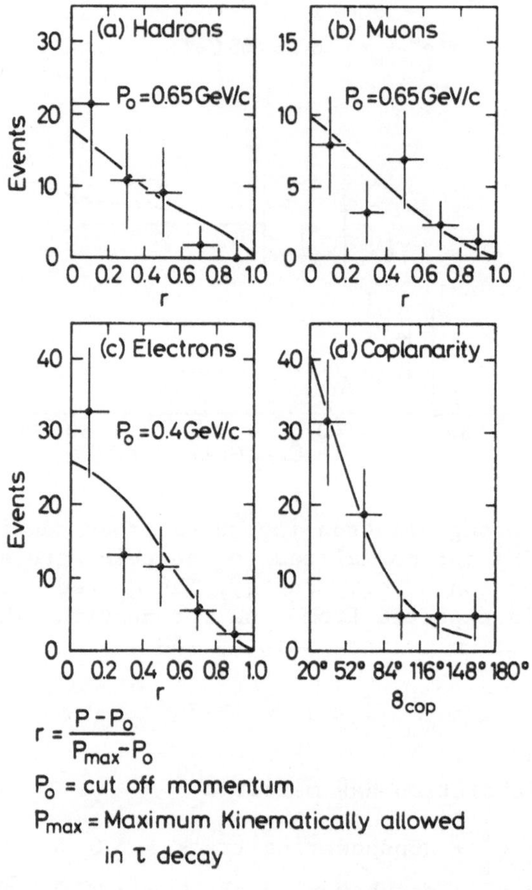

Fig. 4.8 The ratio $r = \sigma(e^+e^- \rightarrow e + 1$ charged + ≥ 0 photons$)/\sigma_{\mu\mu}$ as measured by LBL-SLAC (Ref. 80).

A total of 80 events were found at c.m. energies above 3.9 GeV, 17 events at the ψ' resonance and 1 event at 3.6 GeV. The signal at the ψ' proves conclusively that the τ particle is not associated with charm and we will discuss these data first.

The electron momentum spectrum measured at the ψ' and plotted in Fig. 4.9 shows two clear clusters of events, one centered around 1.5 GeV/c and the other with momenta between 0.4 GeV/c and 0.9 GeV/c. The first cluster can be associated with the cascade decay $\psi' \rightarrow J/\Psi\ X \rightarrow e^+e^-X$.

The electrons in the second cluster have a relatively flat momentum distribution. The background from the reaction $e^+e^- \rightarrow e^+e^-\mu^+\mu^-$ has been estimated to contribute (0.6 ± 0.2) events. An estimate using data at other energies shows that we expect less than 0.1 event from beam-gas interactions. Indeed, all the events originate within the nominal interaction volume. A twobody hadron final state can fake events of type (1) if the charged hadron traversing the magnet is misidentified as an electron. For hadrons with momenta above 0.35 GeV/c the measured probability P_{he} for this to happen is 4×10^{-4}. At the ψ' resonance DASP observed 2113 events of the type $e^+e^- \rightarrow h^\pm$ + nonshowering track + ≥ 0 photons where h is either a kaon or a pion traversing the magnet and the nonshowering track is observed in the inner or outer detector. This class of events therefore contributes a background of (0.84 ± 0.02) events. Dalitz decays of π^0 and η and photons converting in the beam pipe were estimated using the two prong sample above. A total of (0.2 ± 0.1) events were estimated compared to 9 events observed. The computation of background associated with multihadron events was checked by searching for inclusive electron events at the J/ψ resonance. One event of the type $e^+e^- \rightarrow e^\pm$ + nonshowering track + ≥ 0 photon was found which is to be compared with an estimated background of 1.3 events.

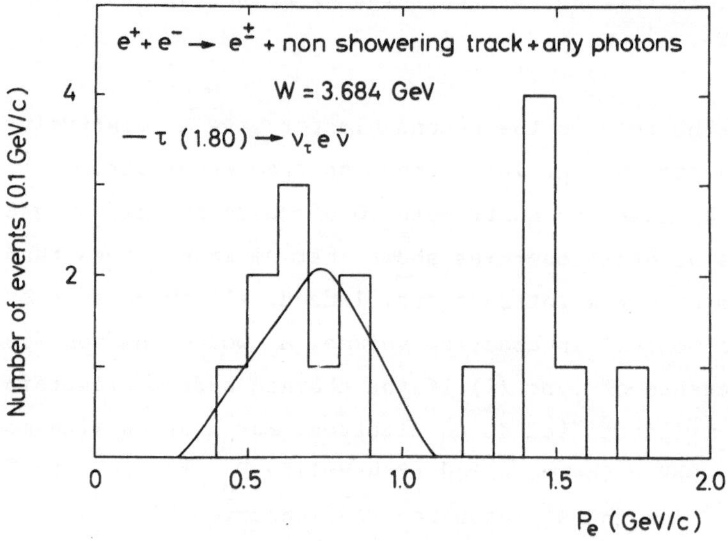

Fig. 4.9 Raw electron momentum distributions observed by DASP[69] at 3.684 GeV. The events here are identified as electron, one nonshowering particle and any number of photons.

Further evidence that the electron events observed at the ψ' are not true hadron events comes from an inspection of the photon multiplicity. Table 4.3 shows a comparison of the photon multiplicities for twoprong electrons (0.4 GeV/c < p_e < 0.9 GeV/c) and two prong hadron events (p_h > 0.4 GeV/c) from ψ' decay.

Table 4.3 Photon multiplicity distributions

Number of photons	0	1	2	3	4	5	6	7
$e^{\pm}$ + nonshowering track at the ψ'(0.4 < p_e < 0.9 GeV/c	4	3	1	1	0	0	0	0
$h^{\pm}$ + nonshowering track at the ψ'(p_h > 0.4 GeV/c)	270	370	440	428	312	199	99	32
$e^{\pm}$ + nonshowering track $\sqrt{s}$: 4-5.2 GeV(p_e > 0.2 GeV/c)	49	17	10	1	2	0	1	0

The distributions are strikingly different. The electron events are accompanied by a few tracks as expected for τ decay whereas the hadron events have a large multiplcitiy.

The authors of Ref. 69 therefore concluded that they have observed an anomalous electron signal at a c.m. energy of 3.684 GeV which is below the charm threshold. This signal is then assumed to come from $\tau\bar{\tau}$ production. The electron momentum spectrum predicted for a τ of mass 1.80 GeV and a zero mass neutrino fits the data well (see curve in Fig. 4.9).

We turn now to the DASP data on lepton two prongs at higher energies. A fraction of the twoprong electron events observed above 3.9 GeV might result from associated production and semileptonic decays of charmed particles. An upper limit can be obtained by assuming that all inclusive electron events with more than two prongs are due to charm production. From the measured multiplicity distributions of these events (Fig. 3.15) and the known detection efficiency a total of (5 ± 2) events has been estimated from this source. The direct decay of a pair of charmed hadrons into a final state with one electron and one nonshowering track is expected to contribute less than one event. The background from all other cources has been estimated to (9 ± 3) events, in agreement with (7 ± 7) events extrapolated from the 3.6 GeV data.

4.2.4 τ mass and spin

The quantity $2\sigma_{\tau\bar{\tau}} B_e \cdot B_{ns}$ is plotted in Fig. 4.10 as a function of c.m. energy. Radiative corrections were applied and the data were corrected for the enhancement at the ψ' due to vacuum polarization. Note the rapid rise near threshold which is characteristic for s-wave production. The data shown in Fig. 4.10 were used to determine the mass of the τ and its spin[69]. The cross section for $\tau\bar{\tau}$ production for a τ spin of 0, 1/2 and 1 reads as follows:

spin 0:

$$\sigma_{\tau\bar{\tau}} = 1/4\ \sigma_{\mu\mu}\ \beta_\tau^3\ |F|^2\ B_e \cdot B_{ns} \tag{4.5}$$

where $\sigma_{\mu\mu} = \frac{4\pi\ \alpha}{3\ s}$ and F is the τ formfactor.

spin 1/2:

$$\sigma_{\tau\bar{\tau}} = \sigma_{\mu\mu}\ \beta_\tau \left\{1 + 1/2(1-\beta_\tau^2)\right\}\ B_e \cdot B_{ns} \tag{4.6}$$

The τ is assumed to be pointlike.

spin 1:

$$\sigma_{\tau\bar{\tau}} = \sigma_{\mu\mu}\ \beta_\tau^3\ \left\{\left(\frac{s}{4M_\tau^2}\right)^2 + 5\ \frac{s}{4M_\tau^2} + 3/4\right\} \cdot B_e \cdot B_{ns} \tag{4.7}$$

The τ is assumed to have the same electromagnetic properties as the W beson[84].

For spin 0 the upper limit on $2\sigma_{\tau\bar{\tau}} B_e B_{ns}$ was calculated with F = 1 and the conservative assumption that the τ has only leptonic decays and $B_e = B_\mu$. This upper limit is plotted in Fig. 4.10 and is seen to be lower than the data by an order of magnitude. For

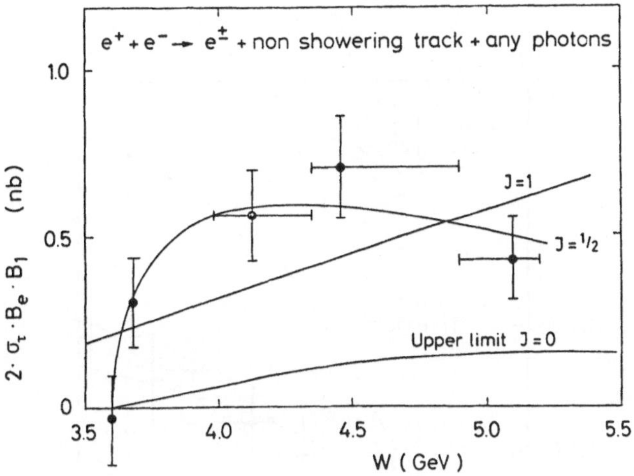

Fig. 4.10 Integrated inclusive cross section for events having an identified electron, a nonshowering particle, and any number of photons as a function of c.m. energy. The data are from DASP[69]. The solid curves show fits to the data assuming pairproduction of pint particles with spin 0, 1/2 and 1.

spin 1/2 and 1 a fit was made treating the τ mass and the products of the branching ratios $B_e \cdot B_{ns}$ as free parameters. The spin 1 curve (see Fig. 4.10) does not describe the data; including the data obtained at higher energies at SPEAR excludes spin 1. The data are well described by a pointlike fermion of spin 1/2. The fit yielded for the τ mass

$$m_\tau = 1.807 \pm 0.02 \text{ GeV}.$$

Similar results though with higher statistics were obtained recently by the DELCO group[85]. Their cross section measurements for

$$e^+e^- \to e^\pm + \text{nonshowering track} + \text{any number of photons}$$

divided by $\sigma_{\mu\mu}$ are plotted in Fig. 4.11. They yield a τ mass of $1.782 \begin{smallmatrix} + 0.002 \\ - 0.007 \end{smallmatrix}$ GeV in agreement with the DASP data.

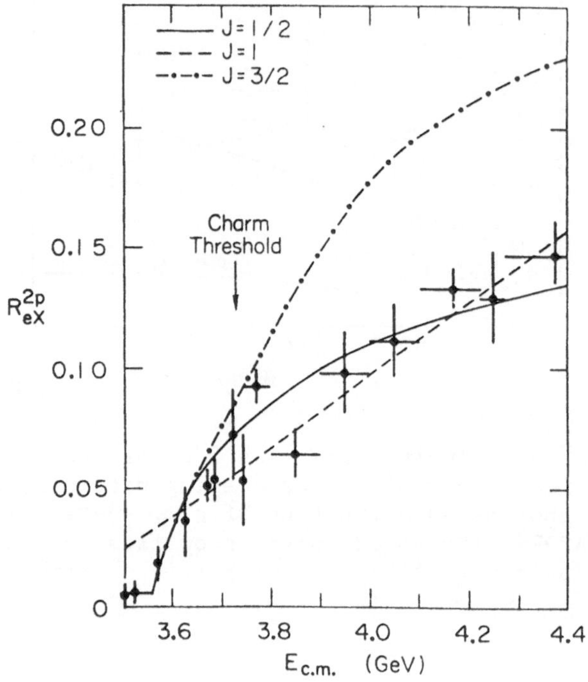

Fig. 4.11 The ratio $\sigma(e^+e^- \rightarrow e^{\pm}X^{\mp}, X \neq e^-)/\sigma_{\mu\mu}$ as measured by DELCO (Ref. 85)

A similar measurement performed by the DESY-Heidelberg group[86] gave

$$m_\tau = 1.790 \begin{smallmatrix} + 0.007 \\ - 0.010 \end{smallmatrix} \text{ GeV.}$$

The fit used by DASP to evaluate the τ mass from the electron inclusive events yields $B_e \cdot B_{ns} = 0.086 \pm 0.012$. Using $B_e = 0.182 \pm 0.028$ the DASP group derived the branching ratio for

$\tau \to \nu_\tau$ + nonshowering particle + $\geq$ 0 photons, B_{ns} = 0.47 ± 0.10. The branching ratio B_{1h} for $\tau \to \nu_\tau$ + hadron + $\geq$ 0 photons is given by $B_{1h} = B_{ns} - B_\mu$ = (0.29 ± 0.11). The systematic errors are small compared to the statistical error. The average number of photons associated with $\tau \to \nu_\tau$ + hadron + $\geq$ 0 photons can be obtained from Table 4.3 after making background corrections. Averaging the observed photon multiplicity over all two prong events in the higher energy data and correcting for the photon detection efficiency the decay of the type $\tau \to \nu_\tau$ + charged hadron + any number of photons was found to yield on the average 2.8 ± 0.7 photons.

The branching ratio B_{3h} for the τ to decay into final states with at least three charged particles can be obtained from $B_{3h} = 1-B_e-B_{ns}$. (The number of electron events with p_e > 1 GeV/c and 5 or more charged tracks were found to be small.) The result is B_{3h} = 0.35 ± 0.11 in agreement with the PLUTO measurement, B($\geq$ 3 prong) = 0.30 ± 0.10.

The DASP group studied also μ inclusive events[69]. Candidates for muon inclusive events had to have one muon track in the spectrometer, a second nonshowering track and any number of photons observed either in the inner detector or in the spectrometer arms. A charged particle was called a muon if it had a momentum greater than 1.0 GeV/c, gave no signal in the threshold Cerenkov counter, suffered an energy loss consistent with that of a minimum ionizing particle in the shower counter and penetrated at least 60 cm of iron. A total of 25 events with a background of 3.8 events was found.

After all corrections DASP observed (21 ± 5) muon inclusive events and (18.5 ± 4.6) electron inclusive events with momenta above 1.0 GeV/c. The ratio of the leptonic widths evaluated directly from these data are independent of the form of the coupling. This yields B_μ/B_e = 0.92 ± 0.32 with a systematic uncertainty of 0.07. The result is consistent with eμ universality.

The lepton spectrum permits a study of the space-time structure of the weak current mediating the decay of the τ. If only V and A type couplings are considered the Hamiltonian for $\tau \to \nu_\tau \ell \bar{\nu}_e$ is of the form[87]

$$H_{int} = \frac{G_F}{\sqrt{2}} \left[\bar{\psi}_{\nu_e} \gamma_\mu (1-\gamma_5)\psi_e\right] \left[\bar{\psi}_\tau \left\{g_+ \gamma_\mu(1+\gamma_5) + g_- \gamma_\mu(1-\gamma_5)\right\} \bar{\psi}_{\nu_\tau}\right] \tag{4.8}$$

where $g_\pm$ are the coupling strengths for V ± A couplings. In the τ rest system the shape of the lepton spectrum can be expressed in terms of the lepton momentum p, energy E, maximum energy $E_{max} = \frac{m_e^2 - m_{\nu_\tau}^2}{m_\tau}$ and $x = E/E_{max}$:

$$\frac{dN(x)}{d^3p} = \text{const} \left\{3(1 - x) + 2\rho(\tfrac{4}{3}\, x-1)\right\} \tag{4.9}$$

where terms of order m_e/m_τ have been neglected. The shape of the spectrum is then determined by the Michel parameter ρ [88] where ρ is defined as

$$\rho = \frac{3}{4} \frac{g_-^2}{g_+^2 + g_-^2} \tag{4.10}$$

Special cases are:

pure	V-A :	$g_+ = 0$	$\rho = 3/4$
	V+A :	$g_- = 0$	$\rho = 0$
	V :	$g_+ = g_-$	$\rho = 3/8$
	A :	$g_+ = -g_-$	$\rho = 3/8$

The lepton momentum spectrum, obtained by the DASP group[69] by combining the electron and the muon data, is plotted in Fig.4.12

for c.m. energies between 4.0 GeV and 5.2 GeV. This spectrum extends to much higher momenta than the electron spectrum observed in the semileptonic decays of the charmed hadrons, reflecting the pointlike structure of the τ and the low mass of its neutrino.

The solid line shows a fit to the data assuming $m_\tau = 1.80$ GeV, a massless neutrino and a (V-A) structure of the current. The dashed line is a fit keeping the masses constant but changing the left handed V-A current into a right handed V+A current. Both fits are clearly acceptable.

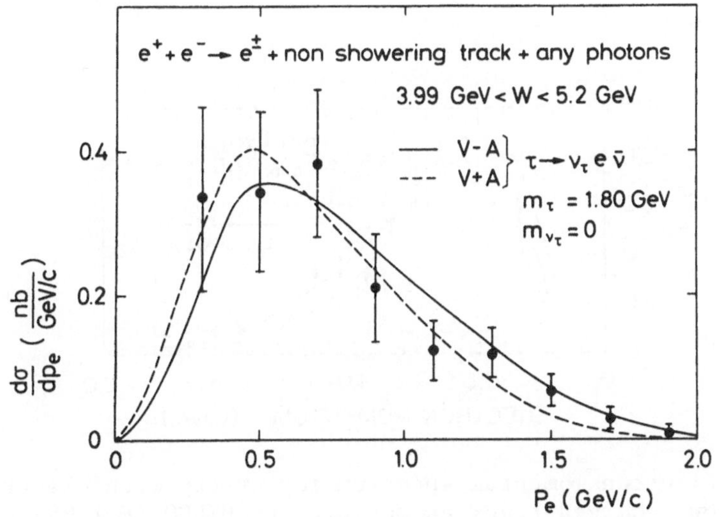

Fig. 4.12 Corrected electron momentum distribution measured by DASP[69] for events having an identified electron, a nonshowering particle, and any number of photons. (Above 1 GeV/c data having a muon instead of an electron are combined with the electron data to form a weighted mean.)

Fits were also made varying the mass of the τ neutrino. The 90 % confidence upper limits on the neutrino mass are $m_{\tau_\nu} < 0.74$ GeV for V-A and $m_{\nu_\tau} < 0.54$ GeV for V+A.
From their high statistics data (see Fig. 4.13) the DELCO group[85] obtained a preliminary value for the Michel parameter of

$$\rho = 0.83 \pm 0.12 \text{ (stat.error)} \pm 0.15 \text{ (system. error)}$$

This result agrees well with V-A and excludes V+A. A pure V or pure A coupling appears to be unlikely. The analysis assumed a massless τ neutrino. If the ν_τ is massive the theoretical spectrum will be

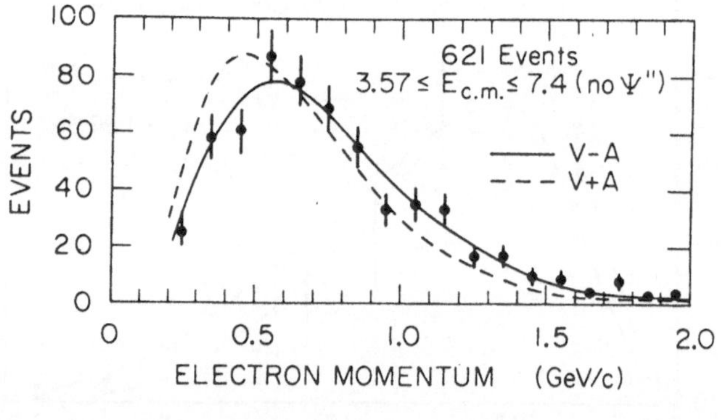

Fig. 4.13 Electron momentum spectrum for events with electron + one charged track as measured by DELCO (Ref.85). The solid and dashed curves represent the spectra expected for V-A and V+A coupling at the $\tau\nu_\tau$ vertex, respectively.

pushed towards smaller x values. As a result the discrepancy between data and the theoretical predictions for V+A, V or A will become even larger.

4.3 Semihadronic Decays of the τ

A characteristic feature of the standard weak interaction is that decays involving strange particles are suppressed relative to decays involving nonstrange final states by $tg^2\Theta_c \approx 0.05$. The DASP group determined[60] the ratio of strange to nonstrange particles in semihadronic τ-decays from a measurement of

$$\frac{\sigma(e^+e^- \to e^\pm + K^\pm + \geq 0 \text{ photons})}{\sigma(e^+e^- \to e^\mp + \pi^\pm + \geq 0 \text{ photons})}$$

It was shown that the two prong cross section including one electron predominantly results from $e^+e^- \to \tau\bar{\tau} \to (\nu_\tau e\bar{\nu}_e)(\nu_\tau + \text{hadrons} + \geq 0$ photons) with only a small contamination from charm decays. The hadrons were identified in either the inner or the outer detector.

The DASP group found:

$$\frac{\sigma(e^\mp + K^\pm + \geq 0 \text{ photons})}{\sigma(e^\mp + \pi^\pm + \geq 0 \text{ photons})} = 0.07 \pm 0.06$$

Therefore on the average only 7 % of all semihadronic τ-decays yield a strange particle in accordance with theory. This should be compared to multiprong events where DASP found[60]:

$$\frac{\sigma(e \;\; + K^\pm + \geq 1 \text{ prong} + \geq 0 \text{ photons})}{\sigma(e \;\; + \pi^\pm + > 1 \text{ prong} + \geq 0 \text{ photons})} = 0.24 \pm 0.05.$$

Since the charged multiplicity is on the order of 4 this is equivalent to (0.9 ± 0.18) charged kaons per multiprong event. (See discussion in section 3).

4.3.1 $\tau \to \pi\nu$

The DASP group[89] searched for the $\tau \to \pi\tau$ decay studying the process

$$e^+e^- \to \tau\bar{\tau} \to (e\nu\nu)(\pi\nu)$$

leading to the final state $e\pi^{\pm}$ 0γ. No positive evidence was found for the existence of the decay into $\pi\nu$. The branching ratio deduced from the data, $B_\pi = 0.02 \begin{smallmatrix} +\ 0.03 \\ -\ 0.02 \end{smallmatrix}$, was about 2.5 s.d. below its theretical value, $B_\mu = 0.10$.
The first clear evidence for the $\pi\nu$ decay was presented by the SLAC-LBL group[90] from an analysis of

$$e^+e^- \to \pi^{\pm}X^{\mp}\ 0\gamma$$

for c.m. energies between 4.8 and 7.4 GeV. Fig. 4.14 shows the pion momentum spectrum in terms of $x = p/p_{max}$ for x above 0.5. The x spectrum is seen to be almost constant in marked difference to the x region below 0.5 where hadronic channels dominate and produce exponentially falling x distribution (see section 6).
The most probable mechanism for producing a constant x distribution at high x is the twobody decay into π + a of a particle with fixed momentum.

The solid line in Fig. 4.14 shows the expected x distribution from $\tau \to \pi\nu$ plus background, the dashed line shows the background (mainly from $\tau^{\pm} \to \rho^{\pm}\nu \to \pi^{\pm}\pi^{0}\nu$) alone. The shape of the spectrum as well as the magnitude of the observed cross section agree with the assumption that most of these high x events come from $\tau \to \pi\nu$ decay.

Further evidence for the $\pi\nu$ decay comes from experiments by PLUTO[91], DELCO[85] and SLAC-LBL (MARK II)[92]. The measured branching ratios are summarized in Table 4.4 (Ref. 93).

Table 4.4 Results on the decay $\tau \to \pi\nu$

experiment	final state	events	back-ground	$B(\tau \to \pi\nu)$ (%)
DASP[89]	$\pi^{\pm}$ + e+0γ	2	1	$2 {+3 \atop -2}$
SLAC-LBL (MARKI)[90]	$\pi^{\pm}$ + ch+0γ	~200	~70	9.3 ± 3.9
PLUTO[91]	$\pi^{\pm}$ + ch+0γ	32	9	9.0 ± 2.9 ± 2.5
DELCO[85]	$\pi^{\pm}$ + e+0γ	15	5.6	$6.0 \pm 1.6 {+1.9 \atop -1.2}$
SLAC-LBL (MARKII)[92]	$\pi^{\pm}$ + ch+0γ	142	46	8.0 ± 1.1 ± 1.5
	$\pi^{\pm}$ + e+0γ	27	10	8.2 ± 2.0 ± 1.5

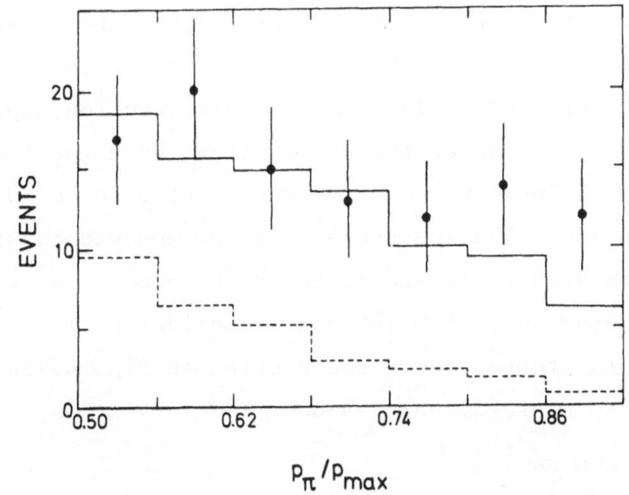

Fig. 4.14 The pion momentum distribution for events of the type $e^+e^- \to \pi^{\pm}$ + 1 charged measured at 4.8-7.4 GeV by the SLAC-LBL group (Ref. 90).

4.3.2 $\tau \to K\nu$

The DASP group has also searched for events of the type

$$e^+e^- \to \tau\bar{\tau} \to \{(K\nu)\ (e\nu\nu) + (\mu\nu\nu) + (\pi\nu)\}$$

$$= K^\pm + \text{charged track} + \text{missing energy}.$$

Only one event with $p_K > 1.0$ GeV/c was found[89]. This yields a 90 % confidence upper limit of $B_K < 0.016$.

4.3.3 $\tau \to \rho\nu$

DASP measured[89] the decay $\tau \to \rho\nu_\tau$ by selecting final states with $\pi^\pm$ + charged track + two photons. Events in which the two photons are compatible with resulting from a π^o decay are retained provided that both computed photon energies are above 50 MeV. The remaining events are plotted versus $M(\pi^\pm\pi^o)$ in Fig. 4.15a. Events with an identified electron are hatched. Events within the ρ-band (0.5 GeV $< M(\pi^\pm\pi^o) < 1.0$ GeV) are plotted versus the momenta of the $\pi^\pm\pi^o$-system in Fig. 4.16. The momentum distribution expected from the decay $\tau \to \rho\nu_\tau$ is shown as the dotted line. Note the flat distribution above 0.9 GeV/c which is characteristic for a two body decay of a moving object. The enhancement at low momenta is presumably due to multihadron events. To reduce the background, only events with a $(\pi^\pm\pi^o)$-momentum above 1.0 GeV/c are considered. The $\pi^\pm\pi^o$ mass distribution for these events are plotted in Fig. 4.15a. These events yield as a preliminary value

$$B_\rho = 0.24 \pm 0.09$$

in good agreement with the theoretical predictions.

4.3.4 $\tau \to \rho\pi\nu$

The PLUTO group has measured[94] the final state $e^+e^- \to e^\pm\pi^\mp\pi^-\pi^+$. They found, after background corrections, that these events result from

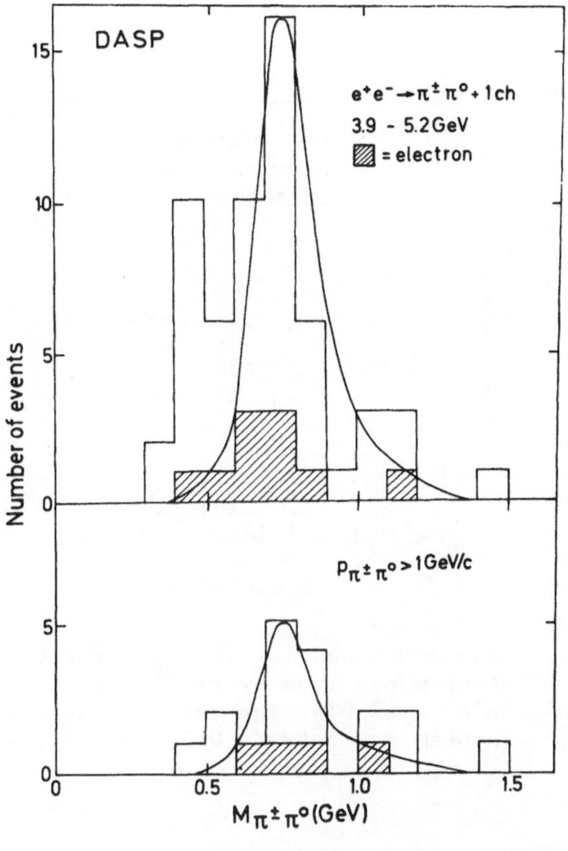

Fig. 4.15 a) The distribution of $M(\pi^{\pm}\pi^{0})$ observed by DASP[89] for events with the topology $e^+e^- \to \pi^{\pm}\pi^{0}$ + charged track.

b) The $M(\pi^{\pm}\pi^{0})$ distribution for events of the same topology as above but with $p_{\pi^{\pm}\pi^{0}} > 1$ GeV/c.

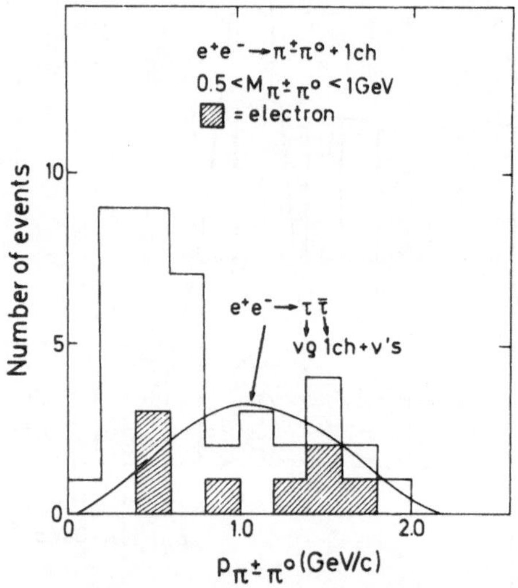

Fig. 4.16 The momentum distribution of the $(\pi^{\pm}\pi^{0})$ system observed by DASP[89] in $e^+e^- \rightarrow \pi^{\pm}\pi^{0}T^{\mp}$. The mass of the $\pi^{\pm}\pi^{0}$ system was between 0.5 and 1.0 GeV.

$$e^+e^- \rightarrow \tau\bar{\tau} \rightarrow (e\nu_\tau\ \bar{\nu}_e)(\rho^0\pi\ \bar{\nu}_\tau).$$

They observe a total of 39 events of the type e + 3 hadrons and assume that the hadrons are all pions. From the sample of 39 events those events are selected for which the 3π-system has an invariant mass and momentum consistent with pairproduction and subsequent decay of a particle of mass $\lesssim 1.9$ GeV; 6 events are rejected. The remaining sample of 33 events contains an estimated background of 13 ± 4 events from hadrons mislabeled as electrons and at most 3 events which had unobserved photons.

The $\pi^+\pi^-$ mass distribution for the 33 events is plotted in Fig. 4.17a. The background is shown by the squares. Each event contributes two mass values and the distribution can be understood as arising mainly from a ρ-signal in one combination and a broad distribution in the other. Plotting only the highest mass combination shows the ρ more clearly. 21 events have a $M(\pi^+\pi^-)$ between 0.70 - 0.84 GeV and are retained. The momentum distribution for the electrons in the remaining sample is plotted in Fig. 4.18b.

The electron spectrum is hard and consistent with $\tau \to e\nu_\tau\bar{\nu}_e$ shown as the solid line but inconsistent with the charm spectrum shown as the dotted line. If the (3π) system results from the twobody decay of a moving object it should uniformly populate momenta between p_{max} and p_{min}. These limits depend on the c.m. energy and the mass of the 3π system. However, the quantity $U = (p_{3\pi} - p_{min})/(p_{max} - p_{min})$ is uniformly populated for any set of c.m. energies and mass values. The events are plotted in Fig. 4.18c versus U. The solid line indicates the event distribution predicted including all the experimental corrections, for a twobody decay $\tau \to (3\pi)\nu_\tau$ of a moving object.

The events can therefore safely be assigned to τ pairproduction. The 3π mass distribution of the 21 events with at least one $\pi^+\pi^-$ combination in the ρ band is shown in Fig. 4.17c. The background expected from misidentified electrons is indicated by the squares. From a spin parity analysis the PLUTO group concluded that the $\rho\pi$ system is preferrentially in a $J^P = 1^+$ state[95]. The mass histogram shows a clustering at low $\rho\pi$ mass as effected for an A_1 signal.

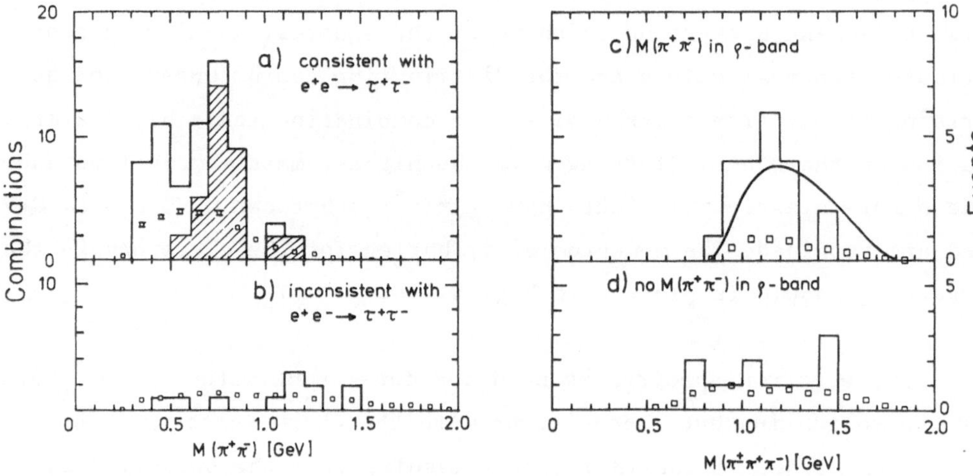

Fig. 4.17 Properties of the two and three pion mass distribution in the $e\pi\pi\pi$ event sample observed by PLUTO[94]. The squares represent the expected distributions for background from identified events. The solid line in c) represents the sum of the background and the $\tau \to \rho^0\pi$ distribution normalized to the data.

a) $M(\pi^+\pi^-)$ for events kinematically consistent with τ pairproduction. There are two combinations per event and the shaded area represents the distribution of the higher $M(\pi^+\pi^-)$ combination.

b) $M(\pi^+\pi^-)$ for events kinematically inconsistent with τ pairproduction.

c) $M(\pi^\pm\pi^+\pi^-)$ distribution for events consistent with pairproduction and having at least one $\pi^+\pi^-$ mass in the ρ band.

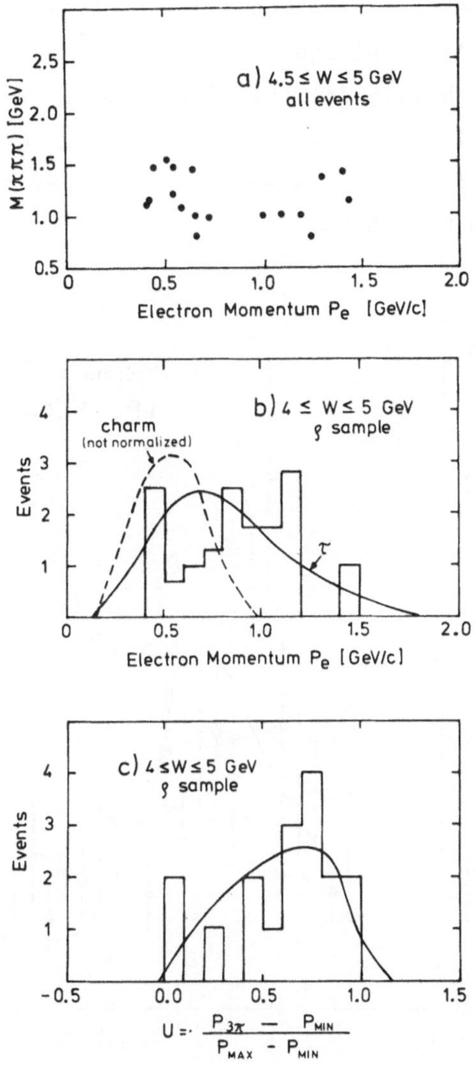

Fig. 4.18 Kinematic properties of the $e\pi\pi\pi$ events observed by PLUTO[94].

a) Invariant mass of the 3π-system plotted against the electron momentum.

b) The electron-momentum distribution for events from the ρ sample (see text). The solid curve is the prediction for τ-decay normalized to the data with $p_e > 0.4$ GeV/c. The dotted curve is the electron spectrum from charm decays measured by DASP. The curves are corrected for the detection efficiency.

c) Distribution of the normalized 3π-momentum (ρ-sample), compared with the expected $\tau \to \nu_\tau \pi\pi\pi$ decay distribution.

The nonresonant mass distribution $\tau \to \rho\pi\nu_\tau$ (solid line) was computed assuming (V-A) couplings, m_τ = 1.8 GeV and a massless τ neutrino. The χ^2 probability for this fit is 5 % compared to 80 % for an A_1 with mass 1.1 GeV and width of 0.25 GeV. The resonance interpretation is thus favoured. Using B_e = 0.16 they find $B(\tau^+ \to \rho^0\pi^+\nu_\tau)$ = 0.05 ± 0.03.

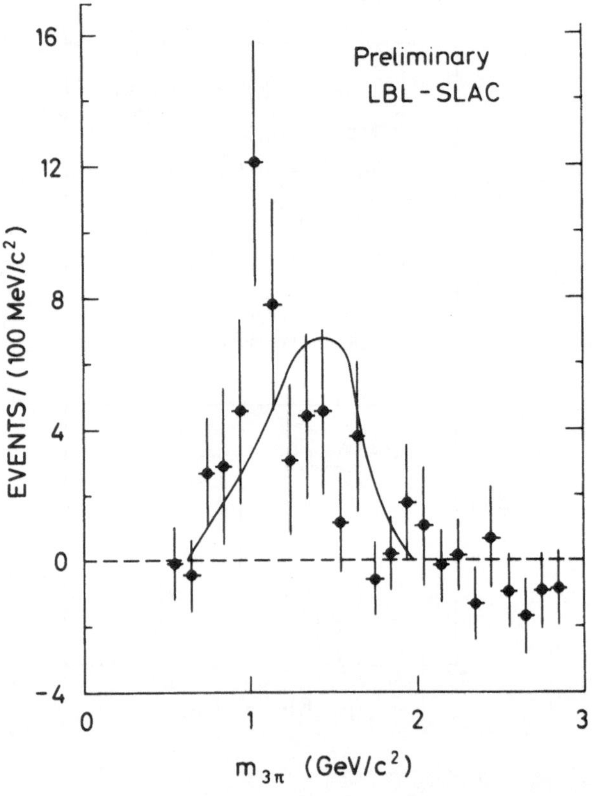

Fig. 4.19 Three-particle invariant mass distribution opposite muons corrected for hadron misidentification as measured by SLAC-LBL[90]. The solid curve is the phase space prediction for $\tau \to \nu\pi\rho$, corrected for acceptance effects. The prediction is normalized to events with a three particle effective mass between 0.7 and 1.8 GeV.

The SLAC-LBL group has also reported[90] evidence for $\tau \rightarrow A_1 \nu_\tau \rightarrow \rho\pi\nu_\tau$. The mass spectrum for the 3π system is plotted in Fig. 4.19. The peak observed is too narrow to result from a phase space like decay $\tau \rightarrow \rho\pi\nu$, plotted as the solid line. A resonance fit to the mass spectrum yields $m_{A_1} \simeq 1.1$ GeV, $\Gamma \approx 0.2$ GeV in agreement with the PLUTO result.

4.4 Summary of the τ Properties

Table 4.6 summarizes the experimental information on mass, leptonic and semihadronic decay modes of the τ.

The observation of the τ at the ψ' below charm threshold conclusively demonstrates that the τ signal has nothing to do with charm. The shape and magnitude of the $\tau\bar{\tau}$ production cross section exclude spin 0 and 1 for the τ: they strongly favor the assignment as a pointlike, spin 1/2 fermion. The best value for the τ mass at present is $1.782 \,^{+\,0.003}_{-\,0.004}$ GeV. The analysis of the decay spectra puts a limit of 0.25 GeV on the mass of the τ neutrino. The lifetime of the τ is less than $9 \cdot 10^{-12}$ sec. The lepton momentum spectrum strongly favors a V-A type coupling of the weak current to the $\tau\bar{\nu}_\tau$ system. The measured leptonic decay rates are consistent with e/μ universality. The leptonic and semihadronic branching ratios agree with those expected from theory for a heavy lepton of mass 1.8 GeV.

The consistency with e/μ universality classifies the τ as either a ortholepton or a sequential lepton. In the first case the τ has the same lepton number as e or μ. In the second case the τ carries a new lepton number. Recent neutrino experiments rule out that the τ is μ-like[97].

Table 4.5 Properties of the τ

	experimental results	predicted by theory	experiment	comments
mass (GeV)	$1.782^{+0.003}_{-0.004}$ GeV		DASP[69], DELCO[85], D-HD[86]	
Spin	1/2			
lifetime (sec)	$<9\cdot10^{-12}$	$2.9\cdot10^{-13}$	PLUTO[94]	
m_{ν_τ} (GeV)	<0.25		DELCO[85]	
$\tau\bar{\nu}_\tau$ coupling to weak current	V-A		DELCO[85]	
B_μ/B_e	0.92 ±0.32	1(seq.lept)	DASP[69]	
B_e	0.186±0.010±0.028	0.18	SLAC-LBL[75]	from eμ; assume $B_e=B_\mu$ and V-A
	0.16 ±0.06		PLUTO[94]	from eμ,μX; assume V-A
	0.224±0.032±0.044		LBL-SLAC[80]	from eμ; assume $B_e=B_\mu$
	0.182±0.028±0.014		DASP[69]	from eμ; assume B_e-B_μ and V-A
	0.16 ±0.04		DELCO[85]	from eX; preliminary
B_μ	0.175±0.072±0.030	0.18	SLAC-LBL[82]	from μX; assume $B_{1prong}=0.85$ and V-A
	0.14 ±0.034		PLUTO[79]	from μX; assume V-A
	$0.22^{+0.07}_{-0.08}$		Iron ball[76]	from μμ
	0.20 ±0.10		Maryland-Princeton-Pavia[76]	from μX

	experimental results	predicted by theory	experiment	comments
$B(\tau \to e\gamma)$	<0.026		LBL-SLAC[96]	
$B(\tau \to \mu\gamma)$	<0.013		LBL-SLAC[96]	
$B(\tau \to$ 3 charged leptons)	<0.006		SLAC-LBL[96]	
	<0.01		PLUTO[68]	
$B(\tau \to$ 1 charged + any photons)	0.70 ± 0.10		PLUTO[79]	
	0.90 ± 0.10		LBL-SLAC[-80]	
	0.65 ± 0.12		DASP[69]	
$B(\tau \to$ 1 charged hadron + any photons)	0.40 ± 0.15		PLUTO[79]	
	0.45 ± 0.19		LBL-SLAC[80]	
	0.29 ± 0.11		DASP[69]	
$B(\tau \to$ 3 charged hadrons + any photons)	0.35 ± 0.11		DASP[69]	
$B(\tau \to \pi\nu)$	0.077 ± 0.013	0.10	SLAC-LBL[90], PLUTO[91], DELCO[85], and Ref.93	
$B(\tau \to K\nu)$	<0.016	0.005	DASP[89]	
$B(\tau \to \rho\nu)$	0.24 ± 0.09	0.22	DASP[89]	
$B(\tau \to \rho^0\pi^\pm\nu)$	0.050 ± 0.015	~0.1	PLUTO[94]	$\rho\pi$ spectrum consistent with A_1 decay. Including neutral decay modes would give $B(\tau \to A_1\nu)$ = 0.10 ± 0.03

5. THE Υ FAMILY

The observation of the Υ gave the impression of a déjà vu phenomenon in more than one respect. It was discovered by a Columbia-FNAL-Stony Brook Collaboration[98,99] studying the $\mu^+\mu^-$ mass spectrum produced via

$$400 \text{ GeV } p + N \rightarrow \mu^+\mu^- + X$$

where Be, Cu and Pt were used as targets. Fig. 5.1 shows the latest data from this experiment[99]. Plotted is the $\mu^+\mu^-$ mass spectrum in terms of the differential cross section $B\frac{d^2\sigma}{dmdy}\Big|_{y=0}$ per nucleon with m = $\mu^+\mu^-$ mass, y = c.m. rapidity of the $\mu^+\mu^-$ system, and B = branching ratio for the $\mu^+\mu^-$ channel. A broad enhancement is seen around 9.5 GeV riding on a falling background. No other structure is observed between 6 and 13 GeV. The background subtracted signal is plotted in Fig. 5.2. One observes a 500 MeV wide (FWHM) peak around 9.4 GeV followed by a second bump near 10 GeV. The r.m.s. mass resolution in this experiment was roughly 500 MeV.

The authors have fitted the mass spectra to a two peak and a three peak hypothesis. The latter was found to produce a somewhat better chi square per d.f. In both cases the resonances were assumed to be narrow which implies that the mass shapes are determined by the experimental resolution. The results from these fits are listed in Table 5.1.

Meanwhile further experiments have observed the Υ in hadron initiated reactions. They are summarized in Ref. 99.

From the experience with the J/ψ it seemed obvious to associate the Υ with a new heavy quark Q. An important piece of information was missing, however. The widths of the Υ states. Are those states indeed

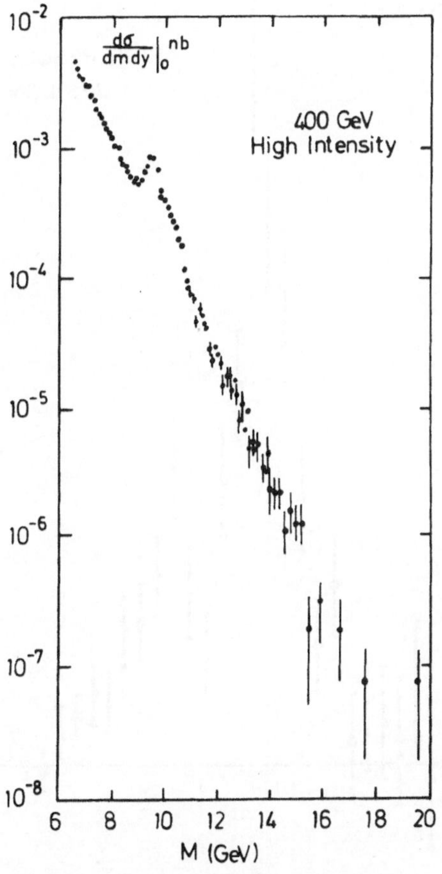

Fig. 5.1 Dimuon production at 400 GeV via $pN \to \mu^+\mu^-X$ as measured by the Columbia-FNAL-Stonybrook Collaboration (Ref. 99). Plotted is the cross section $d\sigma/dmdy$ at $y = 0$ as a function of the dimuon mass m.

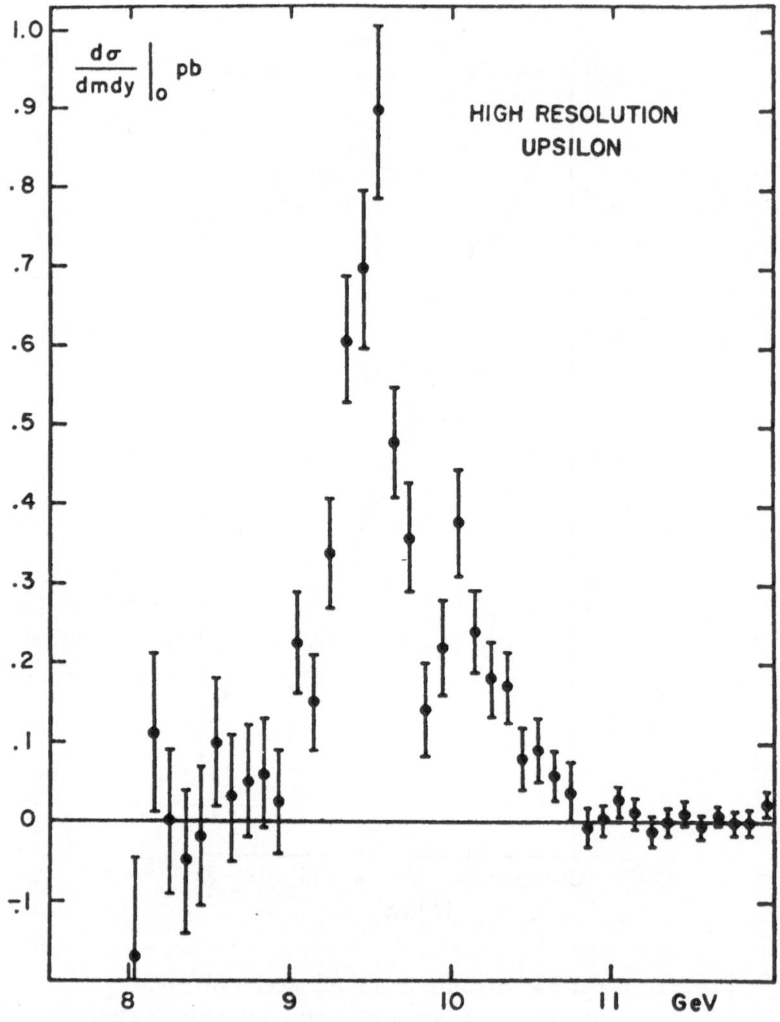

Fig. 5.2 Dimuon production via $pN \rightarrow \mu^+\mu^-X$ as measured by the Columbia-FNAL-Stonybrook Collaboration (Ref. 99). Plotted is the dimuon mass spectrum obtained after subtracting a smooth background.

Table 5.1 Resonance fits to the background subtracted $\mu^+\mu^-$ mass spectrum (Ref. 98)

		2 peaks	3 peaks	
T	M	9.41 ± 0.013	9.40 ± 0.013	GeV
	$B\frac{d\sigma}{dy}\Big\vert_{y=0}$	0.18 ± 0.01	0.18 ± 0.01	pb
T'	M	10.06 ± 0.03	10.01 ± 0.04	GeV
	$B\frac{d\sigma}{dy}\Big\vert_{y=0}$	0.069 ± 0.006	0.065 ± 0.007	pb
T''	M		10.40 ± 0.12	GeV
	$B\frac{d\sigma}{dy}\Big\vert_{y=0}$		0.011 ± 0.007	pb

narrow like the $J/\psi, \psi'$? This question can be studied best by producing the T in an e^+e^- storage ring. DORIS, which at the time could reach a total c.m. energy of 2 × 4.3 = 8.6 GeV, was converted to a single beam machine in the fall of 1977 and was upgraded to 2 × 5 = 10 GeV subsequently.

The heroic efforts of the DORIS machine crew led to the observation of the T by the PLUTO[100] and the DASP2 groups[101] in June of this year. In a subsequent running period at DORIS the T was also seen by a DESY-Hamburg-Heidelber-Munich collaboration[102]. Fig. 5.3 shows the total cross section as measured by these experiments near the T mass region. Only the PLUTO data are corrected for acceptance. The T shows up as a narrow (<10 MeV wide) signal at 9.46 GeV. As in the case of the J/ψ and ψ' the observed width (σ_m = 7.8 ± 0.9 MeV) is determined by the energy spread of the beams which for m = 9.5 GeV was calculated to give an r.m.s. resolution of σ_m = 8 MeV.

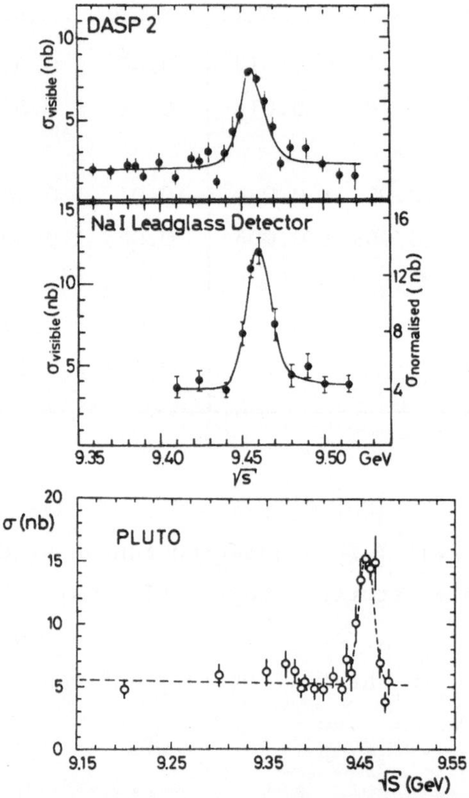

Fig. 5.3 The total cross section for hadron production in the Υ region. The top two measurements by DASP2 (Ref. 101) and DESY-Hamburg-Heidelberg-Munich (Ref. 102) give visible cross sections, the bottom measurement by the Pluto group (Ref. 100) shows the corrected cross section.

Integration of the resonance signal relates the hadronic cross section to the total, the hadronic and the leptonic widths Γ_{tot}, Γ_{had} and Γ_{ee}:

$$\frac{M^2}{6\pi^2}\int \sigma_{had}\, dm = \frac{\Gamma_{ee}\Gamma_{had}}{\Gamma_{tot}}$$

For Γ_{tot} we can write

$$\Gamma_{tot} = \Gamma_{had} + \Gamma_{ee} + \Gamma_{\mu\mu} + \Gamma_{\tau\tau} = \Gamma_{had} + 3\Gamma_{ee}$$

where e,μ,τ universality was assumed. In order to determine Γ_{had} and Γ_{ee} separately a measurement e.g. of $\Upsilon \rightarrow \mu^+\mu^-$ is required which yields

$$\frac{M^2}{6\pi^2}\int \sigma_{\mu\mu}\, dm = \frac{\Gamma_{ee}^2}{\Gamma_{tot}}$$

In addition to the observation of the Υ in the hadronic final states a first attempt was made to measure $B_{\mu\mu} \equiv \Gamma_{\mu\mu}/\Gamma_{tot}$ [103-105] and to deduce Γ_{ee} and Γ_{tot}. The mass and widths parameters as obtained by the three experiments are listed in Table 5.2.

Table 5.2 Mass and widths parameters of the Υ as measured at DORIS

	PLUTO[103]	DASP2[104]	D-H-HD-M[105]
M_y (GeV)	9.46 ± 0.01	9.46 ± 0.01	9.46 ± 0.01
Γ_{ee} (keV)	1.3 ± 0.4	1.5 ± 0.4	1.1 ± 0.3
$B_{\mu\mu}$ (%)	2.7 ± 2.0	2.5 ± 2.1	$1.0^{+3.4}_{-1.0}$
Γ_{tot} (keV)	>20 (2 s.d.)	>20 (2 s.d.)	>15 (2 s.d.)

The mean values are[106]

$M_{T} = 9.46 \pm 0.01$ GeV
(The error reflects the 1 $^{o}/_{oo}$ uncertainty in the energy calibration of DORIS)

$\Gamma_{ee} = 1.3 \pm 0.2$ keV

$B_{\mu\mu} = 2.6 \pm 1.4$ %

$\Gamma_{tot} >$ 25 keV (95 % C.L.)
$<$ 18 MeV

best value: $\Gamma_{tot} = 50$ keV.

The DASP2[104] and DESY-Hamburg-Heidelberg-München[102] collaboration also searched for the T'. The visible cross sections measured by these experiment around 10 GeV is displayed in Fig. 5.4. Both experiments observe a peak at 10.02 GeV which is consistent with the T' mass value quoted by the Columbia-FNAL-Stony Brook collaboration[98]. The T' parameters deduced by the storage ring experiments are listed in Table 5.3.

Table 5.3 Mass and width parameters of the T' as measured at DORIS

	DASP2[104]	D-H-HD-M[102]
$M_{T'}$ (GeV)	10.012 ± 0.01	10.02 ± 0.02
$M_{T'} - M_{T}$ (MeV)	555 ± 11	560 ± 10
Γ_{ee}	0.35 ± 0.14	0.32 ± 0.13
$\Gamma_{ee}(T)/\Gamma_{ee}(T')$	≈ 3	3.3 ± 0.9

Charge of the new quark:

In the nonrelativistic quark model the leptonic decay width for the vector states is directly proportional to the quark charge, Q:

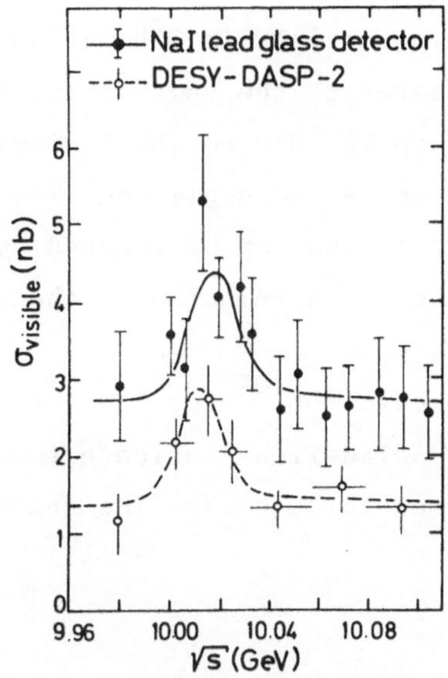

Fig. 5.4 Evidence for the Υ' as measured by the DASP2 (Ref. 104) and DESY-Hamburg-Heidelberg-Munich group (Ref. 102).

$$\Gamma_{vee} = \frac{16\pi\alpha^2 Q^2}{M^2} \left|{}^3S_1(0)\right|^2 \qquad (5.1)$$

It is an empirical fact that the leptonic decay width divided by the average quark charge is approximately the same for all vector ground states, viz. ρ, ω, Φ and J/ψ: $\Gamma_{vee}/\left|\sum c_i Q_i\right|^2 \approx 11$
This is shown by Fig. 5.5. This rule applied to the T yields $Q = 0.34 \pm 0.04$ or $Q = 1/3$.

The leptonic width depends on the wave function which is a function of the quark potential. In QCD the quark potential is flavour independent. Therefore the same potential which describes the J/ψ should also be applicable to the T provided that effects due to the difference in mass can be neglected. Their influence on the extraction of the quark charge can be reduced by comparing the leptonic widths of T and T' in relation to those of J/ψ and ψ'.

The theoretical prediction[107] for $\Gamma(T \to ee)$ and $\Gamma(T' \to ee)$ is shown in Fig. 5.6. The solid lines indicate the lower limits for $Q = 1/3$, the dashed lines for $Q = 2/3$. The shaded region indicates

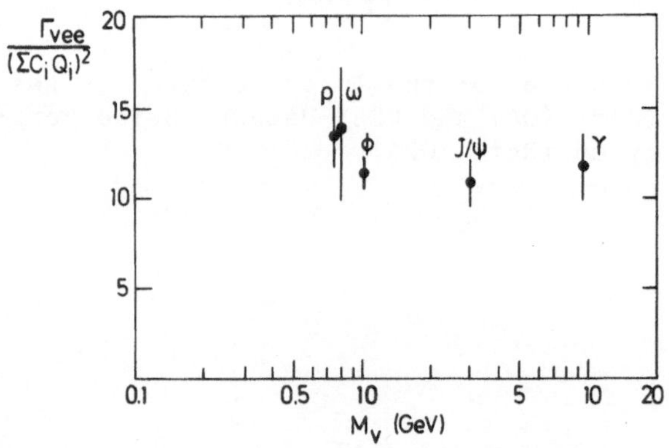

Fig. 5.5 The leptonic decay widths of vector mesons divided by the square of the average quark charge plotted as a function of the vector meson mass.

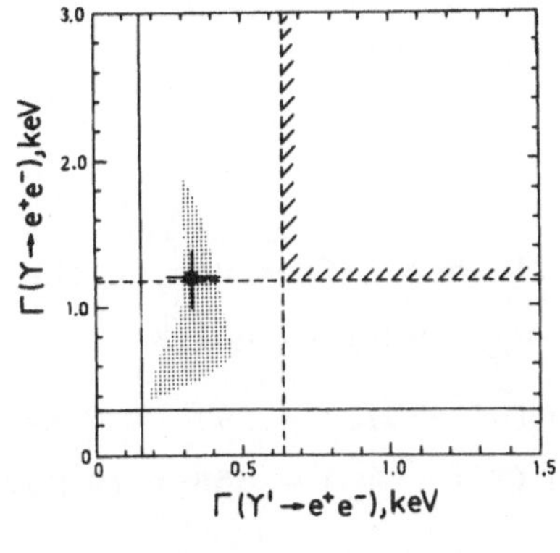

Fig. 5.6 Predictions of the quarkonium model for Γ(Υ → ee) and Γ(Υ' → ee) as computed by Quigg and Rosner (Ref. 107). The solid lines show the lower limits for charge Q = 1/3 of the b quark, the dashed lines for charge Q = 2/3. The hatched area shows the most likely region for Q = 1/3. The data point indicates the average value of the DORIS experiments.

the most probable values for the leptonic widths if Q = 1/3. They were determined from twenty different potentials which reproduce Γ_{ee} for J/ψ and ψ'. The cross shows the position of the average value measured by the DORIS experiments. The data disagree with the Q = 2/3 prediction but are consistent with Q = 1/3. The experiments do not fix the sign of the charge. The comparison with the d and s quarks suggests Q = −1/3, however. It has become customary to name the new quark b for bottom (or beauty).

The Υ - Υ' mass difference

Despite the large difference in mass between J/ψ and Υ the Υ - Υ' mass difference is almost the same as between J/ψ and ψ'. In a nonrelativistic quark model the mass difference depends on the type of the potential. If Υ and Υ' are assigned to the 1^3S_1 and 2^3S_1 states of the $b\bar{b}$ system, respectively, the model predicts the mass difference shown in Table 5.4.

Table 5.4 The T-T' mass difference in the nonrelativistic quark model. m_Q = mass of quark

a) experiment: $m(\psi' - J/\psi) = 591 \pm 1$ MeV

$m(T' - T) = 558 \pm 10$ MeV

b) theory:

type of potential	$\Delta m(2^3S_1 - 1^3S_1)$	$\Delta m(T' - T)^*$
coulomb	$\sim m_Q$	1800 MeV
linear	$\sim m_Q^{-1/3}$	400 MeV
logarithmic	const	591 MeV

* computed from $\Delta m(\psi' - J/\psi)$

A pure logarithmic potential which predicts $\Delta m(T' - T)$ and $\Delta m(\psi' - J/\psi)$ to be the same is at variance with experiment while a superposition of a coulomb and a linear term could give the correct description.

Fig. 5.7 shows the potential

$$V = -\frac{4}{3}\frac{\alpha_s}{r} + ar$$

with $\alpha_s = 0.2$ and $a = 0.25$ GeV as a function of the radius. Indicated are also the compton wave lengths of the pion, the charm quark and the b quark. It seems reasonable to expect the $b\bar{b}$ spectroscopy to be slightly more Coulomb-like than the $c\bar{c}$ spectroscopy.

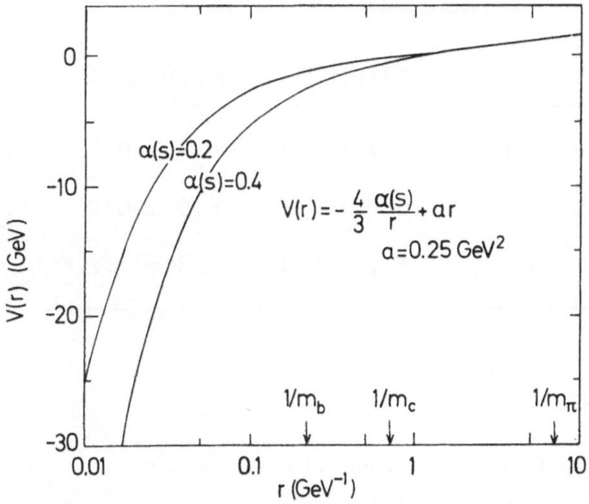

Fig. 5.7 The radius dependence of the quarkonium potential.

Υ final states

Apart from the μ pair decay no other exclusive decay channel of the Υ has yet been identified. The analysis of the overall features of the hadronic final states, however, produced very interesting results, must remarkably the difference in jet formation on and off the Υ. These results will be discussed in section 7.

The average number of charged particles is larger on resonance by roughly one unit. This feature is observed by all three storage ring experiments (see Table 5.5).

Table 5.5 Average charged particle multiplicities at the Υ and off resonance. The values quoted are raw numbers and have not been corrected for acceptance.

	DASP2[104]	D-H-HD-M[102]	PLUTO[106]
$\langle n_{ch} \rangle$ off resonance	5.2 ± 0.1	6.4 ± 0.2	4.9 ± 0.1
$\langle n_{ch} \rangle$ on resonance	5.7 ± 0.1	6.9 ± 0.2	5.4 ± 0.1
$\langle n_{ch} \rangle$ Υ direct decay	6.2 ± 0.1	7.3 ± 0.2	5.9 ± 0.1

The PLUTO group has determined the cross section for inclusive K^o_S production. The ratio

$$\frac{\sigma_{on}(K^o_S)}{\sigma_{off}(K^o_S)} = 4.0 \pm 1.7$$

should be compared to the ratio of the total cross section

$$\frac{\sigma^{tot}_{on}}{\sigma^{tot}_{off}} \simeq 2.5 \ .$$

In both cases an average was taken over the resonance region. No significant change is seen in K^o production on and off the resonance.

A similar conclusion was drawn by the DASP2 group who studied inclusive charged hadron production. The data are shown in terms of $\frac{s}{\beta}\frac{d\sigma}{dx}$ in Fig. 5.8. Within the large errors the same yield is found for $K^\pm$ and $\bar{p}$ relative to $\pi^\pm$ production.

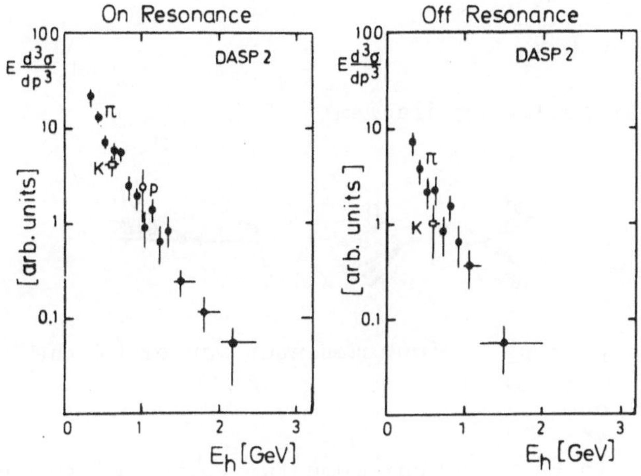

Fig. 5.8 The invariant cross section $E\,\frac{d^3\sigma}{dp^3}$ for the sum of $(\pi^+ + \pi^-)$, $(K^+ + K^-)$ and $2\cdot\bar{p}$ production on and off the T resonance (Ref. 104.

6. INCLUSIVE HADRON PRODUCTION

6.1 Basic Formulae

One of the basic properties of electron hadron scattering is the almost perfect scale invariance exhibited by the structure functions in the deep inelastic region. Inclusive hadron production in e^+e^- annihilation is expected to posses similar properties. First measurements on this subject were carried out by the SLAC-LBL collaboration who observed approximate scaling of the sum over all charged hadrons produced. In the DASP experiment the $\pi^{\pm}$, $K^{\pm}$ and $\bar{p}$ spectra were determined separately which allowed to test scaling for each particle species.

We start with a brief description of the formalism[108] for inclusive hadron production

$$e^+e^- \rightarrow h\ X \tag{6.1}$$

depicted by the following diagram:

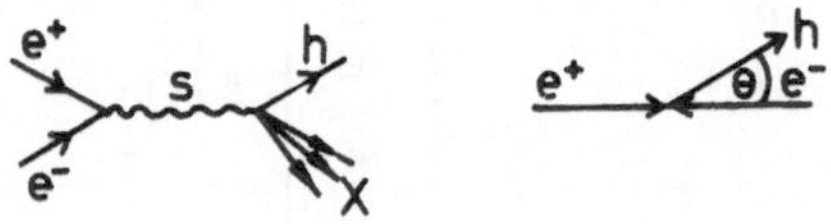

Define: $q = p_+ + p_-$ four momentum vector of the virtual photon

$q^2 = s$

$p = (\vec{p}, E)$ four momentum vector of the hadron h

Θ = production angle of h with respect to the e^+ direction

$x = \frac{2q\cdot p}{s} = \frac{2E}{\sqrt{s}}$ fractional energy of h

$\nu = \frac{q\cdot p}{m} = \frac{E}{m}\sqrt{s}$ energy of γ in h rest system

Note that $\frac{2m\nu}{s} = x$.

The virtual photon, as seen in the rest system of h, has transverse (T) and longitudinal (L) components. As a consequence the process (6.1) is described by two independent structure functions, e.g. $\bar{W}_T\ (s,\nu)$ and $\bar{W}_L\ (s,\nu)$. The differential cross section has the form

$$\frac{d^2\sigma}{dxd\Omega} = \frac{\alpha^2}{s}\frac{|\vec{p}|}{\sqrt{s}}\, m\ \{\bar{W}_T\ (1 + \cos^2\Theta) + \bar{W}_L\ (1 - \cos^2\Theta)\} \tag{6.2}$$

Special cases of (6.1) are pair production of fermions (e.g. $e^+e^- \rightarrow \mu^+\mu^-$) where $\bar{W}_L = 0$ and of scalar or pseudoscalar mesons (e.g. $e^+e^- \rightarrow \pi^+\pi^-$) where $\bar{W}_T = 0$. It is customary to use instead of

$\bar{W}_L, \bar{W}_T$ the structure functions $\bar{W}_1$ and $\bar{W}_2$ which are defined in terms of the tensor $\bar{W}_{\mu\nu}$:

$$\overline{W_{\mu\nu}} \equiv 4\pi^2 \frac{E}{m} \sum_n <0|J_\mu(0)|p,n> <n,p|J_\nu(0)|0> \cdot (2\pi)^4 \delta^4(p - p - p_n)$$

$$= -(g_{\mu\nu} - \frac{q_\mu q_\nu}{s}) \bar{W}_1(s,\nu) + \frac{1}{m^2}(P_\mu - \frac{p\cdot q}{s} q_\mu)(P_\nu - \frac{p\cdot q}{s} q_\nu)\cdot \bar{W}_2(s,\nu) \tag{6.3}$$

and

$$\bar{W}_1(s,\nu) = \bar{W}_T(s,\nu)$$
$$\bar{W}_2(s,\nu) = \frac{m^2}{|\vec{p}|^2} (\bar{W}_L(s,\nu) - \bar{W}_T(s,\nu)) \tag{6.4}$$

Note that in the pure transverse case ($\bar{W}_L = 0$) the equivalent of the Callan-Gross relation reads for $\beta = 1$

$$\nu\bar{W}_2 = - \frac{2m}{x} \bar{W}_1 \tag{6.5}$$

From (6.2) and (6.4) we find

$$\frac{d^2\sigma}{dxd\Omega} = \frac{\alpha^2}{s} \beta x \{m \bar{W}_1 + \frac{1}{4} \beta^2 x \bar{W}_2 \sin^2\Theta)\} \tag{6.6}$$

where $\beta = |\vec{p}|/E$. Note that $m\bar{W}_1 \geq 0$ since the cross section has to be a positive quantity. After integrating over the angles and replacing $4\pi\alpha^2/3s$ by $\sigma_{\mu\mu}$ one has

$$\frac{d\sigma}{dx} = 3\sigma_{\mu\mu}\beta x \{m \bar{W}_1 + \frac{1}{6} \beta^2 x \nu\bar{W}_2\} \tag{6.7}$$

For E >> m this simplifies to

$$\frac{d\sigma}{dx} = 3\sigma_{\mu\mu} x \{m \bar{W}_1 + \frac{1}{6} x \nu\bar{W}_2\} \tag{6.8}$$

If the structure functions $\bar{W}_1$ and $\nu\bar{W}_2$ obey scaling they become functions of the ratio ν/s alone. Using $x = \frac{2m\nu}{s}$ as the scaling variable and substituting

$$\begin{aligned} -m\bar{W}_1(s,\nu) &\equiv \bar{F}_1(x,s) \\ \nu\bar{W}_2(s,\nu) &\equiv \bar{F}_2(x,s) \end{aligned} \tag{6.9}$$

scale invariance is defined as

$$\lim_{\substack{\nu\to\infty \\ x = \text{const}}} -m\bar{W}_1(s,\nu) = \lim_{\substack{s\to\infty \\ x = \text{const}}} \bar{F}_1(x,s) \equiv \bar{F}_1(x) \tag{6.10}$$

and similarly

$$\lim \nu\bar{W}_2(s,\nu) = \bar{F}_2(x)$$

Scale invariance leads to the following expression for the inclusive cross section

$$\frac{d\sigma}{dx} = 3\sigma_{\mu\mu} x \,\{-\bar{F}_1(x) + \frac{1}{6}\, x\, \bar{F}_2(x)\} \tag{6.11}$$

If scale invariance is fulfilled the shape of the particle energy spectra, $d\sigma/dx$, is independent of s. Furthermore, the magnitude of the inclusive cross section behaves like s^{-1}.

Inclusive production of hadrons in e^+e^- annihilation is closely related to electron hadron scattering. We consider electroproduction on protons,

$$ep \to e'X \tag{6.12}$$

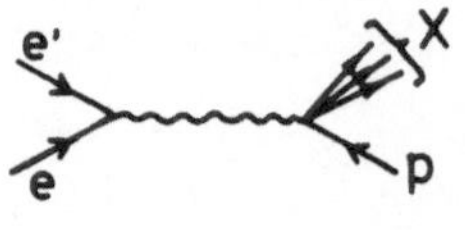

Notation:

$q = p_e - p_{e'}$ = virtual photon, $q^2 < 0$

P = incoming proton

Θ = scattering angle of e' with respect to e

$$\nu \equiv \frac{p\cdot q}{m} \quad = \text{photon energy in the rest system of the incoming proton (= lab. system)}$$

$$\nu = E_e - E_{e'}$$

$$\omega = \frac{2p\cdot q}{-q^2} = \frac{2m\nu}{-q^2}$$

The electroproduction cross section reads

$$\frac{d^2\sigma}{dE'_e \, d\cos\Theta} = \frac{8\pi\alpha^2}{(q^2)^2} E'^2 \{W_2(q^2,\nu)\cos^2\frac{\Theta}{2} + 2W_1(q^2,\nu)\sin^2\frac{\Theta}{2}\} \tag{6.13}$$

where the structure functions W_1 and W_2 are defined through the tensor $W_{\mu\nu}$:

$$\begin{aligned} W_{\mu\nu} &= 4\pi^2 \frac{E}{m} \sum \langle p|J_\mu(0)|n\rangle \langle n|J_\nu(0)|p\rangle (2\pi)^4 \delta^4(q + p - p_n) \\ &= -(g_{\mu\nu} - \frac{q_\mu q_\nu}{q^2})W_1(q^2,\nu) + \frac{1}{M^2}(p_\mu - \frac{p\cdot q}{q^2}q_\mu)(p_\nu - \frac{p\cdot q}{q^2}q_\nu)W_2(q^2,\nu) \end{aligned} \tag{6.14}$$

With the substitutions

$$\begin{aligned} mW_1(q^2,\nu) &\equiv F_1(\omega,q^2) \\ \nu W_2(q^2,\nu) &\equiv F_2(\omega,q^2) \end{aligned} \tag{6.15}$$

Scale invariance leads to

$$\lim_{\substack{\nu\to\infty \\ \omega \text{ fixed}}} mW_1(q^2,\nu) = \lim_{\substack{-q^2\to\infty \\ \omega \text{ fixed}}} F_1(\omega,q^2) \equiv F_1(\omega) \tag{6.16}$$

In field theory the tensors $W_{\mu\nu}$ and $\bar{W}_{\mu\nu}$ are related by crossing symmetry:

$$\overline{W_{\mu\nu}}(q,p) = - W_{\mu\nu}(q,-p) \tag{6.17}$$

(We consider here only the case where h is a fermion. If h would be a boson, $\bar{W}_{\mu\nu}(q,p) = W_{\mu\nu}(q,-p)$)

Consequently

$$m\bar{W}_1(q^2,\nu) = - mW_1(q^2,-\nu)$$
$$\nu\bar{W}_2(q^2,\nu) = (-\nu)W_2(q^2,-\nu) \tag{6.18}$$

The kinematical regions for annihilation and the scattering processes are separated in the q^2,ν plane ($q^2 > 4\ m^2$ and $q^2 < 0$). The analytic continuation does not exist in general. In the scaling limit the two processes have the point $\omega = x = 1$ in common:[109]

$$\bar{F}_1(x = 1) = -F_1(\omega = 1)$$
$$\bar{F}_2(x = 1) = F_2(\omega = 1) \tag{6.19}$$

In certain field theories Gribov and Lipatov[110] found for $e^+e^- \to pX$

$$\bar{F}_1(x) = -\frac{1}{x} F_1(\omega = \frac{1}{x})$$

and

$$\bar{F}_2(x) = -\frac{1}{x^3} F_2(\omega = \frac{1}{x}) \tag{6.20}$$

6.2 Momentum Spectra without Particle Separation

We first discuss spectra of charged particles not separated by mass. In the SLAC-LBL experiment[111] momentum spectra were determined for events with three or more charged particles registered in the detector. In the PLUTO experiment[112] two charged particles were required per event while the DASP group employed a genuine inclusive trigger[113]. Fig. 6.1 shows a plot of $s\,\frac{d\sigma}{dx}$ as measured by SLAC-LBL for $\sqrt{s}$ = 3.0, 4.8 and 7.4 GeV. The data were summed over all charged particle species. Consequently, instead of the scaling variable $x_E = 2E_h/\sqrt{s}$ the variable $x = x_p = 2p_h/\sqrt{s}$ was used (note: $x_p \simeq x_E - 2m_h^2/(x_E\cdot s)$). The PLUTO data obtained at energies between 3.6 and 5.0 GeV are shown in Fig. 6.2. For $x \lesssim 0.6$ the cross section points measured in the three experiments agree within their systematic uncertainties. At higher x values the SLAC-LBL values are higher compared to those from PLUTO (Fig. 6.3). All three experiments observe the same qualitative behaviour: the x spectrum passes through a maximum at small x values which is followed by an exponential fall-off. The latter is reminiscent of hadron-hadron collisions.

As a function of s, $s d\sigma/dx$ shows the following features:
(a) For $x \geq 0.4$ (and $\sqrt{s} \geq 3$ GeV) the spectra are the same to within a factor of ~1.5, i.e. scale invariance is satisfied to within this

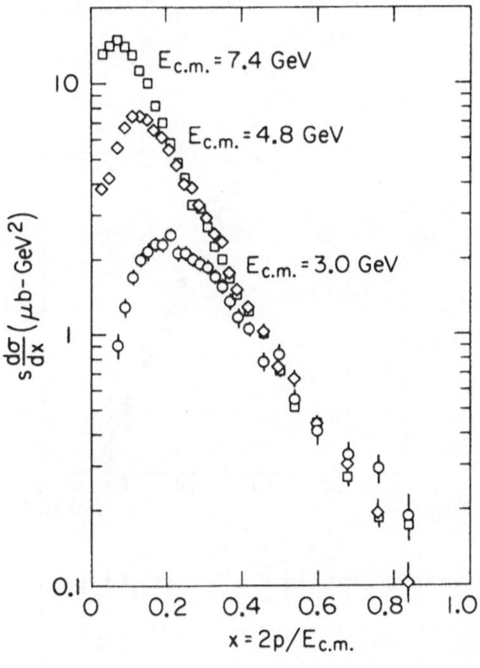

Fig. 6.1 $s d\sigma/dx$ for c.m.s. energies E_{cm} of 3.0, 4.8 and 7.4 GeV; $x \equiv 2p/E_{cm}$ (Ref.).

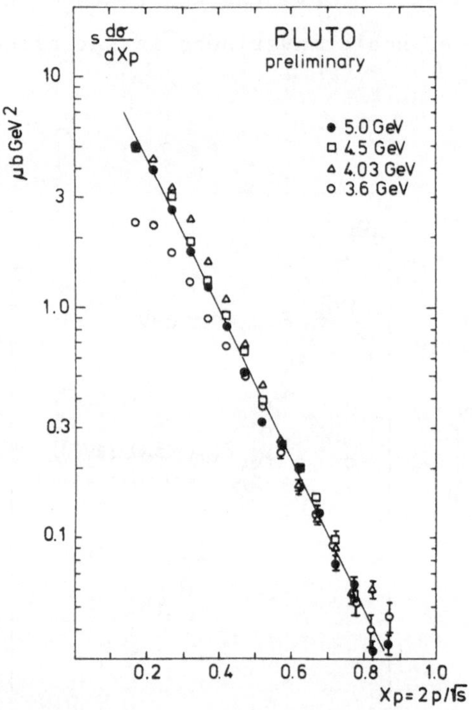

Fig. 6.2 Momentum spectra of charged particles. (Ref.112)

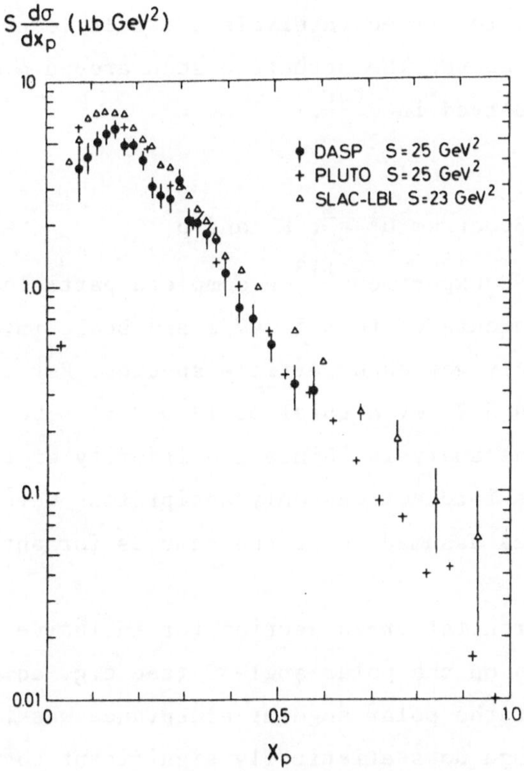

Fig. 6.3 Momentum spectra of charged particles

accuracy. The region in x over which scaling is fulfilled is extended downwards to smaller values of x with increasing s: For $\sqrt{s} \geq 5$ GeV, $x_{scaling} \geq 0.2$. (b) $\sigma^{tot}/\sigma_{\mu\mu}$ rises by a factor of 2 between 3.8 and 4.6 GeV. This additional contribution to σ^{tot} is seen to add only events at small x($x \leq 0.4$).

In Fig. 6.4 scale invariance is tested by plotting $s\,\frac{d\sigma}{dx}$ as a function of s for fixed intervals of x. It leads to the same conclusions as above. The structure seen around 4 GeV corresponds to the bump observed in σ^{tot}.

6.3 Momentum Spectrum of $\pi^{\pm}$, $K^{\pm}$ and $\bar{p}$

In the DASP experiment[113] a complete particle separation was possible for momenta up to 1.5 GeV/c and scale invariance was tested separately for each particle species. For c.m. energies between 3.6 and 5.2 GeV a total of 13 000 $\pi^{\pm}$, 890 $K^{\pm}$ and 130 $\bar{p}$ were used in the analysis. Since the majority of the protons were due to beam gas interactions only antiprotons were considered. The proton yield was assumed to be the same as for antiprotons.

The differential cross section for inclusive production in general depends on the polar angle Θ (see e.g. eq.6.6). In the DASP experiment the polar angular acceptance was $|\cos\Theta| < 0.55$. Within this range no statistically significant $\cos\Theta$ dependence was observed and a constant angular distribution was assumed in order to integrate the cross section over $\cos\Theta$. The possible error introduced by this procedure was estimated by considering the limit that only transverse photons contribute, $W_L = 0$. In this case the DASP cross sections given below would have to be increased by at most 24 %. With the angular dependence observed by SLAC-LBL[111] the estimated increase is less than 5 % for $x \leq 0.5$ and ≈ 13 % above.

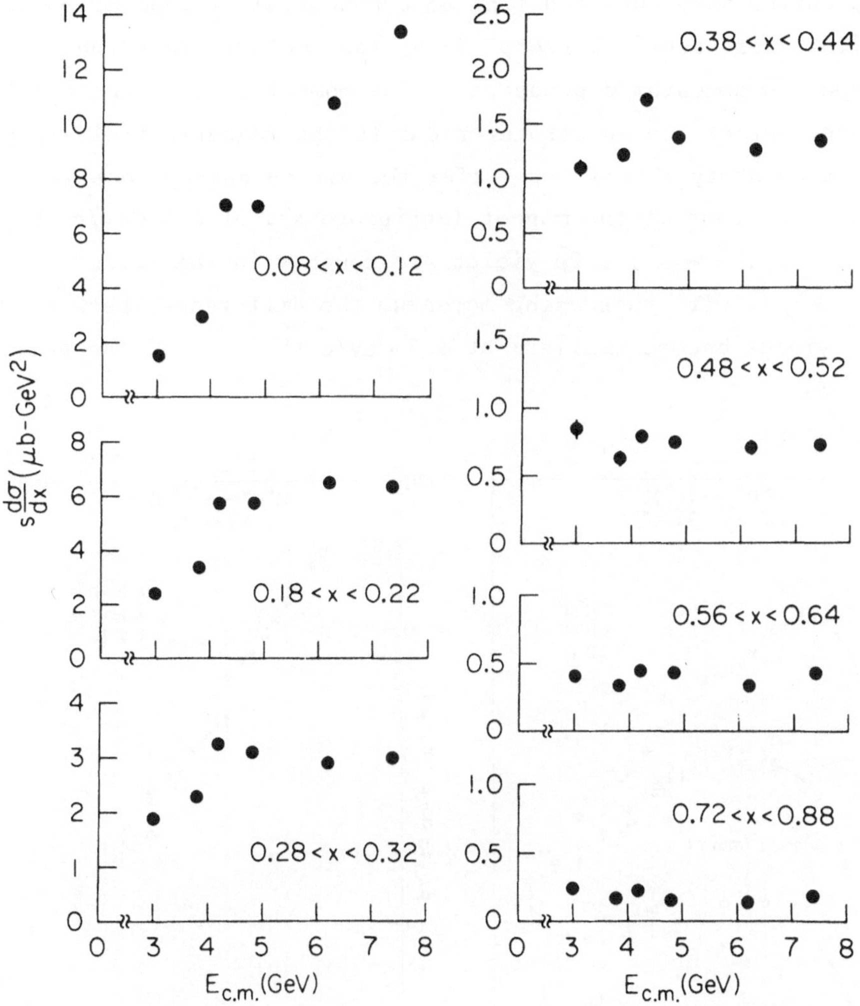

Fig. 6.4 $sd\sigma/dx$ as a function of the c.m. energy $E_{c.m.}$ for various x intervals; $x = 2p/E_{c.m.}$ (Ref. 111)

The relative frequency of pions, kaons and nucleons can be read off from Fig. 6.5 which shows the cross section $d\sigma/dp$ as a function of momentum averaged over c.m. energies from 4 to 5.2 GeV for the sum of π^+ and π^-, K^+ and K^-, and twice the antiproton yield. The dashed curves were obtained from an exponential extrapolation of the invariant cross sections $E/4\pi p^2\, d\sigma/dp$ (see below). They indicate the expected momentum dependence at low momenta, where the particles are swept out of the spectrometer due to the magnetic field (pions), are lost by decay (kaons) or suffer too big an energy loss in the material in front of the magnet (antiprotons). At 0.5 GeV/c the $(\pi^+ + \pi^-) : (K^+ + K^-) : 2\bar{p}$ yields are roughly in the ratio 100 : 10 : 1. With increasing momentum the differences between $\pi^\pm$ and $K^\pm$ yields become smaller: at 1.5 GeV/c they are of the same magnitude.

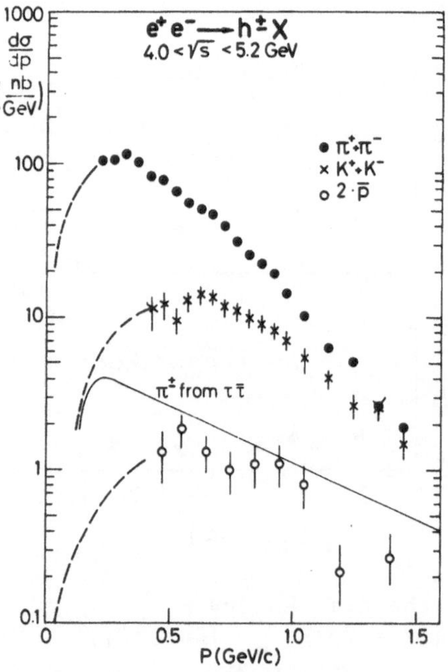

Fig. 6.5 Momentum spectrum of $\pi^\pm$, $K^\pm$ and $\bar{p}$ as measured by DASP

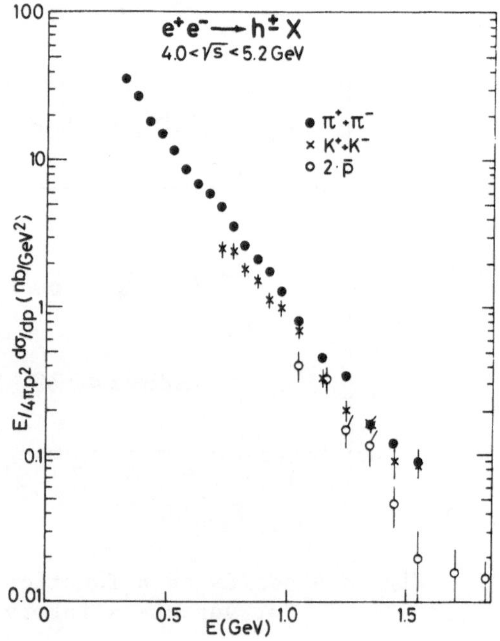

Fig. 6.6 The invariant cross section for $\pi^\pm$, $K^\pm$ and $\bar{p}$ as measured by DASP

The inclusive cross sections include the contributions from τ pair production,

$$e^+e^- \rightarrow \tau\bar{\tau} \rightarrow h^\pm X.$$

The pion yield from τ production was calculated by Monte Carlo using the decay branching ratios predicted by theory for a τ of 1.8 GeV mass. The pion yield is shown by the solid curve in Fig.6.5. It is suppressed below 0.6 GeV/c because of the selection criteria described above. The τ contribution accounts for ~4 % of all pions at p = 0.6 GeV/c and ~25 % at 1.5 GeV/c. The τ contributions to $K^\pm$ and $\bar{p}$ production are negligibly small. The data shown below have not been corrected for the contribution from τ production.

Fig. 6.6 shows the same data plotted in terms of the invariant cross section $Ed^3\sigma/dp^3 \simeq (E/4\pi p^2)\, d\sigma/dp$. To within 20 or 30 % accuracy the $\pi^\pm$, $K^\pm$ and $2\bar{p}$ cross sections fall on the same curve, which is well approximated by an exponential,

$$(E/4\pi p^2)\, d\sigma/dp \sim \exp(-bE). \tag{6.21}$$

E_{cm}=4–5.2 GeV b_π=5.0±0.1 GeV^{-1} b_K=4.9±0.2 GeV^{-1} $b_{\bar{p}}$=5.4±0.5 GeV^{-1}

Inclusive spectra in hadronic collisions behave in a similar manner when plotted as a function of the transverse energy $E_T = \sqrt{m^2 + p_T^2}$

6.4 Particle Multiplicities

From the measured momentum distributions the DASP group[113] calculated the average numbers of charged pions, kaons and nucleons, e.g.

$$\langle n_{\pi^\pm}\rangle = \frac{\int_0^{P_{max}} \frac{d\sigma}{dp}(e^+e^- \rightarrow \pi^\pm X)\, dp}{\sigma_{tot}}$$

The extrapolation of $d\sigma/dp$ to zero momentum was done using the exponential fits described by eq. (6.21). For σ_{tot} the data obtained in the same experiment[20] were used. The fraction of the cross section determined by extrapolation amounted to 10 % for the majority of the pion data, to 20 % for kaons and to 30 % for antiprotons. The particle multiplicities are shown in Fig. 6.7 as a function of the c.m. energy. The errors given are purely statistical. The systematic errors of the integrated inclusive cross sections and of σ_{tot} add a further uncertainty of 20 - 25 % of which ~15 % are due to normalization. Above charm threshold on the average 3 - 4 $\pi^{\pm}$, 0.5 - 0.6 $K^{\pm}$ and 0.05 - 0.08 p and $\bar{p}$ are produced per event.

6.5 Test for Scaling

In general the structure functions $\overline{W}_1$ and $\nu\overline{W}_2$ depend on two variables, e.g. x and s. If scale invariance is fulfilled $\overline{W}_1$ and $\overline{W}_2$ are functions of x alone and the cross section $s/\beta\ d\sigma/dx$, is almost the same for all values of s. (See eqn. 6.6).

In Fig. 6.8 the scaling cross sections $s/\beta\ d\sigma/dx$ as measured by the DASP group[113] for $\pi^{\pm}$, $K^{\pm}$ and $2\bar{p}$ production are plotted as a function of x in eight energy intervals between 3.6 and 5.2 GeV. Most spectacular is the similarity between the three types of particles. Within a factor of two their cross sections seem to fall on a common curve decreasing exponentially for $x \gtrsim 0.2$. No peculiarities are observed when comparing the shapes of the cross sections measured at the 4.04. 4.16 and 4.4 GeV resonances (second, third and fifth interval) and outside.

In order to test the pion data for scaling Fig. 6.9 compares the π cross sections outside the resonance region at energies of $\sqrt{s}$ = 3.6 and 5.2 GeV. Below x = 0.25 the cross section rises by a factor of 1.5 to 2 between $\sqrt{s}$ = 3.6 and 5.2 GeV. At higher x values the two cross section sets agree within errors. This shows that

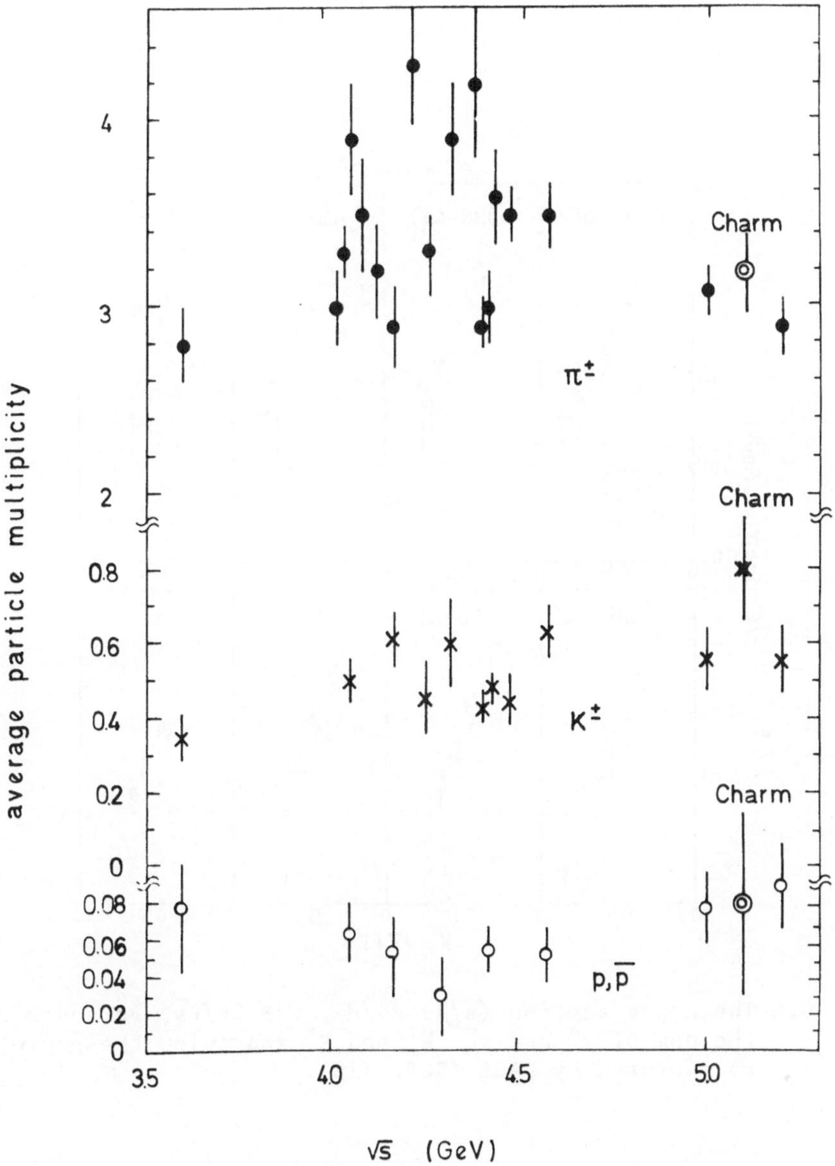

Fig. 6.7 Average $\pi^{\pm}$, $K^{\pm}$ and p,$\bar{p}$ multiplicities per event (Ref. 113).

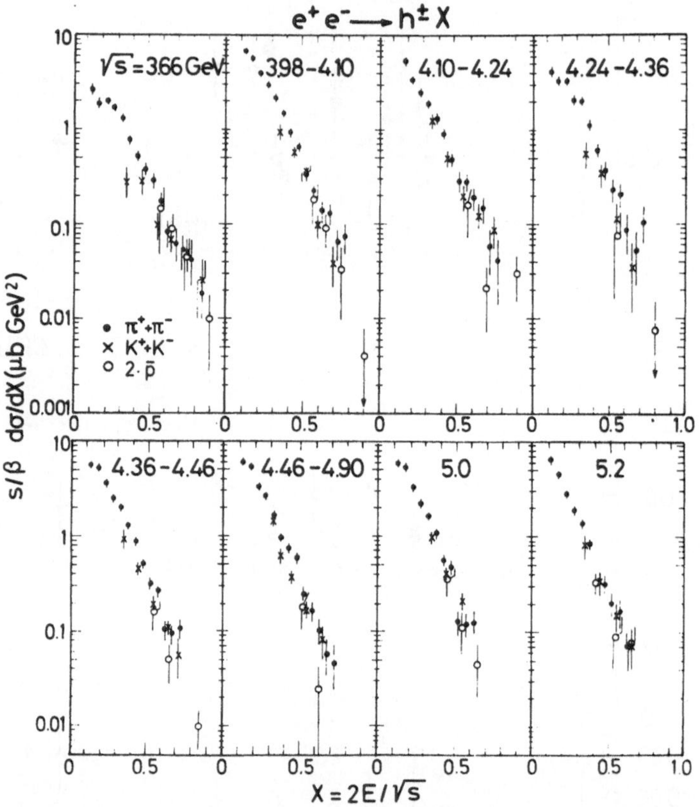

Fig. 6.8 The cross section $(s/\beta)\ d\sigma/dx$, $x \equiv 2E/\sqrt{s}$, versus x for the sum of π^+ and π^-, K^+ and K^- and twice the $\bar{p}$ yield, as measured by DASP (Ref. 113).

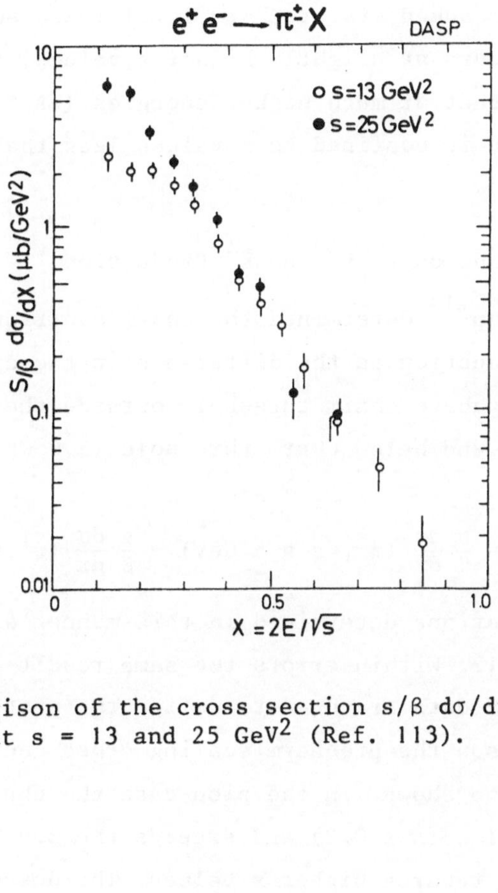

Fig. 6.9 Comparison of the cross section $s/\beta\ d\sigma/dx$ for $\pi^{\pm}$ production at s = 13 and 25 GeV^2 (Ref. 113).

the rise of R from a value of 2.3 at $\sqrt{s}$ = 3.6 GeV to 4.5 above $\sqrt{s}$ = 5 GeV is associated with low x pions.

Fig. 6.10 shows the $\pi^{\pm}$ and $K^{\pm}$ data plotted as a function of s for fixed x. Both cross sections for x < 0.5 rise by a factor of two to three when crossing the charm threshold near $\sqrt{s}$ = 4 GeV. In the pion case for x > 0.2 this added contribution disappears above the resonance region. For x > 0.3 the scaling cross section at $\sqrt{s}$ = 5.2 GeV has reached its precharm level measured at $\sqrt{s}$ = 3.6 GeV. For kaons this occurs at slightly higher x values, x ≈ 0.4. We may therefore expect that at much higher energies ($\sqrt{s}$ >> 5 GeV) the charm contribution is confined to x values less than 0.3.

6.6 Charm Contribution to $\pi^{\pm}$ and $K^{\pm}$ Production

The DASP group[113] determined the charm contribution to charged pion and kaon production as the difference in the cross sections for c.m. energies above charm threshold outside the resonances (5.0 and 5.2 GeV) and below charm threshold (3.6 GeV), viz.

$$\frac{s}{\beta}\frac{d\sigma}{dx}^{\text{charm}}(\pi^{\pm}) = \frac{s}{\beta}\frac{d\sigma}{dx}(\pi^{\pm},\sqrt{s} = 5\text{ GeV}) - \frac{s}{\beta}\frac{d\sigma}{dx}(\pi^{\pm},\sqrt{s} = 3.6\text{ GeV}).$$

The charm contributions determined in this manner are plotted in Figs. 6.11 and 6.12. Within errors the same result is found whether the 5.0 or 5.2 GeV data are used to define the post charm-threshold data. For comparison the precharm scaling cross sections measured at 3.6 GeV are also shown. In the pion case the charm contribution is large for small x (x ≲ 0.2) and exceeds the 3.6 GeV values. It falls off rapidly towards higher x values. The descent is steeper than for the precharm data. For x values above 0.3 the charm contribution is close to zero. The data for kaon production behave in a similar manner although the conclusions are less firm because of the larger statistical errors. For x values between 0.3 and 0.4

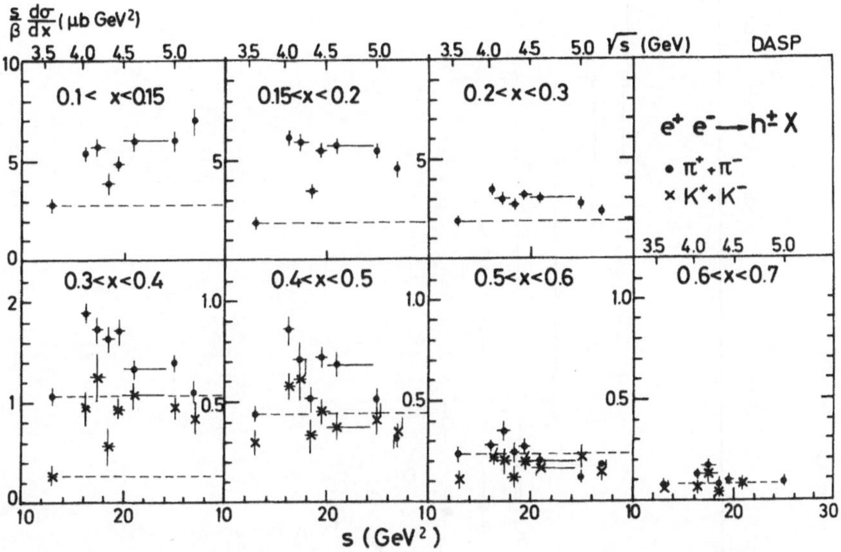

Fig. 6.10 The cross section $(s/\beta)\ d\sigma/dx$ versus s for fixed x for $\pi^\pm$ and $K^\pm$ production as measured by DASP (Ref.113).

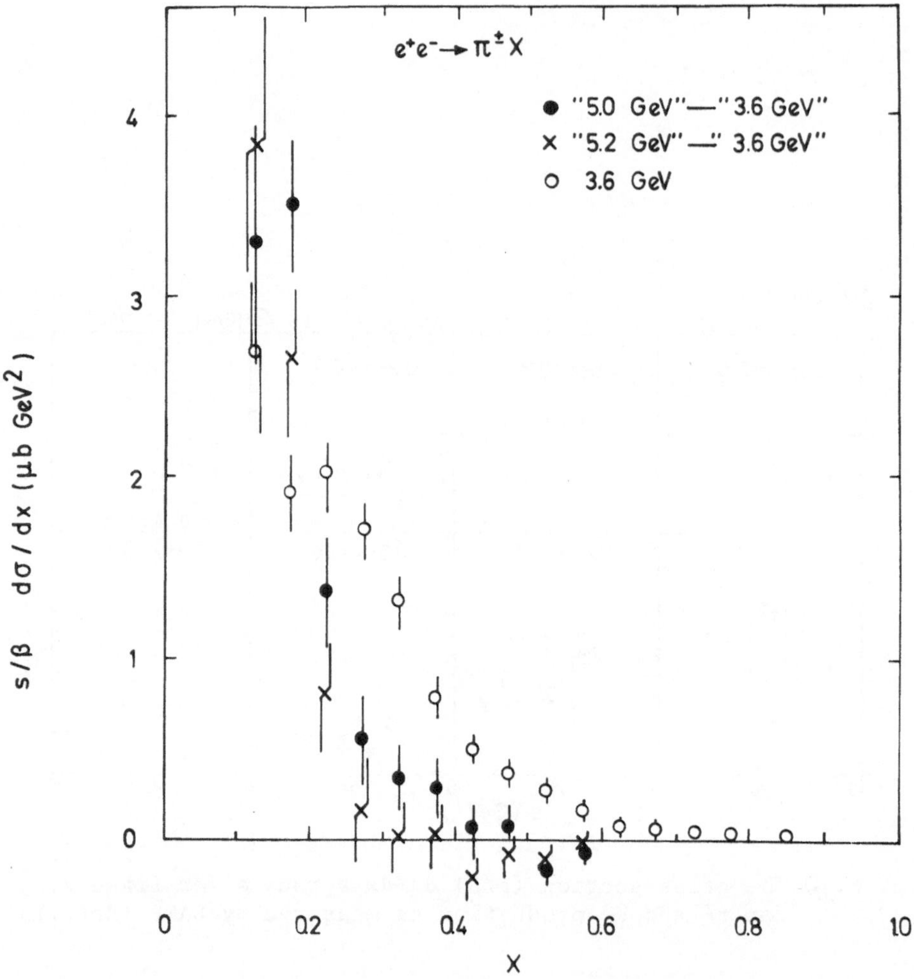

Fig. 6.11 The charm contribution to the scaling cross section of $\pi^\pm$ production determined as the difference between the 3.6 GeV and 5.0 GeV data (●) and as the difference between the 3.6 GeV and 5.2 GeV data (×). The open points show the $\pi^\pm$ scaling cross section measured below charm threshold at 3.6 GeV (Ref. 113).

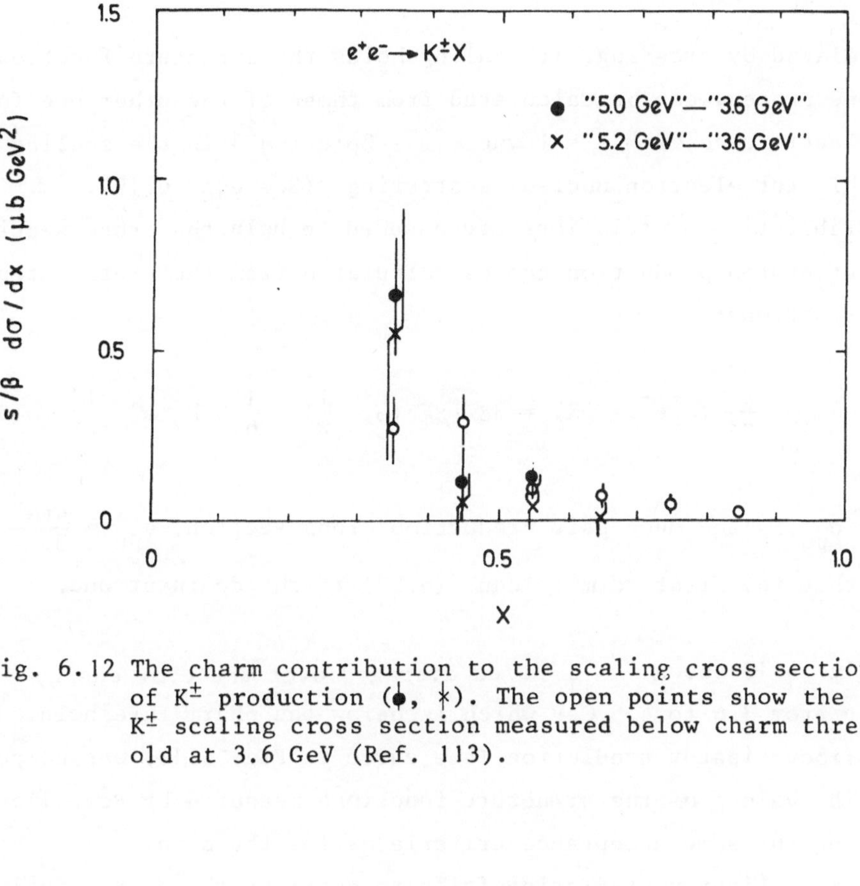

Fig. 6.12 The charm contribution to the scaling cross section of $K^{\pm}$ production (●, ×). The open points show the $K^{\pm}$ scaling cross section measured below charm threshold at 3.6 GeV (Ref. 113).

the charm contribution is larger than the 3.6 GeV data. For x values above 0.4 the charm contribution is small.

6.7 e^+e^- Annihilation and Inelastic ep Scattering

Inclusive antiproton production by e^+e^- annihilation

$$e^+e^- \rightarrow \bar{p}X$$

and electron proton scattering,

$$ep \rightarrow eX$$

are related by crossing. If scaling holds the structure functions for one process can be calculated from those of the other one for the elastic case, $x = \omega = 1$ where $\omega = 2p\cdot q/(-q^2)$ is the scaling variable for electron nucleon scattering (See eqn. 6.19). If the Gribov-Lipatov relations are assumed to hold the cross section for antiproton production can be calculated from the proton structure functions:

$$\frac{x}{\sigma_{\mu\mu}} \frac{d\sigma}{dx} (e^+e^- \rightarrow \bar{p}X) = 3\beta \{xF_1(\omega = \frac{1}{x}) - \frac{1}{6}\beta^2 F_2(\omega = \frac{1}{x})\} \quad (6.22)$$

where $\sigma_{\mu\mu}$ is the muon pair production cross section, $\sigma_{\mu\mu} = \frac{4\pi\alpha^2}{3s}$.
Note that the first term in eqn. (6.22) is the dominant one.

Fig. 6.13 shows $x/\sigma_{\mu\mu}\, d\sigma/dx$ averaged over the c.m. energy region from 3.6 to 4.5 GeV which is below the charm threshold. The Gribov-Lipatov prediction (see curve in Fig. 6.13) was computed from the values of the structure functions measured by Ref. 115 imposing the same acceptance criteria as for the data.
The Gribov-Lipatov prediction fails to describe the data quantitatively. The theoretical curve is always below the measured points. The discrepancy appears to increase towards smaller values of x

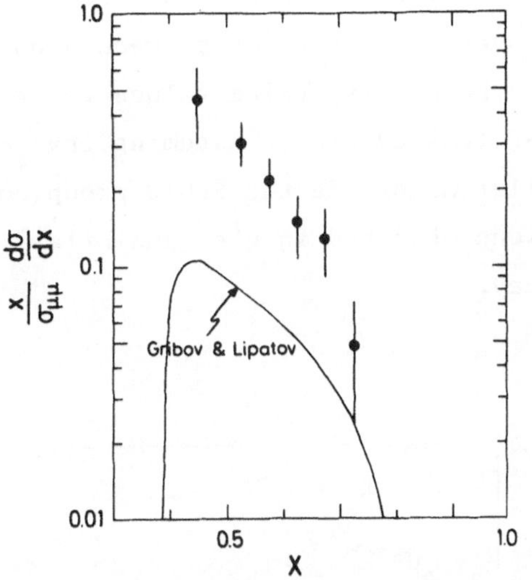

Fig. 6.13 The quantity $x/\sigma_{\mu\mu}\ d\sigma/dx$ versus x for $e^+e^- \to \bar{p}x$ averaged over c.m. energies from 4.0 to 4.5 GeV. The curve shows the prediction of Gribov and Lipatov (from Ref. 113).

reaching a factor of three at x = 0.5. Part of this failure - if not all - may have to be attributed to contributions of the type

$$e^+e^- \to h^*X,\ h^* = \bar{\Lambda} \to \bar{p}..,\ \bar{\Sigma} \to \bar{p}..,\ \bar{N}^* \to \bar{p}... \quad \text{etc.}$$

which should be excluded from the e^+e^- data before a comparison is made[116].

6.8 Inclusive Rho Production

The PLUTO group[117] analysed inclusive ρ^o production. Identifying all charged particles as pions the Pluto group observed a clear ρ^o signal in the $\pi^+\pi^-$ effective mass distribution (Fig. 6.14).

Fig. 6.15 shows the scaling cross section $s/\beta\ d\sigma/dx$ ($x \equiv 2E_\rho/\sqrt{s}$) versus x averaged over c.m. energies between 4 and 5 GeV. The cross section lies above the corresponding values for π^+ <u>or</u> π^- production as measured by DASP (dashed line). Assuming that charged rho production is of similar magnitude the PLUTO group concludes that the majority of the pions observed in e^+e^- annihilation result from rho production and decay.

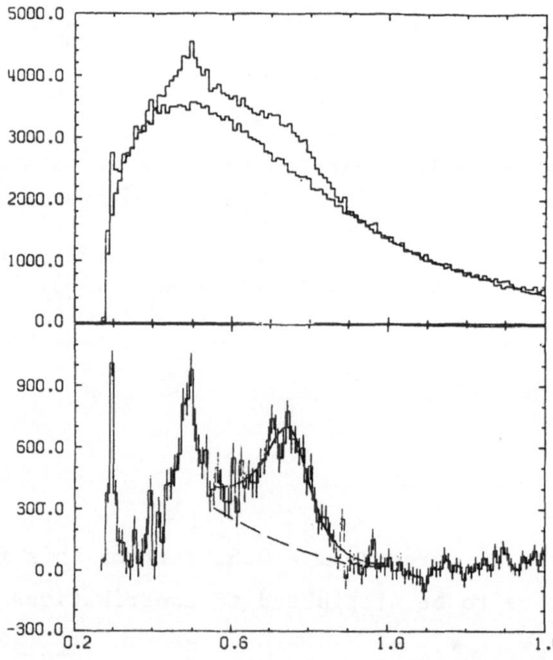

Fig. 6.14 $e^+e^- \rightarrow$ hadrons for c.m. energies between 4 and 5 GeV. The unsubtracted $\pi^+\pi^-$ mass distribution (top) and after subtracting a smooth background (bottom) (Ref. 117).

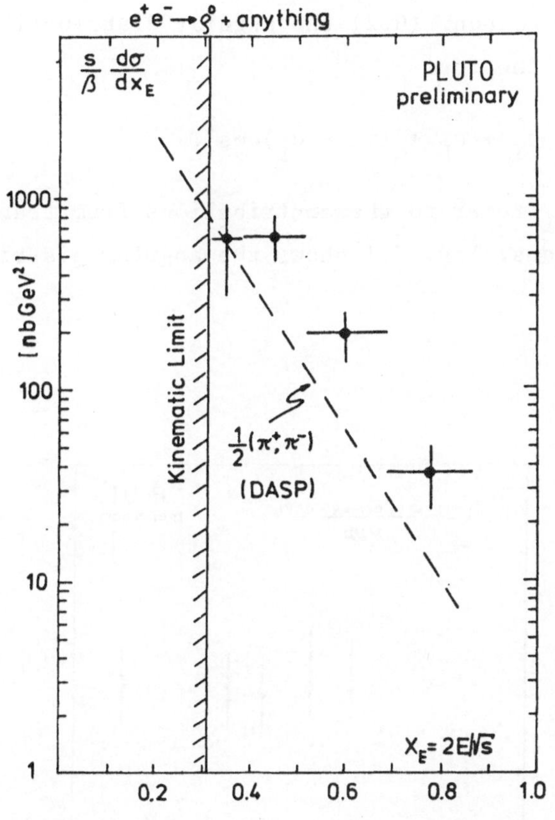

Fig. 6.15 The scaling cross section for inclusive ρ^{0} production between 4 and 5 GeV (Ref. 117).

In Fig. 6.16 the total ρ^{0} cross section is shown relative to $\sigma_{\mu\mu}$; R_{ρ} is seen to rise from a value around 0.8 below charm threshold to ≈ 1.3 above.

7. JET FORMATION

7.1 Angular Distributions

According to eqn. (6.2) the angular distribution of particles produced is of the form

$$\frac{d\sigma}{d\Omega} \sim \sigma_T + \sigma_L + (\sigma_T - \sigma_L)\cos^2\Theta \qquad (7.1)$$

where σ_T and σ_L refer to the contributions from transverse and longitudinal photons. Fig. 7.1 shows the angular distributions for

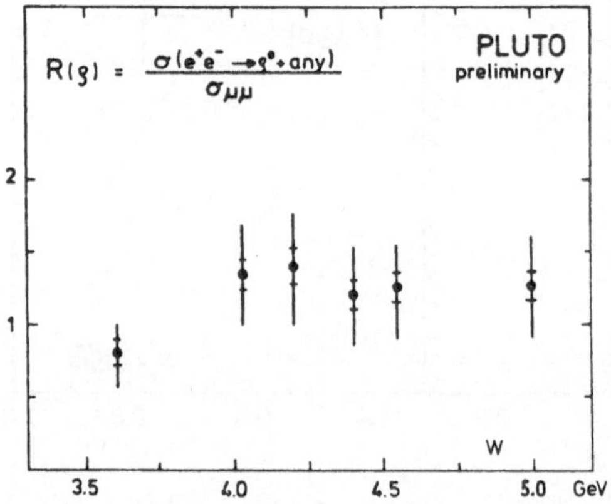

Fig. 6.16 The total cross section for inclusive ρ^{0} production relative to $\sigma_{\mu\mu}$ (Ref.117).

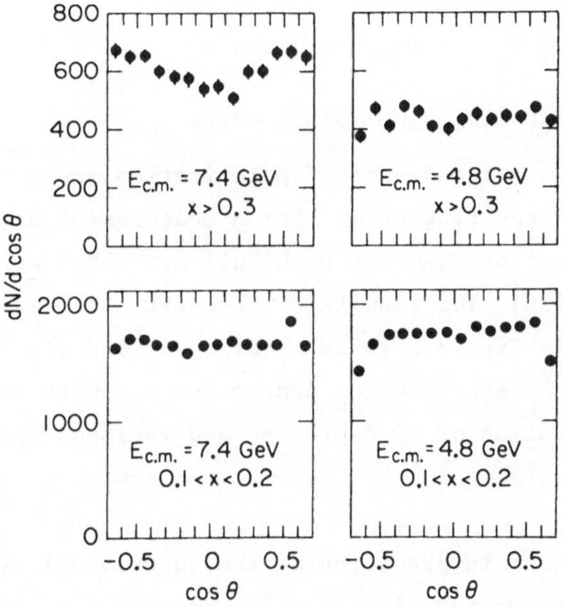

Fig. 7.1 cosΘ distribution for e^+e^- events with three or more charged hadrons at $E_{c.m.}$ = 4.8 and 7.4 GeV for different $x = x_p = 2p/\sqrt{s}$ intervals (Ref. 118).

$e^+e^- \to h^\pm X$ measured by the SLAC-LBL group at 4.8 and 7.4 GeV separately for low x ($0.1 < x < 0.2$) and higher x ($x > 0.3$) values[118]. For low x the $\cos\Theta$ distributions are flat at both energies. At higher x values only the 7.4 GeV data show a $\cos\Theta$ dependence. There, the data imply $\sigma_T > \sigma_L$. Since the acceptance is limited to $|\cos\Theta| < 0.6$ these data are not sensitive to small differences between σ_T and σ_L. The sensitivity is greatly improved by using transverse polarized e^+ and e^- beams. In this case

$$\frac{d\sigma}{d\Omega} \sim \sigma_T + \sigma_L + (\sigma_T - \sigma_L) \cos^2\Theta$$

$$+ P_+P_- (\sigma_T - \sigma_L) \sin^2\Theta \cos 2\phi \qquad (7.2)$$

where P_+, P_- measure the degree of polarization and $\phi = 0$ in the plane of the storage ring beams. For a measurement of this type a detector like that of SLAC-LBL with full azimuthal acceptance near $\cos\Theta = 0$ is optimal. The results of SLAC-LBL obtained at $\sqrt{s} = 7.4$ GeV are summarized in Fig. 7.2 in terms of σ_L/σ_T and $(\sigma_T - \sigma_L)/(\sigma_T + \sigma_L)$[118,119]. As $x \to 0$ σ_L approaches σ_T. With increasing x the relative contribution of σ_L decreases and vanishes near $x = 1$. ($\sigma_L \ll \sigma_T$ at $x = 1$).

This behaviour is predicted by the quark model. The primary process is the formation of a $q\bar{q}$ pair. Because the quark spin $s_q = 1/2$ the production angular distribution of the quark pair is $\sim 1 + \cos^2\Theta$, i.e. $\sigma_L = 0$ (quark spin $s_q = 0$ would lead to $\sigma_T = 0$).

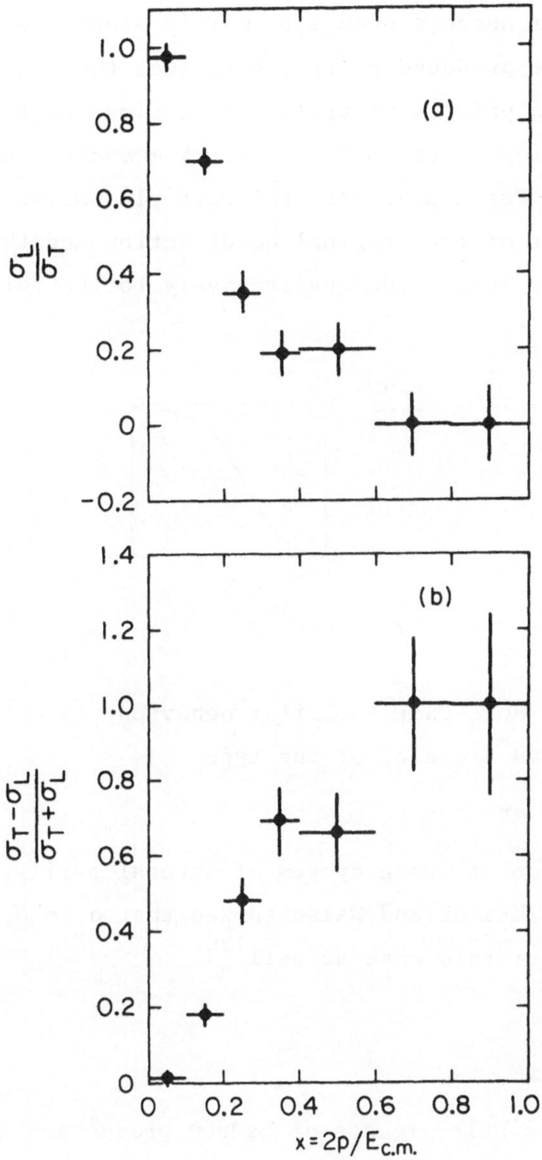

Fig. 7.2 (a) σ_L/σ_T versus $x = x_p = 2p/\sqrt{s}$ at $E_{c.m.} = 7.4$ GeV
(b) $(\sigma_T-\sigma_L)/(\sigma_T+\sigma_L)$ versus x_p at $E_{c.m.} = 7.4$ GeV (Ref. 118).

In a second step physical particle states are formed. The exchange of quantum numbers occurs primarily along the original $q\bar{q}$ direction and the produced particles reflect the cosΘ distribution of the primary $q\bar{q}$ pair. A particle with x close to 1 is emitted along the direction of the original quark momentum and therefore $\sigma_L \to 0$ as $x \to 1$. For x near zero the particle emission is practically independent of the original $q\bar{q}$ direction and therefore $\sigma_L \to \sigma_T$ as $x \to 0$. This leads qualitatively to the following picture:

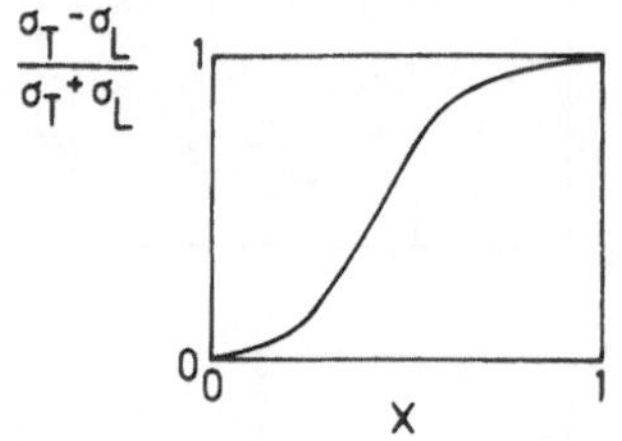

It is amusing to note that a similar behaviour is obtained if the dominant processes are e.g. of the type

$$e^+e^- \to \pi M$$

where M is a meson or meson system of natural parity, $P_M = (-1)^{J_M}$ (e.g. $M = \rho, A_2$). Kramer and Walsh showed that $\sigma_L = 0$ and $\frac{d\sigma}{d\Omega} \sim 1 + \cos^2\Theta$ for this case as well[120].

7.2 Jet Structure

If the quark pair picture of hadron production is valid jet structure must be observed: the hadrons must be emitted in two narrow cones back to back. The axis of the two jets is the direction of the primary $q\bar{q}$ pair. If we assume the hadrons to be emitted in a bremsstrahlung type process the transverse momentum of a hadron with respect to the jet axis will be of the order of the particle mass, $p_\perp \sim m_h$. Perhaps more precisely, the average transverse momentum will be the same as in hadronic collisions, e.g. $\langle p_\perp \rangle \sim 300$ MeV/c for pions. Consequently, jet structure will produce noticeable effects when the average particle energy is large compared to $\sqrt{2} \cdot 0.3$ GeV.

Jet structure was first observed by the SLAC-LBL group, studying e^+e^- collisions up to 7.4 GeV c.m. energy. It was recently confirmed by experiments running at DORIS and extending the measurements up to 10 GeV. In the latter experiments the neutral component was analyzed as well. It was found that the neutral particles (mainly π^0's) are also produced jet-like; furthermore, the neutral jet axis coincides with the one determined from charged particles. We start with a discussion of the SLAC-LBL data[118,119,122].

The SLAC-LBL collaboration has tested their data for jet structure in the following way.

1. For each event determine the axis with respect to which the transverse momenta of the detected particles are minimal,

$$\sum_i p_\perp^2 = \min \tag{7.3}$$

2. Calculate for each event the sphericity,

$$S = \frac{3 \sum_i p_\perp^2}{2 \sum p^2} \tag{7.4}$$

The sphericity is limited to values $0 \leq S \leq 1$. It is instructive to consider two extreme cases:

Isotropic particle production, all particles have the same momentum:

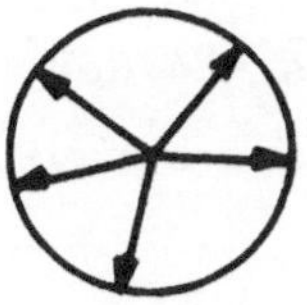

SP = 1

The end points of the momentum vectors lie on a sphere. If the particle multiplicity is large:

$$S = \frac{3 \int p^2 \sin^2\Theta \, d\cos\Theta d\phi}{2 \int p^2 d\cos\Theta d\phi} = 1$$

The other extreme is $p_\perp^2 \to 0$ for all particles. In this case $S \to 0$.

The sensitivity of the sphericity to the formation of jets suffered because only the momenta of charged particles are measured. The axis determined from eqn. (7.3) will in general not be the axis which minimizes the transverse momenta of all particles. In order to interpret the measured sphericity distributions the authors have generated Monte Carlo events according to two models:

- phase space
- jet formation with limited transverse momentum ($\langle p_\perp \rangle = 0.3$ GeV/c).

The artificial events contained charged as well as neutral particles and were subjected to the same selection criteria as the real events. For the analysis only events with three or more charged particles were selected.

Fig. 7.3 shows the S-distributions at $\sqrt{s} = 3.0$, 6.2 and 7.4 GeV and a comparison with these two models. At 3 GeV phase space like behaviour and jet structure are indistinguishable. This was to be expected since $\langle E \rangle \approx \sqrt{2} \; \langle p_\perp \rangle$. For 6.2 and 7.4 GeV the measured sphericities are on the average much smaller than predicted by phase space. This is also seen in Fig. 7.4 where the mean value of S is given versus $\sqrt{s}$. Above $\sqrt{s} = 4$ GeV the observed $\langle S \rangle$ values deviate markedly from a phase space behaviour but agree with the jet picture.

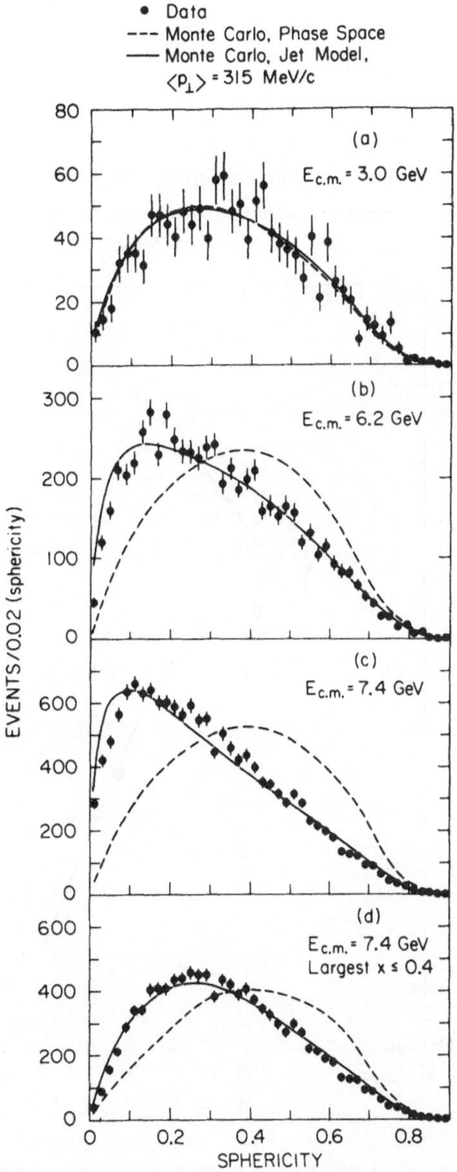

Fig. 7.3 Sphericity distributions for e^+e^- events with three or more charged hadrons at various c.m. energies. The solid curves are the predictions of the jet model ($\langle p_\perp \rangle$ = 315 MeV/c); the dashed curves show fits with invariant phase space. In (d), only those events where all particles have $x = x_p \leq 0.4$ were used (Ref. 118).

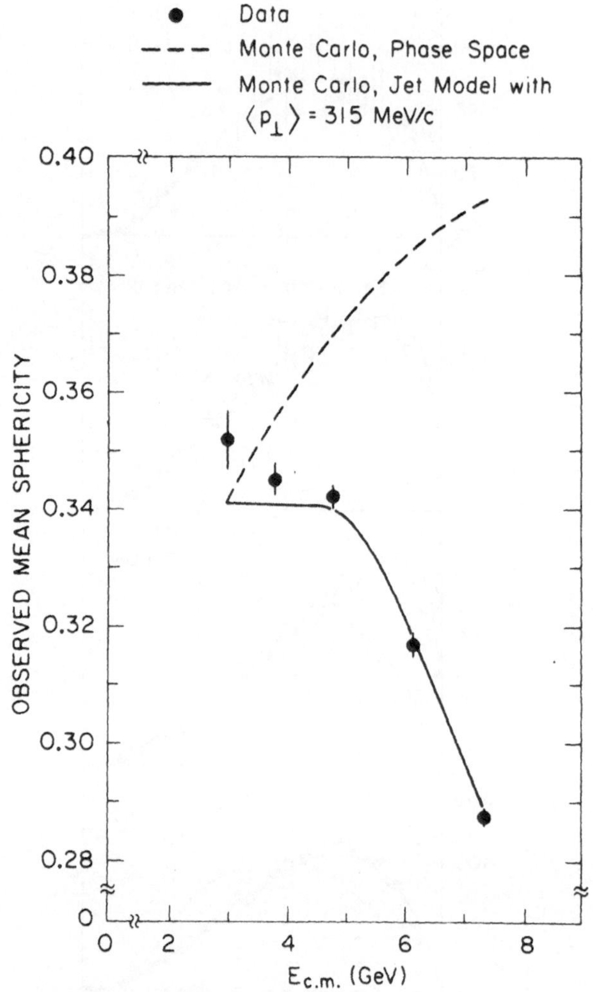

Fig. 7.4 Mean sphericity versus c.m. energy, $E_{c.m.}$ The solid curve is the result of the jet model with $\langle p_\perp \rangle$ = 315 MeV/c. The dashed curve is the invariant phase space prediction (Ref. 118).

In the quark model the jet axis is given by the direction of the primary $q\bar{q}$ pair. In this case, for transversely polarized beams the jet axis must show the same azimuthal anisotropy as the $q\bar{q}$ pair would, namely (see 7.2):

$$\frac{d\sigma}{d\Omega} \sim 1 + \alpha \cos^2\Theta + \alpha P_+P_- \sin^2\Theta \cos^2\phi \qquad (7.5)$$

with $\alpha = 1$. Fig. 7.5 shows the ϕ distribution of the measured jet axis at $\sqrt{s} = 6$ GeV where the beams were unpolarized, and at 7.4 GeV where $P_+P_- \simeq 0.5$.[118,119] The latter data exhibit a clear $\cos 2\phi$ signal. A fit of the observed ϕ anisotropy as a function of x yielded $\alpha = 0.97 \pm 0.1$ again in agreement with the quark picture.

7.3 Particle Emission in the Jet Frame

The observed characteristics of jets as well as the energy dependence and the magnitude of the total hadronic cross section support strongly the view that hadron production in e^+e^- collisions is a two step process where first a pair of quark-antiquarks is formed which then fragment into hadrons. The jet axis on the average measures the quark direction of flight.

If this view is adopted an analysis of hadron production with respect to the jet axis offers the possibility to determine the structure functions for quark fragmentation into hadrons. These may be compared to those deduced from hadron-hadron or lepton-hadron scattering. However, electron-positron annihilation has the advantage that effects due to the quark motion within the nucleon which e.g. may smear the $p_\perp$ distributions are absent.

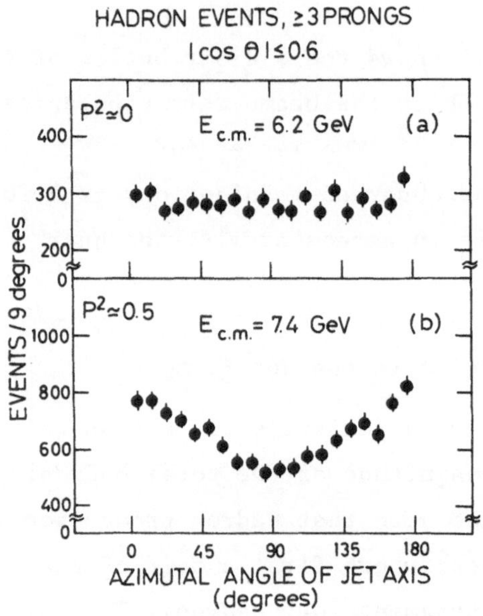

Fig. 7.5 Azimuthal distribution of the reconstructed jet axis; zero degree is in the ring plane = plane of beam polarization:

(a) for a c.m. energy of 6.2 GeV where the beam polarization is zero

(b) for a c.m. energy of 7.4 GeV where the product of the e^+ and e^- beam polarization is $p^2 = 0.5$ (Ref. 118).

The SLAC-LBL group studied the hadron emission in the jet frame[122]. As mentioned before neutral particles were omitted in the analysis. This led to biases in particular for the $p_\perp$ distributions. The biases were corrected for by means of Monte Carlo simulations. The essential input to the Monte Carlo was that only pions are produced and that the frequency of π^0 mesons is given by $N_{\pi^0} = (N_{\pi^+} + N_{\pi^-})/2$. The Lorentz invariant phase space was modified by the following matrix element

$$|M|^2 = \exp \{- (\sum p_{i\perp}^2)/2b^2\} \tag{7.6}$$

Note that $b^2 = \langle p_T^2 \rangle \cdot (N - 1)/N$ where N is the number of particles produced. The b value was chosen such that an average $p_\perp$ of 340 MeV/c was obtained in accordance with the bulk of data.

The data were analysed in terms of the Feynman scaling variable $x_{\parallel} = 2p_{\parallel}/\sqrt{s}$, the transverse momentum $p_\perp$ and the rapidity $y = 1/2 \ln (p_{\parallel} + E)/(p_{\parallel} - E)$ where $p_{\parallel}$ and $p_\perp$ denote the particle momentum components parallel and perpendicular to the jet axis.

In order to have a sample of events with a well defined jet axis, events with at least one high momentum particle were selected. High momentum was defined as $x_{max} > 0.3$. Having found a particle with $x > x_{max}$ the opposite jet (not containing this particle) was studied. This minimizes the biases introduced by the x_{max} cut. The distributions were normalized to the total number of jets studied, σ. The effect of the x_{max} cut is illustrated by Fig. 7.6 which shows the particle density distribution $(1/\sigma)\, d\sigma/dx_{\parallel}$ versus $x_{\parallel}$ around 7.4 GeV for different values of x_{max}. Except for $x_{max} > 0.7$ the "opposite jet" distributions are independent of the particular value of x_{max}.

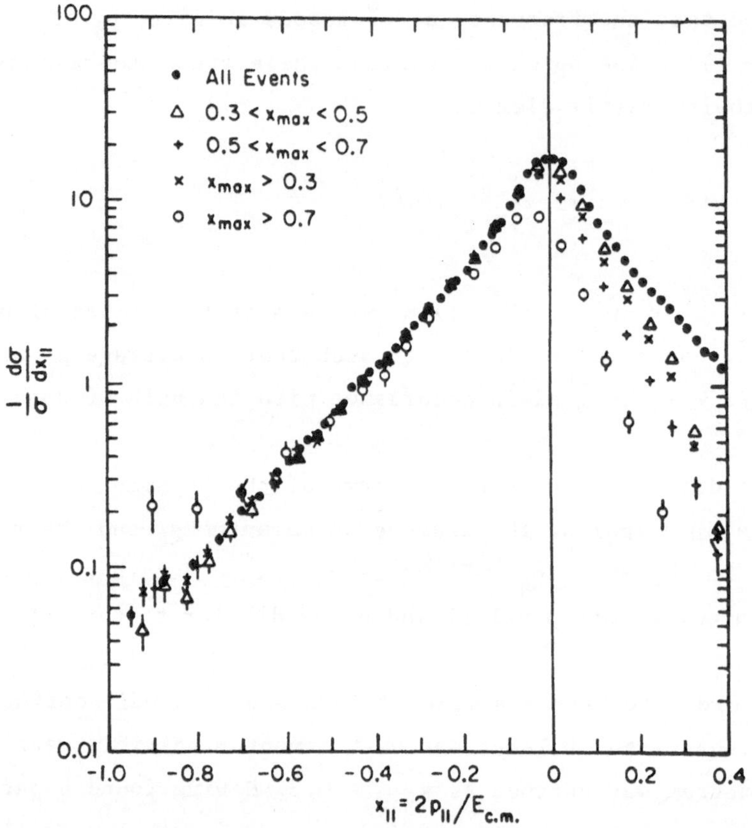

Fig. 7.6 Particle density distributions $(1/\sigma)\ d\sigma/dx_{||}$ vs. $x_{||}$ for various x_{max} cuts for $7.0 < \sqrt{s} < 7.8$ GeV. x_{max} is the highest-x particle on one side of the event and is not plotted. The jet direction is oriented so that x_{max} is at positive $x_{||}$.(Ref.122).

Fig. 7.7 shows the density distributions versus $x_{\parallel}$ for all events (i.e. $x_{max} > 0$) at various c.m. energies between 3 and 7.8 GeV. Here σ is the total cross section. For $x_{\parallel} > 0.1$ approximate scaling is observed.

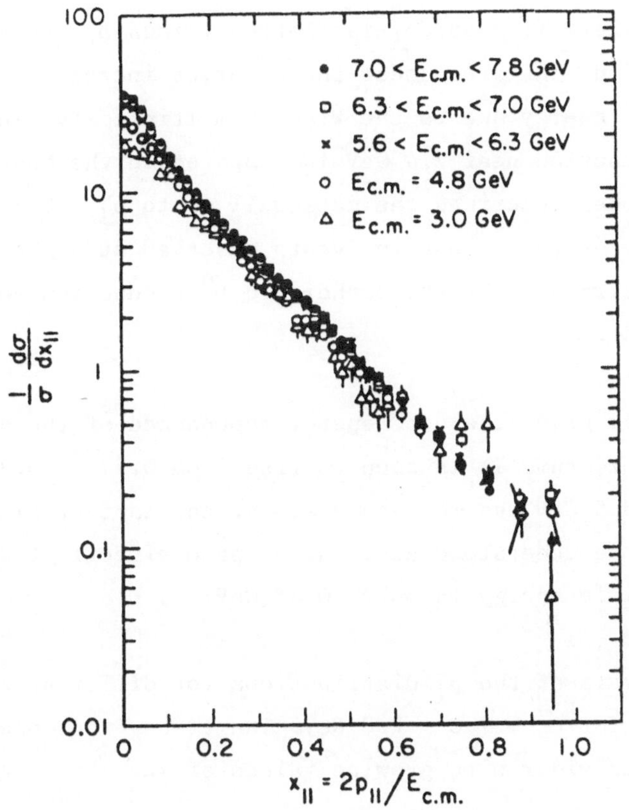

Fig. 7.7 The density distribution $1/\sigma\ d\sigma/dx_{\parallel}$; σ is the total cross section (Ref. 122).

The rapidity distributions are presented in Fig. 7.8. In this case $x_{max} > 0.3$ was required. The x_{max} cut distorts the y distributions for positive y. Two things are noteworthy: with increasing c.m. energy a plateau develops around $y = 0$. The rise of the distribution from y_{max} to the plateau occurs within two units in y. Both features are also common to hadron-hadron scattering.

The behaviour with respect to the transverse momentum is shown in Fig. 7.9 where $(1/\sigma)d\sigma/dp_{\perp}^2$ is plotted versus $p_{\perp}^2$. An x_{max} cut of 0.3 was applied. The area under these curves increases with increasing c.m. energy due to the rise in multiplicity. In Fig. 7.10 the $p_{\perp}^2$ distribution near 7.4 GeV is compared to the Monte Carlo model. The model describes the data well up to $p_{\perp}^2 \approx 0.7$ $(GeV/c)^2$. At least part of the excess in events observed at higher transverse momenta is attributed by the authors to D^o production and decay into $D^o \rightarrow \pi^+ K^-$.

Fig. 7.11 gives the c.m. energy dependence of the average transverse momentum. It is seen to rise from 0.31 to 0.35 GeV/c between 3 and 5 GeV and to become almost constant at higher energies. The rise can be understood as a phase space effect: at 3 GeV the average particle energy is only ~0.45 GeV.

An analysis of the $p_{\perp}^2$ distributions for different $x_{\parallel}$ is shown in Fig. 7.12 for $\sqrt{s} = 7.0 - 7.8$ GeV. The transverse momentum distributions become wider with growing values of $x_{\parallel}$.

The most simple correlation between two opposite jets than can be studied is the charge correlation. Consider e.g. events originating from a u $\bar{u}$ pair. We expect at higher $x_{\parallel}$ to find an excess of positive charges in the direction of the u quark and a corresponding excess of negative charges in the opposite direction.

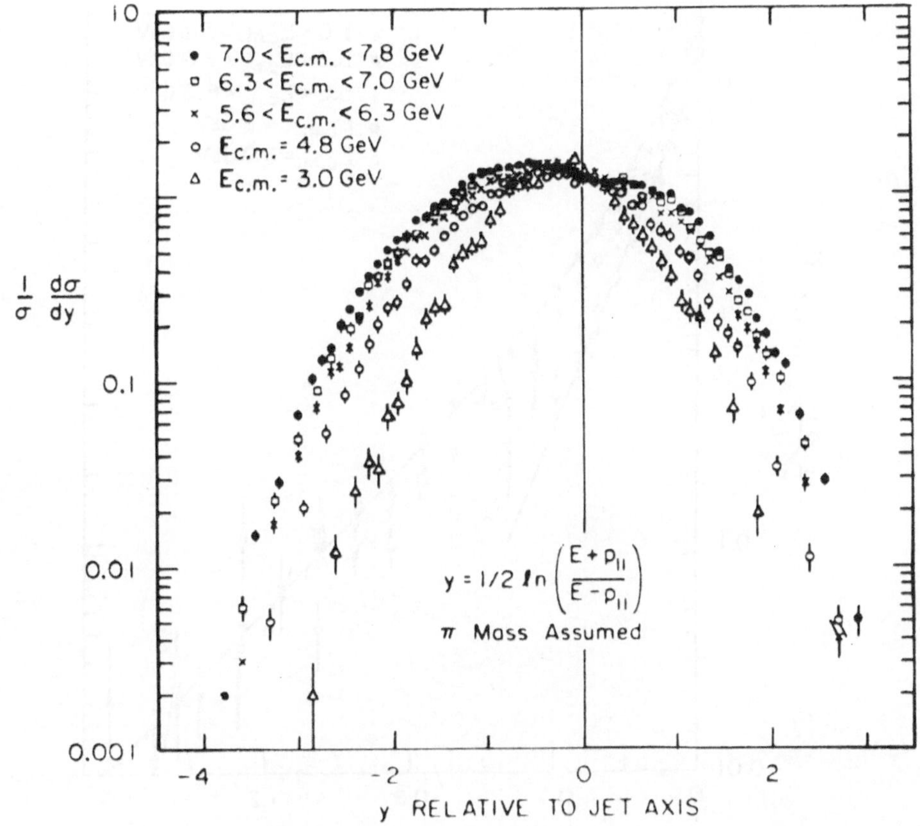

Fig. 7.8 The rapidity distribution (Ref. 122).

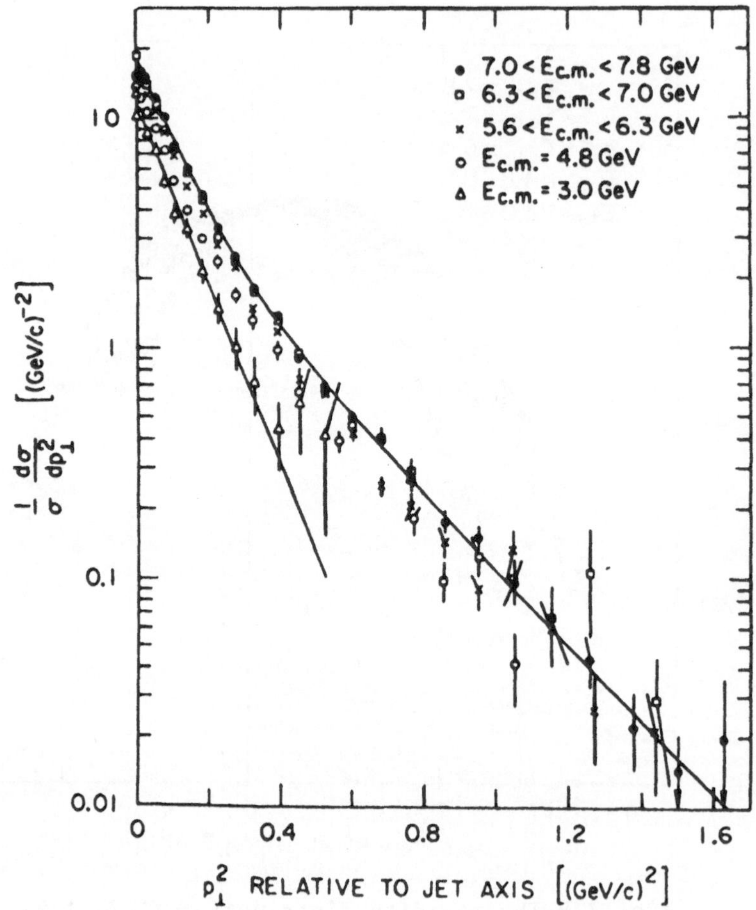

Fig. 7.9 The $p_\perp^2$ distribution (Ref. 122).

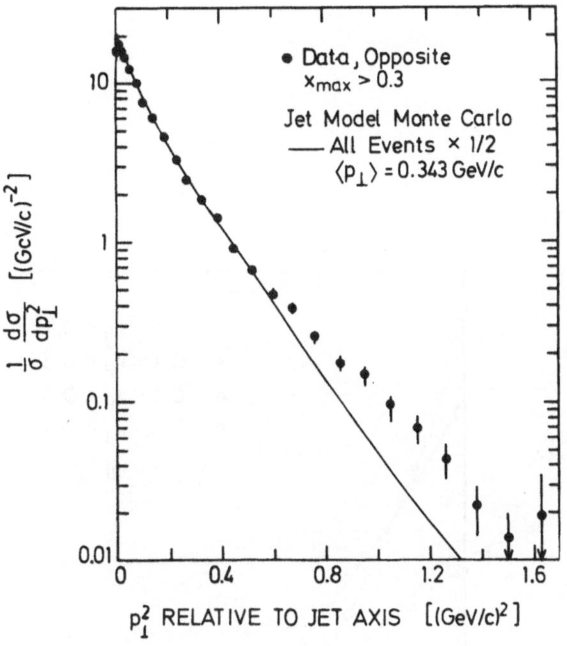

Fig. 7.10 The $p_\perp^2$ distribution at 7.4 GeV (Ref. 122).

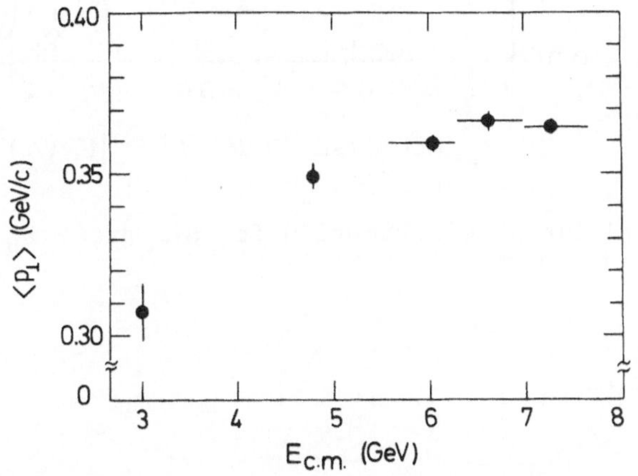

Fig. 7.11 Average $p_\perp$ as a function of c.m. energy (Ref. 122).

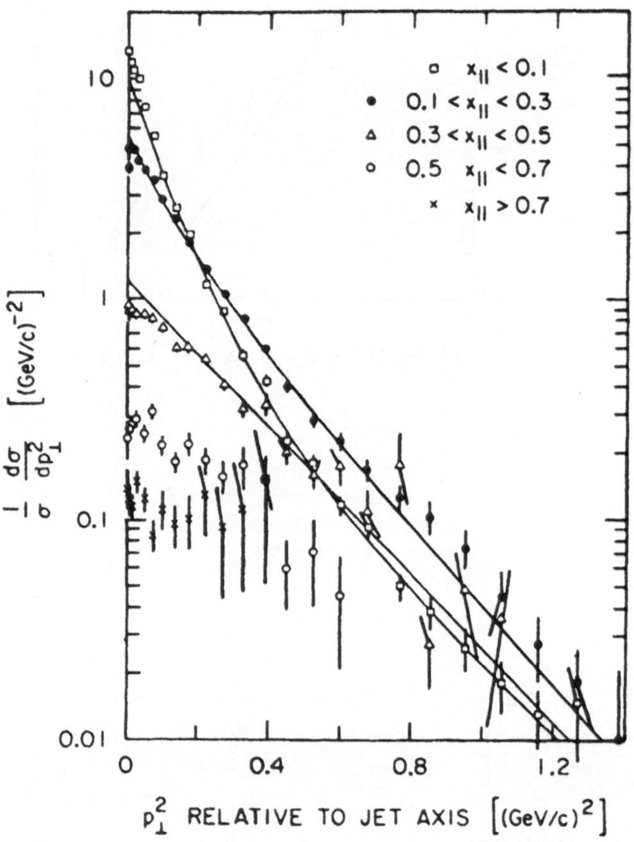

Fig. 7.12 The $p_{\perp}^{2}$ distribution for different $x_{\parallel}$ intervals.

The SLAC-LBL group looked for charge correlations in the following way. The x distributions were determined a) for all particles that had the same charge as the x_{max} particle; b) for those particles that had the opposite charge as the x_{max} particle. Fig. 7.13 shows the observed ratio (opposite charge)/(same charge) of these two distributions for two values of x_{max}, 0.5 and 0.7. The dashed line shows the expectation for no charge correlation. One observes a strong charge correlation on the same side and (within the statistical errors) no charge correlation between opposite jets. The same side correlation most probably is due to low mass pion resonance formation, such as the ρ^o decaying into positive and negative particles (see the strong ρ^o production observed by the PLUTO group, sect. 6). The absence of the expected opposite side charge correlation may have its cause in insufficient statistics and too low c.m. energies.

7.4 Jet Studies in the Υ Region

This section deals with the recent jet studies performed by three DORIS experiments for c.m. energies up to 10 GeV. Jet analyses off and on the Υ resonance are of great importance: they may provide a decisive test on QCD. In QCD the direct hadronic decays of the Υ proceed via a three gluon intermediate state. As a consequence we expect the hadrons to emerge in three rather than two jets (as in nonresonant $q\bar{q}$ formation). The expected properties of the three gluon jets have widely been discussed in recent papers[123-125]. We mention briefly some of the salient features[124].

Consider a $Q\bar{Q}$ system of mass M_{QQ} which decays into three gluons of energies E_i. Define the scaled energies

$$\xi_i = 2E_i/M_{QQ} \qquad (7.7)$$

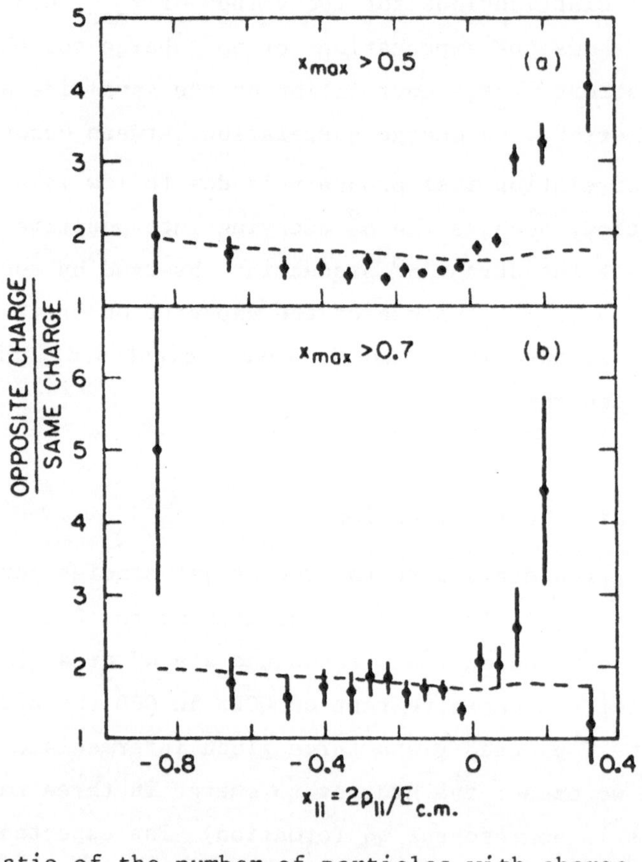

Fig. 7.13 Ratio of the number of particles with charge opposite to the charge of the x_{max} particle, to the number of particles with the same charge: (a) $x_{max} > 0.5$; (b) $x_{max} > 0.7$; (Ref. 122).

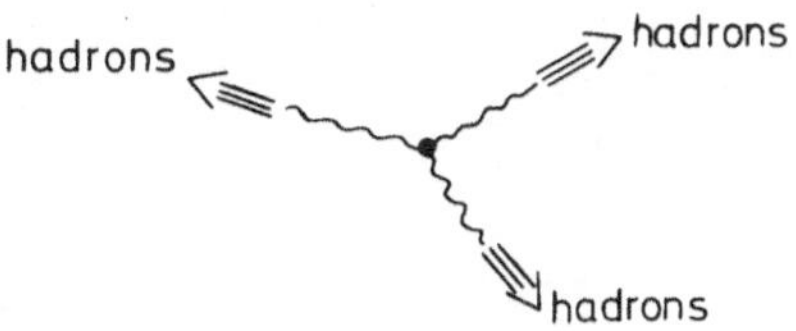

The energy distribution of the gluons is the same as for photons from orthopositronium decay:

$$\frac{1}{\sigma}\frac{d^2\sigma}{d\xi_1\ d\xi_2} = \frac{1}{\pi^2-9}\left\{\frac{1-\xi_3}{\xi_1\xi_2} + \frac{1-\xi_2}{\xi_1\xi_3} + \frac{1-\xi_1}{\xi_2\xi_3}\right\} \tag{7.8}$$

By integrating over e.g. ξ_2 one finds the energy distribution of one gluon (one jet):

$$\frac{1}{\sigma}\frac{d\sigma}{d\xi} = \frac{2}{\pi^2-9}\ F(\xi)$$
$$F(\xi) = \frac{\xi(1-\xi)}{(2-\xi)^2} + \frac{2-\xi}{\xi} + 2\left\{\frac{1-\xi}{\xi^2} - \frac{(1-\xi)^2}{(2-\xi)^3}\right\}\ln(1-\xi) \tag{7.9}$$

The function $F(\xi)$ is shown in Fig. 7.14. The most probable configuration is one where two of the three gluons share basically all of the available energy. Of course, such a configuration will produce two jet events.

Most probable configuration

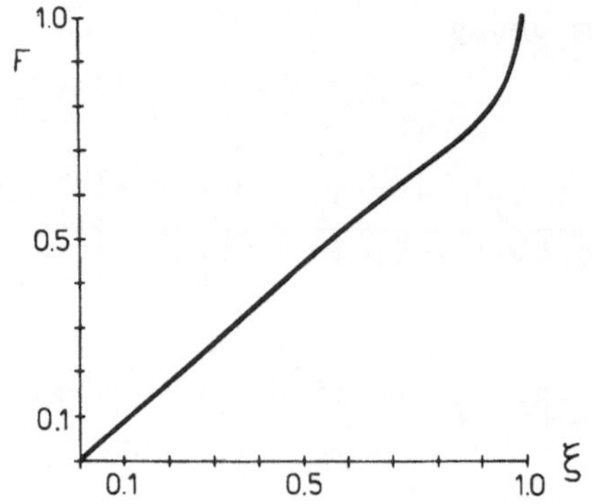

Fig. 7.14 Distribution of the gluon jet energy (Fig. taken from Ref. 124).

The geometrical structure of the three gluons can also be analyzed in terms of the angles Θ and φ between the gluons.

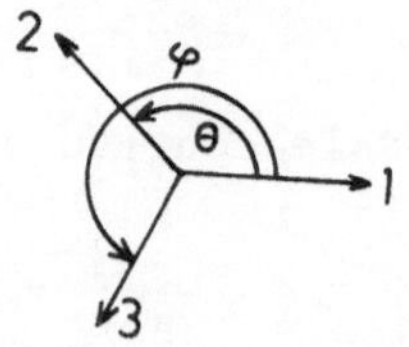

The scaled energies are related to Θ and φ in the following manner:

$$\xi_1 = \frac{2}{A}\sin(\Theta - \varphi)$$

$$\xi_2 = \frac{2}{A}\sin\varphi \qquad (7.10)$$

$$\xi_3 = -\frac{2}{A}\sin\varphi$$

where $A = \sin\varphi - \sin\Theta + \sin(\Theta - \varphi)$

We may ask for the probability of observing clean three-jet events. Theory predicts that in 38 % of the Υ gluonic decays the three gluons are emitted within $\pm 20^{\circ}$ of the symmetric three star directions.

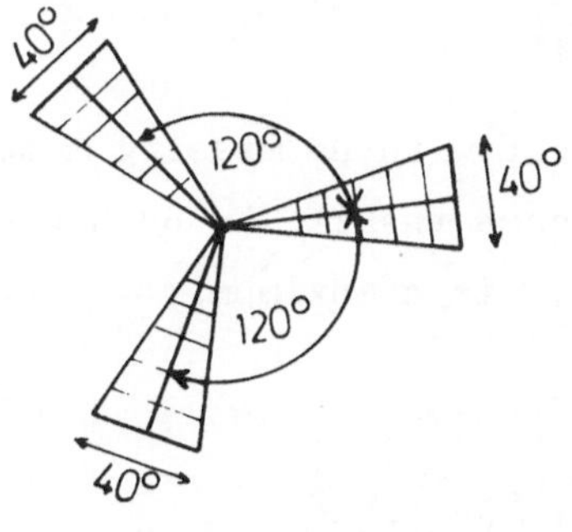

The orientation of the plane defined by the three gluons is a direct consequence of the vector nature of the gluons. If $\hat{n}$ denotes the normal to this plane and Θ_n the angle between $\hat{n}$ and the incoming e^+ (or e^-) then

$$\frac{1}{\sigma}\frac{d\sigma}{d\cos\Theta_n} = \frac{3}{16}(2 + \sin^2\Theta_n) \qquad (7.11)$$

Another test for the vector nature of gluons is provided by the angular distribution between one of the gluon and the incoming e^+. In the limit $\xi_i \rightarrow 1$

$$\frac{1}{\sigma}\frac{d\sigma}{d\cos\Theta_i} \rightarrow \frac{3}{8}(1 + \cos^2\Theta_i) \qquad (7.12)$$

According to present wisdom gluons do not become free but manifest themselves in hadron jets. Besides sphericity two other variables have been devised in the search for jets. These are

thrust[126] $$T = 2\,\frac{\tilde{\sum} p_{\parallel}^{i}}{\sum_i |p_i|} \quad ; \quad \frac{1}{2} \leq T \leq 1 \qquad (7.13)$$

where the summation $\tilde{\sum}$ is to be extended over all particles in one hemisphere;

For practical applications the following definition of T was found to be better suited[127]:

$$T = \frac{\sum_i |p_{\parallel}^{i}|}{\sum_i |p_i|} \qquad (7.13')$$

For an event with no missing momentum the two definitions give the same result. In both cases the jet axis is chosen such that T is a maximum.

spherocity[128]

$$S_o = \left(\frac{4}{\pi}\right)^2 \left(\frac{\sum |p_{\perp}^{i}|}{\sum |p_i|}\right)^2 \qquad (7.14)$$

The jet axis is chosen such that S_o is a minimum. In contrast to sphericity, the momenta are summed linearly. The advantage of S_o over S is that S_o is infrared insensitive and is therefore better suited for QCD calculations.

The PLUTO group[127,130] for their jet analyses selected events with four or more charged particles. The distributions presented below are the observed ones; no corrections were applied for acceptance, cuts or radiative effects. In comparing the observed distributions with theory a Monte Carlo technique was used to impose the same acceptance criteria and cuts onto the theoretical events.

The mean sphericity <S> is plotted in Fig. 7.15 as a function of c.m. energy between 3 and 10 GeV. In agreement with the SLAC-LBL data <S> is seen to decrease with growing energy. The points measured at the J/ψ and ψ' are significantly higher than the value at 3.6 GeV. Perhaps most interesting is the rise in <S> seen directly above charm threshold at 4.03 GeV. The $c\bar{c}$ events do not show a jet structure at all.
At the ϒ the average sphericity is again larger than outside of the resonance at 9.4 GeV. The dashed band shows the expected <S> behaviour for phase space distributed events. At higher energies ($E_{cm} \gtrsim 6$ GeV) the observed sphericity values are well below the phase space prediction. The point measured at 9.4 GeV is in reasonable agreement with the jet modle calculated according to Field and Feynman[129].

The analysis in terms of thrust leads to the same conclusions. Fig. 7.16 shows a plot of 1 - <T> versus E_{cm}. The jet axes determined vis S and T, respectively, are not too different. This is illustrated in Fig. 7.17a where the angle between the two jet axes was plotted. For half of the events this angle is less than 15°. There is a strong correlation between the jet axis and the direction of the particle with the largest momentum. This is shown by Figs. 7.17b and c where the distribution of the angle between the two directions was plotted. The correlation is stronger for the S-jet axis.

The angular distribution of the jet axis with respect to the beam direction is displayed in Fig. 7.18 for E_{cm} = 7.7 and 9.4 GeV. The data are consistent with a $1 + \cos^2\Theta$ behaviour expected for $q\bar{q}$ production.

The energy dependences of the average transverse and longitudinal momenta measured relative to the T-jet axis are plotted in Fig. 7.19a. The longitudinal component is seen to grow much more

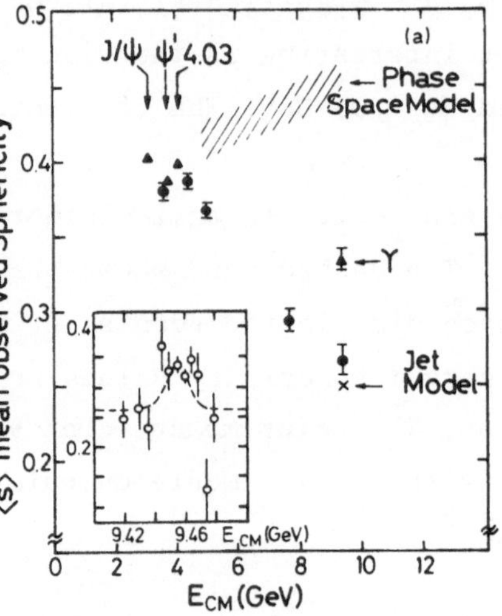

Fig. 7.16 The quantity 1 - <T> as a function of energy where <T> is the average observed thrust (Ref. 127).

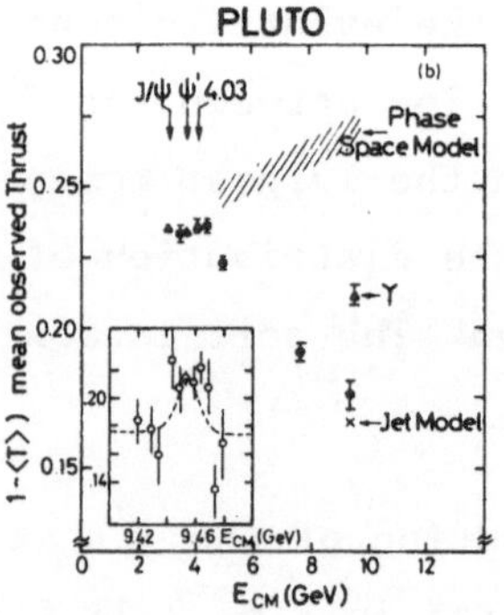

Fig. 7.15 Average observed sphericity as a function of energy (Ref. 127).

rapidly with energy than the transverse momentum. The average half-opening angle of jets computed via $\alpha = \mathrm{atan}\,(\langle p_\perp \rangle / \langle p_\parallel \rangle)$ is 33° at 5 GeV and 28° at 9.4 GeV. We see that the jet cone shrinks with increasing energy but at a relatively slow rate.

Another measure of the jet opening angle is provided by the amount of energy emitted under an angle λ with respect to the jet axis.

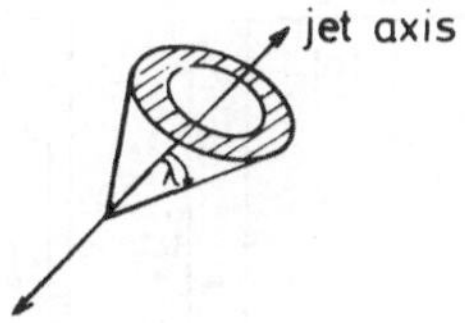

Fig. 7.17d-f shows at 9.4 GeV $\frac{1}{E^c+E^o}\frac{dE^{c,o}}{d\lambda}$ versus λ for different thrust regions; E^c denotes the charged energy (the charged particles were assumed to be pions) and E^o the neutral energy (photons) as measured in the PLUTO shower counters. The histograms show the result for charged particles, the data points measure the neutral component. Note that the jet axis was defined by considering charged particles only. Fig. 7.17d-f tells us that the neutral energy is also concentrated near the jet axis in much the same way as charged particles. The larger width of the E^o distributions may disappear when the neutrals are included in the determination of the jet axis.

The energy detected through charged particles and neutrals which is called visible energy, on the average accounts for 85 % of the total available energy. Fig. 7.19b presents for 9.4 GeV the fraction of the visible energy observed outside a cone of half angle δ. For example 70 % of the visible energy are within a cone of $\delta = 43^\circ$.

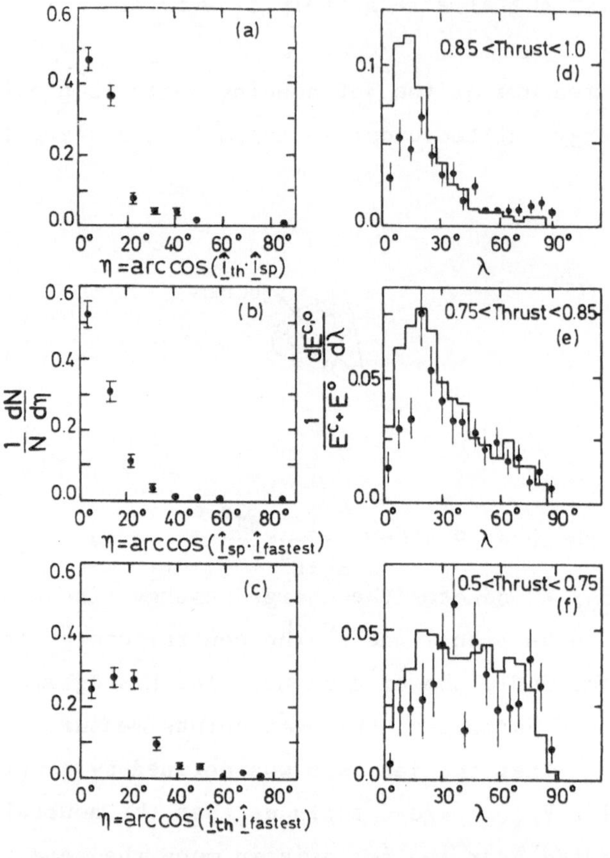

Fig. 7.17 a) The distribution of the angle between the sphericity and the thrust axes at 9.4 GeV;
b,c) The distributions of the angle between the charged particle with the largest momentum and the sphericity and thrust axes, respectively, at 9.4 GeV;
d-f) The angular distributions $1/E\ dE^c/d\lambda$ of charged energy (histograms) and $1/E\ dE^o/d\lambda$ of neutral energy (data points) with respect to the thrust at 9.4 GeV for different T intervals (Ref. 127).

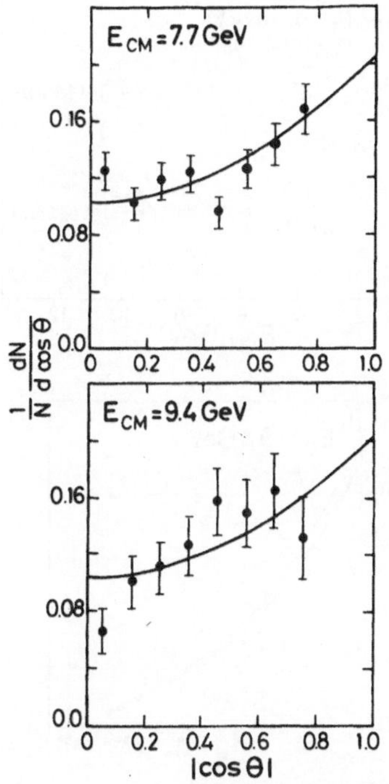

Fig. 7.18 The angular distribution of the thrust axis at 7.7 and 9.4 GeV (Ref. 127).

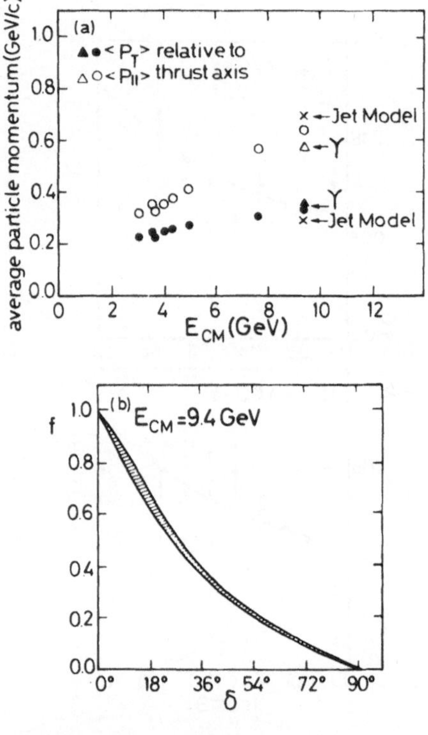

Fig. 7.19 a) The energy dependence of the average transverse and longitudinal momenta;
b) The average fraction of visible energy (f) outside of a cone of halfangle δ at 9.4 GeV (Ref. 127).

We turn now to the PLUTO analysis of the Υ events[130]. Three processes can contribute to events in the Υ region, the direct hadronic decay, the decay through the one-photon channel and the nonresonant continuum:

$$\sigma = \sigma_{dir} + \sigma_{1\gamma} + \sigma_{cont} \qquad (7.14)$$

For the continuum contribution the data at 9.4 GeV were used. Neglecting possible interference effects the one-photon part leads to the same final states as the continuum. The size of $\sigma_{1\gamma}$ relative to σ_{cont} can be computed from the rate of μ pairs observed on and off the Υ:

$$\sigma_{1\gamma} = \frac{\sigma^{on}_{\mu\mu} - \sigma^{off}_{\mu\mu}}{\sigma^{off}_{\mu\mu}} \cdot \sigma_{cont} = (0.24 \pm 0.22)\sigma_{cont}$$

The total number of events available for the analysis were 1418 at the Υ(9.45 - 9.47 GeV) and 420 in the continuum (9.30 - 9.44 GeV). The latter events were used to determine by proper subtraction the distributions for the direct decays of the Υ.
Fig. 7.20a shows the average sphericity plotted versus E_{cm}. The $\langle S \rangle$ value of Υ direct decays is markedly larger than for the continuum. The increase in $\langle S \rangle$ is not caused by the larger multiplicity. This is demonstrated by Fig. 7.20b where $\langle S \rangle$ is plotted for events with 4, 6 and 8 charged particles for 9.4 GeV and for the direct Υ decays. The $\langle S \rangle$ values for the Υ are significantly higher for each one of the topologies.

From the preceding discussion it is clear that the Υ final states are less jet like than those of the continuum. They are incompatible with the two-jet model used to describe the continuum but also with pure phase space (see Fig. 7.20). The PLUTO group compared their data also with the three-gluon model formulated by Koller and Walsh[123]. The gluon jet was assumed to be similar to a quark jet observed in the continuum for the equivalent energy.

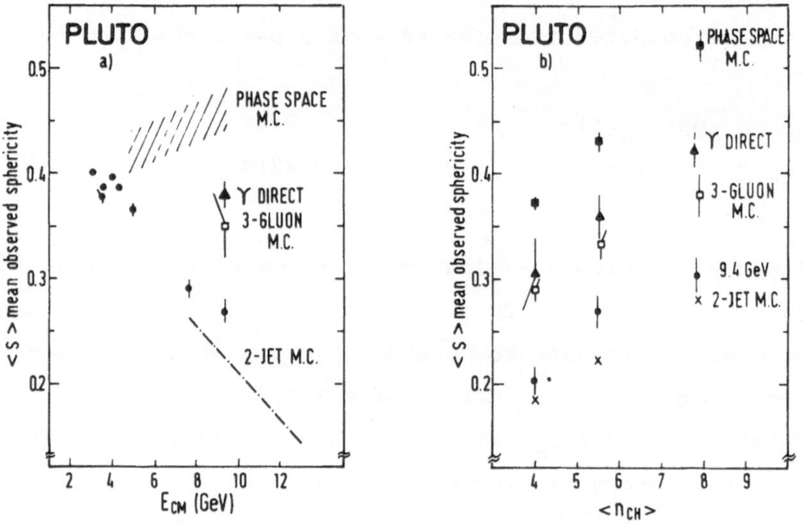

Fig. 7.20 (a) The average sphericity as a function of energy.

(b) The average sphericity for 9.4 GeV from Υ decay for events with different number of charged particles (Ref. 130).

It should be noted that the predicted hadron distributions depend decisively on this assumption. Like for the other model calculations the effects of the detector acceptance and of the experimental cuts were included in the Monte Carlo simulation. Fig. 7.20 shows that the sphericity observed for the T as well as its dependence on the number of charged particles produced is well described by the three-gluon model.

The three gluon decay predicts the hadrons to be concentrated in a plane. The PLUTO group tested the data for flatness in the following manner.

1. Compute for each event the tensor

$$T^{\alpha\beta} = \sum_i (p_i^2 \delta^{\alpha\beta} - p_i^\alpha p_i^\beta) \tag{7.15}$$

where α, β are the coordinate indices.

2. Solve for the eigenvalues λ_i, $i = 1 \ldots 3$ of $T^{\alpha\beta}$ and order them according to $\lambda_1 \geq \lambda_2 \geq \lambda_3$. The sphericity is then given by

$$S = \frac{3\lambda_3}{\lambda_1 + \lambda_2 + \lambda_3} \tag{7.16}$$

3. Compute the quantities

$$Q_K = 1 - \frac{2\lambda_K}{\lambda_1 + \lambda_2 + \lambda_3} = \frac{\sum_i (p^K_{\parallel i})^2}{\sum_i (p_i)^2} \tag{7.17}$$

where p^K is the momentum component along the axis associated with λ_K. Q_1 is a measure of the flatness of the event. $Q_1 = 0$ corresponds to a perfectly planar event, $Q_1 = 1/3$ to a spherical event; of course, an extreme two jet event with all particle momenta along one axis will also lead to $Q_1 = 0$.

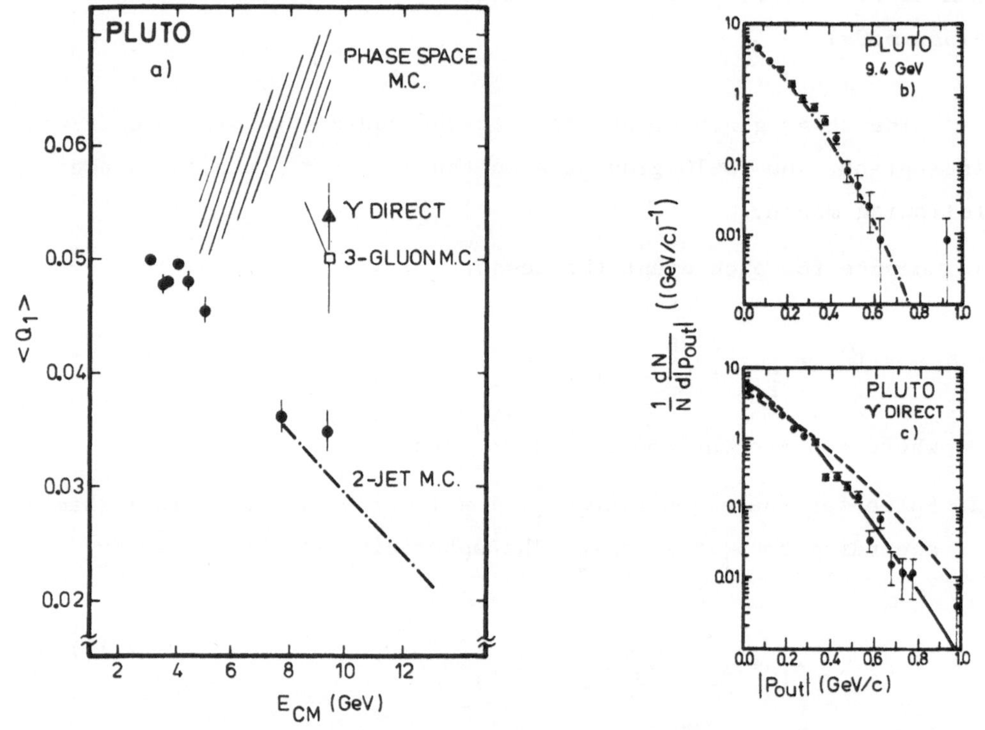

Fig. 7.21 a) The average observed Q_1 as a function of energy.

b),c) The P_{out} distributions for 9.4 GeV and for ϒ direct decay. The dashed-dotted line in b) represents the 2 jet model. The dashed and solid lines in c) represent the phase space and the three-gluon decay, respectively (Ref. 130).

Fig. 7.21a shows the average Q_1 plotted as a function of E_{cm}. It drops with increasing energy in much the same way as e.g. the average sphericity and is consistent with the two jet model. The T decays yield a much larger $\langle Q_1 \rangle$ value in agreement with the three-gluon model.

Figs. 7.21b and c show the p_{out} distributions at 9.4 GeV and from the T decay where p_{out} is the momentum component perpendicular to the plane associated with Q_1. The p_{out} distribution is wider for the T; it agrees well with what a three-gluon decay would predict.

The PLUTO group performed a second test devised by de Rujula et al.[126] to measure the flatness. The test requires to find for each event that plane with respect to which the sum of the momentum components perpendicular to it are a minimum:

$$A = 4 \min \left[\left(\sum_i |p_{out\ i}|\right) / \left(\sum_i |p_i|\right) \right]^2 \qquad (7.18)$$

where A is called the acoplanarity. The average acoplanarity values are given in Table 7.1 together with those measured for sphericity, thrust etc. and with the theoretical predictions. The average acoplanarity observed for the T is larger than for the continuum and is consistent with the three-gluon model.

The DASP2 group also investigated the spatial configuration of events at 9.4 GeV and from T decay[104]. Charged particles were detected in the nonmagnetic inner detector of the apparatus. Since particle momenta would not be determined the particle directions were studied in terms of variables related to sphericity, thrust, sphericity and acoplanarity. Qualitatively, the same behaviour was found as observed by the PLUTO group.

The data presented in this section can be summarized as follows. The hadronic final states produced by nonresonant e^+e^- annihilation possess a two-jet structure which becomes more pronounced as the c.m. energy increases. The origin of the jet structure may well be the production of a pair of quark plus antiquark which fragment into hadrons. One of the pieces yet missing in this puzzle is the observation of correlations between the two opposite jets. The distribution of the hadrons within a jet are remarkably similar to those observed from jets produced by hadron-hadron collisions.

Table 7.1 The observed mean values for sphericity S, Q_1, p_{out}, thrust T and acoplanarity A. The data is compared with three models computed via the Monte-Carlo method to simulate the experimental conditions. The errors given for the values of the phase space and 3-gluon decay models include systematic uncertainties. Taken from Ref. 130.

	M.C.	DATA		M. C.	
	2-jet	9.4 GeV	T DIRECT	3-GLUON	PHASE SPACE
$\langle S\rangle$	0.22	0.27 ±0.01	0.38 ±0.01	0.35 ±0.03	0.46 ±0.02
$\langle Q_1\rangle$	0.030	0.035 ±0.002	0.054 ±0.003	0.050 ±0.005	0.067 ±0.005
$\langle p_{out}\rangle$	0.115	0.122 ±0.003	0.132 ±0.003	0.140 ±0.006	0.177 ±0.006
$\langle T\rangle$	0.84	0.82 ±0.01	0.76 ±0.01	0.76 ±0.01	0.73 ±0.01
$\langle A\rangle$	0.084	0.096 ±0.005	0.14 ±0.01	0.14 ±0.01	0.16 ±0.01

The hadrons emitted in the direct decay of the T show a different spatial structure. They are much less collimated along a common axis. The observed features disagree with the two-jet picture but also with pure phase space. All aspects investigated sofar are in accord with the assumption that the T decays via a three gluon intermediate state. This is in strong support of QCD. However, the basic elements of QCD, such as the vector nature of the gluon, the flavor neutrality, the presence of a three gluon interaction have not yet been established by the data.

Acknowledgements

It is a pleasure to thank Prof. Zichichi and his coworkers for a most enjoyable stay. I am grateful to Mrs. E. Hell for her diligent help in getting the manuscript ready.

References

The following shorthand notations shall be used: 1971 Cornell Conference for the Proceedings of the 1971 Symposium on Electron and Photon Interactions at High Energies, Cornell, ed. by N. Mistry.

1975 Stanford Conference for the Proceedings of the 1975 Symposium on Lepton and Photon Interactions at High Energies, ed. by W.T. Kirk.

1976 Tbilisi Conference for the XVIIIth International Conference on High Energy Physics, Tbilisi, USSR (1976).

1977 Hamburg Conference for the Proceedings of the 1977 International Symposium on Lepton and Photon Interactions at High Energies, Hamburg, edited by F. Gutbrod.

1978 Tokyo Conference for the XIXth International Conference on High Energy Physics, Tokyo, 1978.

1. B.H. Wiik and G. Wolf, DESY Report 78/23 (1978).
2. see e.g. H. Schopper, Lectures given at the International School of Subnuclear Physics, 1977, DESY Report 77/74 (1977).
3. V. Alles-Borelli et al., Phys. Lett. 59B (1975) 201.
4. F.E. Low, Phys. Rev. 120 (1960) 582;
F. Calogero and C. Zemach, Phys. Rev. 120 (1960) 1860.
A. Jaccarini, N. Arteaga - Romero, J. Parisi, P. Kessler, Compt. Rend. B269 (1969) 153, 1129; Lett. Nuovo Cimento 4 (1970) 933;
N. Arteaga-Romero, A. Jaccarini, P. Kessler and J. Parisi, Phys. Rev. D3 (1971) 1569;
see also: S.J. Brodsky, T. Konoshita, H. Terazawa, Phys. Rev. Lett. 25 (1970) 972.
5. S. Orito et al., Phys. Lett. 48B (1974) 380.
6. see e.g. R. Feynman, Photon-Hadron Interactions, W.A. Benjamin (New York) 1972.
7. R.F. Schwitters, 1976 Tbilisi Conference.
8. J.J. Aubert et al., Phys. Rev. Lett. 33 (1974) 1404.
9. J.E. Augustin et al., Phys. Rev. Lett. 33 (1974) 1406.
G.S. Abrams et al., Phys. Rev. Lett. 33 (1974) 1453, ibid. 34 (1975) 1181.
10. see e.g. G.J. Feldman and M.L. Perl, SLAC-PUB-1972 (1977).
11. B. Knapp et al., Phys. Rev. Lett. 34 (1975) 1040.
12. S.L. Glashow, J. Iliopoulos and L. Maiani, Phys. Rev. D2 (1970) 1285.
13. An up to date description of the theoretical framework used to describe the new particles with a complete set of references can be found in K. Gottfried, invited paper at the 1977 Hamburg Conference, p. 667.
see also: J.D. Jackson, Summer Institute of Particle Physics, SLAC, 1976;
J.D. Jackson, invited paper, European Conference on Particle Physics, Budapest, Hungary, 4-9 July, 1977.

V.A. Novikov, L.B. Okun, M.A. Shifman, A.I. Vainshtein, M.B. Voloshin, V.I. Zakharov, ITEP-42,57,58,65,79, 83 (1977).

M. Krammer and H. Krasemann, Lecture given at the Advanced Summer Institute, Karlsruhe (1978); DESY Report 78/66 (1978).

14. T. Appelquist et al., Phys. Rev. Lett. 34 (1975) 43;
C.G. Callan et al., Phys. Rev. Lett. 34 (1975) 52;
E. Eichten et al., Phys. Rev. Lett. 34 (1975) 369;
E. Eichten et al., Phys. Rev. Lett. 36 (1976) 500.
15. J. Pumplin, W. Repko and A. Sato, Phys. Rev. Lett. 35 (1975)1538;
H.J. Schnitzer, Phys. Rev. Lett. 35 (1975) 1540;
H.J. Schnitzer, Phys. Rev. D13 (1976) 74.
16. P.A. Rapidis et al, Phys. Rev. Lett. 39 (1977) 526.
17. DELCO Collaboration, J. Kirkby, rapporteur talk at the 1977 Hamburg Conference, p.3.
18. A.M. Bojarski et al., Phys. Rev. Lett. 34 (1975) 764;
R.F. Schwitters, 1975 Stanford Conference, p. 5.
19. J. Burmester et al., Phys. Lett. 66B (1977) 395; and
G. Knies, rapporteur talk, 1977 Hamburg Conference, p. 93.
20. DASP Collaboration, R. Brandelik et al., Phys. Lett. 76B (1978) 361.
21. J.J. Aubert et al., Phys. Rev. Lett. 33 (1974) 1404.
22. J.-E. Augustin et al., Phys. Rev. Lett. 33 (1974) 1406;
G.S. Abrams et al., Phys. Rev. Lett. 33 (1974) 1453;
A.M. Boyarski et al., Phys. Rev. Lett. 34 (1975) 764.
23. K. Gottfried, invited paper, 1977 Hamburg Conference, p. 667;
M. Caichian and R. Kögerler, Phys. Lett. 80B (1978) 105;

A multichannel analysis of the 4 GeV region has been presented by D. Horn and D.E. Novoseller, preprint, 1978.
24. R. Barbieri, R. Gatto and R. Kögerler, Phys. Lett. 60B (1976) 183.
25. DASP Collaboration, W. Braunschweig et al., Phys. Lett. 57B (1975) 407;

see also B.H. Wiik, rapporteur talk, 1975 Stanford Conference, p. 69;
S. Yamada, rapporteur talk, 1977 Hamburg Conference, p. 69, and E. Gadermann, thesis, Hamburg University, 1978.
26. J.S. Whitaker et al., Phys. Rev. Lett. 37 (1976) 1596;
F.M. Pierre, invited paper, 1976 Tbilisi Conference;
G.H. Trilling in Proceeding of the SLAC Summer Institute on Particle Physics, 1976 SLAC Report 198.
27. C.J. Biddick et al., Phys. Rev. 38 (1977) 1324;
see also H.F.W. Sadrozinski, rapporteur talk, 1977 Hamburg Conference, p. 47.
28. PLUTO Collaboration, V. Blobel, Proc. of the XIIth Rencontre de Moriond, Flaine, March 1977.
29. W. Tannenbaum et al., Phys. Rev. Lett. 35 (1975) 1323.
30. DESY-Heidelberg Collaboration, see J. Olsson, rapporteur talk, 1977 Hamburg Conference, p. 117;
W. Bartel et al., Phys. Lett. 79B (1978) 492.
31. F.M. Pierre, invited paper, 1976 Tbilisi Conference;
G.H. Trilling, Proceedings of the SLAC-Summer Institute on Particle Physics, 1976, SLAC-Report 198.
32. see also G.J. Feldman and M.L. Perl, SLAC-PUB-1972 (1977).
33. R. Barbieri, R. Gatto and E. Remiddi, Phys. Lett. 61B (1976) 465.
34. DASP Collaboration, W. Braunschweig et al., Phys. Lett. 67B (1977) 243 and 249, and
S. Yamada, rapporteur talk, 1977 Hamburg Conference, p. 69.
35. F.A. Berends and R. Gastman, Nucl. Phys. B61 (1973) 414.
36. J. Heintze, rapporteur talk, 1975 Stanford Conference, p.97.
37. W.D. Apel et al., Phys. Lett. 72B (1978) 500.
38. J.S. Whitaker et al., Phys. Rev. Lett. 37 (1976) 1596.
39. F. Buccella, private communication,
see e.g. Y.I. Azimov, L.L. Frankfurt and V.A. Khoze, Leningrad preprint 239 (1976).

40. A. Bradley and F.D. Gault, Manchester Univ. preprint 1978.

41. H.J. Lipkin, H.R. Rubinstein and N. Isgur, Weizmann Institute preprint WIS-78/23 Ph (1978);
G. Eilam, B. Margolis and S. Rudaz, Phys. Lett. 80B (1979) 306.

42. M.S. Chanowitz and F.J. Gilman, Phys. Lett. 63B (1976) 178;
H. Harari, Phys. Lett. 64B (1976) 469.

43. M.K. Gaillard, B.W. Lee and J.L. Rosner, Rev. Mod. Phys. 47 (1975) 277.

44. J. Ellis, M.K. Gaillard and D.V. Nanopoulos, Nucl. Phys. B100 (1975) 313.

45. PLUTO Collaboration, J. Burmester et al., Phys. Lett. 67B (1977) 367.

46. DASP Collaboration, R. Brandelik et al., Phys. Lett. 67B (1977) 363.

47. V. Lüth et al., Phys. Lett. 70B (1977) 132.

48. G. Goldhaber et al., Phys. Rev. Lett. 37 (1976) 255;
I. Peruzzi et al., Phys. Rev. Lett. 37 (1976) 569.

49. E.J. Feldman and M.L. Perl, SLAC-PUB-1972 (1977).

50. G. Goldhaber, lectures given at the International School of Physics, Enrico Fermi, Varenna (1977).

51. L.B. Okun and M. Voloshin, Zh. Eksper. Theor. Fig. 23, (1976) 369;
C. Rosenzweig, Phys. Rev. Lett. 36 (1976) 697;
M. Bander et al., Phys. Rev. Lett. 36 (1976) 695;
A. De Rujula et al., Phys. Rev. Lett. 38 (1977) 317.

52. PLUTO Collaboration. J. Burmester et al., Phys. Lett. 68B (1977) 283.

53. DASP Collaboration, R. Brandelik et al., Phys. Lett. 70B (1977) 132, and DESY Report 78/63 (1978), to be published in Phys. Lett.

54. J.L. Rosner, invited talk given at Orbis Scientiae - 1977, Coral Gables, Fla. 000-2220-102 (1977) and
C. Quigg and J. Rosner, FERMILAB-PUB-77/60-THY (1977).

55. DASP Collaboration, R. Brandelik et al., DESY Report 78/50 and Nucl. Phys., to be published.

56. I. Peruzzi and M. Piccolo, Contribution paper to Festschrift for C. Peyrou, LNF-78/12(P), March, 1978.

57. D. Lüke, invited talk given at the 1977 Meeting of the Division of Particles and Fields of the A.P.S., Argonne, Illinois, October 1977. SLAC-PUB-2086 (February 1978) and A. Barbaro-Galtieri, private communication.

58. D. Fakirov and B. Stech, Heidelberg preprint; Nucl. Phys. B133 (1978) 315;
H. Fritzsch, Phys. Lett. 71B (1977) 429).

59. a) I. Hinchliffe and C.H. Llewellyn Smith, Nucl. Phys. B114 (1976) 45;
b) A. Ali and T.C. Yang, Phys. Lett. 65B (1976) 275;
c) F. Bletzacker, H.T. Nieh and A. Soni, Phys. Rev. D16 (1977) 732;
d) R. Nabari, X.Y. Pham and W. Cottingham, J. Phys. G: Nuclear Physics 3 (1971) 1485;
e) G.L. Kane, Phys. Lett. 70B (1977) 272;
f) X.Y. Pham and J.M. Richard, Preprint PARIS/LPTHE 77.20.

60. DASP Collaboration, W. Braunschweig et al., Phys. Lett. 63B (1976) 47;
DASP Collaboration, R. Brandelik et al., Phys. Lett. 70B (1977) 125 and 70B (1977) 387.

61. DELCO Collaboration, see J. Kirkby, rapporteur talk at the Hamburg Conference 1977, p.3.

62. J.M. Feller et al., Phys. Rev. Lett. 40 (1978) 274 and A. Barbaro-Galtieri, rapporteur talk, 1977 Hamburg Conference, p. 21.

63. P. Rapidis et al., Phys. Rev. Lett. 39 (1977) 526.

64. CERN-Dortmund-Heidelberg-Saclay Collaboration, F. Dydak, private communication.

65. J. Ellis, M.K. Gaillard and D.V. Nanopoulos, Nucl. Phys. B100 (1975) 313.

66. P. Fayet, Nucl. Phys. B78 (1974) 14;
T.D. Cheng and L.F. Li, Phys. Rev. Lett. 38 (1977) 381.

67. M.L. Perl et al., Phys. Rev. Lett. 35 (1975) 1489;
M.L. Perl et al., Phys. Lett. 63B (1976) 466.

68. See the recent review talks by
M.L. Perl, Proceedings of the XII Rencontre de Moriond, Flaine, 1977, to be published, and SLAC-PUB-1923;
and 1977 Hamburg Conference, p. 145;
G. Flügge, invited talk at the Vth International Conference on Experimental Meson Spectrosopy, Northeastern University, Boston, MA, 1977; DESY Report 77/35 (1975).

69. DASP Collaboration, R. Brandelik et al., Phys. Lett. 73B (1978) 109.

70. H.B. Thacker and J.J. Sakurai, Phys. Lett. 36B (1971) 103;
Y.S. Tsai, Phys. Rev. D4 (1971) 2821;
J.D. Bjorken and C.H. Llewellyn-Smith, Phys. Rev. D7 (1973) 887;
K. Fujikawa and N. Kawamoto, Phys. Rev. D14 (1976) 59;
Y.I. Azimov, L.L. Frankfurt and V.A. Khoze, Leningrad Nuclear Physics Institute, Preprint 245, June 1976.

71. N. Kawamoto and A. I. Sanda, DESY 78/14 (1978);
Similar results are reported also by:
F.J. Gilman and D.H. Miller, SLAC-PUB-2046, 1977 and
Y.S. Tsai, private communication.

72. C.H. Llewellyn-Smith, University of Oxford, Preprint 33/76, invited paper presented at the Royal Society Meeting on New Particles and New Quantum Number, London, 11 March, 1976.

73. M. Bernadini et al., Nuovo Cim. 17 (1973) 383.

74. S. Orito et al., Phys. Lett. 48B (1974) 165.

75. M.L. Perl et al., Phys. Lett. 70B (1977) 487;
M.L. Perl et al., Phys. Lett. 63B (1976) 466;
F.B. Heile et al., SLAC-PUB-2059 (1977).

76. Contribution to the Hamburg Conference
see H. Sadrozinski, rapporteur talk, 1977 Hamburg Conference, p. 47.
77. M. Cavalli-Sforza et al., Phys. Rev. Lett. 36 (1976) 558.
78. PLUTO Collaboration, J. Burmester et al., Phys. Lett. 63B (1977) 301.
79. PLUTO Collaboration, J. Burmester et al., Phys. Lett. 68B (1977) 297;
PLUTO Collaboration, G. Alexander et al., DESY Report 78/70 (1978).
80. A. Barbaro-Galtieri et al., Phys. Rev. Lett. 39 (1977) 1058.
81. see S. Yamada, rapporteur talk, 1977 Hamburg Conference, p. 69.
82. G.J. Feldman et al., Phys. Rev. Lett. 38 (1977) 117.
83. F. Gutbrod and Z. Rek, DESY Report 77/45 (1977).
84. W. Alles, Ch. Boyer and A.J. Buras, CERN-TH-220 (1977).
85. W. Bacino et al., Phys. Rev. Lett. 41 (1978) 13 and
J. Kirz, data reported at the 1978 Tokyo Conference.
86. W. Bartel et al., Phys. Lett. 77B (1978) 331.
87. T. Kinoshita and A. Sirlin, Phys. Rev. 108 (1957) 844.
88. see also So-Young Pi and A.I. Sanda, Annals of Physics 106 (1977) 171.
89. DASP Collaboration, S. Yamada, rapporteur talk, 1977 Hamburg Conference, p. 69.
90. G. Hanson, results presented at the 13th Rencontre de Moriond (1978).
91. PLUTO Collaboration, G. Alexander et al., Phys. Lett. 78B (1978) 162.
92. D. Hitlin, talk at the 1978 Tokyo Conference.
93. G. Feldman, review talk at the 1978 Tokyo Conference.
94. PLUTO Collaboration, G. Alexander et al., Phys. Lett. 73B (1978) 99.
95. W. Wagner, Technische Hochschule Aachen, thesis; report PITHA 7801 (1978);

G. Knies, Proc. of the Sixth Trieste Conference on Particle Physics, 1978, ICTP report IC/78/76, p. 142.

96. M.L. Perl, rapporteur talk, 1977 Hamburg Conference, p. 145.

97. R.B. Palmer, results presented at the 13th Rencontre de Moriond (1978), Vol. II, Gauge Theories and Leptons, edited by J. Tran Thanh Van, p. 361.

98. S.W. Herb et al., Phys. Rev. Lett. 39 (1977) 252; W. Innes et al., ibid., page 1240.

99. L.M. Lederman, review talk at the 1978 Tokyo Conference, Columbia University preprint (1978).

100. PLUTO Collaboration, Ch. Berger et al., Phys. Lett. 76B (1978) 243.

101. C.W. Darden et al., Phys. Lett. 76B (1978) 246.

102. J.K. Bienlein et al., DESY Report 78/45 (1978).

103. H. Spitzer, data presented at the 1978 Tokyo Conference.

104. W. Schmidt-Parzefall, data presented at the 1978 Tokyo Conference.

105. G. Heinzelmann, data presented at the 1978 Tokyo Conference.

106. G. Flügge, rapporteur talk at the 1978 Tokyo Conference, DESY Report
H. Spitzer, talk given at the 1978 Kyoto Conference, DESY Report

107. J.L. Rosner, C. Quigg, and H.B. Thacker, Phys. Lett. 74B (1978) 350.

108. see e.g. S.D. Drell, D. Levy and T.M. Yan, Phys. Rev. 187 (1969) 2159; D1 (1970) 1035, 1617, 2402.

109. R. Gatto, P. Menotti and I. Vendramin, Nuovo Cimento Lett. 4 (1972) 35;
R. Gatto and G. Preparata, Nucl. Phys. B47 (1972) 313.

110. V.N. Gribov and L.N. Lipatov, Phys. Lett. 37B (1971) 78.

111. G.G. Hanson, talk given at the 1976 Tbilisi Conference, SLAC-PUB-1814 (1976).

112. PLUTO Collaboration, see U. Timm, rapporteur talk given at the 1977 Budapest Conference;
A. Bäcker, thesis, Gesamthochschule Siegen, 1977, and Internal Report DESY F33-77/03.

113. DASP Collaboration, R. Brandelik et al., Phys. Lett. 67B (1977) 358;
DASP Collaboration, R. Brandelik et al., Nucl. Phys., to be published and DESY Report 78/50 (1978).

114. British Scandinavian Collaboration, B. Alper et al., Nucl. Phys. B87 (1975) 19.

115. E.M. Riordan et al., SLAC-PUB-1634 (1975).

116. G. Schierholz and M.G. Schmidt, Nucl. Phys. B101 (1975) 429.

117. PLUTO Collaboration, data presented by G. Knies, 1977 Hamburg Conference, p. 93.

118. R.F. Schwitters, 1975 Stanford Conference, p. 5;
R.F. Schwitters et al., Phys. Rev. Lett. 35 (1975) 1230;
G. Hanson et al., Phys. Rev. Lett. 35 (1975) 1609.

119. R.F. Schwitters, rapporteur talk at the 1976 Tbilisi Conference;
G.G. Hanson, talk ibid, and SLAC-PUB-1814 (1976).

120. G. Kramer and T.F. Walsh, DESY Report 72/46 (1972), and Z. Phys. 263 (1973) 316.

121. J.D. Bjorken and S.J. Brodsky, Phys. Rev. D1 (1970) 1416.

122. G.G. Hanson, talk given at the 13th Rencontre de Moriond, March 1978 and SLAC-PUB-2118.

123. K. Koller and T.F. Walsh, Phys. Lett. B72 (1977) 227; B73 (1978) 504 and Nucl. Phys. B140 (1978) 449;
T.A. Grand et al., Phys. Rev. D16 (1977) 325;
S. Brodsky et al., Phys. Lett. B73 (1978) 203.

124. see e.g. H. Fritzsch, Schladming Lectures 1978, Acta Physica Austriaca, Suppl. XIX, p. 249 and
H. Fritzsch and K.H. Streng, Phys. Lett. B74 (1978) 90.

125. K. Hagiwara, Nucl. Phys. B137 (1978) 164;
A. de Rujula et al., Nucl. Phys. B138 (1978) 387.

126. S. Brandt et al., Phys. Lett. 12 (1964) 57;
E. Fahri, Phys. Rev. Lett. 39 (1977) 1587;
A. de Rujula et al., Nucl. Phys. B138 (1978) 387.

127. PLUTO Collaboration, Ch. Berger et al., Phys. Lett. B78 (1978) 176.

128. H. Georgi and M. Machacek, Phys. Rev. Lett. 39 (1977) 1237.

129. R.D. Field and R.P. Feynman, Nucl. Phys. B136 (1978) 1.

130. PLUTO Collaboration, Ch. Berger et al., DESY Report 78/71 (1978), to be published in Phys. Lett.

131. C.W. Darden et al., Paper contributed to the 1978 Tokyo Conference, Internal Report DESY F15-78/1 (1978).

D I S C U S S I O N

CHAIRMAN: G. Wolf

Scientific Secretaries: G. Bhanot, C. Newman

DISCUSSION No. 1

- GROSSE:

There is a general result (due to A. Martin and myself) stating that the ratio

$$R = \left|\frac{\psi_{1S}(0)}{\psi_{2S}(0)}\right|^2 = \begin{cases} >1 & \text{if } \left(\frac{\partial}{\partial r}\right)^2 V(r) < 0 \\ <1 & \text{"} \quad \text{"} \quad >0 \\ 1 & \text{if } V(r) = kr\,; \quad 8 \text{ if } V(r) = \alpha/r \end{cases}$$

Clearly to get R = 1.6 (observed experimentally) the coefficient of the Coulomb term must be small.

- THIRRING:

It is true for all S states that

$$|\psi(0)|^2 = \frac{\mu}{2\pi}\langle \partial V/\partial r \rangle$$

Hence, for a linear potential, R = 1 for any two S states.

- BATTISTONI:

Have you any data on the production of mesons with C = 1 through two-photon interactions at different energies?

- WOLF:

The experimental information on two-photon processes is still very sparse. At Frascati the process $e^+e^- \to e^+e^- \mu^+\mu^-$ * and three events on $e^+e^- \to e^+e^- \pi^+\pi^- n\pi^0$ * have been observed.

---

*S. Orito et al. PL 48B(1974)380

Novosibirsk has studied $e^+e^- \to e^+e^- e^+e^-$†. The DASP group recently observed $e^+e^- \to e^+e^- \mu^+\mu^-$; events of this type contributed a sizeable background to inclusive π production, $e^+e^- \to \pi^\pm X$.#

- BARBARO-GALTIERI:

Did PLUTO measure the resonance parameters for the 4·04 and 4·16 GeV states?

- WOLF:

Yes, these parameters agree within errors with those of DASP.

- MONTGOMERY:

Can the differences in radiative corrections account for the disagreement between DASP, SLAC-LBL & PLUTO data?

- WOLF:

All three groups use the same standard (Bonneau & Martin) method for radiative corrections. At DASP we subtract ∿ 8% below 4 GeV, add ∿ 11% at 4·03 GeV, and subtract ∿ 4% above 4·3 GeV. Without these corrections the valley between 4·04 and 4·16 GeV will be filled in and the peaks at these energies decreased.

- ZICHICHI:

How does the acceptance of DASP compare with those of PLUTO and SLAC-LBL?

- WOLF:

SLAC-LBL and PLUTO use solenoids to measure charged particle momentum. They both require at least two charged particles in the trigger. The acceptances of PLUTO and SLAC-LBL are 86% and 67%, respectively. The efficiency of SLAC-LBL for total cross-section events is 50% near 4 GeV. They use a phase-space like Monte Carlo to correct for the events lost and they include in their Monte Carlo pions but no neutrinos from Charm decay. Pluto has total cross-section efficiency of 65%. They use a jetlike Monte Carlo with p_T = 350 MeV. The DASP inner detector has no magnet, so charged particle momenta are not measured. The total cross-section is determined from events with three or more particles where one or more of these particles may be photons. The acceptance is 70% of 4π while the efficiency is 40%.

---

†V.E. Balakia et al. PL 34B(1971)663
#DASP collab. in press.

DASP measures the efficiency directly by identifying one particle as a hadron in the outer detector and checking whether this event would have been seen in the inner detector. Monte Carlo is needed only to correct for the requirement that one more charged particle is produced. This correction is of the order of 5% of σ_{tot}.

- ZICHICHI:

Even after acceptance corrections, the individual characteristics of the three experiments make the final states only semi-inclusive. This may prevent a dip seen in one case from being seen in another, because their sensitivities are different for different exclusive states.

- WOLF:

I agree.

- BARBARO-GALTIERI:

In a 4π detector, QED effects may be easily removed from two-prong events. It is not possible to do this if the solid angles are limited, consequently it is very difficult to compare the three detectors.

- WINTER:

You said that because of the $1 + \cos^2\theta$ angular distribution e^+e^- annihilation products are easier to measure than the strongly forward peaked products of proton proton collisions. Nevertheless, $\sigma_T(pp)$ has been measured to ±1% at the ISR and $\sigma_T(e^+e^- \to \text{hadrons})$ only to ±15%. Can a better experiment for $\sigma(e^+e^- \to h)$ be devised?

- WOLF:

$\sigma_T(e^+e^- \to h)$ is smaller by several orders of magnitude than $\sigma_T(pp)$, so it takes much longer to get a comparable number of events and make cross checks. Furthermore, the trigger to good event ratio is of the order of 100 : 1 to 1000 : 1. The uncertainties in the suppression of the background constitute an important source for the large systematic errors. As an answer to Zichichi's question in the morning on α_s, I can say the following:

One can compare the experimental R value above Charm threshold (after removing 0·7 units for the τ) with the theory prediction of $R = \sum_i Q_i^2 (1 + \frac{\alpha_s}{\pi})$ to calculate α_s. One gets,

$$\left.\begin{array}{l} R = 2 \cdot 3 \pm \cdot 2 \\ \alpha_s = 0 \cdot 5 \pm \cdot 3 \end{array}\right\} \text{ Below charm threshold}$$

$$\left.\begin{array}{l} R = 4 \cdot 3 \pm \cdot 4 \\ \alpha_s = 0 \cdot 9 \pm \cdot 4 \end{array}\right\} \text{ At 5 GeV}$$

- PETERSON:

The new SLAC-LBL data at high energy show a slightly larger value of R.

- ZICHICHI:

So the trouble is worse than you say. There is a one unit discrepancy in R above charm even with all that cooking with α_s.

- BARBARO-GALTIERI:

The SLAC-LBL level of R is 2.6 below charm - so that the discrepancy of 1.0 unit exists even below threshold.

- WOLF:

The systematic error may be different at 2 and 5 GeV, because the type of event changes at charm threshold.

- BARBARO-GALTIERI:

It is also possible that the models (jetlike, phase space etc.) for Monte Carlo may be different leading to variations in systematic error.

- MONTGOMERY:

Do you ever make a comparison of the $\mu^+ \mu^-$ cross-section between the three experiments? To what extent does the systematic error take account of luminosity differences?

- WOLF:

At DASP, the luminosity is determined from 8^0 Bhabha scattering. We have determined large angle Bhabha scattering and μ-pair production and found agreement with the QED to within 7% (with the luminosity determined as discussed above). SLAC-LBL have made similar tests.

- THIRRING:

Does the value of α_s go up or down as s increases?

- KLEINERT:

For a fixed number of flavors, it goes down with increasing s by asymptotic freedom. But when a flavor threshold is crossed, the factor 33 - $2N_f$ in the denominator decreases and α_s jumps upwards.

- BARBARO-GALTIERI:

If you change α_s, the Charmonium model may not fit the data as it does now.

- WOLF:

The Potential model as calculated is non-relativistic, has no gluon corrections etc. After such refinement, one could change α_s to 0.4 and still fit the data.

- BHANOT:

What is the meaning of your Charmonium level nomenclature above charm threshold? The states there mix with the continuum so that, when one is calculating transition amplitudes, one does not know the composition of the state and that makes comparison between theory and experiment very questionable. This is an important effect, because the wave functions of the states have nodes whose location is a function of the mixing.

- WOLF:

That is correct. But the effect of mixing is not that great if you accept the level assignment I showed. The disagreement is 100 MeV in 4 GeV which is a 2.5% effect. In fact, one is amazed why everything comes out so well.

- LACKNER:

The theory calculation of α_s is valid only for vanishing quark mass. What do mass effects do to R?

- ISGUR:

The calculation is valid not for zero quark mass but whenever $\sqrt{s} \gg$ all quark masses.

DISCUSSION No. 2 (Scientific Secretaries: G. Bhanot, C. Newman)

- BHANOT:

Could you suggest an alternative interpretation for the η_c and η_c' ?

- WOLF:

Lipkin, Rubinstein, and Isgur claim the $\chi(2\cdot8)$ to be a $q\bar{q}c\bar{c}$ system analogous to the S^*/φ case while the η_c is close by the J/ψ, but the arguments in favor of this interpretation are not very strong. I would argue against this interpretation in the following way. In the data from Serpukhov on $\pi^-p \rightarrow \gamma\gamma n$, the center of the bump agrees with the mass of the $\chi(2\cdot8)$. Firstly, it would be surprising to find the S^* produced and not the η_c. Secondly, the authors point out that the rates for $\pi^-p \rightarrow \pi^0 n$

ηn

$\chi(2\cdot8)n$

indicate that all seem to come from the same charge exchange mechanism.

- ISGUR:

What about the width of the enhancement at 2·8 GeV seen in $\pi^-p \rightarrow \gamma\gamma n$?

- WOLF:

According to the authors, the observed width is consistent with a zero width object. On the other hand, one can argue that there is some room for an η_c contribution at a mass of 3·0 GeV.

- THIRRING:

You said the D and F are pseudoscalar particles and the D^* and F^* are vector particles. What is the experimental evidence?

- WOLF:

For the D and D^* the evidence for 0^- and 1^- respectively is quite good. The decay analysis of $D \rightarrow K\pi$, $K\pi\pi$ together with a study of $D^* \rightarrow D\gamma$, shows that $J^P(D) = 0^-$ and $J^P(D^*) = 1^-$.

The F has not yet been analyzed in that way. But if what we call F - the 2·03 state - were a 1^- rather than a 0^- state, it could go into $e\nu$ with a large branching ratio. This would produce a peak in the electron spectrum at high momenta which is not seen.

- DE LA VAISSIERE:

Was there any attempt to see the η_c in hadronic modes? What are these hadronic modes?

- WOLF:

Hadronic decay modes have been looked for, but no such decay modes have been found. For a while there was weak evidence

- 2 events - for the decay : $\chi(2\cdot 8) \rightarrow p\bar{p}$. That evidence has disappeared and there is an upper limit:

$$BR(J/\psi \rightarrow \gamma\eta_c) * BR(\eta_c \rightarrow p\bar{p}) < 4x10^{-5}$$

- JACOB:

Do you meet difficulties in charmonium when you include spin effects?

- WOLF:

When one looks at the decay rates for P states, like $\psi' \rightarrow P$ states, one finds that the decay rates predicted by theory are a factor of 2 larger - in amplitude, a factor of 1·4 - than the experimental results. That can be considered good agreement. The level spacing is not very well predicted in the simple charmonium model. The early prediction for the level splitting for the P states was 10 MeV as compared to the 150 MeV found experimentally. This discrepancy could be removed by including spin effects.

For the η_c (that is if you assume the $\chi(2\cdot 8)$ is the η_c) the branching ratios are off by a factor of 50 or so. So far it has not been possible to understand at the same time the large $\chi - J/\psi$ mass splitting and the small transition rate for $J/\psi \rightarrow \chi(2\cdot 8)$.

- WINTER:

Could you please comment once more on the contradiction between the 60% branching ratio for $J/\psi \rightarrow \gamma\eta_c$ derived theoretically and the experimental upper limit of 1·7%?

- WOLF:

Using the mass of the η_c, $m\eta_c = 2\cdot 82$ GeV, and the mass of the charmed quark, $m_c = 1\cdot 55$ GeV, one predicts from QED a rate of 28 keV, while the experimental upper limit is 1·2 keV. The discrepancy may be due to various effects:

1) The overlap integral for the η_c and J/ψ is assumed to be unity in the absence of spin effects. However, the hyperfine splitting is $\sim$ 250 MeV which is a large percentage of the $\psi' - J/\psi$ splitting. So spin effects may be large.

2) The non-relativistic approximation may not be valid for the η_c.

3) The quark magnetic moment may not be of the Dirac type.

DISCUSSION No. 3 (Scientific Secretaries: G. Bhanot, C. Newman)

- ZICHICHI:

Can the τ events originate from the production of $q(\frac{2}{3})\ \bar{q}(-\frac{2}{3})$ followed by the decay $q(\frac{2}{3}) \rightarrow q(-\frac{1}{3}) + e^{+}\nu$ etc.?

- WOLF:

For a charge $2/3$ the cross section is $4/9$ the μ pair cross section. One has measured the cross sections:

$$\sigma(e^{+}e^{-} \rightarrow eX) = \sigma_{qq}\, B_{e}$$

$$\sigma(e^{+}e^{-} \rightarrow e\mu) = \sigma_{qq}\, B_{e}\, B_{\mu} = \sigma_{qq}\, B_{e}^{2} \text{ from } \mu\text{-}e \text{ universality}$$

$$= 4/9\, \sigma_{\tau\tau}\, B_{e}^{2}$$

Now from $e^{+}e^{-} \rightarrow eX$ and $e^{+}e^{-} \rightarrow \mu e$ one can get B_{e} and then check whether the cross section is $\sigma_{\tau\tau}\, B_{e}^{2}$ or $4/9\, \sigma_{\tau\tau}\, B_{e}^{2}$. The errors on B_{e} and $\sigma(ee \rightarrow \mu e)$ are small enough to exclude a cross section smaller by a factor of $4/9$.

A charge of $1/3$ would in general not be detected in the track chambers used or by the scintillation counters, since the threshold is set to $\sim 1/3$ minimum ionization. If we misidentify the decay $q(\frac{2}{3}) \rightarrow q(-\frac{1}{3})\, e\nu$ as $e\nu X$ then from the limits on the $\nu\tau$ mass, one can put a limit of $< .24$ GeV on the missing mass X, i.e. on the mass of the $q^{1/3}$. A quark of this mass would have been detected in the quark searches conducted at accelerators.

- ZICHICHI:

Do the data allow the τ to be a paralepton?

- WOLF:

The strongest argument against this assignment comes from neutrino experiments. The lower limit obtained on the E^{+} and the M^{+} is > 8 GeV (see also lectures by K. Winter).

- ZICHICHI:

But in $e^{+}e^{-}$ physics can you exclude the production of two quasidegenerate gauge leptons?

- BARBARO-GALTIERI:

In principle this can be done from measurement of the ratios:

$$\frac{\sigma(e^{+}e^{-} \rightarrow e^{+}e^{-})}{\sigma(e^{+}e^{-} \rightarrow e\mu)} \quad \text{and} \quad \frac{\sigma(e^{+}e^{-} \rightarrow \mu^{+}\mu^{-})}{\sigma(e^{+}e^{-} \rightarrow e\mu)}$$

There are some such results from SLAC/LBL, but the error bars are large.

- WOLF:

With two gauge leptons, the cross section is twice as large, unless there is some strange interference effect or if their magnetic moments are less than that of a Dirac particle.

- WINTER:

What is the neutrino mass used in the fit for the τ to V-A, V+A?

- WOLF:

The fit was made with a zero τ neutrino mass. If the neutrino is massive, the V-A curve will be shifted to lower electron momenta and will not agree with the data. The V+A disagreement with the data will become worse. From the DELCO group results we also can exclude pure V and pure A.

- THIRRING:

Is there some direct evidence that the τ has no strong interaction?

- WINTER:

One way to get an answer is to photoproduce the τ coherently off nuclei and to measure the A dependence.

DISCUSSION No. 4 (Scientific Secretaries: G. Bhanot, C. Newman)

- MERRITT:

What is the effect of ρ decay on the p_T distribution of the pion? When you start passing particle thresholds do you produce more particles in the jets, and does this affect the p_T distribution?

- WOLF:

Roughly speaking, in $\rho^0 \to \pi^+\pi^-$ each pion gets half the ρ energy and hence $x_\pi \approx \frac{1}{2}x_\rho$. In other words, the slope of the pion x distribution is twice that of the rho.

- PETERSEN:

What is the reason that the average number of particles produced rises in the region of the resonances? Is the heavy lepton contribution subtracted?

- WOLF:

No quantitative study has been made for the first question. However, the following may give the correct explanation. The resonances, at least the 4.04 state, are dominated by $D\bar{D}^*$ and $D^*\bar{D}^*$ production. The average number of charged particles per D decay is around two. In addition, the $D^* \rightarrow D\pi^\pm$ transition will add another charged particle. Hence the average number of charged particles produced in DD^*, D^*D^* production is approximately 5, which is one unit above the charged multiplicity outside the resonances.

The heavy lepton contribution to the multiplicities from DASP is roughly 5%.

- DI IANNI:

Can we measure the total charge of jets?

- WOLF:

The answer is yes. Actually Feynman has proposed a method which will improve the accuracy of such a determination. Consider the particles in one jet, labelled according to their x values, x_i. Weight these particles with a test function f(x) which is large at x = 1 and decreases towards x = 0. For a single event :

$$Q_{jet} = \Sigma f(x_i) Q_i / \Sigma f(x_i)$$

The purpose of the weight function is to emphasize the quark fragmentation region which remembers the quark quantum numbers.

- PLOTHOW:

Regarding your discussion of

$$D \rightarrow e\nu X$$

the value for the branching ratio predicted by theory was 20%, while $\sim$ 8% was observed experimentally. This shows that SU(4) symmetry is broken.

Our group (CERN Hyperon Experiment WA2*) measures the branching ratios:

$$\frac{\Gamma(\Omega^- \to \Xi^- \pi^0)}{\Gamma(\Omega^- \to \Xi^0 \pi^-)} = 2.93 \pm .5$$

The $\Delta I=\frac{1}{2}$ rule would predict 2 for this ratio. Thus we see evidence of SU(3) symmetry breaking.

- WINTER:

In K decay, the $\Delta I=\frac{1}{2}$ rule was found to be satisfied to within a few per cent.

- BARBARO-GALTIERI:

So is the case for the Σ and Ξ decays.

- WINTER:

However, in both cases the Q value is small as compared to the Ω^- decay.

- WOLF:

In D decay the Q value is also large. Therefore the octet and 20-plet enhancements may be a function of the Q value.

---

*"Measurement of the Lifetime and Branching Ratios of the Ω^-." contributed to the 1978 Tokyo conference.

POINT-LIKE EFFECTS IN STRONG INTERACTIONS

Maurice Jacob

CERN

Geneva, Switzerland

Abstract

Particle production at large transverse momentum is found to exceed by a large amount what is expected from the collision of two extended objects about 1 Fermi across. The pertinent effects are associated with collisions among hadron constituents which materialize as jets of particles.

Experimental evidence for a jet configuration is reviewed. A phenomenological analysis of the key features of jet fragmentation is then presented. It is based on the scaling properties of hadronic interactions. Theoretical models are discussed and inparticular the relevance of Quantum Chromodynamics is assessed. We also discuss future prospects at present machines and also consider the use of present synchrotrons in their collider versions.

This series of lectures is based on a review paper " Large transverse momentum and jet studies" written by M. Jacob and P. Landshoff and which will be published in Physics Reports. The main emphasis is on jets in hadron collisions but, whenever possible, connections are made with jets as seen in other processes, and in particular in electron-positron annihilations.

At present jet studies are in a particlularly interesting stage. Jets as expected in the parton model have now been ascertained. However, the special features associated with quantum chromodynaics, which is the great contender for the understanding of the successes of the parton approach, still escape specific tests. The situation is reviewed as it appears after the Copenhagen meeting on "Jets in High Energy Collisions".

FOREWORD

Hadron collision at very high energy usually results in the production of a large number of secondary particles which are mainly pions. The transverse momentum (p_t) of these secondary particles is limited to rather low values ($p_t \approx 0.35$ GeV/c) with an exponential distribution which is most generally associated with the transverse dimension of the colliding hadrons ($\approx$ 1 fermi), in a direction which is not affected by the Lorentz contraction. It was then surprising to find that yields at very large p_t ($p_t > 1$GeV/c say) were far larger than those expected through the naive extrapolation of the low p_t behaviour. A simple quantum mechanical interpretation of such an effect can be obtained by using the concept of scattering among point-like hadron constituents acting incoherently among themselves. Large p_t production then becomes a modern extension of the famous Rutherford experiment, revealing now a coarse-grain structure within the hadron, an idea prompted earlier by the behaviour of deep inelastic lepton scattering. If such an interpretation is now widely accepted, a long experimental and phenomenological effort was required before it could be reasonably well ascertained. Evidence, as it now stands, is reviewed.

Hadron constituents once scattered at wide angle do not escape as such. They materialize as jets of hadrons, a jet being a set of particles, usually pions, all produced in a particular direction. Insofar as a large p_t secondary may signal the occurrence of scattering at the constituent level, jets of hadrons are expected. Evidence for such a jet structure is now compelling but many problems have still to be faced before a detailed picture is available. Present phenomenology for jet formation and fragmentation is discussed.

The apparently puzzling situation whereby the reaction proceeds as if point-like constituents were directly involved, but where no such constituent can be found among the reaction products, now finds a possible interpretation in the framework of quantum chromodynamics, a field theory for hadronic interactions at the quark level. The relevance of such an approach is reviewed.

It was long thought that hadron interactions could become relatively simple at very high energies. The required condition was usually stated as $s >> m^2$, where s and m are the center-of-mass energy squared and the typical hadron mass, respectively. This was motivated by the power behaviour of Regge amplitudes. Research at the CERN ISR quickly found that such an asymptotic simplicity was but an elusive concept. If some simple asymptotic regime may eventually prevail, it should now be met for $\ell n\ s/m^2 >> 1$ rather than for $s/m^2 >> 1$, that is far beyond the present ISR energy range! It now, however, appears that at large transverse momentum ($p_t >> m$) hadron interactions may become simple and there are good reasons to expect that a perturbative approach could even be justified. This is a fascinating prospect which explains all the interest which has already been attached to the study of large transverse momentum phenomena, ever since first evidence was found at the CERN ISR in 1972.

1. Introduction

The study of particle production at large transverse momentum is of great topical interest. It offers a possibility to probe short-range hadron interactions with evidence for substructures which may be compared with those revealed by high-energy processes involving leptons. It has also become of importance recently in the prediction of backgrounds in the search for the W and other particles which could be produced in very high energy hadron interactions. It is however our hope that this review will convince the reader that the actual property of such a hadronic background at large transverse momentum may be as interesting as the long searched-for particles which could possibly appear.

The subject was born at the CERN ISR in 1972 with the discovery of anomalously high particle yields at large p_t, compared with the naive extrapolation of those observed at low p_t. This is illustrated in Figure 1.1, where data on 90° π° production from the CRS collaboration are compared with the form e^{-6p_t}. The latter form fits the data well for $p_t \lesssim 1$ GeV/c but, as may be seen in the figure, above this value of p_t the data depart from the exponential and lie substantially above it. Such behaviour was not totally unexpected, having been hinted at in earlier theoretical papers[1,2] which were prompted by the observation of Bjorken scaling in deep inelastic lepton scattering. However, the actual behaviour agreed only qualitatively with the theoretical expectations.

Although a final understanding of large p_t-phenomena has not yet been achieved, their study is now in an advanced stage. This study has involved an unusually close interplay between theory and experiment. As is clear from Figure 1.1, particle production at a given large value of transverse momentum rises markedly with increasing energy. Hence the ISR has played a dominant part in the study, although its full potential is only now about to be realized because of a previous lack of suitable detectors. Results from Fermilab have also been important, and many of them are complementary to those from the ISR.

Most of the recent theoretical activity in large transverse momentum physics has been within the framework of hard-scattering models[2,3]. These models relate large-p_t phenomena to various types of reaction involving leptons. They predicted the jet structure

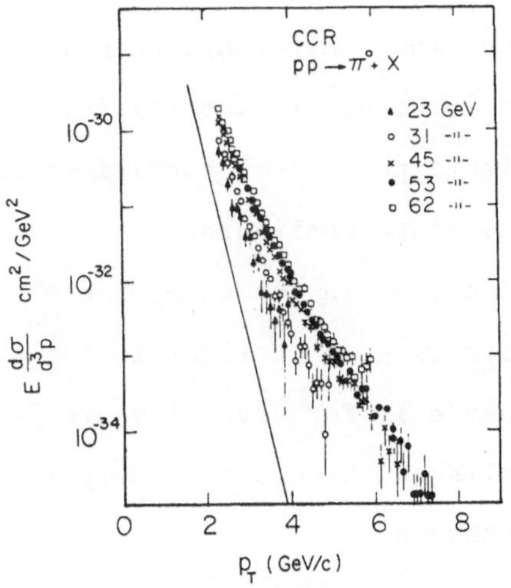

Figure 1.1 Inclusive production of π^0 at 90^0 in pp collisions over the ISR energy range. The solid line is the extrapolation of the fit e^{-6p_T} to data at $p_T \leq 1$ GeV/c. Data from CCRS.

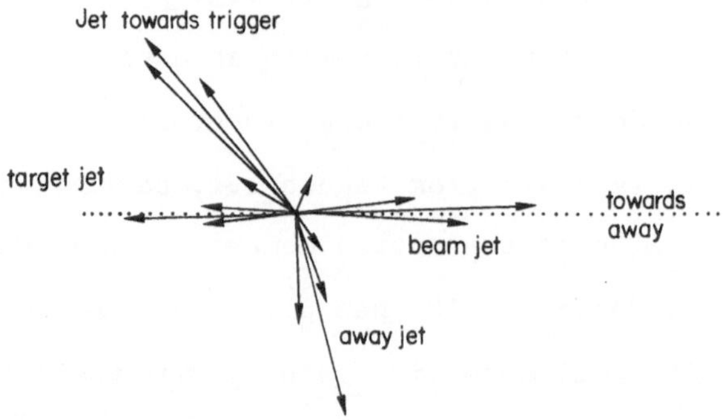

Figure 1.2.a The two jet structures expected from hard scattering among hadron constituents.

which, as we describe in Section 3, seems to be a feature of large-p_t events. As it appears at present, a typical event consists of a pair of transverse jets, whose transverse momenta approximately balance one another, and two longitudinal jets; see in Figure 1.2.a.

At present one sees clear jet effects among hadrons produced in different types of reactions: e^+e^- annihilation, large p_t reaction and deep inelastic neutrino scattering. Examples of each type of process are displayed in Figure 1.2.b-(a),(b) and (c). This is the situation in 1978, with experimental conditions such that the jet structure of the final state configuration leaves no doubt However this requires large energies in the case of e^+e^- annihilation ($\sqrt{s} \gtrsim 8$ GeV) and very large p_t trigger in the case of hadron scattering ($p_t \gtrsim 5$ GeV/c). For several years jet structure had to be gradually established through the analysis of correlation effects, as later discussed.

Other types of models, such as the very-high-spin massive fireball model[4] and the independent-emission model[5], can also accomodate such a jet structure, but it does not occur in them quite so

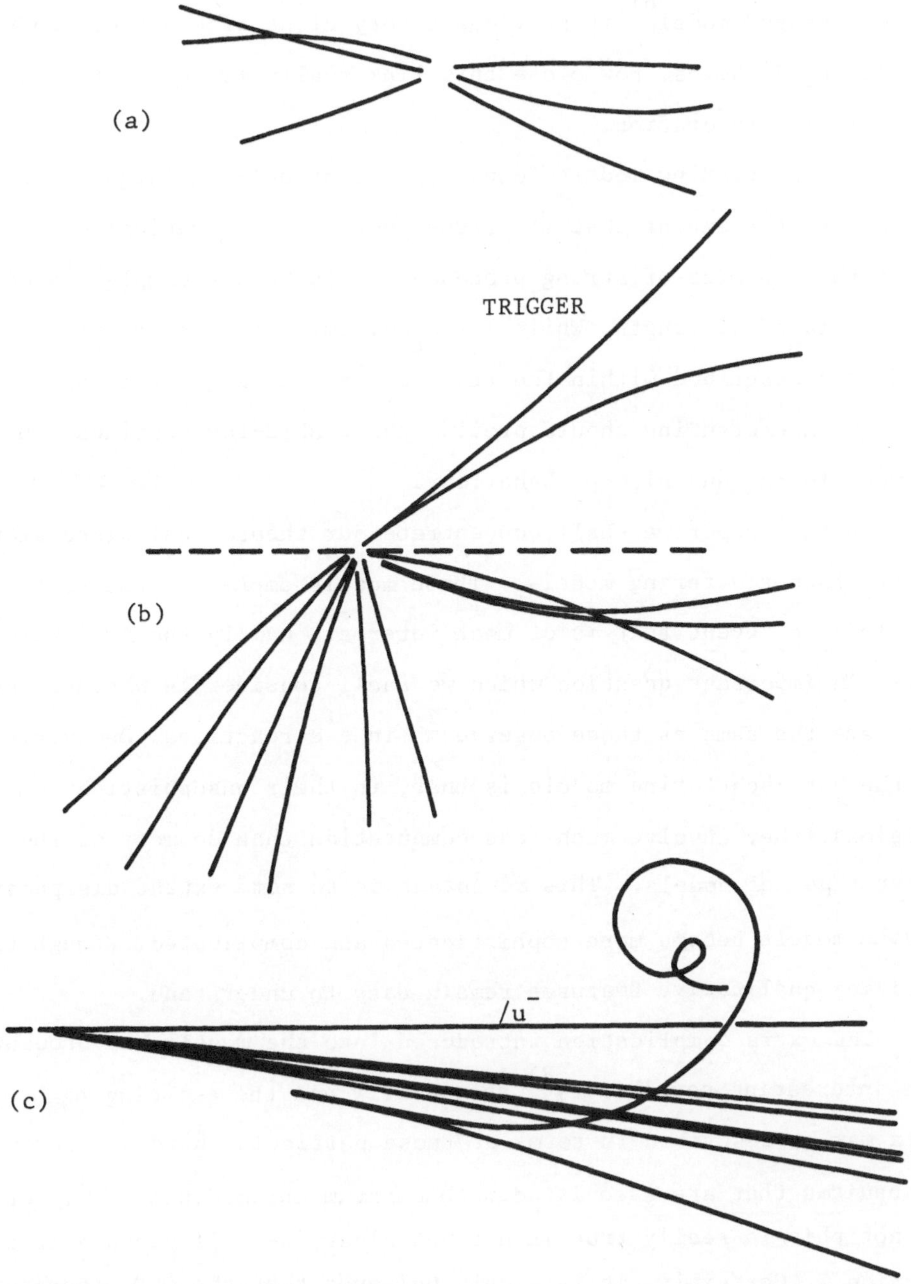

Figure 1.2b Jets as now seen in e^+e^- annihilation at DESY (a); in reactions triggered by a large p_T particle at the CERN-ISR (b); in deep inelastic neutrino scattering in BEBC (c).

naturally. The fireball model does have the feature that, like the "fire-sausage" model[6)] it provides a very close link between high-p_t and low-p_t dynamics. How close this link really is in nature deserves further exploration.

Hard-scattering models focus at present only on large p_t behaviour to the extent that it corresponds to a kinematical domain where the dynamics of strong processes could become simple, as discussed later at length. While low-p_t dynamics is temporarily left aside and described within the framework of the Regge pole approach, further understanding should provide the wished-for continuation between low-p_t and high-p_t behaviour.

In this paper we shall concentrate our theoretical discussions on the hard-scattering models. These models emphasize the part of the large-p_t event that is of most interest, namely the transverse jets. An important question which we shall consider is whether these jets are the same as those observed[7)] in e^+e^- reactions. One merit of the hard-scattering models is that, in their unsophisticated versions, they involve much less computation than do most of the other types of models. This advantage is to some extent disappearing as the models become more sophisticated and complicated, though their striking qualitative features remain easy to understand.

The extra complication introduced into the models has brought the interesting possibility[8)] that nearly all the existing high-p_t data may be described in terms of those particular hard-scattering mechanisms that are associated with quantum chromodynamics[9)]. Whether or not this is really true is not yet clear; we will discuss it in Section 7. Certainly, it is widely believed that the QCD processes should at least be apparent at very large values of p_t. Furthermore, at very large p_t the extra complications introduced into the hard-scattering models are expected to become less important, so that it is hoped that the analysis of data will then become simpler and its interpretation more certain.

For these two reasons, a main thrust of present experimental activity is towards higher values of p_t. This is made possible by the greater acceptance of new detectors, a feature that is necessary also for a more adequate exploration of correlations. Another development that makes possible the exploration of higher values of p_t is the development of calorimeters that allow the use of a whole jet as a trigger rather than just a single high-p_t particle. From an analysis of the ISR correlation data, it was predicted[10,11)] that, for a given p_t, the triggering rate with a suitable calorimeter should be some two orders of magnitude greater than with a single particle. The recent data from Fermilab seem to verify this.

Nearly all the correlation data available so far is kinematic in nature: even if the trigger particle is identified, the associated particles usually are not, except for their respective charges. The information about quantum number correlations that will be accessible with the improved detectors may be expected to play a crucial rôle in sorting out what is happening at large p_t, and in determining the nature of the jets. Hopefully, it may be possible to establish that the jets correspond to large momentum quarks fragmenting into streams of hadrons, as probably occurs in high energy e^+e^- annihilations. But it may be that, at least in some events, more complicated phenomena will be found.

The study of large transverse momentum phenomena has been both exciting and frustrating. It has been exciting because of its association with hard scattering among hadron consituents, something which found supporting evidence at a relatively early stage. It appears at present as the only way to study rather directly quark-quark elastic scattering. It has, however, been frustrating because of the great difficulty in trying to unravel the hard-scattering effects proper from an important and still poorly understood background. Angular correlations are very loose, and they are actually expected to be so in view of the wide momentum spread of the hadron constituents. Their analysis requires large solid angle detectors which have only slowly become available. Worse, at medium p_t values ($3 < p_t < 5$ GeV/c say), where most of the present data are, it is clear that a large fraction of what is observed corresponds more to kinematical constraints than to specific dynamical effects. Ascertaining new dynamical features required much effort and, in particular, the experimental definition of a jet still requires refinement. Furthermore, although one may now reasonably expect that very large p_t triggers ($p_t > 10$ GeV/c say) could show a clear association with quark-quark scattering, the nature of the constituents involved for the bulk of the (lower p_t) data is still a matter of hypothesis. Constituents of a less elementary nature could play an important rôle. Quark-quark scattering might in any case not dominate over other processes expected in the framework of quantum chromodynamics, with an important rôle attributed to gluons.

Such complications make it clear why progress has been relatively slow despite a very substantial experimental effort at the CERN ISR and at Fermilab, over a period of 6 years already. The groundwork seems at long last solid, but much remains to be done.

Considering future prospects one should consider large p_t phenomena in a more general perspective. This is done in Figure 1.3 which combines 4 different processes which, as often hinted at already are expected to show some definite kinship. The first one is e^+e^-

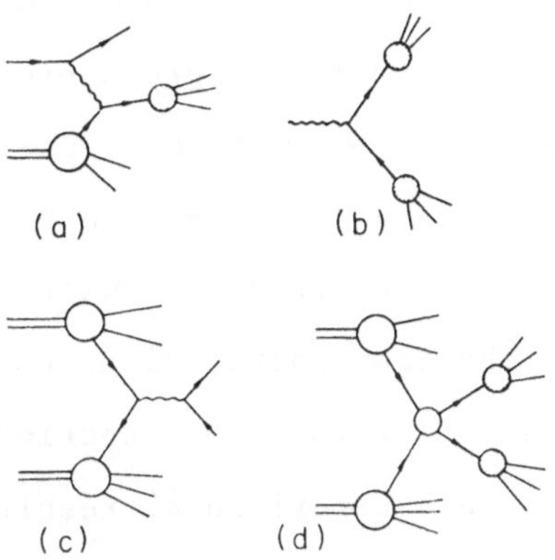

Figure 1.3 Four processes related in the parton model.

annihilation at high energy. Annihilation is analyzed in terms of a quark-antiquark pair production through one photon exchange. Two jets should result in a simple back-to-back configuration. This should become more and more obvious experimentally as the energy increases. The second process corresponds to deep inelastic lepton scattering, analyzed in terms of scattering off a constituent quark. The scattered quark may result in a jet at wide angle while the remainder of the struck nucleon appears as a backward jet. The jet structure should become clearer as the momentum transfer (and therefore the incident energy) increases. The third process corresponds to large mass lepton pair production in hadron-hadron interaction. The reaction is here analyzed in terms of the now familiar Drell-Yan process, whereby a quark and an antiquark annihilate into a lepton pair. The remainder of the incoming hadrons should give forward and backward jets. Finally the fourth process is large-p_t hadron production, analyzed in terms of quark-quark scattering. Large angle scattering results in the production of two wide angle jets which should also have some longitudinal momentum. The remainder of the two hardons should give forward and backward jets.

Such an analysis corresponds to the standard parton approach, identifying partonsas quarks. It has great predictive value insofar as factorization properties are implied in the parton approach. The same parton fragmentation distribution into hadrons enters in the calculation of cross-sections associated with the first, second and fourth processes. The same parton distribution in hadron structure enters the calculation of the cross-sections for the second, third and fourth processes. Experimental results corresponding to four very different types of reactions can thus be analyzed in terms of two distibution functions. The quark-quark differential cross-section enters the calculation of large transverse momentum yields in a simple way (5.1). Conversely, the study of large p_t phenomena provides a direct clue to quark-quark scattering. As discussed in Section 5 this is the approach followed in particular by Feynman and Field in their first analysis of inclusive and correlation data: the "black box" approach.

Quantum chromodynamics provides a theoretical framework for the parton model. The property of asympotic freedom leads us to expect that the four types of phenomena should be analyzable by perturbation theory. However, while the single hard process in the first three reactions may define the dominant kinematical features of the final state, corrections corresponding in particular to one extra gluon emission may not be negligible. Indeed, they are understood to be responsible for part of the large mean transverse momentum measured in high mass lepton pair production. Moreover, in the case of the large transverse momentum production processes of special interest here, several basic reactions are now on the same footing and, even in two jet reactions, gluon jets may dominate over quark jets. Figure 1.3 corresponds therefore to an oversimplification. Process 9.1d) involves constituents which are not sensitive to electromagnetic interactions. As a result, the study of large p_t phenomena appears as intrinsically more complicated but also richer than if it involved quark jets only.

Indeed, the study of quark jets is far simpler in e^+e^- annihilation. As previously mentioned, the general features of jets have been well displayed first in the SPEAR data. The first results from DORIS above 9 GeV show even more dramatic structures. Figure 3.1 gives a typical event, as observed in the PLUTO detector. The distribution of the charged tracks clearly displays a two-jet structure.

One may be tempted to conclude that jet studies should from now on be pursued at PETRA and at PEP rather than at the ISR. Jet production now seems to be the typical reaction in e^+e^- annhilation at high energy, whereas it has to be looked for with a very selective trigger in hadron collisions. Moreover, all hadrons produced through e^+e^- annihilation can be associated with the interesting jet system, whereas separating jet particles from background particles raises difficulties in hadron collisions. One should certainly expect very interesting results from the first round of experiments at PETRA (PEP) and in particular one could obtain evidence for extra jets associated with radiated hard gluons. Nevertheless, further jet studies in hadron collisions remain highly worthwhile. Different types of jets may exist and quantum number correlations may reveal the importance of quark-gluon jet configuration as opposed to the mainly quark-antiquark jets of e^+e^- annihilation. Secondly, progress in jet definitions resulting from e^+e^- studies should help in refining jet triggers and, with the relatively high yields now ascertained, a serious study of the basic processes should be achieved.

Despite the apparently overwhelming advantages of electron-positron storage rings, proton machines can therefore still contribute in a unique way.

2. Single-particle Inclusive Data

Parametrization of the Data

In so far as Bjorken scaling is valid in deep inelastic lepton scattering, this corresponds to the property that dependence on any fixed dimensional parameter has disappeared. A similar property in large-p_t physics would correspond[1)] to a single-particle inclusive cross-section of the form

$$E \frac{d\sigma}{d^3p} \sim p_t^{-n} \quad F(x_t, \theta) \qquad (2.1)$$

with n = 4. Here θ is the centre-of-mass production angle and x_t is the dimensionless scaling quantity

$$x_t = 2p_t \ / \ \sqrt{s} \qquad (2.2)$$

so that F is a dimensionless function.

The data on pion production from the ISR and Fermilab in the range 2 GeV/c < p_t < 9 GeV/c fit very well to the form (2.1), but with n = 8 (see Figure 2.1). This means that the function F is not dimensionless, but rather contains some fixed dimensional parameter. Very recent data from the ISR indicate that when one includes larger values of p_t the effective fitted value of n becomes smaller; for instance, the quoted value from the CERN-Saclay experiment with data up to $p_t \simeq 12$ GeV/c, is $n = 6.6 \pm 0.8$. The CERN-Saclay data are

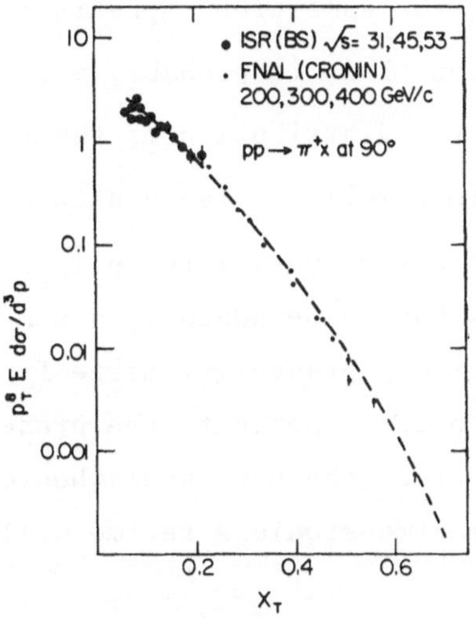

Figure 2.1 Inclusive production of π^+ at 90^0. The figure combines data from the ISR (full dots) and Fermilab (dots). It shows that at fixed x_T the distribution behaves as p_T^{-8}. The dashed curve corresponds to $(1-x_T)^9$.

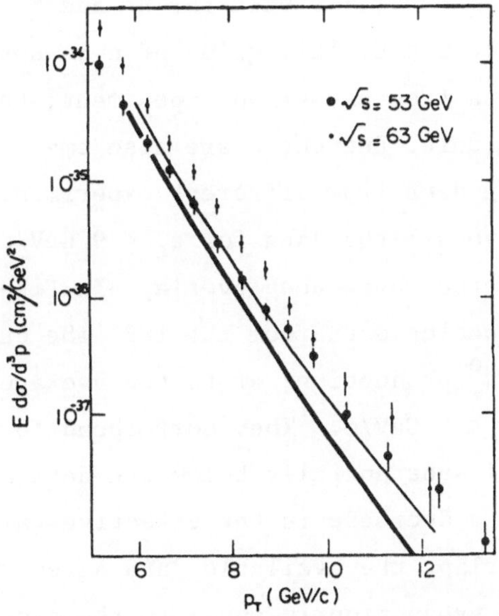

Figure 2.2 Inclusive π^0 production at 90^0. Data from the CERN-Saclay collaboration at the ISR. The curves correspond to lower p_T fits according to the p_T and x_T behaviour in Figure 2.1

shown in Figure 2.2; the same trend appears in the recent data from the CCOR collaboration. This is probably more realistically interpreted as saying that a distribution of the type (2.1) with $n = 8$, which reproduces rather well available data up to $p_t \simeq 9$ GeV/c say, no longer fits by itself at very large p_t ($p_t > 10$ GeV/c). A new type of contribution has to be added or a simpler and more fundamental behaviour may gradually emerge from the more complicated régime which corresponds to the presently studied ranges in p_t and $\sqrt{s}$. This raises the hope that when even larger values of p_t can be explored a dimensionless régime will reveal itself, with $n = 4$.

In Figure 2.1 we plot $p_t^8 Ed\sigma/d^3p$ for π^+ production at 90° in pp collisions, with $p_t < 9$ GeV/c. The data extend from $\sqrt{s} = 20$ at Fermilab to $\sqrt{s} = 53$ at the ISR, and they seem to fit well to a single curve as required by (2.1) when $n = 8$. It is found that the θ dependence is rather weak, with very little variation in the whole central region $45^o < \theta < 135^o$. Data for π^- and π^0 production likewise fit well to $n = 8$. This value of n is not necessarily the best fit to the data from any given experiment: there are variations up to about half a unit. But there are also small disagreements between overlapping data from different experiments, and $n = 8$ gives an acceptable fit to all the data for $p_t < 9$ GeV/c. The data shown in Figure 2.1 do agree where they overlap. In Figure 2.2 we show the recent π^0 production data from the ISR. The curves are fits of the form (2.1) to π^0 production, or to the average of π^+ and π^- production, for $p_t < 9$ GeV/c. They correspond to $n = 8$, and the fact that they fall substantially below the data at larger values of p_t may indicate a decrease in the effective value of n.

When they overlap, the available data agree well with a hypothesis that π^0 production is equal to the average of π^+ and π^- production. However, π^+ an π^- production differ in rate, particularly at larger values of p_t. This is shown in Figure 2.3, which displays data from the same two experiments as in Figure

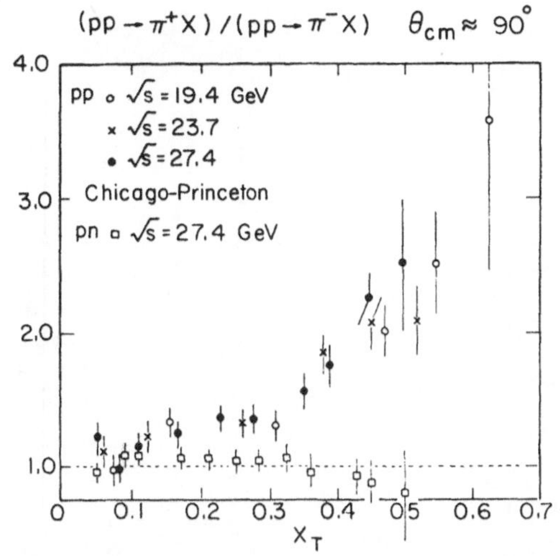

Figure 2.3.a Charged effects for large p_T yields.

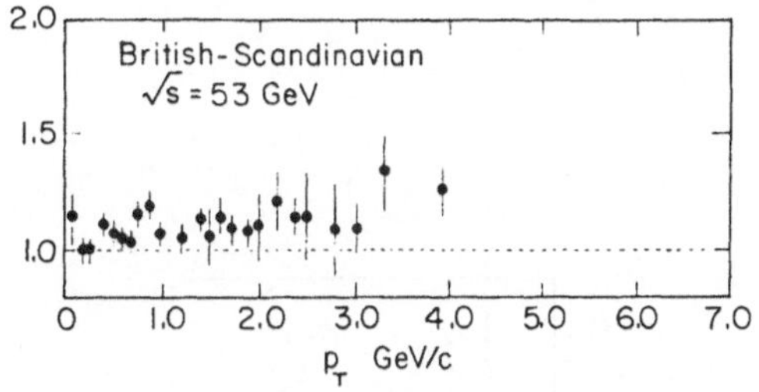

Figure 2.3.b Charged effects for large p_T yields.

2.1. As is indicated by the fact that the π^+/π^- ratio at different energies falls more or less on a single curve when plotted against x_t, both π^+ and π^- correspond quite well to (2.1) with n = 8. The same is true of K^+ and, to a lesser extent, K^-. Data for the ratios K^+/π^+ and K^-/π^- at 90° are shown in Figure 2.4, which contains also results on η production. For p production, on the other hand, the best value of n is close to 12. Data for this from Fermilab are shown in Figure 2.5, which also contains $\bar{p}$ data. In the case of $\bar{p}$, there

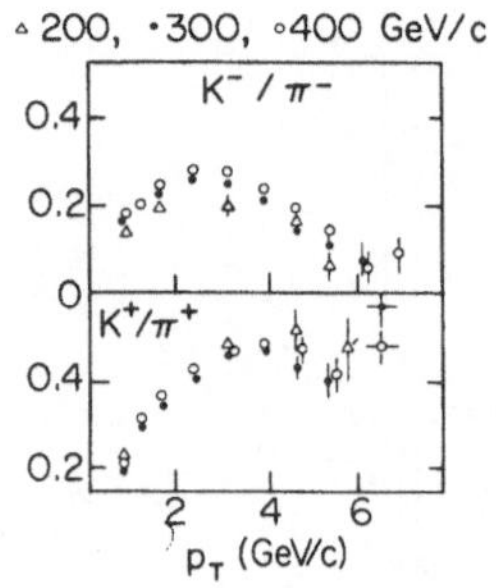

Figure 2.4.a
Yield ratios at large p_T.

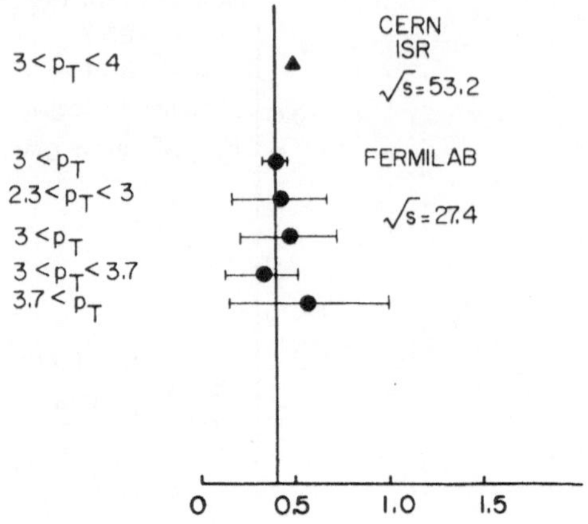

Figure 2.4.b
Yield ratios at large p_T.

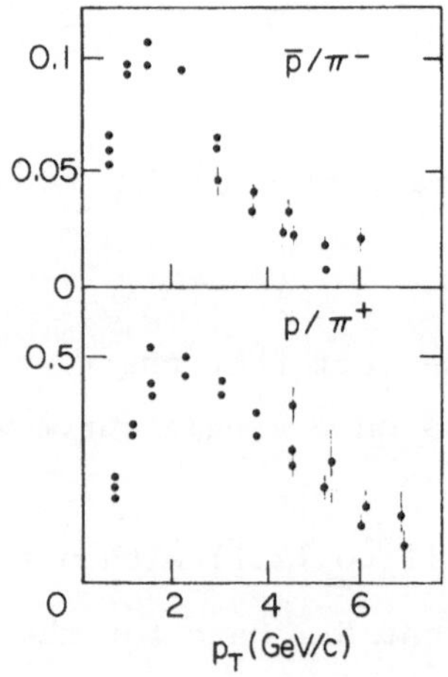

Figure 2.5 Yield ratios at large p_T. There is a relatively important proton yield. It decreases faster with p_T (p_T^{-12}).

is some discrepancy with overlapping data from the BS experiment at the ISR. The relatively large value of the p/π ratio at moderate p_t values ($1 < p_t < 4$ GeV/c) and the sharper fall-off of the proton yield with increasing p_t are surely both related to the fact that a large p_t proton is likely to originate through one of the incident protons being scattered at wide angle. This is further supported by recent data on fast low-p_t leading protons in reactions that contain large p_t secondaries, as later discussed. At present, data on large p_t reactions with baryon triggers are extremely limited as compared to what one would wish to have. We have remarked that, in the case of π production, the data break away from the exponential form that fits at low p_t when $p_t > 1$ GeV/c. In the case of p and $\bar{p}$ the break-away occurs at rather a larger value of p_t, 3 GeV/c or more. This possibly indicates that larger values of p_t are needed in p and $\bar{p}$ production before one can be reasonably sure of being in the régime of high-p_t dynamics.

At low p_t much of the difference between pion and proton distributions may be eliminated using variable $m_t = \sqrt{p_t^2 + m^2}$ rather than p_t. We are however considering a domain where difference in mass is no longer very important. The steeper fall-off of the proton distribution may rather reflect the disappearance with increasing p_t of a specific mechanism whereby the incident proton is scattered at wide angle.

Nuclear Effects

The Fermilab data that we have shown are for a hydrogen target. To achieve higher event rates, earlier Fermilab experiments used heavier targets, but revealed somewhat surprising nuclear effects. With the parametrization

$$E \frac{d\sigma}{d^3p} \sim A^{\alpha} \qquad (2.3)$$

the parameter α is shown in Figure 2.6. These data are for single-particle inclusive experiments. It has been found recently in experiment E 494 at Fermilab that if one requires a high-p_t particle to emerge at 90° on each side of the incident beam, the value of α stays close to 1. This suggests[12)] that the effects of Figure 2.6 may have their origin in the transverse Fermi motion of the nucleons within the nuclear target: the single-particle experiments select events where the Fermi motion is aligned towards the detected particle and so the cross-section is enhanced, while in the double-arm experiment the symmetry of the configuration removes this bias.[13)]

The value $\alpha = 1$ is understood in hard-scattering models as resulting from a parton of the beam particle penetrating the nucleus and retaining its identity as it does so. It then scatters at wide angle on one of the nucleons and begins to fragment into hadrons only after it has escaped from the nucleus. In ordinary, low-p_t reactions rather smaller values of α are found.

Other Beams

The data in Figures 2.1 to 2.6 all concern collisions of a proton. Of considerable theoretical importance are corresponding data (Figure 2.7) for different types of beam particles,particularly. the comparison of π and p - see Section 6 .The $\bar{p}$, p comparison is also important in connection with plans for the construction of $\bar{p}p$ colliding beam facilities, but the data are evidently not yet good enough to make any confident predictions about the necessary extrapolations to higher values of p_t and $\sqrt{s}$

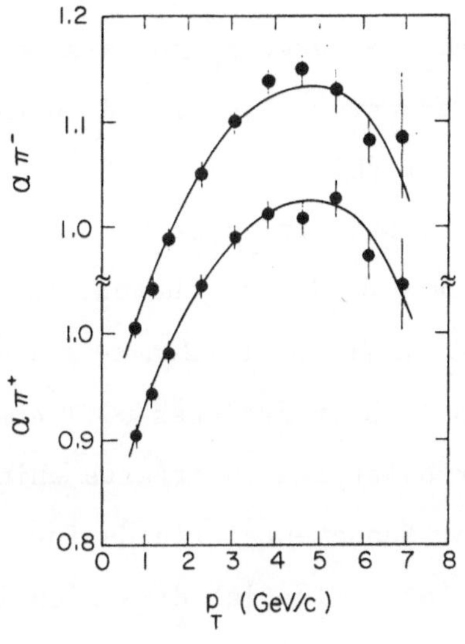

Figure 2.6 Nuclear effect for large p_T reaction.

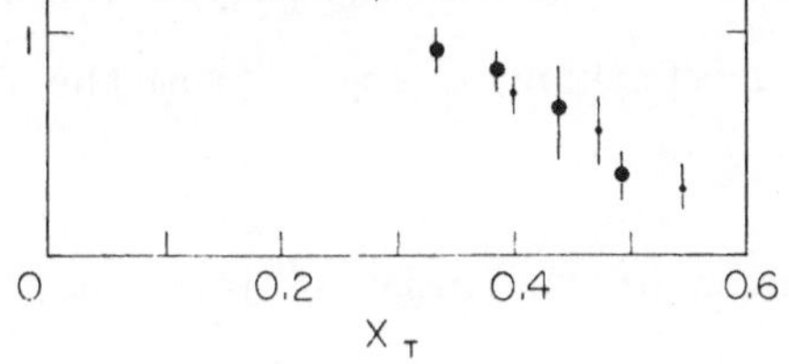

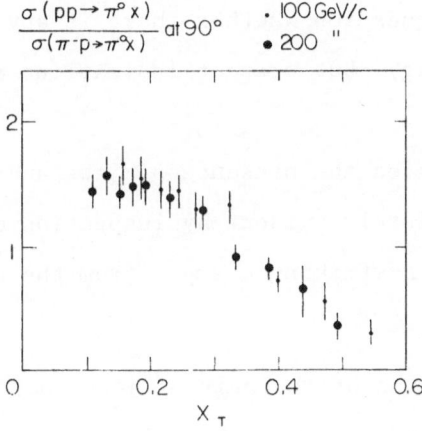

Figure 2.7 Beam ratios. Pions are more efficient than protons at producing large p_T particles.

3. Correlation Data: Jet Structure

The SPEAR Jets

We have said that the large-p_t correlation experiments reveal the existence of two transverse jets, and in this Section we review some of the evidence for this.

Recall first a few properties of the jets observed[7)] at SPEAR. In e^+e^- annihilation via a virtual photon, the photon is at rest in the laboratory and so there is no advance expectation that in a particular event the two jets it produces will emerge in any particular direction, except for polarization effects which can be associated with the nature of the fundamental fields involved in the process. We come back to this later. The jet direction is identified by choosing an axis in some direction, calculating p_t^2 for all the particles relative to that direction, and then varying the direction of the axis so as to minimize this quantity. It is then found that the average p_t of the particles relative to this preferred direction remains limited to a few hundred MeV/c, even as the energy increases. It remains an open question whether there is any rise in this $\langle p_t \rangle$ with increasing energy, but present indications are that any such rise is very slow.

At higher energies the presence of jets and the direction of their axis is immediately evident by inspection of the events. In Figure 1.2.b we saw a striking example from the PLUTO experiment at DORIS.

When the direction of the pair of jets has been identified event by event, one may examine that distribution of its direction relative to the initial colliding beams when the events are summed. Because the virtual photon has spin 1, this distribution must take the form $(1 + \alpha \cos^2\theta)$, for some fixed α. It is found[7] that $\alpha = 0.97 \pm 0.14$. The value $\alpha = 1$ is characteristic of the production of a pair of jets each of spin ½. With spin 0, the value would be $\alpha = -1$. This encourages the hypothesis that the jets are in fact a fragmenting quark-antiquark pair, as expected from the parton-model considerations. This is shown in Figure 3.1. The figure does not show the final-state interaction that must occur if there is to be confinement of fractional quantum numbers. The produced hadrons are expected to have only limited transverse momenta with respect to the direction associated with the quarks, thus appearing as two jets.

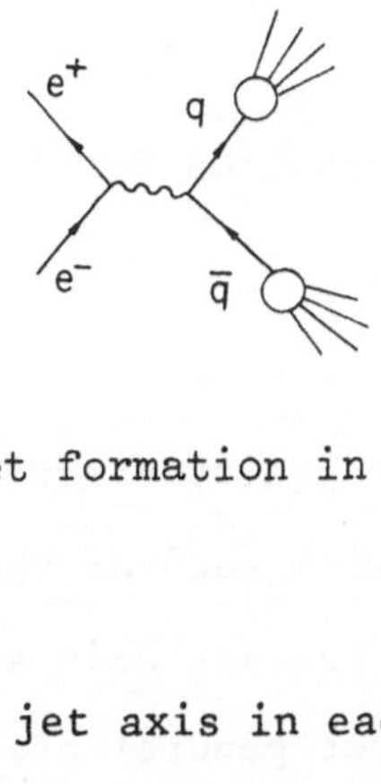

Figure 3.1 Quark jet formation in e^+e^- annihilation.

Having identified the jet axis in each event, one may again sum over events and obtain the distribution of longitudinal momentum of particles within the jets. To a good approximation, this is found to scale: for a given type of particle it depends only on the ratio z of longitudinal momentum of the particle to the total jet momentum. There is some disagreement between published data from SPEAR and from DESY[13] on the precise shape of this distribution; we return to this in Section 4.

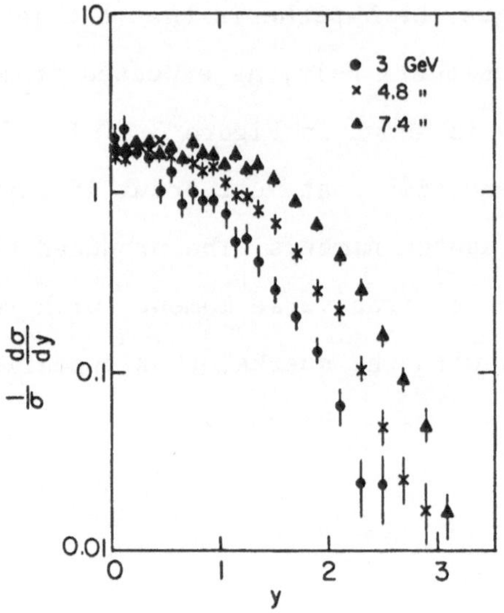

Figure 3.2 Rapidity distribution for jets as observed at SPEAR.

Figure 3.2 gives the rapidity distribution of the particles (mainly pions) associated with each of the two jets. The rapidity is defined with respect to the jet axis as previously identified. It is interesting to see that general properties of particle production similar to those seen in the longitudinal jets produced in early experiments at the ISR and at Fermilab are found again in jet fragmentation in e^+e^- annihilation. The different distributions in Figure 3.2 simply translate from one another in so far as the large rapidity behaviour is concerned, as implied by Feynman scaling. A rapidity plateau starts to develop as the jet momentum becomes large enough.

Transverse Jets in High-p_t Events: The Trigger Side

Superficially, it seems that the transverse jets seen in large-p_t reactions are the same as those at SPEAR. A central question in this review is whether they are exactly the same. A difficulty in the experimental study of this question has been that so far the transverse momentum possible to have been investigated has been limited to about 3 GeV/c, or at most 5 GeV/c. The problem which this causes may be seen in Figure 3.3. This shows rough plan drawings of 2.5 GeV/c and 5 GeV/c jets, based on SPEAR measurements of multiplicities and transverse momentum distributions. The broken lines represent neutral particles. At these momenta, the jets are not at all well collimated and they are unlikely to contain more than one charged particle with momentum much greater than 500 MeV/c, so that the transverse jets cannot easily be separated from the longitudinal jets (Figure 1.2). Hopefully this problem will be much reduced when higher-momentum jets can be studied.

Nevertheless, even with a 5 GeV/c jet produced at 90°, the charged multiplicity found in a cone of half-angle 45°, say, should be about three times that in an ordinary, low-p_t event. So one way to search for jets is to look for unusually large multiplicity fluctuation.

Almost all the available correlation data come from experiments that have used a single high-p_t particle as a trigger. As we investigate in more detail in Section 4, this imposes a substantial trigger bias on the structure of the trigger-side jet: it selects the somewhat unusual events where the trigger-side jet gives most of its momentum to the single trigger particle, and little energy and momentum are left to be shared among other fragments of the

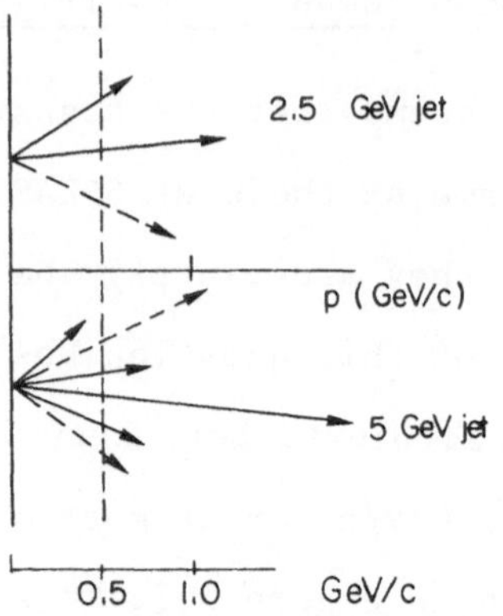

Figure 3.3 Typical jets. The full and dotted lines stand for charged and neutral particles, respectively.

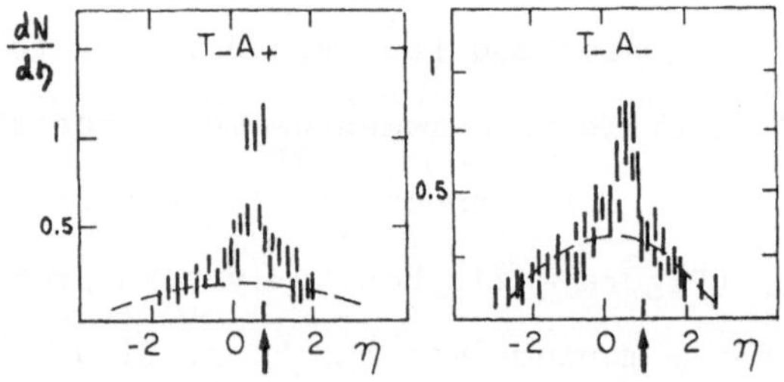

Figure 3.4 Same side jet. Rapidity correlation for large p_T particles ($p_T > 1$ GeV/c) observed on the same side as the trigger particle. Whenever another large p_T particle is found it is in the direction of the trigger particle ($p_T > 2$GeV/c). It belongs to the same jet. Data from the CCHK Collaboration.

jet. This means that, usually, these further fragments are few in number, if they are present at all, and when they are present they are not readily separated from the "background" of low-p_t particles associated with the longitudinal jets. However, there is clear evidence that the trigger-side multiplicity is enhanced close to the trigger particle, both nearby in rapidity (Figure 3.4) and in azimuthal angle. This enhancement is more marked for the rare events where there are secondaries of larger transverse momentum, because such secondaries are more likely genuinely to belong to the transverse jets rather than to the background.

The recent results of the CERN-Saclay-Zurich collaboration provide a test of such jet features with much larger p_t triggers (typically 8 GeV/c). A large lead glass array allows for triggering on the energy of one (a set of) neutral particle(s). Associated charged secondaries belonging to the same jet candidate are required to have $p_t > 0.8$ GeV/c. Figure 3.5 shows the rapidity correlation then obtained. Any large p_t charged particle is strongly correlated in direction with the large-p_t neutral trigger.

If one takes those comparatively rare events with a high-p_t secondary, and assumes that the sum of its momentum vector and that of the trigger give a good indication of the direction of the trigger-side jet, one can measure the transverse momentum k_t of the two jet fragments relative to the resulting jet axis. The distribution is shown in Figure 3.6. The average value for k_t of a few hundred MeV/c agrees well with that found in SPEAR jets.

Figure 3.6 displays recent results obtained with very large p_t neutral trigger particles. The transverse momentum of the associated charged secondaries with $p_t > 0.8$ GeV/c, here measured in the triggering plane, show a sharp cut-off.

Mass plots of the pair of particles show a clear ρ signal. It may be deduced that production of a ρ at a given p_t is comparable with that of a π at the same p_t. On the other hand, the rate of ϕ production measured at Fermilab (E 357) is some ten times less.

As is obvious from Figure 3.4, two body resonances do not explain all of the correlation on the trigger side. There are also clear correlations between pairs of π°.

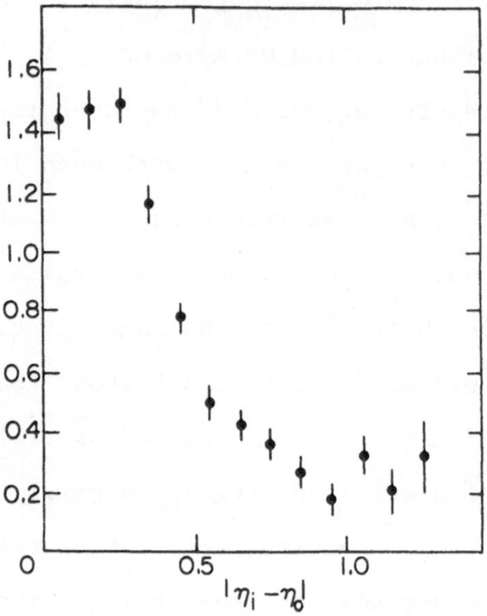

Figure 3.5 Rapidity correlation for large p_T particles ($p_T > 0{,}8$ GeV/c) seen on the same side as a very large p_T π^0 trigger ($p_T > 8$ GeV/c). Data from the CERN-Saclay collaboration.

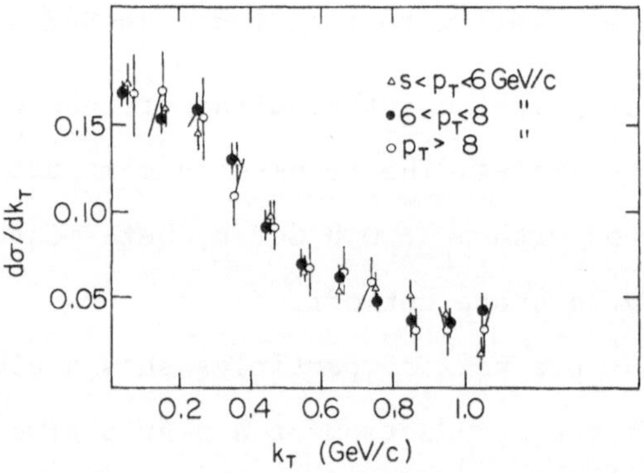

Figure 3.6 Transverse momentum distribution with respect to the jet axis. Data from the CERN- Saclay Collaboration.

It is worthwhile at this stage to give at least a provisional (and phenomenological) definition of what is meant by a jet. A jet of hadrons is a set of particles with limited transverse momentum with respect to the total momentum of the set. The mean transverse momentum (with, say, an exponential distribution) is of the order of 0.35 GeV/c, which is readily associated with the size of the produced hadrons as seen in a direction which is not affected by the Lorentz contraction. A typical resonance (ρ or ω say) has a Q value which, when distributed among the daughter particles, corresponds to the observed transverse momentum cut-off. A typical resonance satisfies the phenomenological definition for a jet, but it is probable that the notion of a jet is far more general. It may be that in some circumstances a jet is just a superposition of resonances, a hadronic system most generally associated with a quark-antiquark (mesonic) state.

As discussed later, the prominence of the ρ signal in the toward jet results from a trigger bias effect. No such strong signal appears in the away side jet which is a priori unbiased.

The Away-side Jet

The experimental study of the correlations on the side of the beam opposite from the trigger particle is harder. This is because, even if the direction θ_1 of the trigger particle is fixed, the direction θ_2 of the opposite-side jet varies substantially from event to event. However, by taking two particles on the opposite side that have transverse momentum large enough to give a reasonable chance that they do not belong to the background, one finds event by event clear signs of a correlation in their rapidities and azimuthal angles. An example is the data of Figure 3.7a, taken with a trigger particle having $p_t \approx 2.5$ GeV/c and rapidity $y \approx 1$. Events are chosen where on the oppposite side the "fastest" (i.e. largest p_t) particle has $p_t > 1$ GeV/C, and the next fastest $p_t > \frac{1}{2}$ GeV/c. The distribution of the rapidity of the second fastest when, respectively, the fastest falls in the rapidity intervals $\frac{1}{2} < y < 2$ and $-2 < y < -\frac{1}{2}$ are shown in the two parts of the Figure. The problem in such studies is always how to make an estimate of the background, contributed from particles that belong to the longitudinal jets. In Figure 3.7a this is done by drawing a curve through the data in $y < 0$ in the upper part of the Figure, and through the data in $y > 0$ in the lower part, and in each case completing the curve by making it symmetric about $y = 0$. The fact that this procedure results in the same background curve in each case leads one to have some confidence in it. The figure shows a clear rapidity correlation above the background, as expected for a transverse away-side jet. However, one cannot yet be at all certain that the picture of Figure 1.2, with a single pair of transverse jets, gives an exhaustive description of all events.

The clear jet structure also appears on the BFS data shown in Figure 3.7b. The trigger particle is now a $\pi^{\pm}$ at 90°. The Δy distribution for large-p_t particles on the opposite side shows a sharp peak in both the opposite and same charge case.

It seems that the trigger-side and away-side jet axes are almost coplanar with the initial colliding beams. At present, this is verified by measurements of p^{out}, the component of momentum of the away-side particles perpendicular to the plane defined by the beams and the

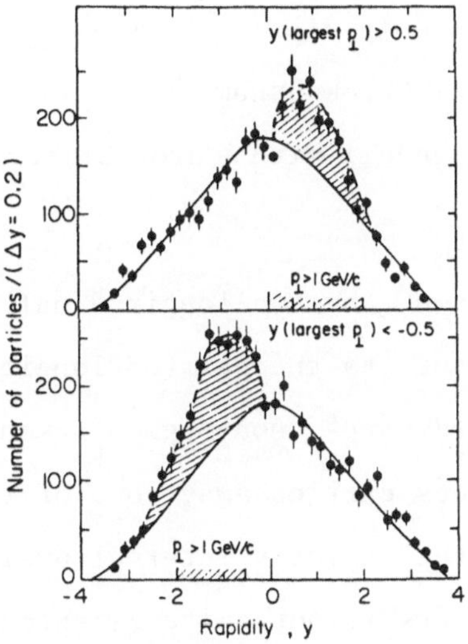

Figure 3.7.a Correlations among two large p_T particles on the away side. Data from the CCHK collaboration at the ISR.

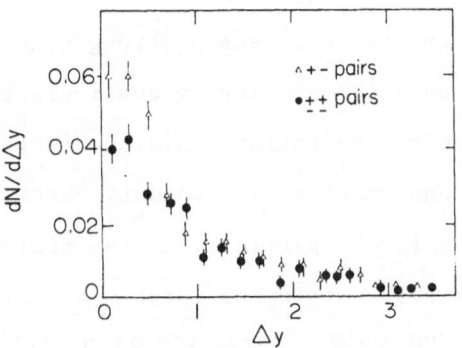

Figure 3.7.b Correlation among two large p_T particles on the away side. Data from the BFS Collaboration at the ISR.

momentum of the trigger particle (Figure 3.8a). The average value $\langle p^{out} \rangle$ remains fixed at about ½ GeV/c, even for high trigger-particle momenta (Figure 3.8b). More precisely speaking $\langle p_{out} \rangle$ increases with x_e and p_t, though only slowly. This is qualitatively in agreement with Quantum Chromodynamics expectations, a gluon being able to add an extra recoil with a probability increasing with p_t.

The Longitudinal Jets

So far as is known, the two longitudinal jets of Figure 1.2 are similar in structure to the pair of longitudinal jets that are characteristic of "ordinary", non-high-p_t events. Of course, the pair of transverse jets carries away some of the available momentum and energy, so that the rapidity interval available to particles in the longitudinal jets is not quite as extensive as in ordinary events. As a verification, the BFS collaboration finds the leading particle structure shown in Figure 3.9. What is presented is the x distribution for fast positive particles, where $x = 2p_t/\sqrt{s}$, as observed for different triggering requirements. Considering first the sum over all events, one sees a strong quasi-elastic peak associated with protons which are merely quasi-elastically scattered in single diffractive excitation. This quasi-elastic peak disappears as expected when one requires a secondary particle at wide angle, which is usually a low-p_t particle. As the minimum required value for p_t is increased the yield for large x positives (most likely protons) decreases, but only slowly. Requiring rather a large p_t value ($p_t \simeq 3$ GeV/c) does not strongly suppress the yield for rather fast protons associated with the forward and backward jets of Figure 1.2. The observed fall-off of the fast proton yield is actually a little stronger than what is expected from energy balance alone, granting

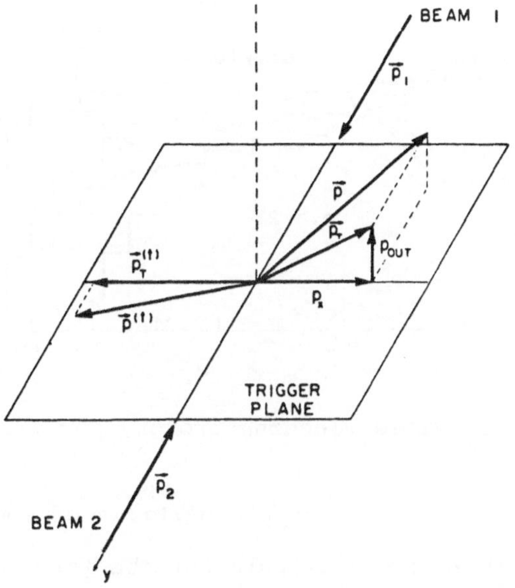

Figure 3.8.a Kinematical variables for the study of large p_T correlations. One defines x_e as $p_x=x_e p_T$.

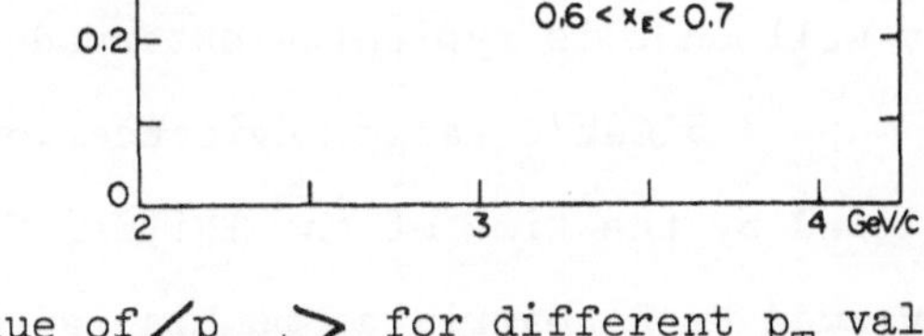

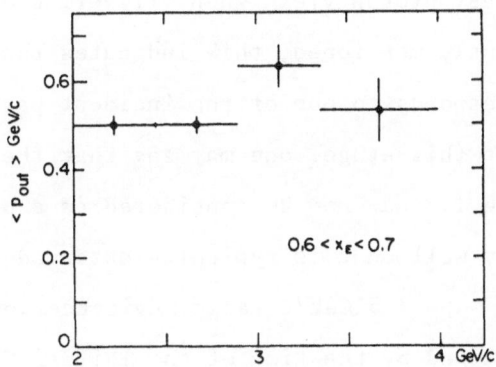

Figure 3.8.b Value of $\langle p_{out} \rangle$ for different p_T values. Actually one observes a slow rise with x_e (and p_T). This is reported by the CCOR collaboration with triggers up to 12 GeV/c.

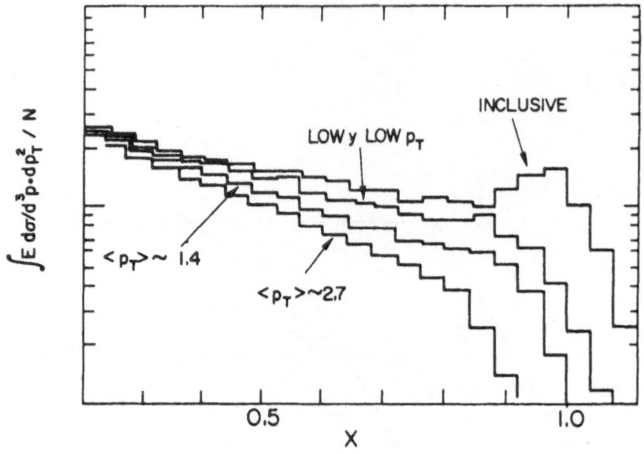

Figure 3.9 Large p_T yield (leading proton) for different trigger specifications.

the fact that observing a wide-angle particle of momentum p_t decreases the total energy available for the forward and backward jets from $\sqrt{s}$ to $\sqrt{s} - 2p_t$.

One observes specific correlations between large p_t particles and leading protons. As expected, π^-'s are more frequently associated with a fast proton than π^+'s. The most striking feature is the sharp decrease of the fast proton yield when triggering on a large p_t proton. As previously mentioned, this indicates that the trigger proton often corresponds to one of the incident particles.

Concluding at this stage, one may say that the jet structure sketched in Figure 1.2 can now be considered as a prominent feature. It summarizes very well what is typically observed with a trigger particle in the $2 < p_t < 5$ GeV/c range. Evidence for it could be considered as obtained by the time of the Tbilisi Conference in 1976; this was reviewed by P. Darriulat in his rapporteur talk. We shall consider it as the starting point for our discussion of jets. Whether it becomes far clearer experimentally, with "pencil-like" jets as a rule, or becomes more complicated with perhaps further jets also showing up as the trigger momentum is much increased (p_t > 10 GeV/c say), is one of the important open questions. We shall come back to this later on.

4. Phenomenological Study of the Transverse Jets

A Very Simple Model

We describe a very simple calculation[10]. It assumes that two transverse jets are produced somehow it does not matter what dynamical mechanism is responsible for this. Once they have been produced, the jets are supposed to fragment into hadrons in a manner that is independent of their production. In particular, their fragmentation is independent of the total energy of the process and of the angle at which the jet emerges. For simplicity, we ignore all quantum numbers and suppose that there is only one type of jet and one type of hadron; it is not too hard to make the model more realistic by incorporating different flavours into the calculation.

As a first approximation, it is usual to assume that, as seems to be true at least approximately of the SPEAR jets, the fragmentation of a jet into hadrons scales. By this, we mean that number density F(z) of hadron fragments having fractional longitudinal momentum z is a function only of z, and not explicitly of the total jet momentum P. Also as a first approximation, one may ignore the transverse momentum of the hadron fragments relative to P. Then the inclusive cross-section for the production of a hadron of large transverse momentum p_t is

$$\frac{d\sigma}{dp_t} = \int_{P_t > p_t} dP_t \, \frac{d\sigma^{Jet}}{dP_t} \int_0^1 dz \, F(z) \, \delta(p_t - zP_t) \tag{4.1}$$

where $d\sigma^{Jet}/dP_t$ is the cross-section for the production of a transverse jet with transverse momentum P_t. Note that both $d\sigma/dp_t$ and $d\sigma^{Jet}/dP_t$ are integrated over angles.

It is convenient to parametrise the jet-production cross-section in the form

$$\frac{d\sigma^{Jet}}{dP_t} = \frac{A}{P_t^{N-1}} \tag{4.2}$$

where, at given energy $\sqrt{s}$, both A and N are assumed to vary only slowly with P_T. Then (4.1) becomes

$$\frac{d\sigma}{dp_t} = \frac{A}{p_t^{N-1}} \int dz\, z^{N-2}\, F(z) \tag{4.3}$$

Trigger Bias

Notice two features of this formula. The first is Bjorken's[10] parent-child relationship: the single-particle cross-section has the same power dependence as the jet cross-section. This similarity of shape follows only if the fragmentation function F does indeed scale, and it is true only to the extent that N is indeed constant (For a discussion of the consequences of relaxing these assumptions, see the review by Craigie[14]). The second important feature of (4.3) is the factor z^{N-2} under the integral: this implies that large values of z contribute most to the single-particle cross-section. This property, that the trigger-side jet gives most of its momentum to the trigger particle, is known as trigger bias. The larger the value of N is, the more marked is the trigger bias.

If one parametrises the inclusive single-hadron data in the form

$$E\frac{d\sigma}{d^3p} \sim p_t^{-n}\, e^{-bp_t/\sqrt{s}} \tag{4.4}$$

and supposes, as seems to be the case, that there is little variation with angle, then

$$N = -p_t \frac{\partial}{\partial p_t} \log (E \frac{d\sigma}{d^3p})$$

$$= n + bp_t/\sqrt{s} \qquad (4.5)$$

The CCRS collaboration quotes values $n \approx 8.5$ and $b \approx 25$ for pion production. Hence when p_t changes from 3 GeV/c to 6 GeV/c, at the upper energies N varies from about 10 to 11.5, while at the lower ISR energies it varies from about 11.5 to 14.5. Thus N is large, and so the trigger bias is substantial. Also, at least at the upper ISR energies, N is indeed reasonably independent of p_t.

Jet Fragmentation

To carry the simple model further, it is necessary to make an assumption about the form of the jet fragmentation F(z). A simple form that is commonly taken is

$$F(z) = B (1-z)^m / z \qquad (4.6)$$

where B is a constant. The factor z^{-1} in this expression deserves some discussion. According to the definition of the function F(z), the average multiplicity of jet fragments having $z > z_o$ is

$$n(z_o) = \int_{z_o}^{1} dz \ F(z) \qquad (4.7a)$$

and the average fraction of the total jet momentum that these fragments take is

$$Z(z_o) = \int_{z_o}^{1} dz \ z \ F(z) \qquad (4.7b)$$

If we insert (4.6) into the two formulae (4.7), we see that $Z(z_o)$ remains bounded as $z \to 0$. Indeed, the constant B must be chosen such that $Z(o) = 1$:

$$B = (m + 1) \tag{4.8}$$

so that the total momentum of all the jet fragments is just equal to the momentum P of the jet. (If there are several types of jet fragment, one must arrange that the sum of these contributions to the momentum is correctly normalized). On the other hand, the multiplicity $n(z_o)$ diverges logarithmically as $z_o \to 0$, so that the jet has a large number of particles carrying low (longitudinal) momentum. Obviously, the number is not truly divergent except in the limit $P \to \infty$. One way to arrange this is to define z in a different fashion from that which we have described, but in such a way that asymptotically, as $P \to \infty$, it does become equal to the fractional longitudinal momentum. For example, one could take z to be the fractional energy instead of momentum. Then z has a natural lower cut-off, proportional to 1/P, and the total jet multiplicity contains a term proportional to log P. This matter has been discussed by various authors, and recently by Field and Feynman[15)], but has not really been resolved in any satisfactory manner. It is probable that there is no satisfactory solution: if a particle has very small z, then both in principle and as a matter of experimental practice, one cannot be sure that it belongs to a given jet in the final-state configuration of hadrons, rather than to some other jet or to the low-p_t background. For particles having very small z, quantum-mechanical interference effects between these different parts of the final state will surely be important and the concept of a jet fragmentation function F(z), which is a probability function rather than an amplitude, is not really valid. The lesson is that one should study events where the jet momentum P is as large as possible, so that the range of values of z for which there is real doubt how the particles should be attributed among the various components of the final state is as small as possible.

The z^{-1} term will lead to a rapidity plateau for the distribution of secondaries along the jet axis. This is compatible with the distribution observed for SPEAR jets. At present it is not yet possible to acertain such a behaviour with ISR jets, since the relevant low-p_t particles cannot be clearly distributed among the jets in any unambiguous way. One sees, however, an increase of the associated multiplicity which is compatible with what is expected from jet fragments with (4.6). As shown in Figure 4.1, which gives some early Pisa-Stony Brook results, the associated multiplicity on the away side depends on the transverse momentum of the trigger particle, and hence on the jet momentum, but not on the centre-of-mass energy. Combining charged and neutrals, at moderate p_t one typically gets one more extra particle per GeV/c. This is a significant excess since within the angular range $\Delta y = 1$, $\Delta\phi = 60°$, which one may tentatively attribute to a jet at wide angle, one expects on the average half a particle only.

Jet Triggers

If we insert (4.6) and (4.8) into (4.3), we obtain

$$\frac{d\sigma}{dp_t} = \frac{A}{p_t^{(m+1)}} \left(\frac{m!\,(N-3)!}{(N+m-2)!} \right) \qquad (4.9)$$

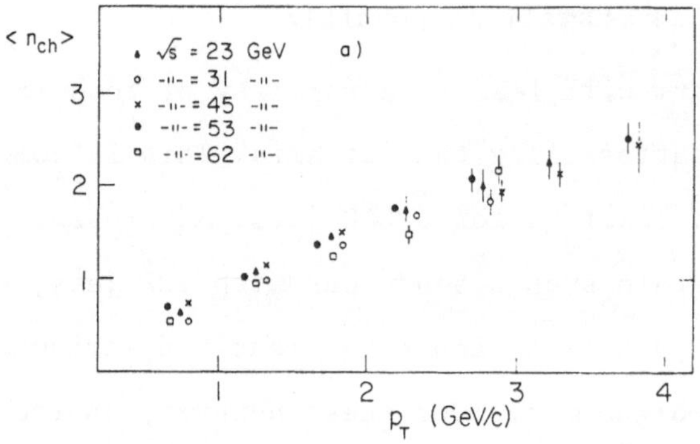

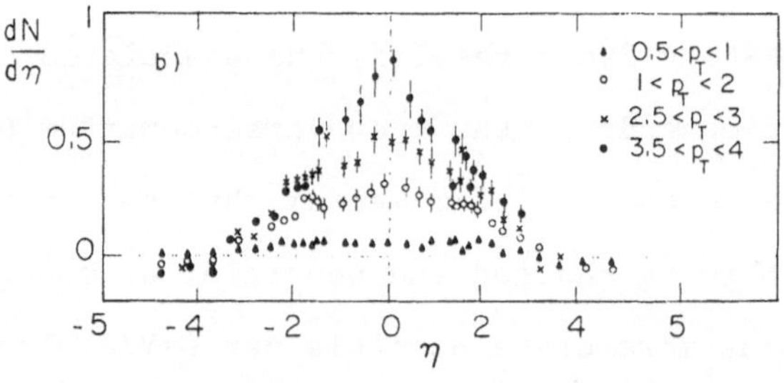

<u>Figure 4.1</u> Jet multiplicity and angular distribution of the associated particles to a large p_T trigger (on the away side). Data from the Stony Brook Collaboration.

Acoording to (4.2), the factor in curly brackets is the ratio of the cross-section for the production of a particle of a given transverse momentum to that for a jet of the same transverse momentum. This factor is small: for m = 2 and N = 10 it is equal to 10^{-2}. Notice that because, according to (4.5), N varies with p_t and $\sqrt{s}$, the factor is not quite constant.

As we discuss below, the simple form (4.6) for the jet fragmentation function F(z) is not necessarily correct. By choosing forms that agreed with correlation data from the ISR, it was predicted[10,11] that the ratio of the jet cross-section to the single particle cross-section should be some two orders of magnitude. The data from Fermilab seem to verify this prediction: Figure 4.2 shows data from the E 260 and E 395 experiments. Such experiments use a calorimeter to trigger on the transverse jet, and there is a severe problem in

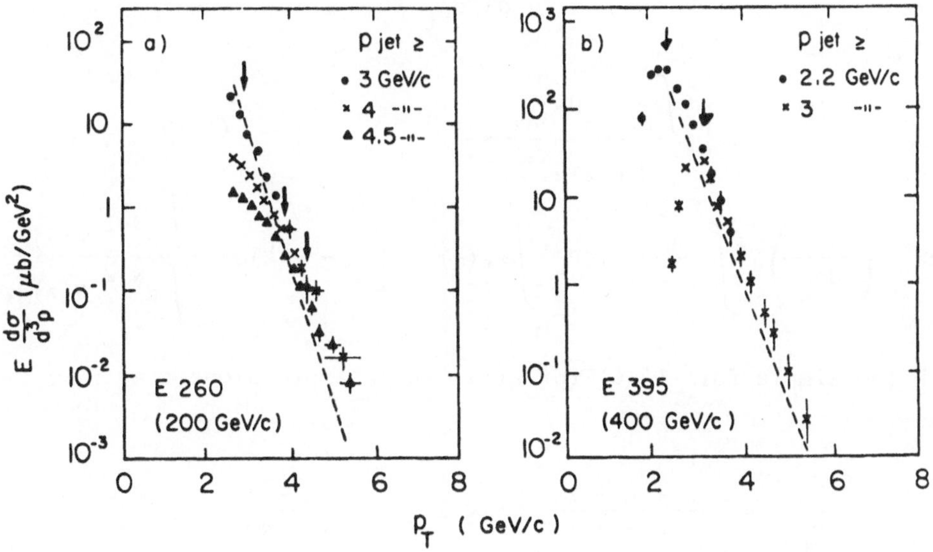

Figure 4.2 Jet cross sections as observed at Fermilab. The arrows indicate the minimum momentum imposed on the Calorimeter trigger with a faithful inclusive distribution to the right. The dashed curve is 10^2 times the single charged particle yield.

ensuring that the calorimeter receives the whole jet, but only the jet. Nevertheless, the results strongly encourage the hope that calorimeter triggers will make possible the study of jets up to very much larger values of transverse momentum than single-particle triggers.

Momentum in the Trigger-side Jet

We have said that, with a single high-p_t particle as trigger, the trigger bias has the effect that most of the momentum in the trigger-side jet will be given to the trigger particle. It is simple to calculate the average total momentum $<P_t>$ of the trigger-side jet, or the average fractional momentum $<z>$ of the trigger particle:

$$<P_t> = \frac{\int P_t \frac{A}{P_t^{N-1}} dP_t \int dz\, F(z)\, \delta\,(p_t - zP_t)}{dT/dp_t}$$

$$= p_t \frac{\int F(z) z^{N-3} dz}{\int F(z) z^{N-2} dz} \tag{4.10}$$

$$<z> = \left(\frac{d\sigma}{dp_t}\right)^{-1} \int \frac{A}{P_t^{N-1}} dP_t \int zF(z)\, \delta\,(p_t - z^{pt}) dz = \frac{\int z^{N-1} F(z)dz}{\int z^{N-2} F(z)dz}$$

With the simple form (4.6) for F(z), these expressione are

$$\frac{< P_t >}{P_t} = \frac{m + N - 2}{N - 3} \quad ; \quad < z > = \frac{N - 2}{m + N - 1}$$

With m = 2 and N = 10, these quantities take values 1.4 and 0.73, respectively.

What is the Trigger-side Jet?

We explained in Section 3 that there seems good reason to believe that two jets observed in e^+e^- annihilation are fragments of a quark and an antiquark, respectively, with some sort of interaction between the two systems in order to remove the unobserved fractional baryon number and charge. It has also been hypothesized[6,16] that the longitudinal jets produced in low-p_t hadronic collision have their origin in fragmenting quarks, though the details of this are not yet agreed. The obvious question is whether the two transverse jets produced in high-p_t collisions are fragmenting quarks. An alternative proposal[6,10,17] that deserves serious attention is that at least one of the jets is a quark-antiquark system, which emerges either as a stable meson, or a resonance, or perhaps as a continuum. Another suggestion, which has come to the fore recently[8,18], is that one of the jets seen in high-p_t reactions may, at least sometimes, be a fragmenting gluon. This will be discussed later.

One has therefore to face the possibility that the type of jet seen in high-p_t collisions may vary with the precise triggering conditions and with the range of energy and transverse momentum being explored. It is very hard, if not impossible, to establish the identity of a jet on an event-by-event basis, since a given type of jet will choose to fragment in a different way in each event, and also there are always almost insuperable problems in completely separating the components of a jet from the background.

There are two conflicting pieces of evidence about the identity of the trigger-side jet; maybe the conflict is due to the different triggering conditions. The E 260 experiment uses a calorimeter trigger and measures the z distribution of charged hadrons within the jet: the result is that it seems rather similar to that found in an e^+e^- jet, and so the working hypothesis is that the jet is quarklike. The British-French-Scandinavian experiment uses a single-particle trigger, and measures the total momentum of the nearby charged particles. As we now explain, their results do not favour the proposition that the trigger-side jet is quarklike.

In order to discuss this, we must first emphasize that there is a problem about what one should take for the fragmentation function F(z) of a quarklike jet. On the basis of the early SPEAR data[7)], Feynman and collaborators[11)] take a form for F(z) that is rather more complicated than (4.6), and which goes to a constant as $z \to 1$. At $\sqrt{s} = 53$ and for p_t in the range 2 - 4 GeV/c, their form gives $\langle P_T \rangle / p_t \approx 1.3$. On the other hand, the form (4.6), with m = 2, seems to be rather a better fit to the more recent DESY data from DASP and PLUTO 13 and corresponds to $\langle P_T \rangle / p_t \simeq 1.4$.

The BFS data do not favour a value for $\langle P_T \rangle / p_t$ as large as either of these results; indeed, for a pion trigger a value less than 1.1 seems to occur. This value is arrived at by measuring the momentum of the charged particles close to the trigger particle (Figure 4.3), and subtracting a background contribution which is assumed to be independent of p_t. Allowance is then made for neutrals and, of course, for limited acceptance. Although the conclusion cannot be regarded as completely reliable, it seems rather unlikely that a value as large as 1.3 is tenable.

A value of $\langle P_T \rangle / p_t$ less than 1.1 would seem to favour the hypothesis that the trigger-side jet is a $q\bar{q}$ system. In this case the trigger-side jet is able to consist of the trigger particle alone, so that in such events $P_t = p_t$. In a small fraction of the

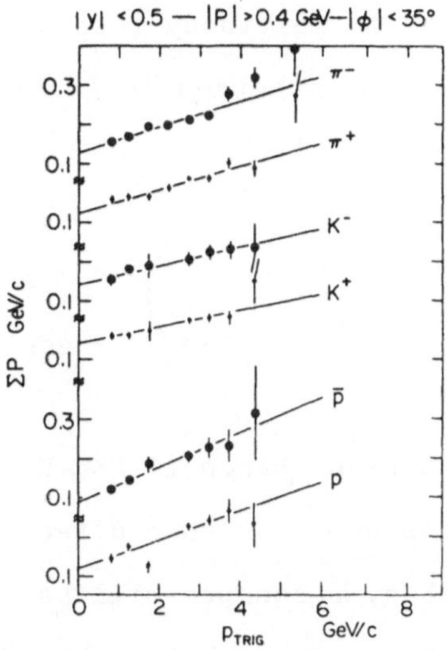

Figure 4.3 Momentum flow on the trigger side. Data from the BFS collaboration at the ISR. The trigger particle takes most of the jet momentum.

events, perhaps about 25%, the qq system emerges as an excited system and provides additional jet fragments[F)]. By choosing[10,17)] an appropriate mixture of the single-particle and multiparticle possibilities, one can readily fit the same-side correlation date from the ISR. In such a case, the fragmentation function of the jet into pions would take the form.

$$F(z) = K\,\delta(1 - z) + L + \ldots\ldots \qquad (4.11)$$

The first term represents the possibility that the pion forms the whole jet, and the second term is the contribution from two-body resonances, such as the ρ. Data from the R 412 experiment (Figure 3.8) indicate that ρ production at a given p_t is roughly equal to π production at the same p_t, so $L \approx K$. The terms not written explicitly in (4.11) correspond to multibody resonances and perhaps also to a continuum. If (4.11) is inserted in (4.3), the two terms shown give the contribution

$$\frac{d\sigma}{dp_t} = \left(K + \frac{L}{N-1} + \ldots \right) \frac{A}{p_t^{N-1}} \tag{4.12}$$

to $d\sigma/dp_t$ for pion production. Hence if $L \approx K$, pions coming from ρ are about 10 times less numerous than direct pions. The contribution from a given multibody resonance will be even more suppressed: compare the small term between brackets in (4.9). We have said that the same-side correlation data suggest that the K term should contribute about 25% of $d\sigma/dp_t$; to achieve this, one must take K (and also L) to be very small. This means that if the trigger bias is removed, for example using a calorimeter trigger, the contribution from the K term is only at the 1% level; that is, it is then rather unlikely that the jet is found to consist of just one particle.

Multi-particle Triggers

We should stress at this stage the importance of scaling on the trigger side. Our present approach starts from jet formation with eventually different fragmentation modes. Imposing scaling, one predicts that while measured yields may and do change by a great deal depending on what one triggers upon (one particle, two particles, whatever may be received in a certain solid angle...), the p_t dependence should always show the same inverse power, as appearing in (4.2). This is not at all a trivial statement. One could a priori consider mechanisms which would lead to different power behaviour for single particle and jet triggers. This was actually proposed as an interpretation of the observed p_t^{-8} behaviour for single-particle triggers, when p_t^{-4} was expected on mere scaling grounds. As shown by Figure 4.2, present results are nevertheless compatible with a single power behaviour holding for widely different types of triggers. This certainly supports the approach which was followed in this section. It should be stressed, though, that further tests of the assumed scaling behaviour are very much needed.

We may for instance work out explicitly the two-particle case. The trigger in this case could be the total p_t of two π^o, summing over the internal degrees of freedom of the two-pion system.

For reasons discussed previously, we include explicitly a $(z_1\ z_2)^{-1}$ factor in the two-particle fragmentation function, which we take as

$$F^{ij}(z_1, z_2) = \frac{G^{ij}(z_1, z_2)}{z_1\ z_2} \tag{4.13}$$

where i and j are indices referring to the type (s) of particles associated with the pair.

For reasons of clarity, we do not include explicitly the dependence on the transverse momentum of each particle with respect to the jet axis (this can be done easily, as in the single-particle trigger case). The two-particle cross-section then reads (compare (4.1))

$$\frac{d^2\sigma^{ij}}{dp_{t1}\ dp_{t2}} = \tag{4.14}$$

$$\int_{P_t > p_{t1}+p_{t2}} dP_t \frac{d\sigma^{Jet}}{dP_t} \int_0^1 dz_1 \int_0^{z_1} dz_2 \frac{G^{ij}(z_1, z_2)}{z_1\ z_2}\ \delta(p_{t1}-z_1P_t)\ \delta(p_{t2}-z_2P_t)$$

We introduce the variables $p_t = p_{t_1} + p_{t_2}$ (which defines the trigger) and $z = p_t / P_t$. We write $z_1 = \nu z$, $z_2 = (1 - \nu)z$. The cross-section then reads:

$$\frac{d^2 \sigma^{ij}}{p_t \, dp_t \, d\nu} = \frac{1}{p_t} \int \Phi \left(\frac{p_t}{z}\right) G^{ij} (z, \nu) \frac{dz}{z^2 \nu (1 - \nu)} \tag{4.15}$$

with $\Phi (p_t) = \frac{d\sigma^{Jet}}{dp_t}$. The selected cross-section is then:

$$\int \frac{d^2 \sigma^{ij}}{dp_t \, d\nu} d\nu = \int \Phi\left(\frac{p_t}{z}\right) \frac{1}{z^2} \frac{G^{ij} (z, \nu)}{\nu (1 - \nu)} dz \, d\nu \tag{4.16}$$

whereas the one-particle inclusive cross-section is given by (4.1)

$$\frac{d\sigma^i}{dp_t} = \int \Phi\left(\frac{p_t}{z}\right) \frac{G^i (z)}{z^2} dz \tag{4.17}$$

now writing $F^i (z) = \frac{G^i (z)}{z}$ We see that the two distributions are clearly proportional, as imposed by scaling. A simple (first approximation) form for $G^{ij} (z)$ is $C^{ij}(1-z)^m$, which is chosen in analogy with (4.6). This should a priori be suitable for π^o pairs. The ν dependence is then simply integrated over the acceptance limits imposed by the triggering device. Figure 4.9 displays data from ACHM at the ISR for 2 π^o production at large p_t. When the 2 π^o distribution is suitably integrated over the internal degrees of freedom of the dipion system, it is found to be proportional to the single π^o cross-section. Further studies should extend such a comparison over a wider p_t range and vary the type of triggers. One may for instance use a "software" trigger, summing the total transverse momentum of charged particles observed within a certain angular acceptance or, when considering calorimeter triggering, compare the inclusive yields observed with the full calorimeter to those observed with separate fractions of the calorimeter. All available data are compatible with scaling, as we have assumed to hold. Tests are not yet extensive or accurate enough though. Recent results from the CERN-Saclay-Zürich collaboration provide nevertheless a very good illustration of scaling on the trigger side. The distribution of associated charged secondaries with $p_t > 0.8$ GeV/c scales according to the trigger momentum (neutral system) set at very large values. This is displayed in Figure (4.4).

The Away-side Jets

We have seen in Section 3 there to be evidence that the transverse momentum of the trigger-side jets is, at least in part, balanced by a jet on the opposite side. For simplicity, we add to our simple model the further assumption that the transverse momenta of the two transverse jets are exactly equal. As we shall explain in Section 5, there is some reason to believe that this may be a bad approximation when the trigger momentum is not very large ($p_t \lesssim 3$ GeV/c say). With this assumption, the cross-section for the production of a pair of particles with transverse momenta p_t, P_t' on opposite sides is, for $p_t > p_t'$,

$$\frac{d^2\sigma}{dp_t\, dp_t'} = \int_{P_t > p_t} \frac{A}{p_t^{N-1}}\, dP_t \int F(z)\, \delta(p_t - zP_t)\, dz \int F(z')\delta(p_t' - z'P_t) dz'$$

$$= \frac{A}{p_t^{N}} \int z^{N-1}\, F(z)\, F(z\, x_e)\, dz \qquad (4.18)$$

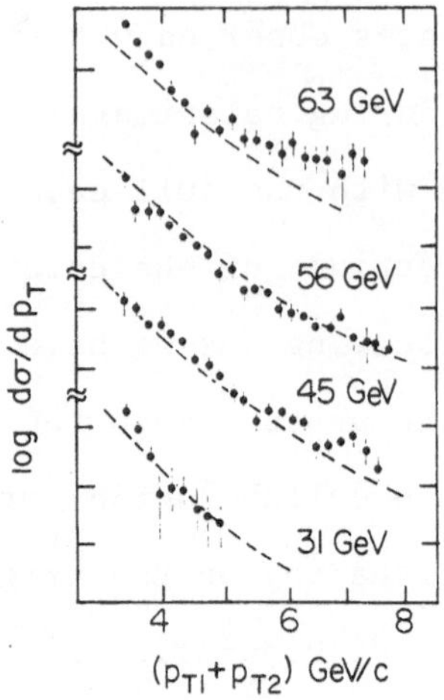

Figure 4.4.a Two π^0 distribution (averaged over the difference between π^0 momenta). The distribution is proportional to the single π^0 distribution (dashed curve). Data from Experiment 412 at the ISR.

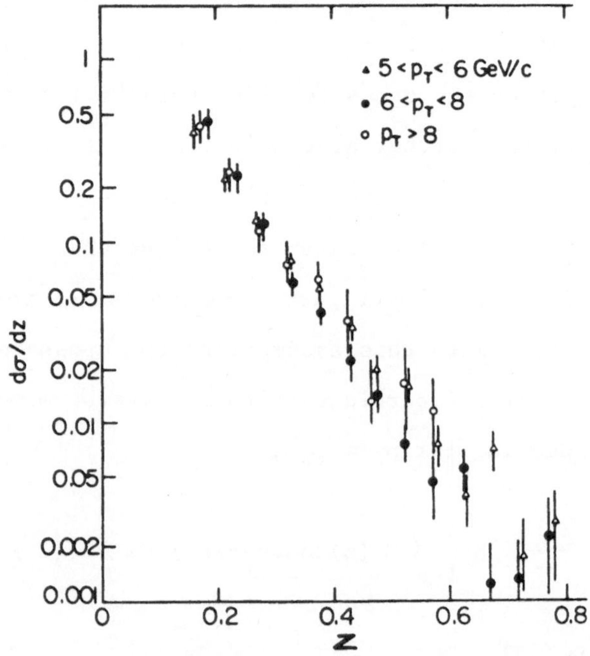

Figure 4.4.b Scaling for the trigger side jet. Inclusive distribution of associated charged particles on the same side as a very large p_T π^0. Data from the CERN-Saclay collaboration.

where $x_e = p_t'/p_t$. We have supposed that both jets have the same fragmentation function F; the necessary correction if this is not the case is trivial. From this formula we may calculate the distribution of away-side particles in the variable x_e:

$$\frac{dN}{dx_e} = \left(\frac{d\sigma}{dp_t}\right)^{-1} \int \frac{d^2\sigma}{dp_t\, dp_t'} \delta\left(x_e - \frac{p_t'}{p_t}\right) dp_t'$$

$$= \frac{\int z^{N-1} F(z)\, F(z\, x_e)\, dz}{\int z^{N-2} F(z)\, dz} \qquad (x_e < 1) \tag{4.19}$$

the corresponding expression for $x_e > 1$ is obtained by replacing x_e by x_e^{-1} and multiplying by x_e^{-N}.

Evidently the result (4.19) for dN/dx_e scales: it depends only on x_e. The data (Figure 4.5) show that scaling is not at all good when the p_t of the trigger particle is less than 3 GeV/c, but it becomes rather better for higher trigger-particle momenta. There are a number of possible explanations[19)] for this:

a) The scaling is only predicted by (4.19) to the extent that N is constant. In fact this is not exactly true though, according to (4.5), it will become a better approximation at higher energies $\sqrt{s}$. The variation of N causes dN/dx_e to decrease with increasing trigger p_t, at least for large values of x_e;

b) There may be intrinsic scale-breaking effects in the fragmentation function F(z), though in e^+e^- annihilation these appear to be fairly small. More imortant is the neglect of the transverse momenta of the jet fragments relative to the total jet momentum;

c) Measurements of dN/dx_e cannot distinguish the particles in

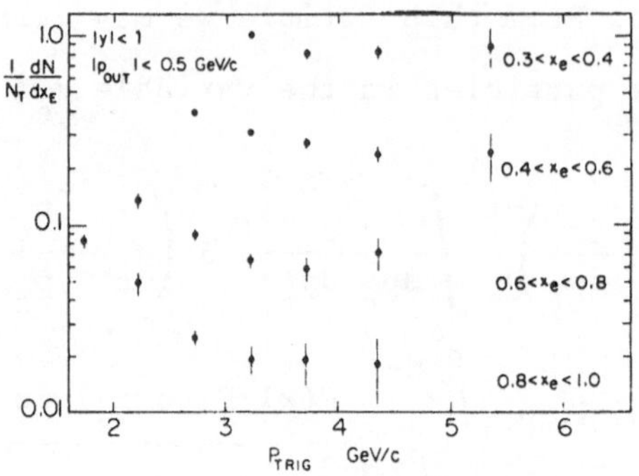

Figure 4.5.a Scaling for the opposite side jet. The yield at fixed x_e should not depend on p_T. Data from the BFS collaboration.

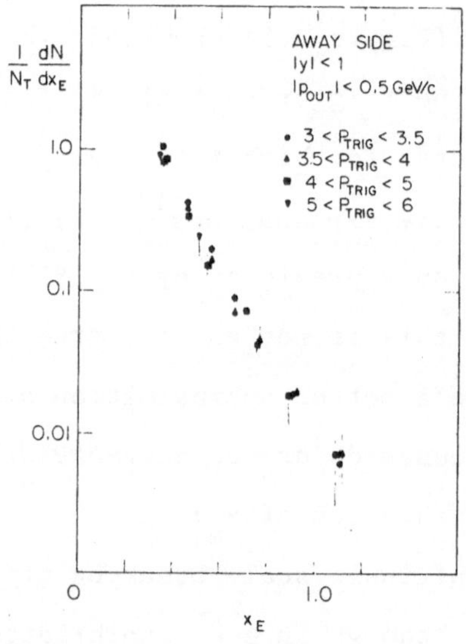

Figure 4.5.b Scaling on the away side (same data as those of figure 4.5.a).

the away-side jet from the unwanted background particles. The background particles will make the apparent dN/dx_e decrease with increasing trigger p_t;

d) If there is a constant, or slowly increasing, difference between the transverse momenta of the two jets, dN/dx_e will increase with increasing trigger p_t;

e) As the trigger p_t increases, the character of the away-side jet may change. For example[8,18], in QCD models it changes from a fragmenting gluon to a fragmenting quark. Apparent non-scaling will then result from these having different fragmentation functions F (z).

Taking these effects into account, one does expect the scaling of dN/dx_e to improve as the trigger p_t increases, but it may never become perfect.

At present one may consider that scaling for the away-side jet is reasonably well established. The recent Fermilab results (E 260) and the recent results of the British-French-Scandinavian collaboration at the ISR provide an already strong support. Scaling has certainly reached the status of a good working hypothesis open to further tests. Changing triggering conditions provide different expected changes which should be verified. For instance, observing a large p_t particle on the opposite side of the trigger should from quite general arguments, enhance the probability of observing another large p_t particle on the trigger side. Again, requesting another large p_t particle beside the trigger particle may perhaps enhance the inclusive p_t yield on the away side as compared to what it is with a single-particle trigger with the sum of the two momenta.

5. Hard-scattering Models: General Features

General Form

All the qualitative features of high-p_t events agree well with the predictions of hard-scattering models. In this section we review some general features of such models, and then in the next two sections we consider particular examples of the models.

In a sense, the observation that high-p_t events result in a pair of transverse jets C, D is directly equivalent to the statement that a hard scattering has occurred: some part A of one of the initial hadrons must have scattered through wide angle on some part B of the other, so as to produce the transversely-moving systems C, D (see Figure 5.1). The question that experimentalists and theorists are seeking to answer is what is the nature of each of the systems A, B, C, D.

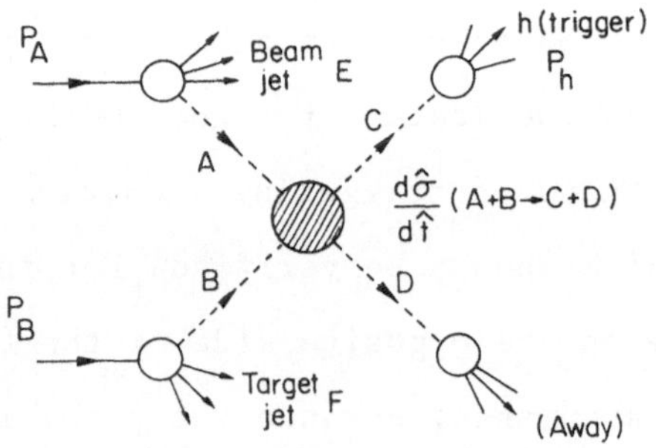

Figure 5.1 Hard scattering process in the parton model. The reaction amplitude is supposed to fall into different specific terms.

The inclusive cross-section corresponding to Figure 5.1 is:

$$E \frac{d\sigma}{d^3 p} = \int d^2 k_{tA}\, d^2 k_{tB}\, d^2 k_{tC}\, dx_A\, dx_B \tag{5.1}$$

$$G_A/p\,(x_A\, k_{tA})\, G_B/p(x_B, k_{tB})\, \frac{1}{\pi}\, \frac{d\hat{\sigma}}{d\hat{t}}\, (\hat{s}, \hat{t})\, \frac{1}{z}\, F_{h/C}\, (z, k_{tc})$$

This is the single-particle cross-section for the process $pp \to hX$. The functions G_A/p, G_B/p measure the probability distributions for finding constituents A, B in the proton, with fractional longitudinal momenta x_A, x_B and transverse momenta k_{tA} and k_{tB}. The function $F_{h/C}$ is the fragmentation function for the jet C into the hadron of type h; it is defined as in Section 4, except that we have allowed it to vary with the transverse momentum k_{tc} of h relative to the direction of C as well as with its fractional longitudinal momentum z. The differential cross-section $d\hat{\sigma}/d\hat{t}$ describes the central wide-angle scattering $AB \to CD$. The variable $\hat{s}$ is the energy variable for this scattering, and $\hat{t}$ is the momentum transfer between A and C. These variables are given by

$$\begin{aligned} \hat{s} &= x_A x_B s \\ \hat{t} &= x_A t/z \end{aligned} \tag{5.2}$$

where s is the energy variable for the overall process and $t = (p_A - p_h)^2$ is the momentum transfer to the trigger hadron h. The constraints (5.2) are imposed inside the integral (5.1), together with

$$z = -\frac{1}{s}\left(\frac{t}{x_B} + \frac{u}{x_A}\right) \tag{5.3}$$

where $u = (p_B - p_h)^2$. Further related formulae are given in Section 8. The form (5.1) is common to all hard-scattering models; the various models differ in how they identify A, B, C and D, and in the explicit expressions for the functions under the integral. This is described in Sections 6 and 7.

Initial and Final State Interactions

In most of the models, neither A nor B is the whole of its parent hadron, but rather some fragment of it. The remaining parts E and F of the two initial hadrons break up and emerge as longitudinal jets. The observation of the presence of leading particles (Figure 3.9) is evidence for the existence of these. In addition to the hard-scattering, $A + B \rightarrow C + D$, there will be ordinary, soft interactions between the hadrons in the initial state, between the systems E and F in the final state, and cross-interactions between the initial and final state. These ordinary interactions are presumably mediated by pomeron exchange. They are partly quasi-elastic and partly inelastic, producing extra particles by pionization. These ordinary interactions are to be distinguished from any quark or colour confining interactions that may operate in addition. It may be shown that, asymptotically, they do not affect the two hard-scattered systems C and D.

These ordinary interactions, then, produce extra small-p_t particles. They are responsible for generating the background particles in the central region. The mechanism that generates them is just the same as that which generates the central particles in ordinary, non-high-p_t interactions, and so the central background should in many respects be similar to the central "plateau" region of ordinary interactions. For example, it should contain about ten times as many pions as kaons. Although the interactions in question generate additional hadrons in the final state, there is a theorem[20] that, asymptotically, they may be ignored in the calculation of the inclusive cross-section for the production of high-p_t particles or jets. This is because it is found that in such a cross-section the effects of the various additional, soft interactions destructively interfere with one another and just cancel in the inclusive cross-section.

It is generally assumed that any quark or colour-confining interactions may similarly be ignored in the calculation of inclusive cross-sections. However, these also will produce additional final-state particles. The nature of the confining interactions, is not, o course, understood, but it is generally assumed[21] that when the confining force acts between two coloured objects the multiplicity of particles produced as a result of the interaction is proportional to the rapidity separation between those objects. The particles are expected to appear with small transverse momentum in the centre-of-mass frame of the two objects, and to fill in the rapidity gap between them in that frame. The consequent deployment of particles in the overall centre-of-mass frame has been discussed by Morton[22]. Notice that in the case where the two transverse jets C and D are quarks (rather than one a quark and the other an antiquark), an interaction between them would not serve to confine the unwanted quantum numbers. Rather, the confining force will act between each of C, D and the two longtudinal jets E, F. So the situation is then not the same as in e^+e^- annihilation, where the confining force just acts between the two jets.

Effects of Transverse Momentum of Constituents

In the hard-scattering models, the cross-section for large-p_t

processes is small mainly because the cross-section for the central wide-angle scattering A + B → C + D is small. Combridge[23)] pointed out that the need for this wide-angle scattering is reduced, and so the magnitude of the cross-section is enhanced, if the constituents A, B of the initial hadrons already have some transverse momentum. If either a single-particle or a single jet trigger is used, their transverse momenta will tend to line up in the direction of the trigger. If the central wide angle scattering occurs through a $\hat{t}$-channel exchange, where $\hat{t}$ is the momentum transfer between A and C, then[24)] if C provides the trigger it will be mainly the transverse momentum of the constituent A that is aligned towards the trigger, though momentum conservation results in this being the case also for B, but to a lesser extent. For the case of $\hat{u}$-channel exchange, where $\hat{u}$ is the momentum transfer between B and C, it is the transverse momentum of B that receives the greater alignment.

The consequence of the transverse momentum of either A or B being aligned towards the trigger is that some of the transverse momentum of the trigger-side jet is balanced by a recoil of the longitudinal jets slightly away from the trigger. This has been verified by the BFS collaboration, who examine those events in which there is a fast longitudinal particle. They find that, on average, it does recoil away from the trigger but that the average value of its recoil transverse momentum is not very large. One finds that, the mean transverse momentum opposite to the trigger direction rises with p_t, to practically saturate (within the p_t range studied) below 0.1 GeV/c per fast particle $0.4 < x_F < 0.8$. When the value of x_F is larger than 0.8, the mean value per particle is found to be larger. One should remark however that by then, the total multiplicity is likely to be significantly smaller than what is found in events with leading particles of lower x_F values.

The enhancement of the cross-section from this effect is obviously greatest when the cross-section is falling most steeply, that is when N in (4.2) or (4.3) is largest. According to (4.5), for given p_t, N is largest at the lower energies $\sqrt{s}$. On the other hand, the enhancement decreases with increasing p_t, because the effect then results in a small proportionate reduction in $|\hat{t}|$ or $|\hat{u}|$. This argument assumes that the transverse momentum available to the constituents is fixed; the conclusions may be modified if the transverse momentum of the constituents varies with their fractional longitudinal momentum or with the momentum transfer. A number of theoretical papers have argued[25)] that the transverse momentum of the constituents is not fixed, but there is no agreed theoretical approach to this question and there are not yet enough experimental data to decide the matter. However, the observation[26)] in dilepton production that the lepton pair is produced with a surprisingly large transverse momentum (about 1.2 GeV/c for dilepton masses greater than 4 GeV) indicates that the constituent transverse momentum may be quite large. The same experiment measures the average transverse momentum of dihadron pairs produced in a similar configuration and finds that $\langle p_t \rangle$ achieves even greater values and seems to increase linearly with the dihadron mass. The interpretation of this result is straightforward[27)] in the two-transverse-jet picture of high-p_t hadronic events, and is not directly related to the corresponding dileption result.

Several authors[8,11,24,28-30)] have incorporated the effects to the transverse momentum of the constituents A, B into calculations of the hard-scattering mechanism. They fold into the calculation some distribution function for the constituent transverse momentum, typically either an exponential or a gaussian. Their quantitative conclusions vary. The most dramatic effect is obtained by those[30)] who take the central wide-angle scattering to correspond to the exchange of a massless gluon, with cross-section $d\hat{\sigma}/d\hat{t}$ proportional to $1/\hat{t}^2$. In such a case, it is obviously advantageous to the dynamics to use the tail of the transverse momentum distributions, so as to be able to operate on the singularity at $\hat{t} = 0$. Most authors agree that this is not realistic, since it corresponds to a soft-scattering, rather than a hard-scattering, model. One way to avoid this problem is to allow the constituents A and B to depart from the mass shell in the integration in (5.1). Many authors impose a mass-shell constraint, but in fact it is more correct not to do so.

Effects of Breaking of Bjorken Scaling

In the early calculations of hard-scattering models, it was always customary to assume that the distributions of the constituents A, B in their parent hadrons satisfied Bjorken scaling. That is, not only was their transverse momentum neglected, but also their distribution was assumed to be a function only of their fractional longitudinal momentum. Theoretical arguments, based on the use of the renormalization group equations in perturbation theory (see, for example, the reviews in reference 31) nowadays lead one to expect that the distribution should vary also with the momentum transfer with which the distribution is probed. Various authors[8,18,28,32)] have incorporated such non-scaling effects into calculations of large p_t processes, and also a similar non-scaling of the jet fragmentation function F(z). An example is shown in Figure 5.2.

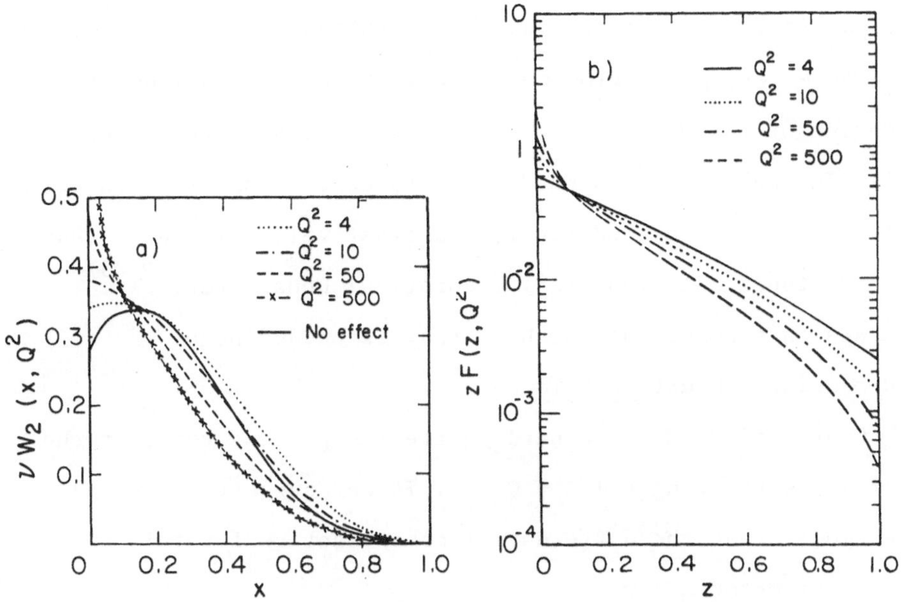

Figure 5.2 Fragmentation function with scaling violation effects as expected according to QCD.

It is not clear how much non-scaling should be put in. Although non-scaling of quark distributions within hadrons has been observed[33)] in deep inelastic lepton scattering, its magnitude remains to be settled, as does the amount of the non-scaling that should be attributed to effects not directly connected with the renormalization group (see, for example, the papers in reference 34). For application to large p_t processes, the non-scaling effects observed in lepton scattering at momentum transfers less than 20 GeV^2 or so have to be extrapolated to values greater than 100 GeV^2. There is a further theoretical problem, that in the large p_t application it is not always clear which momentum transfer is the appropriate one in this context, $\hat{t}$ or $\hat{u}$. (This uncertainty is most acute when the wide-angle scattering $A + B \rightarrow C + D$ is an $\hat{s}$-channel interaction, rather than a $\hat{t}$ or $\hat{u}$ exchange.) As for the jet fragmentation function F(z), measurements in other reactions such as e^+e^- annihilation or electroproduction do not give a sufficiently accurate indication of the amount of non-scaling, if indeed they show any at all. The best that can be done is to assume that the non-scaling in the fragmentation function is quantitatively related to that in the constituent distributions. Theoretical considerations[35)] encourage this hope, but further study is needed here.

The Dimensional Counting Rule

Contained within the hard-scattering process there is the wide-angle scattering A + B → C + D. It was suggested several years ago[36)] that the differential cross-section for this should take the asymptotic form

$$\frac{d\hat{\sigma}}{d\hat{t}} = \hat{s}^{-m} f(\hat{\theta}) \qquad (5.4)$$

where m is a fixed constant and $f(\hat{\Theta})$ is some function of the scattering angle $\hat{\Theta}$ in the centre-of-mass frame of the sub-process or, equivalently, of $\hat{t}/\hat{s}$. It was further suggested[37)] that the value of m is given by

$$m = n_A + n_B + n_C - 2 \qquad (5.5)$$

where n_H is the number of (valence) quarks plus antiquarks in hadron H. This is one part of what is known as the dimensional counting rule.

One can try and check this rule experimentally by studying exclusive processes. The data are reviewed in reference 3. While they are in agreement with (5.4) and (5.5) they are not accurate enough to reach any firm conclusion about the validity of the rule.

Until recently, the best data were those for wide-angle proton-proton scattering at rather low energies, $s < 60\ GeV^2$. These data fit (5.4) and (5.5) very well[38], but they also fit at least as well[39] to functional forms quite different from (5.1). It will be of great interest to have data for exclusive processes at wide angle at high energies with a variety of beams.

There are numerous difficulties in trying to derive the results (5.4) and (5.5) from field theory. Theoretical progress on this problem has been rather little since it was reviewed[40] in 1974. If one works in terms of Feynman diagrams, the diagrams must be selected[41] with some care if the desired results are to follow, and even then extraneous softening factors have to be imposed on the vertex functions within the diagrams if additional logarithmic factors are to be avoided[42]. In particular, it was shown some time ago[43] that a theory that contains vector gluons is unlikely to satisfy (5.4) with m fixed. The counting rule has certainly had some experimental success. Nevertheless the theoretical situation is unclear.

The other part of the dimensional counting rule concerns the behaviour at $x \to 1$ of the constituent distribution functions, or the $z \to 1$ behaviour of the jet fragmentation function. For the distribution function $G_{A/H}(x)$ of the constituent A in the hadron H, the behaviour near x=1 is supposed to be $(1-x)^{2n_{\bar{A}H}-1}$, where $n_{\bar{A}H}$ is the minimum number of quarks in the spectator system $\bar{A}H$. Likewise for the fragmentation of a jet with the quantum numbers of H into a fragment A (plus anything), the fragmentation function $F_{A/H}(z)$ is supposed to behave like the same power of (1-z) as $z \to 1$. The application of this rule has been extensively discussed in references 3 and 17. The theoretical status of the rule is that it is an extension of the Drell-Yan-West threshold relation in electroproduction or the Bloom Gilman relation. No entirely satisfactory derivation of these relations has been given[46] and their validity does not seem to be confirmed by the data[47] - though the situation here is not clear-cut. Hence the generalization to $G_{A/H}(x)$ or $F_{A/H}(z)$ must be regarded as having rather an uncertain status. Furthermore, the dimensional counting rule is supposed to apply only near x = 1 or z = 1, and most of the available data are for smaller values of x or z. Notice that even if this application of the dimensional counting rule does not turn out to be correct, the application in (5.4) and (5.5) may still be valid as it requires fewer theoretical assumptions.

If the form (5.4) for $\frac{d\hat{\sigma}}{d\hat{t}}$ holds with a simple power, and if Bjorken scaling is strictly valid and parton transverse momenta are negligibly small, then relation (5.1) readily leads to the form (2.1) for the inclusive cross-section at large p_t.

6. Constituent Interchange and Other Models

CIM and Quark Fusion

The central question of high-p_t physics is what are the objects A, B, C and D of the hard-scattering mechanism in Figure 5.1. We shall conclude in this Section and in the next that we do not yet have a definite answer to this.

To begin with, suppose that the calculation of Figure 5.1 is as simple as it can be. Specifically, assume that the following assumptions are correct: (i) the transverse momenta of the constituents A, B are negligible; (ii) their distributions in their parent hadrons scale, as do the jet fragmentation functions; (iii) the wide-angle scattering $A + B \to C + D$ has the functional form (5.4). In this case the single-particle inclusive cross-section has the form given in (2.1):

$$E \frac{d\sigma}{d^3p} = p_t^{-n} \; F\,(x_t, \theta) \tag{6.1}$$

$$x_T = 2p_t/\sqrt{s}$$

θ = production angle in centre-of-mass frame.

In terms of the parameter m in (5.4), n = 2 m.

As we reviewed in Section 2, data for π^+, π^- and π^0 production at various angles θ agree well with (6.1) throughout the Fermilab/ISR energy range for $p_t \gtrsim 1$ GeV/c, at least up to $p_t = 9$ GeV/c. The value found for n is consistent with 8, though decent fits cover the range 7.5 to 9.5.

If one makes the assumptions (i) to (iii) listed above, and assumes also the dimensional counting rule (5.5), the choice $q + q \to q + q$ for the wide-angle scattering $A + B \to C + D$ corresponds to $n = 4$. The data clearly rule out $n = 4$, so that this choice for the wide-angle scattering is tenable only if at least one of the assumptions is relaxed. We discuss this below.

On the other hand, either of the choices $q + \bar{q} \to M + M$ or $M + q \to M + q$ corresponds to $n = 8$. These choices are known respectively as quark fusion and constituent interchange (CIM). The symbol M denotes a quark-antiquark system; when it emerges as a jet in the final state, it can be either a stable meson, or a two-body resonance or a higher excited $q\bar{q}$ state - the jet fragmentation function is as in (4.11). These two choices for the wide-angle scattering, then, each have the merit that, with the simplest assumptions, they correctly predict the value of n. (The possible slight preference for $n \approx 8.5$ in the data, rather than $n = 8$, is most naturally accommodated by giving the initial constituents some transverse momentum.)

A full scaling behaviour would correspond to $n = 4$. This exhausts the dimensions of the inclusive cross-section. It corresponds to a primary interaction involving elementary fields (gluon exchange between quarks, say). A larger power involves either more complicated basic information or, in particular, a primary interaction among non-basic constituents. The structure of such constituents ($q\bar{q}$ pair for instance) brings an extra dimension which manifests itself through a larger value of n.

A detailed confrontation of all the available single-particle data with the quark fusion and CIM models has been described by Chase and Stirling[48]. Their conclusion is that, while each model separately is successful in many respects, a mixture of the two models is really needed.

Quark-Quark Scattering

The original expectation[1] was that large-p_t hadron production would be scale-free, so that n = 4 in (6.1). If one makes the simplest assumptions, as we have described, the quark-quark scattering model [2] provides a concrete realization of this expectation. The discovery that the data rather correspond to n = 8 for a long time led people to believe that some other dynamical mechanism must be operating, though there has always been the hope and expectation that the quark-quark scattering would become apparent at higher values of p_t.

Such an idea collected more strength recently when the success of calculations based on quark-gluon coupling in the analysis of Charmonium provided confidence in the value to expect for quark-quark scattering at large momentum transfer and hence jet production at large p_t[9]. In the framework of a perturbative calculation in terms of a running coupling constant, one for instance predicts a jet production cross-section which is far below that observed (by over an order of magnitude) at a presently typical p_t value of 5 GeV/c. While what is observed cannot yet correspond to quark-quark scattering, as most simply expected to occur, the quark-quark scattering jet production, with p_t^{-4} behaviour (up to logarithmic terms) should eventually win over an initially overwhelming background which drops down as p_t^{-8}. One may thus estimate that at ISR energies a change of régime could occur around 9 GeV/c, with quark-quark effects eventually dominating in the production process.

Quark-quark scattering may however be far more involved than what is associated with simple gluon exchange. Indeed, more recently, there have been various attempts to describe the existing, N = 8, data within the framework of the quark-quark scattering model. One approach to this is that of the Caltech Group[11)], who abandon the idea that wide-angle quark-quark scattering should be scale-free. They therefore discard the predictions of dimensional counting, and assume that the scattering occurs through some unknown mechanism that is more complicated than single gluon exchange. Instead, they postulate that the quark-quark scattering cross-section is such as to give n = 8 and also the right angular dependence for large-p_t production:

$$\frac{d\hat{\sigma}}{d\hat{t}} = \frac{2300 \text{ mb GeV}^6}{\hat{s}\hat{t}^3} \tag{6.2}$$

As is clear from the calculation of the inclusive cross-section in terms of the differential cross-section at the constituent level, such a choice readily gives a p_t^{-8} behaviour. So would however any form of the type $(\hat{t}^4 (\hat{s}/\hat{t})^m)^{-1}$. With m = 0 the distribution would be far too peripheral, or too strongly peaked at forward and backward angles. With m = 1, Feynman, Field and Fox get the proper type of angular dependence, namely a weak angular dependence of the inclusive cross-section, at least over the range $45^\circ < \theta < 135^\circ$, and, the proper (and weak) amount of angular correlations between the toward and away jets, as later discussed.

By a suitable choice of the quark-jet fragmentation functions, good agreement is found for much of the single-particle and correlation data. In particular, the major problem that confronts the quark-fusion/CIM approach is largely overcome, that of the ratio $(pp \to \pi^{0}X)/(\pi^{-}p \to \pi^{0}X)$. However, there are some difficulties. We have explained that there is some reason to believe that, in the existing data, the trigger-side jet is not quark-like. The reason for this is that the evidence from $e^{+}e^{-}$ annihilation experiments at DESY is that the quark-jet fragmentation function F(z) vanishes something like $(1-z)^{2}$ as $z \to 1$. In order to increase the average fraction z of the jet momentum taken by the trigger particle, Field and Feynman[11)] assume that instead $F(z) \to$ const. as $z \to 1$. This helps with the problem which we have described concerning momentum measurements in the trigger-side jet, but does not completely solve it. It also leaves a problem with the momentum distribution observed in the away-side jet. The greater the momentum in the trigger-side jet is, the greater is that in the away-side jet. Even if one reduces the latter by assuming that the background is biased so as to balance some of the transverse momentum of the trigger-side jet, the observed away-side distribution does not correspond in normalization to that expected from a quark jet with as large a momentum as they need to postulate. A meson-like jet on the trigger side, with a smaller total average transverse momentum and, therefore, correspondingly less momentum for the away-side jet, does better here. Of course, the problem can be cured by allowing a very large transverse momentum for the quark constituents in the proton, so that the background balances more of the p_t of the trigger-side jet; but an average transverse momentum of the quark constituents of 500 MeV/c is not enough. It will be important to have correlation data with rather larger trigger p_t, so that the effects of the transverse momentum of the constituents become less significant.

The other approach used in attempts to make the quark-quark scattering model reproduce the n = 8 data is to relax the assumptions of scaling for the quark distributions in the nucleon and for the quark fragmentation function, to introduce fairly large transverse momentum for the quark constituents before they undergo wide-angle scattering [44)] (even as large as an average of 850 MeV/c), and to use the QCD running coupling constant for the vector gluon exchange in the wide-angle quark-quark scattering. We have already explained that each of these modifications is subject to arbitrariness and uncertainty at the present time. Nevertheless, the consensus of opinion[8,18,28,32)] is that, with reasonable inputs, the best that can be done is to bring the quark-quark scattering model into agreement with the data at p_t values greater than about 6 GeV/c. This conclusion, of course, increases the need for correlation data at such large values of p_t[J)].

One further modification has been introduced by Duke[28)]; this enables him to fit the data also at lower values of p_t. He modifies the wide-angle quark-quark scattering as in (5.4). However, we have explained that this may be criticized on the grounds that, in this context, the quark-quark scattering is not required to be an exclusive process, and for inclusive processes the form (5.5) is probably invalid.

All approaches discussed so far are mainly phenomenological in character. There is now an a priori reliable theoretical frame-

TABLE 7.1

Cross-sections for the various constituent quark-quark, quark-gluon and gluon-gluon subprocesses as calculated according to QCD. The differential cross-section is given by $d\hat{\sigma}/d\hat{t} = \pi\alpha_s^2(Q^2)\ |A|^2/\hat{s}^2$, where $\alpha_s(Q^2)$ is the effective coupling. This table is taken from Reference 57. The original derivation is by B. Combridge et al., Ref.18.

	Subprocess	$\lvert A\rvert^2$
1.	$q_i q_j \to q_i q_j$ $q_i \bar{q}_j \to q_i \bar{q}_j$ $(i \neq j)$	$\frac{4}{9}\,\frac{\hat{s}^2+\hat{u}^2}{\hat{t}^2}$
2.	$q_i q_i \to q_i q_i$	$\frac{4}{9}\left(\frac{\hat{s}^2+\hat{u}^2}{\hat{t}^2}+\frac{\hat{s}^2+\hat{t}^2}{\hat{u}^2}\right)-\frac{8}{27}\,\frac{\hat{s}^2}{\hat{u}\hat{t}}$
3.	$q_i \bar{q}_i \to q_i \bar{q}_i$	$\frac{4}{9}\left(\frac{\hat{s}^2+\hat{u}^2}{\hat{t}^2}+\frac{\hat{t}^2+\hat{u}^2}{\hat{s}^2}\right)-\frac{8}{27}\,\frac{\hat{u}^2}{\hat{s}\hat{t}}$
4.	$q_i \bar{q}_i \to gg$	$\frac{32}{27}\left(\frac{\hat{u}^2+\hat{t}^2}{\hat{u}\hat{t}}\right)-\frac{8}{3}\left(\frac{\hat{u}^2+\hat{t}^2}{\hat{s}^2}\right)$
5.	$gg \to q_i \bar{q}_i$	$\frac{1}{6}\left(\frac{\hat{u}^2+\hat{t}^2}{\hat{u}\hat{t}}\right)-\frac{3}{8}\left(\frac{\hat{u}^2+\hat{t}^2}{\hat{s}^2}\right)$
6.	$q_i g \to q_i g$	$-\frac{4}{9}\left(\frac{\hat{u}^2+\hat{s}^2}{\hat{u}\hat{s}}\right)+\left(\frac{\hat{u}^2+\hat{s}^2}{\hat{t}^2}\right)$
7.	$gg \to gg$	$\frac{9}{2}\left(3-\frac{\hat{u}\hat{t}}{\hat{s}^2}-\frac{\hat{u}\hat{s}}{\hat{t}^2}-\frac{\hat{s}\hat{t}}{\hat{u}^2}\right)$

The numerical coefficients correspond to the sum (average) over final (initial) colors, hence the difference between processes 4 and 5.

work in which to cast present ideas about quark-quark interactions. This is quantum chromodynamics. But quantum chromodynamics involves several terms and large effects could come from gluon production. This is what we now turn to with the question: is QCD actually relevant?

7. Quantum Chromodynamics

QCD Mechanisms

Field and Feynman[11)] fit the large p_t inclusive data by choosing a quark-quark scattering of the form (6.2). At the moderate values of $\hat{t}$ this is, as we previously said, an order of magnitude greater than the form that corresponds to the exchange of a single colour-octet gluon:

$$\frac{d\sigma}{d\hat{t}} = \frac{4\pi\alpha_s^2}{9}\left(\frac{\hat{s}^2 + \hat{u}^2}{\hat{s}^2\hat{t}^2}\right) \tag{7.1}$$

with $\alpha_s \approx 1/3$. This is why the attempts to make the quark-quark scattering with single gluon exchange fit the inclusive data fall short at moderate values of p_t. The value $\alpha_s \approx 1/3$ is suggested by the analysis of scale-breaking effects in deep inelastic lepton scattering and charmonium decay rates, but one cannot yet be entirely confident of it.

However, once one admits the presence of vector gluons there are various additional hard-scattering processes that must be considered[8,18)]. These are quark-gluon scattering, gluon-gluon scattering and $q\bar{q} \to gg$ (Figure 7.1). These terms have been calculated, using appropriate guesses for the gluon distribution within the proton and for the gluon fragmentation function. The conclusion is that the additional processes dominate over the quark-quark scattering at small values of x_t. They are still not sufficient to

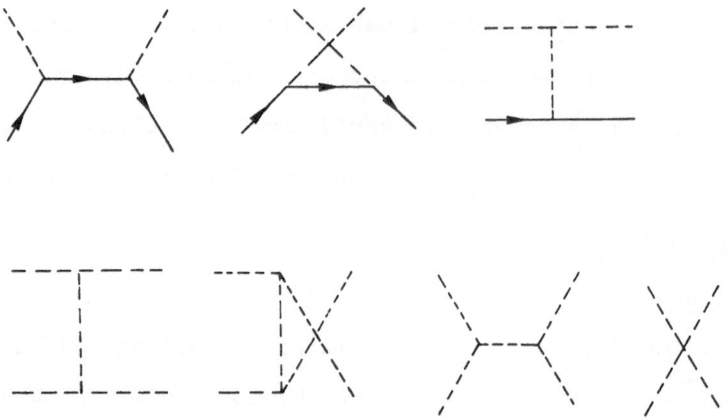

Figure 7.1 Quark gluon and gluon gluon scattering to lowest order in Quantum Chromodynamics.

bring the calculated contributions up to the data, unless one includes rather a large transverse momentum for the constituents[8], probably as much as 850 MeV/c. We have already explained that having such a large transverse momentum for the constituents helps also in explaining the away-side correlations. For this there is now also the additional helpful feature that the away-side jet is quite likely to be a gluon. As we shall explain below, it is possible that the gluon fragmentation function is such that gluons are more reluctant than quarks to fragment into fast hadrons, and we have said that indeed there is a shortage of fast hadrons observed in the away-side jet.

The importance of gluon effects is a very important point of the QCD approach. While at large x_t one is more likely to trigger on a quark jet, gluons are otherwise more important as a source of hadrons. A particularly interesting process is of course $gg \to q\bar{q}$ (5 in table 7.1) in view of the fact that gluons are insensitive to flavor. Large p_t production of hadrons according to this process should give as many charmed quark pairs as others. It turns out that the corresponding cross-section is not relatively important. It is at the < 1% level. Nevertheless, with jet cross-section at the mb level at 2 GeV/c, this contribute charmed particle production at a level much above that expected on statistical grounds. It should therefore be very interesting to look for charmed particles among large p_t hadrons. One could see (if the QCD approach is correct) yields which may be compatible with the Beam dump results (10 -100 μb say).

The additional QCD terms do not help with the possible problem of the total momentum measured in the trigger-side jet. It also remains to be decided whether it is reasonable to include so much transverse momentum for the constituents. The large value that is used may be motivated by the large values of transverse momentum observed in dimuon production[26], but it is not yet clear that in fact dimuon production serves as a direct probe of constituent transverse momentum when that transverse momentum q_t is large[45]. Essentially, the issue is whether the simple Drell-Yan mechanism that seems to describe the dimuon production at small q_t has to be modified or even replaced for larger values of q_t.
In the Drell-Yan mechanism, the transverse momentum of the dimuon arises directly from the internal motion of the constituent quarks in the incident hadrons. One modification of the Drell-Yan mechanism is the radiation of gluons from the annihilating quark and antiquark. When a gluon is radiated with large transverse momentum, it causes the dimuon to recoil with a larger transverse momentum than results from the intrinsic quark motion. See Figure 7.2. Such radiation must be allowed for also when quarks or gluons participate in large-p_t hadron production; for example, in $qq \to qq$

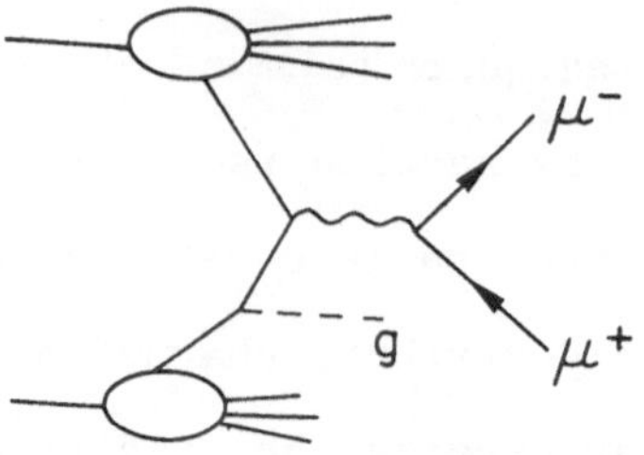

Figure 7.2 Scaling violation through gluon exchange in large mass lepton pair production.

any of the four quarks may radiate a gluon. These effects have been calculated[46)] in leading order, together with corrections from internal, virtual gluons. There are cancellations among the different effects, as is usual in theories that display a gauge invariance. Related effects occur in deep inelastic lepton scattering and e^+e^- annihilation, and the conclusion for large-p_t production seems to be that one should use the net apparent parton distribution within hadrons as measured in deep inelastic scattering, the net apparent quark fragmentation function as measured in e^+e^- annihilation, and the running coupling constant. These include gluon effects, as shown in Figure 7.3. Then only the simplest diagrams need be considered for the large-p_t process, that is the ones shown in Figure 7.1.

Direct Photon Production at Large p_t

A number of authors have predicted[47)] that the cross-section for direct photon production at large p_t should be quite large. In QCD models this comes about because for every diagram in which a high-p_t gluon is emitted, there is a diagram in which the gluon is replaced by a high-p_t photon. This second diagram is smaller than the first by a factor $\alpha/\alpha_s(Q^2)$. This is the ratio of direct photon production to jet production. However, if one uses the first

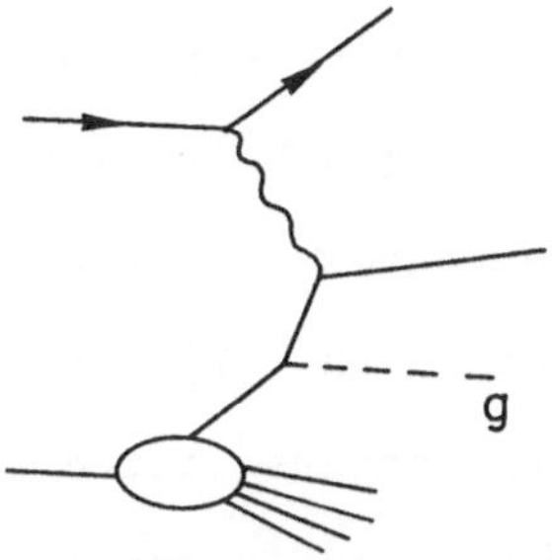

Figure 7.3 Scaling violation through gluon exchange in deep inelastic scattering.

diagram to calculate single pion production at large p_t, whether this pion is a fragment of the gluon or of a jet on the opposite side one has to pay the price of a trigger-bias factor (see the term in brackets in (4.9)). The conclusion is that at $p_t = 4$ GeV/c one should expect a ratio γ/π at the 10% level, rising with p_t to maybe as much as 100% at $p_t = 10$ GeV/c. The ratio also rises slowly with energy.

Similar results are found in models where the trigger-side jet in π production is a $q\bar{q}$ system. Whatever dynamics is responsible for producing such a jet can instead produce a single high-p_t photon. Again the coupling of the photon is smaller, but again there is no trigger-bias factor for the photon. Because, unlike the pion, the photon is elementary, dimensional counting predicts that the ratio γ/π is equal to $p_t\sqrt{s}$ times a function of x_t; in an explicit model, Escobar[54)] finds that this function is rather constant.

The experimental search for direct photons is very difficult, and the data now available are probably best regarded as an upper limit.

Quark Fragmentation

There is a widespread feeling that the fastest fragment of a jet retains some information about the quantum numbers of the jet. In particular, when the jet is a quark it is believed that, in events where it fragments to give rather a fast meson, that meson is most likely to be such as to retain the original quarks as a valence quark. See, for example, the papers listed in reference 55. So a fragmenting u quark can more readily result in a fast π^+ or π^0 than a fast π^-.

In order to predict the details of large-p_t events in which there are quark jets, a more complete model of quark fragmentation is needed. Such a model has recently been developed rather fully[44)]. When a coloured quark emerges from a hadronic system, a colour field is set up between it and the system. If this field is somehow more or less confined to a tube-like region, rather than spreading throughout space, its energy increases with the distance of the quark from the system, until there is enough energy in the field for quark-antiquark pairs to be created. So the initial quark a can combine with an antiquark $\bar{b}$ from a new pair, so forming a meson $a\bar{b}$ and leaving the other quark b of the pair to combine with further antiquarks. This process continues until the quark has shed nearly all of its energy and momentum. The details of what happens then are obscure, but the quark somehow gets annihilated by the residual hadronic system, so that no fractional charge or baryon number emerges. Baryon production from quark jets is not at all understood.

SU(2) symmetry is assumed to require that at each stage the quark has equal probability of encountering either a $u\bar{u}$ or a $d\bar{d}$ pair, no matter what type of quark it may be. (However, the validity of this assumption has been questioned by Sukhatme[48)]. SU(3) symmetry is broken by assuming that the probability of encountering an $s\bar{s}$ pair is somewhat smaller.

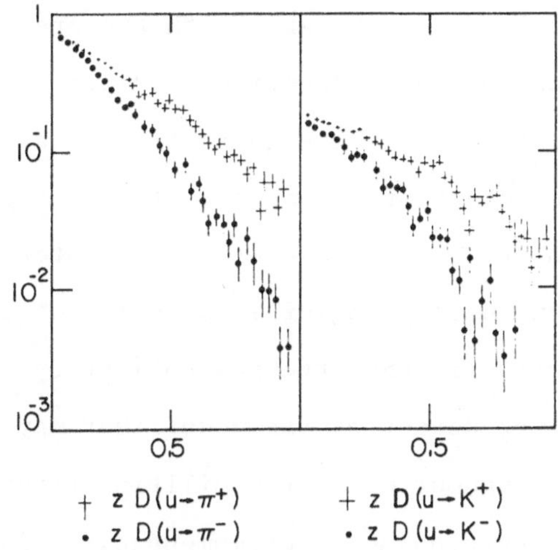

Figure 7.4 Quark fragmentation into pions. Monte Carlo calculation of Feynman and Field.

The first mesonic system $a\bar{b}$ evidently does contain the original quark a as a valence quark. However, this does not necessarily mean that the fastest-detected meson contains a as a valence quark, for two reasons. Firstly, it is not necessarily true that the first system emitted is the fastest; and secondly, the system $a\bar{b}$ may be an unstable resonance that rapidly decays into the stable mesons that are actually detected. Making reasonable input assumptions, Field and Feynman[44)] have generated 40,000 quark decays in the model, using a Monte Carlo procedure. Some of the results are shown in Figure 7.4. Their conclusion is that, on average, the notion of quantum number retention for the fastest hadron works fairly well; for example, they find

$$\frac{F_{\pi^+/u}(z)}{F_{\pi^-/u}(z)} \approx 3 \text{ at } z = \tfrac{1}{2} \tag{7.2}$$

As $z \to 1$, this ratio diverges as $(1-z)^{-1}$.

This last result does not agree with the predictions of dimensional counting, which we described earlier; dimensional counting would make the ratio diverge as $(1-z)^{-4}$ as $z \to 1$. This may or may not be a defect of the model; it arises probably because no account is taken of the possible effect of intermediate quarks in the fragmentation chain going off mass shell, while this is an important consideration in the attempts to justify the dimensional counting rule. Another defect of the model that may be important is that it deals throughout with probabilities rather than with amplitudes, and so contains no quantum-mechanical interference effects. Nevertheless, the model does provide the only available framework for calculating all the expected properties of quark fragmentation, and many of these properties are listed in the Field-Feynman paper[44)], in addition to the simple ones that we have described.

For the case of a gluon jet, one possible assumption is [49)] that it initially materializes as a $q\bar{q}$ pair; the q and $\bar{q}$ then each fragment as we have described. In this case, $F_{H/g}/F_{H/q}$ vanishes like $(1-z)$ as $z \to 1$, so that a gluon jet is less likely to contain fast hadrons than is a quark jet. Since a gluon has neutral flavour, the π^+ and π^- distributions that it produces are identical.

In the case of $q\bar{q}$ jets M, such as are needed in CIM, no corresponding model has been described in the literature.

8. Future prospects

Refining the jet picture

At present jets in large p_t reactions can be considered as established. The jet structure, as expected according to the parton picture (figure 1.3), corresponds to correlation studies and to the analysis of jet fragmentation which could be done so far. While

QCD may provide a rationale for present parton model calculations, considered as a first order estimate, specific QCD features cannot be ascertained yet. There are indeed qualitative effects which are found to follow QCD prediction but an actual test is not yet available.

Jets should become clearer still as one increases the trigger momentum (in practice going from 3 to 10 GeV/c at the ISR). Nevertheless, some parameters like $\langle k_t \rangle$ and $\langle p_{out} \rangle$, which should be fixed in the parton model, should be found to slowly increase with p_t according to QCD. Jets should be more spread out than initially expected and eventually branch into several jets.

The probability of an extra gluon being radiated at wide enough an angle from the main jet and therefore appearing as another jet should increase with p_t. However, this is still a difficult calculational problem with QCD and much theoretical attention is presently devoted to this question. Going up to triggers with $p_t > 10$ GeV/c one should find typical jets as discussed in Section 3. One should also find more isotropic configurations. A double-arm calorimeter detector may introduce too strong a bias in favour of two-jet configurations and a detector with full azimuthal coverage may reveal interesting new features.

It is therefore very important to explore jet structure much further than has been done so far. This is an extremely important test of ideas based on the parton picture. It is also very important to check whether, once the existence of two-jet configurations is on very firm ground, there are also configurations of a different kind which correspond to reactions with several jets.

At present, jet studies justify intensive experimental effort on proton machines and therefore at CERN and at Fermilab[50].

Large p_t Production at Collider Energy

While many questions connected with hadronic jets are still open for research at present energies, the most challenging question remaining is the relevance of QCD. As previously discussed, this calls for still higher energies than those available at the CERN ISR, if one wishes to get a clear-cut answer. Indeed, while the trend now found in very large p_t data (Figure 2.2) complies with what is expected according to QCD, the discussion in Section 7 shows that it is only at much higher energies (and much larger p_t values) that the complications associated with scaling violations, constituent transverse momenta and the presence of different types of jets might eventually disappear.

Such studies are not to be considered as for the remote future. As recently emphasized, in particular by C. Rubbia, colliding beams at very high energy can be achieved with the present machines. One possibility consists in using the main ring and the Doubler at Fermilab in a colliding beam version (the Tevatron project). One could thus envisage experimentation with $\sqrt{s} \approx 1000$ GeV (1000 GeV for the Doubler beam and 250 GeV for the main ring beam) and a luminosity of the order of 10^{30} cm^{-2} s^{-1} [51]. One could also use recent developments with cooling techniques[52] to obtain an intense source of antiprotons and inject and accelerate at the same time protons and antiprotons in the CERN SPS or the Fermilab machine[53]. At CERN one may thus consider experimentation at $\sqrt{s} \approx 500$ GeV with a luminosity of the order of 10^{30} cm^{-2} s^{-1}. This should be achieved in 1981. Later on, Isabelle should provide a much higher luminosity (of the order of 10^{33}) for experimentation at such energies[54].

The main motivation for such an endeavour is searching for

the weak boson(s). Future prospeots have been recently discussed in detail by Quigg[55]. The weak bosons are expected to decay most frequently into hadron jets with leptonic decays at the 10% level, say. One may search for very large p_t jets ($p_t \gtrsim 30$ GeV) or very large p_t leptons. While jet search might seem easier in view of the expected larger yields, present ideas about large p_t hadron production lead one, however, to expect a background which could be forbiddingly large. If this is the case, W(Z) search would be more difficult, and possible only through lepton search, but the jet study in hadronic interaction would be very rewarding for its own sake.

Feynman and Field have recently explored in detail expectations based on QCD[44]. As discussed in Section 7, one then expects behaviour to match present data as a result of many correcting terms, the influence of which would gradually disappear. As a result $p_t^8 Ed\sigma/d^3p$ at fixed x_t should not be a constant as p_t (and $\sqrt{s}$) increases, but instead rapidly increase. Figure 9.1 shows the corresponding effect for the expected π^0 yield at 90^0, as calculated by Feynman and Field. As correction terms include in a QCD calculation become less important, one departs from the p_t^{-8} behaviour to eventually reach a p_t^{-4} behaviour. At $\sqrt{s}$ = 500 GeV this corresponds to a yield which is two orders of magnitude larger than what the naive extrapolation of the observed large p_t behaviour (Figure 2.1) would lead one to predict. The approach to the asymptotic behaviour remains however rather slow. This is illustrated byFigure 9.2 which gives the extrapolated single particle yield at fixed x_t as calculated by Feynman and Field. The quantity $p_t^4\ Ed\sigma/d^3p$ should eventually show a logarithmic behaviour. This is the case at collider energy but one needs to go far beyond the present ISR energy range to see it appearing. "Smear", the

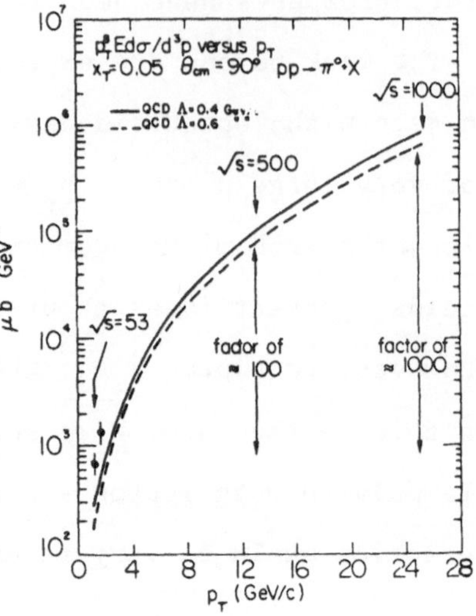

Figure 9.1 Large p_T pion yield at collider energy showing departure to the p_T^{8} fit to medium p_T ISR results.

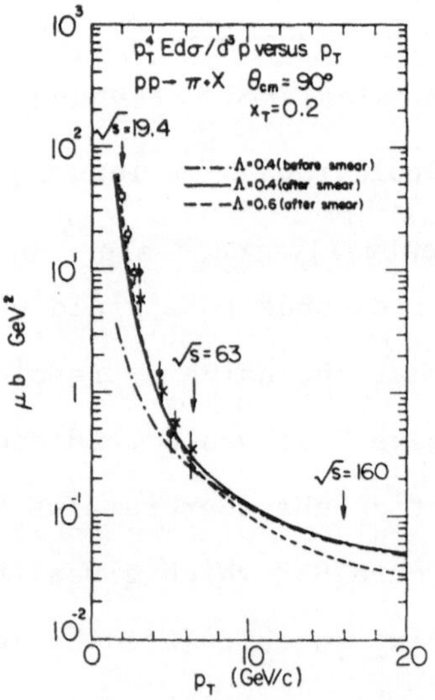

Figure 9.2 Large p_T pion yield at collider energy showing the approach to the full scaling limit.

effect of which is seen in Figure 9.2 corresponds to the internal motion of quarks. At $\sqrt{s}$ = 500 GeV, the jet yield is calculated to be almost 3 orders of magnitude above the single particle yield. If present ideas about the QCD interpretation of large-p_t phenomena are correct, spectacular effects are to be found at collider energies and prominent jets should be the rule. At $\sqrt{s}$ = 500 GeV (a typical collider energy) the cross-section for the production of a 20 GeV jet should be of the order of 5×10^{-3} $\mu b/GeV^2$. It should be as high as 1 $\mu b/GeV^2$ for an 8 GeV jet. This is two orders of magnitude above what should be observed at the ISR (a value which is close to 10^3 larger than the already measured π^o yield at the same p_t).

Taking the (conservative) p_t^{-8} behaviour as holding to larger p_t's and larger energies, Quigg has estimated that the jet yield associated with the production of the weak bosons at collider energy should be slightly above the hadronic background[55]. Figure 9.3 shows the integrated yield (yield above a specific p_t value) associated with the W, the Z^o and the large-p_t hadronic interaction. The yieldsin weak boson decay should still be multiplied by the relevant branching ratio (0.9 say). A signal could be seen though uncertainties in such extrapolation should not give too much weight to a jet search for the weak bosons[53].

Extrapolating according to QCD the picture is very different and the expected signal becomes much weaker than the expected background. Figure 9.4 shows the differential cross-section for jet production due to W^+ decay (the yield has still to be multiplied by the corresponding branching ratio). One notices the strong Jacobian peak at $p_t \approx M_W/2$. It could be further smeared by the transverse motion of the produced W if this is large. The signal stands above the background calculated according to the "black box"

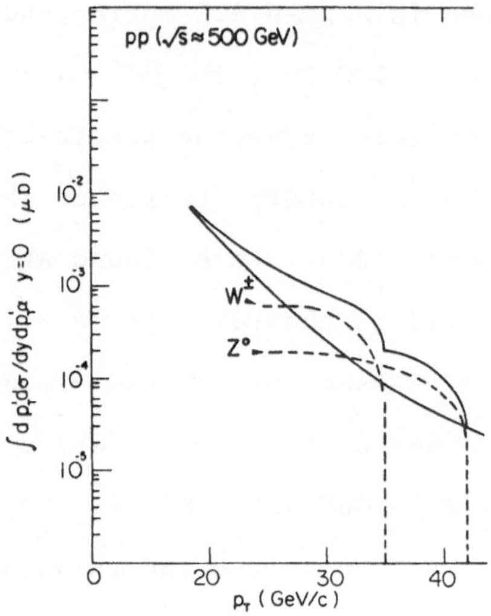

<u>Figure 9.3</u> Jet yields at collider energy. The W and Z signals stand above a p_T^{-8} background.

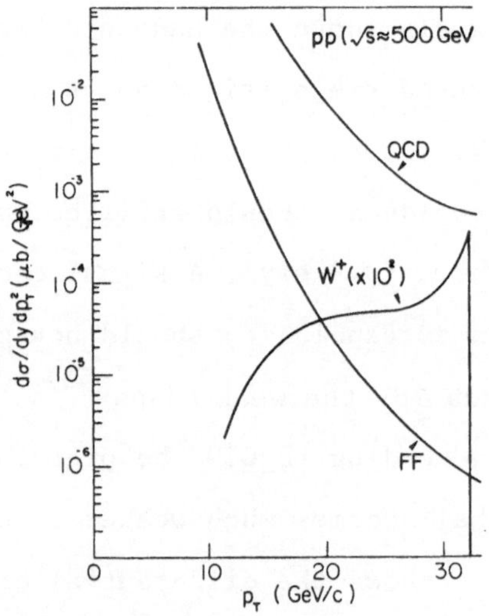

<u>Figure 9.4</u> The jet signal from the W is far below the QCD jet yield.

model of Feynman and Field (FF) with an asymptotic p_t^{-8} behaviour. It stands however below the background expected from QCD. The signal over background ratio should be better in $\bar{p}p$ than in pp collisions. At such energies however, the expected signals differ by less than an order of magnitude[55)].

W(Z) search is still possible through its leptonic modes, in particular the $e^+ e^-$ ($\mu^+ \mu^-$) decay of the Z^o. Extrapolating from present trends and with present wisdom one may however not reasonably expect any sizeable hadronic signal over the relatively strong hadronic background. While evidence for the Weak bosons may thus be more difficult to collect, our present understanding of large-p_t phenomena leads to expect spectacular jet effects at Collider energy. This should be a very interesting facet of particle physics in the early eighties. It would prove QCD correct and show that strong interactions become simple at large momentum transfer.

Concluding we are at present at a very interesting stage. On the one hand evidence for effects expected according to the parton approach are now ascertained and, while hadron constituents do not appear directly they result in jets of hadrons with prominent kinematical features. On the other hand an underlying field theory for strong interactions appears as a serious contender and could provide the proper interpretation for the observed effects. It is at present too early to conclude. Nevertheless, prediction for collider energies, which should be available within 2 to 3 years, are of great interest. One may expect very prominent effects with spectacular jets of hadrons.

ACKNOWLEDGEMENT

These lecture notes borrow a very great deal from a review paper written by P.V. Landshoff and the author. It will be published in Physics Reports and could be consulted for a more extensive discussion. They also borrow from the introductory review to the Conference on "Jets in High Energy Collisions" held at the Niels Bohr Institute and Nordita in July 78, written by the author and to be published in the Conference proceedings (Academica Scripta). This second paper could also be consulted for a review of the latest developments.

I am indebted to Prof. A. Zichichi for giving me the opportunity of presenting this review on jets in hadronic interactions at the Erice Summer School.

TABLE*

LARGE-P_T EXPERIMENTS AT THE ISR

Saclay-Strasbourg R 102	Magnet	Single charged particles at 90°	M. Banner et al., Phys. Lett. 41B, 547 (1972) and 44B, 537 (1973)
CERN-Columbia-Rockefeller (CCR) R 103	Lead glass	Inclusive photons (π^o) at 90°	F.W. Büsser et al., Phys. Lett. 46B, 471 (1973), 51B, 306 (1974) and 55B, 233 (1975)
British-Scandinavian R 203	Magnetic spectrometer	Single charged particles at wide angle	B. Alper et al., Phys. Lett. 44B, 521 and 527 (1973); Nuclear Physics B 100, 237 (1975)
Pisa-Stony Brook R 801	Lead glass and 4π scintillator hodoscope	Rapidity distributions of charged hadrons accompanying a large-p_t π^o	G. Finochiaro et al., Phys. Lett. 50B, 396 (1974); R. Kephart et al., Phys. Rev. D14, 2909 (1976)
CERN-Columbia-Rockefeller-Saclay (CCRS) R 105	Two magnets and lead glass	Photons, electrons and charged hadrons at large p_t (single and double inclusive)	F.W. Büsser, Phys. Lett. 56B, 482 (1975); Nuclear Physics B106, 1 (1976)
CERN-Daresbury-Liverpool-Rutherford R 205	Mgnetic spectrometer	Rapidity distributions of charged hadrons accompanying identified high-p_t charged hadrons	B. Alper et al., Nuclear Phys. B114, 1 (1976)
Aachen-CERN-Heidelberg-Munich (ACHM) R 701	Streamer chamber and lead glass	Rapidity distributions of charged hadrons accompanying high-p_t π^o's	K. Eggert et al., Nuclear Phys. B98, 49 and 73 (1975)

CERN-Collège de France-Heidelberg-Karlsruhe (CCHK) R 407-408	Split Field Magnet	Correlations involving large-p_t charged particle(s)	R. Cottrell et al., Phys. Lett. 55B, 341 (1975); M. Della Negra et al., Phys. Lett. 59B, 401 and 481 (1975); Nuclear Physics B104, 365 (1976) and B127, 1 (1977)
CERN R 412	SFM and lead glass	Rapidity distributions of charged hadrons accompanying a high-$p_t\pi^0$	P. Darriulat et al., Nuclear Physics B107, 429 (1976) and B110, 365 (1976)
British-French-Scandinavian (BFS) R 413	SFM and wide-angle spectrometer	Correlations involving large-p_t charged particle(s)	M. Albrow, et al.
CERN-Columbia-Oxford-Rockefeller (CCOR) R 108	Solenoid and lead glass	Correlations involving large-p_t π^0's and charged particles	
CERN-Saclay R 702	Two magnets and lead glass	Production of electrons, photons and hadrons at large p_t	P. Darriulat et al.
Brookhaven-CERN-Syracuse-Yale (BCSY) R 806	Argon calorimeter and transition radiation	Production of electrons and photons at large p_t	J.H. Cobb et al.
In preparation			
CERN-Bologna R 415	Upgraded SFM with electron trigger		
Annecy-CERN-Collège de France-Heidelberg-Karlsruhe-Warsaw R 416	Upgraded SFM		
British-Scandinavian-CERN R 807	Calorimeter and toroidal magnet		
Elastic Scattering at Large t			
CERN-Hamburg-Heidelberg-Annecy-Vienna (CHHAV) R 401	SFM	H. de Kerret, E. Nagy, R.S. Orr, W. Schmidt-Parzefall, K. Winter, A. Brandt, F.W. Büsser, G. Flügge, F. Niebergall, P.E. Schumacher, H. Eichinger, K.R. Schubert, J.J. Aubert, C. Broll, G. Coignet, J. Favier, L. Massonet, M. Vivargent, W. Bartl, H. Dibon, C.H. Gottfried, G. Neuhofer, M. Regler, Phys. Lett. 68B, 374 (1977) and submitted to Nuclear Physics	

* Table Throughout the text data are referred to according to the experimental collaboration from which they originate. The collaborations and some of their publications are listed in this table.

LARGE P_T EXPERIMENTS AT FERMILAB

Single-arm Experiments			
E 100 A Chicago-Princeton	Single-arm spectrometer	Protons on heavy targets	J.W. Cronin et al., Phys. Rev. D11, 3105 (1975)
E 300 Chicago-Princeton	Single-arm spectrometer	Protons on hydrogen and heavy targets	D. Antreasyan et al., Phys. Rev. Lett. 38, 112 and 115 (1977)
E 268 Brookhaven- E 350 Caltech-LBL	γ-ray detector (π^o)	Various Beams on hydrogen	G. Donaldson et al., Phys. Rev. Lett. 36, 1110 (1976) and 40, 917 (1978)
Jet Studies			
E 260 Caltech-UCLA-Fermilab-Chicago Circle-Indiana-MPI	Multi-particle spectrometer and calorimeter	Protons and pions on hydrogen and nuclear targets	C. Bromberg et al., Phys. Rev. Lett. 38, 1447 (1977); Nucl. Physics B134, 189 (1978)
E 236 Fermilab-Tufts-Washington	Calorimeter, single-particle spectrometer and Cerenkovs		
E 395 Lehigh-Penn-Wisconsin	Two calorimeters and magnet	Protons and pions on hydrogen	M.D. Corcoran et al.
Double-arm Experiments			
E 494 Columbia-FNAL-Stony-Brook	Double-arm spectrometer	Protons on hydrogen and heavy targets	R.D. Kephart et al., Phys. Rev. Lett. 39, 1440 (1977); R.L. McCarthy et al., Phys. Rev. Lett. 40, 213 (1978); R.J. Fisk et al., Phys. Rev. Lett. 40, 984 (1978)
E 357 FNAL-Michigan Purdue	Double-arm spectrometer	Protons on nuclei target	C.W. Akerlof et al., Phys. Rev. Lett. 39, 861 (1977)
Elastic Scattering at Large t			
E 177 A Cornell-Lebedev-McGill-North Eastern		Protons on hydrogen	J.L. Hartmann et al., Phys. Ref. Lett. 39, 975 (1977)

REFERENCES

The list of reference papers contains mainly theoretical or phenomenological papers which are quoted in the text. It continues with a table giving a list of experiments on large p_t reactions at the CERN ISR and at Fermilab. Experimental results quoted in the text are normally referred to according to the collaboration which obtained them. The table should allow an easy tracking of the relevant pubblication.

References

1. S.M. Berman and M. Jacob, Phys. Rev. Lett. 25, 1687 (1970)

2. S.M. Berman, J. D. Bjorken and J. B. Kogut, Phys. Rev. D4, 3388 (1971). See also S.D. Ellis and M. Kislinger, Phys. Rev. D9; 2027 (1974)

3. D.I. Sivers, R. Blankenbecler and S.J. Brodsky, Physics Reports 23C, 1 (1976)

4. R. Hagedorn and U. Wambach, Nuclear Physics B123, 382 (1977)

5. S.D. Ellis and J. Gasser, Nuovo Cimento 39A, 279 (1977); C. Michael and L. Vanryckeghem, J. Phys. G3, L151 (1977) and G4, 683 (1978)

6. G. Preparata and G. Rossi, Nuclear Physics B111, 111 (1976)

7. G. Hanson, Proc. 7th International Colloquium on Multi-particle Reactions, Tutzing (1976)

8. R.D. Field, Phys. Rev. Lett. 40, 197 (1978); A.P. Contogouis, R. Gaskell and S. Papadopoulos, McGill preprint

9. V. Marciano and H. Pagels, Physics Reports 36, 137 (1978)

10. J.D. Bjorken, Phys. Rev. D8, 4098 (1973); S.D. Ellis, M. Jacob and P.V. Landshoff, Nuclear Physics B108, 93 (1976); M. Jacob and P.V. Landshoff, Nuclear Physics B113, 395 (1976); S.D. Ellis and R. Stroynowski, Rev. Mod. Phys. 49, 753 (1977)

11. R.D. Field and R.P. Feynman, Phys. Rev. D15, 2590 (1977); R.P. Feynman, R.D. Field and G.C. Fox, Nuclear Physics B128, 1 (1977)

12. G.C. Fox, talk at 1977 Argonne meeting, CALT-68-630

13. R. Brandelik et al., Phys. Lett. 67B, 358 (1977); G. Knies, talk at International Symposium on Lepton-Photon Interactions, Hamburg (1977). A large-p_t trigger could also select a double scattering process. This would imply two (softer) jets on the other side. Requesting a large-p_t particle on each side disfavours on the contrary such a process with as result $\alpha \approx 1$.

14. N.S. Craigie, Physics Reports (in press)

15. R.D. Field and R.P. Feynman, preprint CALT-68-618

16. W. Ochs, Nuclear Physics B118, 397 (1977); B. Andersson, G. Gustafson and C. Peterson, Phys. Lett. 71B, 337 (1977); K.P. Das and R.C. Hwa, Phys. Lett. 68B, 459 (1977); D.W. Duke and F.E. Taylor, Fermilab preprint 77/95; S. Pokorski and L. Van Hove, CERN preprint TH 2427; V.V. Knyazev et al., Serpukhov preprint IHEP 77-106

17. R. Blankenbecler, S.J. Brodsky and J.F. Gunion, preprint SLAC-PUB-2057

18. R.F. Cahalan, K.A. Geer, J. Kogut and L. Susskind, Phys. Rev. D11, 1199 (1975); R. Cutler and D.I. Sivers, Phys. Rev. D16, 679 (1977); B.L. Combridge, J.Kripfganz and J. Ranft, Phys. Lett. 70B, 234 (78), preprint CERN-TH 2343; J.F. Owens, E. Reya and M. Glück, Florida preprint FSU-HEP-77-09-07 and erratum

19. G.A. Ringland, talk at Bielefeld Workshop on High-p_t Physics (1977)

20. C.E. Detar, S.D. Ellis and P.V. Landshoff, Nuclear Physics B87, 176 (1975); J.L. Cardy and G.A. Winbow, Phys. Lett. 52B, 95 (1974)

21. J.C. Polkinghorne, Nuclear Physics B93, 515 (1975); S.J. Brodsky and J.F. Gunion, Phys. Rev. Lett. 37, 402 (1976)

22. W. Morton, Cambridge preprint DAMTP 77/4

23. B.L. Combridge, Phys. Rev. D12, 2893 (1975)

24. M.K. Chase, Cambridge preprint DAMTP 77/29

25. J.B. Kogut, Phys. Lett. 65B, 337 (1977); I. Hinchcliffe and C.H. Llewellyn-Smith, Phys. Lett. B66, 281 (1977); P.V. Landshoff, Phys. Lett. 66B, 452 (1977); F.E. Close, F. Halzen

and D.M. Scott, Phys. Lett. 68B, 447 (1977); J.F. Gunion, Phys. Rev. D15, 3317 (1977); A.C. Davis and E.J. Squires, Phys. Lett. 69B, 249 (1977); D.E. Soper, Phys. Rev. Lett. 38, 461 (1977); J.C. Polkinghorne, N. Phys. B128, 537 (1977); K.J. Kim, Phys. Lett. 73B, 45 (1978); C.S. Lam and T.M. Yan, Phys. Lett. 71B, 165 (1977); K. Kinoshita et al., Phys. Lett. 63B, 355 (1977); R.J. Hughes, MIT preprint CTP 638; T. Kawabe, Kyushu preprint 77-HE-9; J. Bartelski, Aachen preprint

26. D.M. Kaplan et al., Phys. Rev. Lett. 40, 435 (1978)

27. C.J. Burrows, Cambridge preprint DAMTP 78/10

28. R. Baier and B. Peterson, Bielefeld preprints BI-TP-77/08 and 77/10; D.W. Duke, Fermilab preprint 77/35; R. Raitio and R. Sosnowsky, Bielefeld Workshop on High-p_t Physics (1977)

29. J. Ranft and G. Ranft, CERN preprint TH-2363; M. Fontannaz and D.I. Schiff, Nuclear Physics B132, 457 (1978); F. Halzen, G.A. Ringland and R.G. Roberts, Phys. Rev. Lett. 40, 991 (1978)

30. M. Della Negra et al., Nuclear Physics B127, 1 (1977)

31. H.D. Politzer, Physics Reports 14C, 129 (1974); P.V. Landshoff and H. Osborn in Electromagnetic Interactions of Hadrons (Plenum, ed. A. Donnachie and G. Shaw)

32. A.P. Contogouris, R. Gaskell and A. Nicolaidis, Nuclear Physics B126, 157 (1977) and Prog. Theor. Phys. 58, 1238 (1977); R. Hwa, A. Spiessbach and M. Teper, Phys. Rev. Lett. 36, 1418 (1976); E. Fischbach and G.W. Look, Phys. Rev. D15, 2576 (1977); W.J. Stirling, Nuclear Physycs B124, 330 (1977); T. Uematsu, Kyoto preprint RIFP-309

33. L. Hand, rapporteur talk at 1977 International Symposium on Lepton and Photon Interactions at High Energies, Hamburg

34. R. Barbieri et al., Nuclear Physics B117, 50 (1976); P.V. Landshoff and D.M. Scott, Nuclear Physics B131, 172 (1977); F. Halzen and D.M. Scott, Wisconsin preprint

35. A. Mueller, Phys. Rev. D9, 963 (1974); C.G. Callan and M.L. Goldberger, Phys. Rev. D11, 1542 (1975); N. Coote, Phys. Rev. D11, 1611 (1975); V.N. Gribov and L.N. Lipatov, Sov. J. Nuc. Phys. 15, 675 (1972)

36. R. Blankenbecler, S.J. Brodsky and J.F. Gunion, Phys. Lett. 39B, 649 (1972)

37. V.A. Matveev, R.M. Muradyan and A.N. Tavkelidze, Lett. Nuov. Cim. 7, 719 (1973); S.J. Brodsky and G.R. Farrar, Phys. Rev. Lett. 31, 1153 (1973)

38. P.V. Landshoff and J.C. Polkinghorne, Phys. Lett. 44B, 293 (1973)

39. B. Schrempp and F. Schrempp Phys. Lett. 55B, 203 (1975)

40. P.V. Landshoff, rapporteur talk at XVII International Conference on High Energy Physics, London 1974

41. P.V. Landshoff, Phys. Rev. D3, 1024 (1974), G. Farrar and C.C. Woo, Nuclear Physics B85, 50 (1975)

42. P.V. Landshoff and J. C. Polkinghorne, Phys. Rev. D8, 927 (1973); P.M. Fishbane and I.J. Muzinich, Phys. Rev. D8, 4015 (1973)

43. H.M. Fried, B. Kirby and T.K. Gaisser, Phys. Rev. D8, 3210 (1973); I.G. Halliday, J. Huskins and C.T. Sachrajda, Phys. Lett. 47B, 509 (1973)

44. R.D. Field and R.P. Feynman, preprint CALT-68-618

45. G. Altarelli; G. Parisi and R. Petronzio CERN preprint TH-2413

46. C.T. Sachnajda CERN preprint 2459

47. C. Escobar Phys. Rev. D15, 355 (1977); G.R. Farrar and S. Frautschi Phys. Rev. Lett. 36, 1017 (1976)

48. U.P. Sukhatime Phys. Lett. 73B, 478 (1978)

49. K. Koller and T. Walsh, Phys. Lett. 72B, 227 (1977)

50. For a more detailed discussion see ISR WOTKSHOP/2-7 and ISR WORKSHOP/2-14, CERN reports (1977)

51. The Tevatron Project, Fermilab Report (1977)

52. See for instance ISR WORKSHOP/2-12 and 2-13, CERN reports (1977)

53. See for instance Physics at Collider Energies, M. Jacob, Fermilab Summer Study, 1977

54. A Proton-proton Colliding Beam Facility, BNL Report 50648 (1977)

55. C. Quigg, Rev. Mod. Phys. 49, 287 (1977)

DISCUSSION

CHAIRMAN: M. Jacob

Scientific Secretaries: G. Anastaze, I. Dadic, J. Lindfors, R. Roth

DISCUSSION

- ISGUR:

How do the multiplicities behave in high p_T events? In particular, is there any variation of the beam jet multiplicities with p_T?

- JACOB:

The beam jet multiplicity is about what you would expect after allowing for the missing energy which has gone into the high p_T jets. It hardly changes with p_T, since it is only a logarithmic effect. On the contrary, a large multiplicity is associated with the transverse jets.

- MOTZ:

What do you think of the physical nature of the jets? Is there any indication of the structure of the elementary particles so as we have large p_T, or can they also be explained independently from such structures as a result of final state interactions of the constituents of the jet? It seems to me that there is some parallel to the cluster-problem in nuclear physics?

- JACOB:

The association between large p_T particles and the "tail" of an otherwise standard production mechanism has indeed long been advocated as an alternative to the jet picture. Nevertheless, once correlation results turned out to be in agreement with the hard scattering approach it became the "standard model". At present one may contrast the predictive value of the hard scattering approach with the difficulties which the cluster model has in adapting itself to the large p_T data. The hard scattering approach remains a model. Nevertheless, its general features are no longer challenged.

- MOUZOURAKIS:

What does the variation of $\langle p_{out} \rangle$ as a function of x_e imply in terms of constituents?

- JACOB:

There is first a trivial effect. As x_e increases we are dealing with two large p_T particles and they can both be used in defining the reaction plane. This will decrease the observed value of $\langle p_{out} \rangle$. There is, however, a more important effect and the former and trivial one is not enough. This new effect is associated with the fact that, as we consider larger transverse momenta the probability for the radiation of an extra hard gluon should increase. This will result in less coplanar configurations and the effective value of $\langle p_{out} \rangle$ will increase. The observed effect is in qualitative agreement with what is expected according to QCD. Nevertheless, it is too early to attempt a quantitative test.

The increase in $\langle p_{out} \rangle$ would then be related to the increase with energy of the mean transverse momentum of a large mass lepton pair, as found at Fermilab.

- KOCA:

What have we learned so far from the high energy total and inclusive cross sections of hadrons? Can properties like rising total cross sections, scaling in inclusive cross sections and the regularities such as $\sigma_{\pi N}/\sigma_{NN} \approx 2/3$ or much better ones à la Lipkin be explained by hard QCD?

- JACOB:

At present one may separately consider large cross section processes where what appears to be relevant is the available rapidity range, which increases as ln s, rather than the center of mass energy proper, and low cross section processes (large p_T production and large mass lepton pair production) where hadron constituents appear to interact incoherently from one another and where more rapid changes with energy occur. The relative simplicity of the latter processes justifies the interest which they presently generate. This is where one hopes to get a simple and sufficiently general understanding of strong interaction mechanisms. The former ones should eventually be understood in terms of QCD, if QCD meets the hopes which it at present arouses. Nevertheless, complicated coherent effects are involved and in a domain where perturbative

QCD should a priori not apply.

Perturbative QCD is therefore not expected to apply in any simple manner to the phenomena which you listed in your question. Nevertheless, if successful with phenomena discussed in this series of lectures, it should trigger further effort towards a theory of soft processes, with ln s_χ behaviour, which are a priori far more complicated to analyze within this framework. At present Regge models are best at providing rather a satisfactory description of the observed behaviour. They could eventually result from an underlying field theory.

- ROTH:

Don't jets formed from colored particles have to exchange further colored particles in order to result in colorless hadrons? If so, won't this extra exchange change the final jet structure?

- JACOB:

Jet fragmentation involves particles with widely different momenta. There is a close connection between the quantum numbers associated with the jets and those of the leading particles which originate from them. The slow particles are in the same region of phase space as the many other slow particles associated with other constituents. The proper balancing of quantum numbers occurs among these slow particles which softly (and strongly) interact among themselves.

- WOLF:

Which QCD diagrams are expected to dominate quark-quark scattering at very large p_T?

- JACOB:

At very large p_T perturbation theory should apply and the dominant diagram corresponds to single gluon exchange. Nevertheless, the observed hadrons may originate from different types of constituents (quarks or gluons). One has to consider several diagrams which occur at the same order in α_s namely quark-quark, quark-gluon and gluon-gluon scattering.

- SURSOCK:

A basic result of axiomatic S-Matrix theory is the one to one correspondence between pole of transition amplitude and asymptotic state. How can we reconcile this result with confinement on the one hand and the diagram where quark appears as a pole in the propagator on the other?

- JACOB:

Confinement does not hold within perturbative QCD where the present calculations are made. It should be stressed that what is calculated is not the transition probability to a particular final state defined in terms of quarks and gluons, but rather the energy and momentum flow corresponding to a specific solid angle. This is the physical quantity which is expected to remain the same after confinement has been imposed. Perturbative calculations are indeed conducted as if quark existed as a free particle.

- NEWMAN:

From our FERMILAB dimuon experiment using a pion beam, we see the mean p_T rise with μ pair mass and then flatten off as the $\langle p_T \rangle$ vs mass curve does for proton-induced dimuons. But $\langle p_T \rangle$ for pions flattens at a higher value:

$$\sim 1{\cdot}2 \text{ GeV compared to } \sim 1{\cdot}0 \text{ for proton}$$

Do you know why?

- JACOB:

No. As the lepton pair mass is fixed and also its center of momentum, the values of X_a and X_b, namely the fractions of the incident particle momenta taken by the active constituents have fixed values whether one deals with a pion or a proton incident particle.

The mean transverse -(primordial)- momentum of the constituents could be different in a pion and in a proton.

I do not see why one should get an extra transverse momentum of 0·2 GeV/c with an incident pion.

- GERHOLD:

Could you comment on the comparison between large p_T jets and low p_T jets, especially from the experimental point of view?

It is more natural to consider inclusive large p_T scattering, but we are not so sure whether this also is the case for low p_T scattering. What do you think about this?

- JACOB:

Firstly, there are actually many similarities. Indeed the $X \to 1$ behaviour of inclusive production is fruitfully analyzed in terms of parametrization obtained from the analysis of large p_T jet, namely, forms of the type $(1 - X)^m$.

This has been discussed by Ochs, and the CHLM and the UCLA groups at the ISR have both analyzed their data in such a way with much success. There are expected, (and observed), similarities between constituent fragmentation at wide angle and in the forward direction.

Secondly, there is much emphasis on inclusive large p_T scattering but exclusive scattering should also give very interesting clues about hadron structure. The differential cross section at large $|t|$, (fixed angle), is expected to behave as S^{-N}, (N = 10 for pp scattering), as opposed to the $S^{2(\alpha(t)-1)}$ Regge behaviour at low $|t|$. However, the elastic cross section is very small and data do not extend in S and $|t|$ as much as one would hope for.

If the emphasis is on inclusive production, it is first of all because of the relatively large cross section which one finds.

- ZICHICHI:

In order to understand $\alpha>1$, it is not necessary to have coherent effect. Rescattering is enough. This is supported by the fact that massive lepton pair production in nuclei saturates at $\alpha=1$. In fact, the leptons do not rescatter inside the nucleus. Low mass lepton pairs have $\alpha\simeq 2/3$, while for $m_{\ell\ell} \gg 3$ GeV, $\alpha\simeq 1$. The circumstance that high p_T "hadrons" have $\alpha>1$ can be understood in terms of the fact, for example, that a π is made of a $q\bar{q}$ pair, and the quark can rescatter inside the nucleus, thus producing $\alpha>1$. This matter was discussed last year at the Erice school. For details see my closing lecture 1977. (Status of the Subnuclear Whys.)

- MOLSON:

Does the radiation of hard gluons initiate jets?

- JACOB:

Yes, they should. One expects quantum number effects to be different (as many π^+'s as π^-'s), a larger multiplicity and a distribution more peaked at low X, than for quark jets. However, all this is still speculative.

The best place where to look for gluon jets is probably $e^+e^- \to gg\,\gamma$, the photon in the final state imposes a C= +1 state with an expected two gluons decay. Such gluon jets should be seen at PETRA.

- MOUZOURAKIS:

Is the value $<k_T> \sim 850$ MeV used by Feynman and Field for partons in their QCD model compatible with the average $<p_T> \sim$ 350 MeV found in typical hadronic collisions?

- JACOB:

The value of $<p_T>$ of Feynman and Field is a parameter which includes recoil associated with gluon production. It includes a primordial value which could be of the order of 350 MeV/c and an extra term associated with scaling violation.

- MOUZOURAKIS:

Are there data on the p_T dependence of the jet cross section when double arm triggering on the total p_T (= no trigger bias) is used?

Is this compatible with the Feynman prediction of p_T^{-6}?

- JACOB:

Data are being obtained in particular at Fermilab with the double arm calorimeter. It is, however, too early to draw any conclusion.

- WINTER:

What is the correlation between the flavor of the quark initiating a jet and the final state? What is the fraction of jets on the average in which we can identify the flavor correctly on the basis of the leading particle charge and flavor?

- JACOB:

There is a connection. The charge (quantum numbers) of the quark reacts on the mean charge observed at rather large z ($z > 0 \cdot 2$) There is no certain flavor identification except at $z>1$. Nevertheless, there certainly is a statistical identification. Charged distributions have been obtained by Feynman and Field using a Monte Carlo calculation. They are in agreement with observations for jets produced in neutrino scattering where the incident $\nu(\bar{\nu})$ picks the quark flavor.

- MONTGOMERY:

Another place to look for gluon jets is in muon-nucleon scattering where one normally expects the quark jet along the virtual photon direction. A gluon jet would be characterized by high p_T relative to the virtual photon and a particular ϕ dependence in the final state hadrons.

- MOUZOURAKIS:

Is there any connection between the jet/π^0 ratio and the γ/π^0 ratio? Could the non-observation of a large direct γ yield at high p_T constitute a severe blow to QCD?

- JACOB:

In the theory of quarks and gluons, photons must couple and one may say that the ratio of gluon to photon coupling goes as $\sqrt{\alpha_s/\alpha}$. Large p_T photons should be produced. The low yield is compensated for to some extent by the fact that only a fraction of the jet momentum is found on the leading π^0. This is the reason why one quotes values of the order of 5-10% for the γ/π^0 ratio at large p_T. There is some freedom about the value of the ratio. Nevertheless, one does expect large p_T photons at least at the few percent level. Such a conclusion does not require QCD. Nevertheless in QCD it would be very difficult to explain away the absence of large p_T photons.

- MOUZOURAKIS:

R806 (ISR) finds, by using the virtual photon continuum going into e^+e^- pairs in the p_T region 2 - 3 GeV and the mass region $200 < m_{e^+e^-} < 500$, a value of $\gamma/\pi^0 = 0\cdot55 \pm 0\cdot9$, i.e., consistent with zero, and also an equivalent upper limit of 2% with 90% confidence.

PRODUCTION AND DECAY OF CHARMED PARTICLES IN e^+e^- COLLISIONS

A. Barbaro-Galtieri

Lawrence Berkeley Laboratory
University of California
Berkeley, California 94720

ABSTRACT

This is a review of all the data available on production and decay of charmed particles in e^+e^- collisions. Production and decay of D^*, D, F mesons and charmed baryons are discussed. Comparisons with theoretical predictions, where available, are made.

I. INTRODUCTION

Since the discovery of J/ψ in 1974, a lot of data has been accumulated on charmed particles. In these lectures I will review all that we have learned so far on charmed particles from e^+e^- collisions.

Section II will give naive quark model predictions on production and decays of charmed particles. In particular, the expectation for R, the total hadronic cross section divided by $\sigma_{\mu\mu}$, and for vector meson production will be discussed. Section III will deal with evidence for D, D^*, F and charmed baryon production and their cross sections as well as with the resonant states above $c\bar{c}$ threshold. In Section IV all the results on masses of D and D^* and decay properties of the D mesons will be presented and compared with expectation.

II. NAIVE QUARK MODEL PREDICTIONS FOR CHARMED PARTICLES

There are many excellent review articles on theoretical predictions for production and decay of charmed particles. Here, while discussing the experimental data, we will only review the most basic expectations of the quark model. For a more complete treatment of the subject we refer the reader, for example, to Refs. 1-3.

A. Expected States with Charm Quantum Number

With the addition of a fourth quark, charm, a number of new states are expected by combining it with the old quarks.

For the pseudoscalar mesons we can construct all the states shown in the following 4×4 matrix:

$$\begin{array}{c} \\ u \\ d \\ s \\ c \end{array} \begin{array}{cccc} \bar{u} & \bar{d} & \bar{s} & \bar{c} \end{array}$$

$$\begin{pmatrix} \frac{\eta' + \eta + \sqrt{2}\pi^0}{2} & \pi^+ & K^+ & \bar{D}^0 \\ \pi^- & \frac{\eta' + \eta - \sqrt{2}\pi^0}{2} & K^0 & D^- \\ K^- & \bar{K}^0 & \frac{\eta' - \eta}{\sqrt{2}} & F^- \\ D^0 & D^+ & F^+ & \eta_c \end{pmatrix}$$

that is, we add at least seven more pseudoscalars to the ones of the old spectroscopy. The new states are

$$\begin{array}{ll} c\bar{u} & D^0 \\ c\bar{d} & D^+ \\ c\bar{s} & F^+ \\ c\bar{c} & \eta_c \end{array}$$

and the charge conjugates of the first three. We also expect the vector mesons to increase by the same number, and so on for the other nonets. In addition to the η_c we expect to find an η_c', because the first excitation level of the $c\bar{c}$ system is expected to be very close in mass to the ground level. The same consideration is valid for the ψ. As we will see in Section II.B.2, many of these excitation states have been found (see Table I).

As for the baryons (qqq states) the increase is even more dramatic:

u d s		→	u d s c
$(\frac{1}{2})^+$	octet	→	20-plet
$(\frac{3}{2})^+$	decuplet	→	20-plet

The $J^P = (\frac{1}{2})^+$ 20-plet is made of the old octet with $c = 0$, nine new states with $c = 1$, and three with $c = 2$. As for the 20-plet with $J^P = (\frac{3}{2})^+$ we expect, in addition to the old decuplet, six states with $c = 1$, three with $c = 2$, and one with $c = 3$.

In these lectures we will talk about the production and properties of D and D* states and discuss the experimental evidence for F meson and charmed baryon production in e^+e^- annihilation.

B. Hadron Production in e^+e^- Collisions

Figure 1 shows some schematic diagrams of the major phenomena occurring in e^+e^- collisions at the presently available energies (total energy $E < 10$ GeV). Diagrams (a) and (b) represent the most copious QED processes producing lepton pairs (the one photon processes); diagram (c) is one of the two possible diagrams for the annihilation into two photons; diagram (d) represents the so called "two photon" production, it only contributes a few percent of the hadronic cross section at these energies and will not be mentioned any further in these lectures. Finally, diagrams (e) and (f) represent the one photon hadron production and are the diagrams relevant to charmed particle production.

1. Total Hadronic Cross Section. The cross section for $q\bar{q}$ pair production can be calculated from QED, treating the quarks as point-like objects. In this case the cross section can be calculated in the same way as for a QED process involving leptons [diagram (b) in Fig. 1]. For μ pairs, at a total energy of $\sqrt{s}$, it is

$$\sigma_{\mu\mu} = \sigma(e^+e^- \to \mu^+\mu^-) = \frac{4\pi\alpha^2}{3s} = \frac{21.7}{E_b^2} \text{ nb} \qquad (1)$$

where E_b (in GeV) is the energy of one of the beams. The difference between diagrams (e) and (b) in Fig. 1 is that μ and q have different charges. The cross section for hadron production, assuming that the probability for $q\bar{q}$ pairs to go into hadrons is unity, is

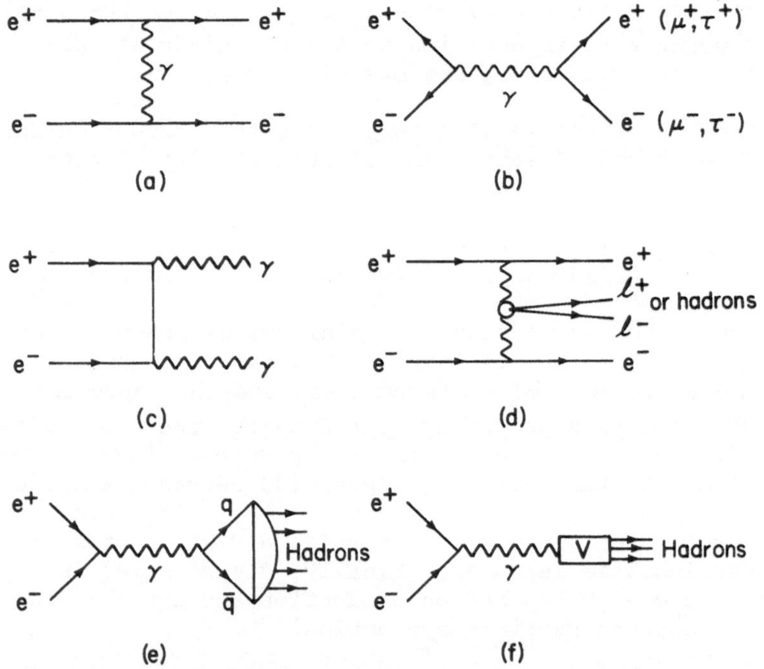

Fig. 1. Some schematic diagrams of processes taking place in e^+e^- collisions. The wavy lines represent the photon, the full lines are leptons and hadrons.

$$\sigma_h = \sigma(e^+e^- \to \text{hadrons}) = \sigma_{\mu\mu} \cdot 3 \cdot \sum_i Q_i^2 \qquad (2)$$

where the $\sum_i$ includes the charges of all the types of quarks involved and the factor 3 comes from quantum chromodynamics (QCD), that is, from the hypothesis that quarks come in three colors. It is customary to analyze the experimental data in terms of

$$R = \frac{\sigma_h}{\sigma_{\mu\mu}} = 3 \sum Q_i^2 \qquad (3)$$

in order to be able to detect the deviation from this simple hypothesis. Of course, the expression (3) is not expected to be valid where resonant processes, like the diagram of Fig. 1(f), occur. We will discuss that case in the next section.

At low energies only u, d, and s quarks are involved. Away from resonances we expect

$$R = 3\left(\frac{1}{9} + \frac{1}{9} + \frac{4}{9}\right) = 2 \text{ , below charm threshold} \qquad (4a)$$

and

$$R = 3\left(\frac{1}{9} + \frac{1}{9} + \frac{4}{9} + \frac{4}{9}\right) = \frac{10}{3} \text{ , above charm threshold .} \qquad (4b)$$

Figure 2 shows the value of R measured at SPEAR in the 3 to 8 GeV energy regions.[4,5] Below and above charm threshold for the hadronic component of R, we find approximately

$$R_a = 2.5 \text{ , below charm threshold}$$

$$R_b = 5.2 - 1 = 4.2 \text{ , above 6 GeV}$$

where we have subtracted one unit of R for heavy lepton production. In the region just above charm threshold resonance structures are present; we will discuss these in the next section.

In general, the total hadronic cross section behaves as expected for the onset of production of a new type of particle,

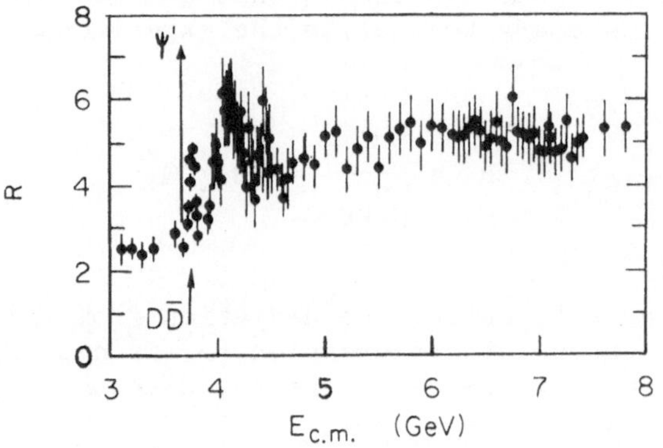

Fig. 2. The ratio of the total hadronic cross section to the μ pair production cross section, $R = \sigma_h/\sigma_{\mu\mu}$, as a function of the center of mass energy. The plot is taken from Ref. 4, the data[5] at the ψ(3772) have been added to it.

as a simple quark counting model would predict. To check if particles with a new quantum number are produced and that, in fact, they correspond to the expectation from a charmed quark, we have to go into more details and study the different hadronic final states. This is the subject of these lectures.

The experimental values of R_a and R_b deviate from the naive quark model expectation [Eqs. (4a) and (4b)]. In QCD one would expect corrections due to quark-gluon interactions. Away from thresholds and resonant structures these corrections, in an asymptotically free theory, have been estimated to give[6]

$$R = 3 \sum Q_i^2 \left(1 + \frac{\alpha_S(E)}{\pi}\right) \qquad (5)$$

where $\alpha_S(E)$, defined in analogy to α, the fine structure constant, is the running coupling constant of asymptotically free theories. In a particular $SU(4) \times SU(3)$ model with only four quarks, α_S has the form[7]

$$\alpha_S(E) = \frac{\alpha_S(E_o)}{1 + \frac{25}{12\pi}\,\alpha_S(E_o)\,\ln\frac{E}{E_o}} \quad . \tag{6}$$

For a detailed review of these ideas, see Appelquist et al.[3] The experimental value of R differs from the expected value (Eq. 4) by about 25% above charm threshold, which would imply a value of $\alpha_S \sim 0.7$ in the 5-8 GeV region. This value of α_S is in gross disagreement with the value of $\alpha_S \sim 0.2$ evaluated at the ψ, as we will see in the next section. However, as discussed in Section III.D, the systematic errors on the measured R are quite large.

2. Vector Meson Formation. Close to a resonance in the $q\bar{q}$ system, the cross section (Eq. 2) has to be modified to account for the resonant matrix element. The cross section for vector meson production (diagram (f) in Fig. 1) and decay into the final state i, takes the form

$$\sigma(e^+e^- \to V \to i) = \frac{\pi(2J+1)}{s}\;\frac{\Gamma_e\,\Gamma_i}{(E-M)^2 + \Gamma^2/4} \tag{7}$$

where M is the mass of the vector meson, Γ its total width, Γ_e and Γ_i the partial widths into electron pairs (the incoming channel) and into the ith final state respectively. The factors of α and Q are now included in the-partial widths, as we will see shortly. For vector mesons, that of course couple to the e^+e^- system through the photon [Fig. 1(f)], J = 1, so at resonance Eq. (7) becomes

$$(\sigma_0)_i^{max} = \frac{12\pi}{M^2}\;\frac{\Gamma_e\,\Gamma_i}{\Gamma^2} \tag{8}$$

the subscript σ_0 has been added to remind the reader that this expression is valid only if radiative effects were not present. In order to compare with the data, radiative corrections have to be applied either to the formula or to the data. These corrections have been discussed in the literature[8] and summarized recently by

Jackson and Scharre.[9] In addition to radiative corrections, often the uncertainty in the energy of the e^+e^- beams contributes to alter the resonant shape (Eq. 7) and has to be taken into account, especially when Γ is small compared with the beam resolution. Techniques used to take into account these effects are extensively described in Ref. 9. Because of these corrections the relevant measurement is the area of the resonance curve instead of the height. For the cross section in the ith final state we have:

$$\int \sigma_i \, dE = (\mathrm{Area})_o^i = \pi \frac{\Gamma}{2} (\sigma_0)_i^{max}$$

so it is

$$\frac{\Gamma_e \Gamma_i}{\Gamma} = \frac{M^2}{6\pi^2} \int \sigma_i \, dE \quad . \qquad (9)$$

Through this expression and similar ones for the different final states we can determine experimentally Γ_e, Γ_i, and the total width $\Gamma = \sum_i \Gamma_i$.

The vector mesons produced in e^+e^- interactions in the 2-8 GeV energy region appear as peaks in a plot of R (Fig. 2). In Table I we tabulate some parameters for the vector mesons of the new spectroscopy (states of $c\bar{c}$) and the ρ, ω, and ϕ of the old spectroscopy. For completeness we have added the Υ and Υ' recently observed at high energies,[10-12] they are $b\bar{b}$ bound states (b is the bottom quark of charge $Q = -1/3$). For a discussion of the relation between Γ_e and the charge of the quark responsible for the Υ and Υ' as well as for a discussion of different choices of potential to describe $q\bar{q}$ systems, see Quigg and Rosner.[13] The states of $c\bar{c}$ above ψ and ψ' are above threshold for charmed particle production and will be discussed in Section III.A.2. Note that

$$M_{\psi'} - M_{\psi} = 589 \pm 1 \text{ MeV}$$

$$M_{\Upsilon'} - M_{\Upsilon} = 556 \pm 3 \text{ MeV}$$

as quoted in Ref. 16 and 20 respectively. These mass differences are relevant to the choice of a potential to describe the $q\bar{q}$ system.[3,13]

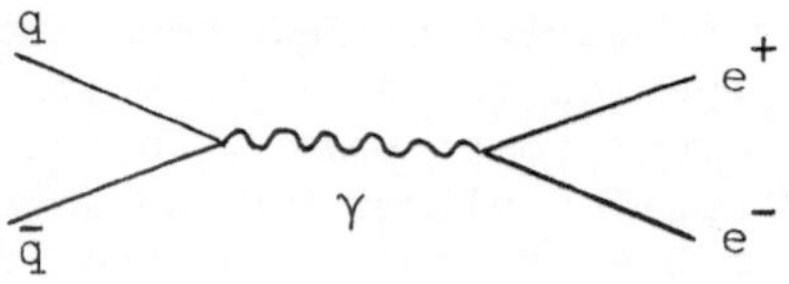

The theoretical expressions for Γ_e and the other partial widths depend upon the model used to calculate them. In general for a $q\bar{q}$ bound state going into e^+e^- as in the diagram at right, the

TABLE I. Resonance parameters for vector mesons.[a] Γ is the total width, Γ_e is the partial width to electron pairs, and B_e is the branching fraction to electron pairs.

State	Mass (MeV)	Γ (MeV)	Γ_e (keV)	B_e	Ref.
ρ	776±3	155±3	6.7 ±0.8	$(4.3\pm0.5)10^{-5}$	14
ω	782.6±0.3	10.1±0.3	0.76±0.17	$(7.6\pm1.7)10^{-5}$	14
ϕ	1019.6±0.2	4.1±0.2	1.31±0.10	$(31\pm1)10^{-5}$	14
ψ	3095±4	0.069±0.015	4.8 ±0.6	$(69\pm9)10^{-3}$	SLAC-LBL[15]
ψ'	3684±5	0.228±0.056	2.1 ±0.3	$(9.3\pm1.6)10^{-3}$	SLAC-LBL[16]
ψ''	3772±6	28 ±5	0.35±0.09	$(1.2\pm0.3)10^{-5}$	Rapidis et al.[5]
	3770±6	24 ±5	0.18±0.06	$(0.7\pm0.2)10^{-5}$	DELCO[17]
4.04[b]	4040±10	52 ±10	0.75±0.10	$(1.4\pm0.4)10^{-5}$	DASP[18]
4.16[b]	4159±20	78 ±10	0.77±0.20	$(0.9\pm0.3)10^{-5}$	DASP[18]
4.41	4414±7	33 ±10	0.44±0.14	$(1.3\pm0.3)10^{-5}$	SLAC-LBL[19]
Υ	9460±10	~0.05	1.2 ±0.2	$(2.6\pm1.4)10^{-2}$	20[c]
Υ'	10016±10	--	0.33±0.10	--	20[c]

[a] Other states have been reported between the ϕ and the ψ by experiments at Frascati and Orsay; we do not include them here.

[b] The SLAC-LBL and DELCO data do not separate this region into two states (see Figs. 2 and 14).

[c] Values for Υ and Υ' are averages as quoted by Flügge.[20]

leptonic width has been calculated to be[21]

$$\Gamma_e = \frac{16\pi\ \alpha^2\ Q^2}{M^2}\ |\psi(0)|^2 \qquad (10)$$

where M is the mass of the vector meson, Q the charge of the quark and $|\psi(0)|^2$ is the square of the wave function at the origin and it depends upon the potential chosen to represent the $q\bar{q}$ interaction. Extensive work has been done to understand the $c\bar{c}$ interactions. Appelquist and Politzer[22] conjectured the existence of $c\bar{c}$ bound states, that is, states below the threshold for producing particles with charm, just about at the same time that the ψ was discovered.[23]

Eichten et al[24] have used a short range Coulomb potential and a long range linear potential to describe $c\bar{c}$ interactions and have been successful in predicting some of the features of the charmonium spectrum, including the existence of $\psi''(3772)$, later discovered.[5] The experimental data on charmonium spectroscopy are discussed in Prof. Wolf's lectures and will not be discussed here. For a review of the theoretical work on charmonium see Refs. 1, 3 and 25, where expressions for Γ_h can be found. We will only discuss here briefly Γ_h for the ψ, as derived by Appelquist and Politzer.[22] The $\psi \to h$ process (for $\psi \to 3\pi$ a schematic diagram is shown in Fig. 3c) can take place through three-gluon exchange. In analogy with the three-gamma process in positronium, by substituting α^3 with $5/18\ \alpha_S^3$, they derived the expression

$$\Gamma(\psi \to h) = \frac{40}{81\pi}\ (\pi^2 - 9)\ \frac{\alpha_S^3}{M^2}\ |R(0)|^2 \qquad (11)$$

where $|R(0)|^2 = 4\pi\ |\psi(0)|^2$. By measuring Γ_e and Γ_h (see Table I) and using Eqs. (10) and (11), one can calculate the value for α_S at the ψ to be

$$\alpha_S(3.1) = 0.19 \quad . \qquad (12)$$

We comment here that charmonium models using a more sophisticated potential than that of Ref. 24 require $\alpha_S \sim 0.4 - 0.5$ to fit the charmonium spectrum. We refer the reader to Ref. 3 and 81 for discussion of this point.

One final observation on the parameters of Table I is with regard to total widths. The ψ and ψ' that are $c\bar{c}$ bound states, and Υ and Υ' that are $b\bar{b}$ bound states, have a very narrow total width. This is expected by the <u>OZI rule</u>,[26] found empirically years ago to explain the observed rates in meson decays. This rule says that transitions described by diagrams where the initial quarks annihilate each other and do not appear in the final state are suppressed. The relevant diagrams for ϕ, ψ and ψ'' decays are shown in Fig. 3. Diagrams (a) and (c) are forbidden by the OZI rule, diagrams (b) and (d) are allowed. For the ϕ decay, the ratio of rates for diagrams (b) and (a) is not very large due to kinematic factors. There are various possible dynamical approaches to the theoretical understanding of the OZI rule, but a quantitative formulation of this rule has not been achieved yet. See Jackson's review of this point.[1] In any case, the OZI rule tells us that above threshold for $D\bar{D}$ production we should expect Γ_{tot} to be larger than those observed for ψ and ψ', which is what we see in Table I.

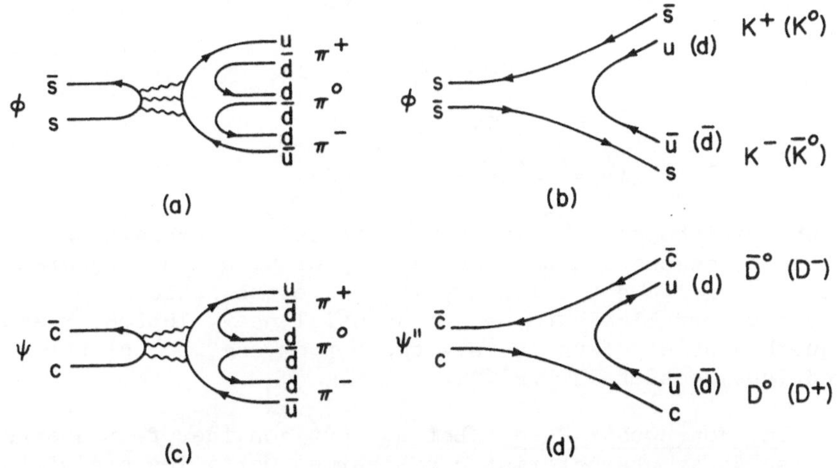

Fig. 3. Illustration of the OZI rule. Diagrams (a) and (c) are forbidden, (b) and (d) are allowed. The decay rates for the four diagrams are $\Gamma_a = 0.6$ MeV, $\Gamma_b = 3.4$ MeV, $\Gamma_c = 0.7$ keV, and $\Gamma_d = 28$ MeV. The wavy lines represent gluons.

C. Expected Decays of Charmed Particles

The charmed quark, c, was introduced by Glashow, Iliopoulos and Maiani (GIM)[27] to explain the absence of strangeness changing neutral currents; that is, the nonobservation of decays like $K^0 \to \mu^+\mu^-$ (now measured to have a branching fraction of 9×10^{-9}) and $K^- \to \pi^-\nu\bar{\nu}$. The presence of a fourth quark would produce the cancellation of the $\Delta S = 1$ piece of the neutral current.

The expected decays of charmed particles can be derived using a conventional Weinberg and Salam theory[28] with left-handed weak isodoublets and right-handed weak isosinglets, along with the GIM quark structure.[27] One possible model with six leptons and six quarks (see for example the review of Harari on a variety of possible models[2]) includes the following isodoublets:

$$\begin{pmatrix} \nu_e \\ e^- \end{pmatrix}_L \begin{pmatrix} \nu_\mu \\ \mu^- \end{pmatrix}_L \begin{pmatrix} \nu_\tau \\ \tau^- \end{pmatrix}_L \qquad \begin{pmatrix} u \\ d' \end{pmatrix}_L \begin{pmatrix} c \\ s' \end{pmatrix}_L \begin{pmatrix} t \\ b \end{pmatrix}_L \tag{13}$$

where

$$\begin{aligned} d' &= d\cos\theta + s\sin\theta \\ s' &= -d\sin\theta + s\cos\theta \end{aligned} \tag{14}$$

with θ the Cabibbo angle[29] introduced in 1963 to explain the decay properties of mesons and baryons made up of u, d and s quarks. Here we have included the τ heavy lepton, established as a new particle and very likely to be a sequential heavy lepton,[30] and a new quark doublet of which only the bottom one, b, has been observed (see Section II.B.2).

1. <u>The Four Quark Case</u>. Let us just consider four quarks first. The decay characteristic of charmed particles are dictated from whatever charged current we can form in this framework. The charged current has the form:

$$J_h^c = \bar{u}d' + \bar{c}s' = \bar{u}(d\cos\theta + s\sin\theta) + \bar{c}(s\cos\theta - d\sin\theta)$$

which can also be written as

$$J_h^c = \cos\theta\,(\bar{u}d + \bar{c}s) + \sin\theta\,(\bar{u}s - \bar{c}d) \tag{15}$$

where the quark symbols have been used instead of the complete expression for the current, i.e., $\bar{u}d = \bar{u}\gamma_\mu(1-\gamma_5)d$. The values of the coefficients are $\cos\theta = 0.974$ and $\sin\theta = 0.227$, and for this reason the transitions that can be done with the first term of the current are called "Cabibbo favored" and the others are called "Cabibbo suppressed."

The "favored" decays of the c quark are the transitions with

$$c \to s \,, \quad \text{with} \quad \Delta C = \Delta S = \Delta Q \quad , \quad \Delta I = 0 \quad , \tag{16}$$

the "Cabibbo suppressed" are the transitions

$$c \to d \,, \quad \text{with} \quad \Delta C = \Delta Q \,, \quad \Delta S = 0 \,, \quad \Delta I = \tfrac{1}{2} \quad . \tag{17}$$

Some diagrams for two-body Cabibbo-favored and Cabibbo-suppressed modes of the charmed mesons are shown in Fig. 4. Notice that in order to have a "Cabibbo favored" decay of these mesons, the c quark has to decay according to (16) and the W must have the favored transition $W \to u\bar{d}$ as in Eq. (15). From these diagrams (just on the basis of diagram counting) we notice that:

a. Diagrams of type (b) are nine times more likely than diagrams like (d), because in (b) the $u\bar{d}$ pair can have any of three colors, whereas in (d) the $u\bar{d}$ pair must have the same color as c and $\bar{s}$.

b. Comparing (a) and (c) one can easily derive that (a) is a factor 3 larger because of the three colors of the quarks. Here we comment that on the basis of diagram counting we would expect

$$\begin{aligned} D^0 &\to K^- + \text{hadrons} && 60\% \\ D^0 &\to K^- + e^+ + \ldots && 20\% \\ D^0 &\to K^- + \mu^+ + \ldots && 20\% \end{aligned} \tag{18}$$

We will come back to this point in Section 3 below.

c. The rates for the "Cabibbo suppressed" decays are smaller than the "favored" ones by at least a $\tan^2\theta = 0.55$ factor. Of course, phase space factors are to be properly taken into account.

In summary, just on the basis of the predictions of Eqs. (15) (16) and (17), we would expect that charmed mesons prefer the following hadronic decays:

$$\begin{aligned} D^0 &\to K^-\pi^+, \ \bar{K}^0\pi^0, \ \bar{K}^0\pi^+\pi^-, \ \bar{K}^0\eta, \ \bar{K}^0\eta', \ \text{etc.} \\ D^+ &\to \bar{K}^0\pi^+, \ K^-\pi^+\pi^+, \ \text{etc.} \\ F^+ &\to \bar{K}^0K^+, \ K^+K^-\pi^+, \ \eta\pi^+, \ \eta'\pi^+, \ \text{etc.} \end{aligned}$$

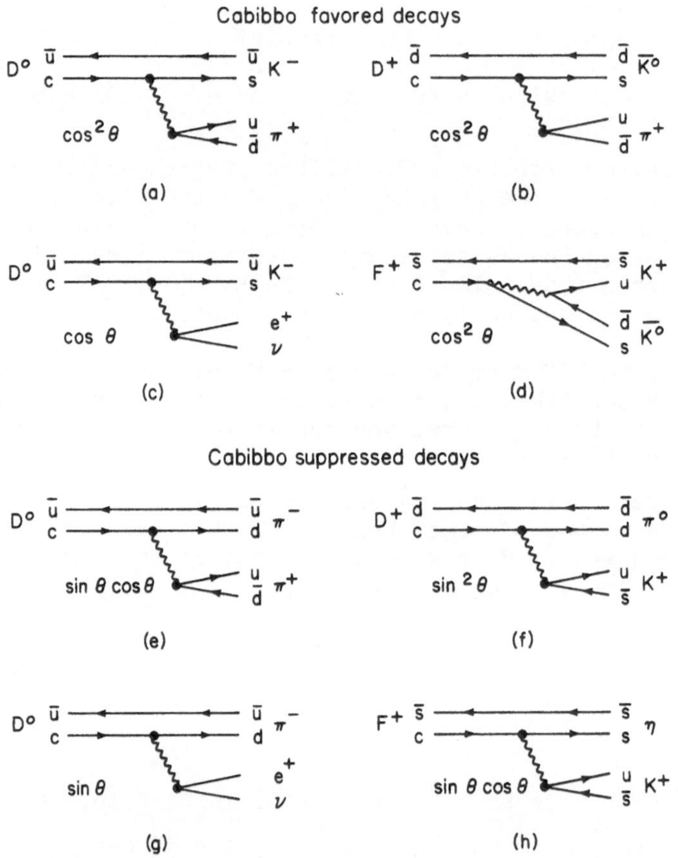

Fig. 4. Schematic diagrams illustrating charmed particle decays. The wavy lines represent the W boson, solid lines are hadrons or leptons. Diagrams (a)-(d) are for Cabibbo favored decays, (e)-(h) are for Cabibbo suppressed decays. The amplitude for each diagram is proportional to cosθ or sinθ factors as shown.

and the following semileptonic decays:

$$D^0 \to K^- e^+ \nu, \quad K^{*-} e^+ \nu, \quad K^- \mu^+ \nu, \quad \text{etc.}$$

$$D^+ \to \bar{K}^0 e^+ \nu, \quad \bar{K}^{*0} e^+ \nu, \quad \text{etc.}$$

$$F^+ \to \eta e^+ \nu, \quad \eta' e^+ \nu, \quad \text{etc.}$$

2. <u>The Six Quark Case</u>. If we introduce mixing of d, s and b quarks we need three angles and a phase to describe such mixing.[31] This means that instead of a 2×2 matrix as in Eq. (14), we have a 3×3 unitary matrix.

$$J_h^c = \bar{u}\ \bar{c}\ \bar{t}\ \gamma_\mu (1 - \gamma_5) U \begin{pmatrix} d \\ s \\ b \end{pmatrix}$$

$$U = \begin{pmatrix} C_1 & -S_1 C_3 & -S_1 S_3 \\ S_1 C_2 & C_1 C_2 C_3 - S_2 S_3 e^{i\delta} & C_1 C_2 S_3 + S_2 C_3 e^{i\delta} \\ S_1 S_2 & C_1 S_2 C_3 + C_2 S_3 e^{i\delta} & C_1 S_2 S_3 - C_2 C_3 e^{i\delta} \end{pmatrix} \quad (19)$$

where $C_i = \cos\theta_i$, $i = 1,2,3$
$S_i = \sin\theta_i$.

In a graphic presentation Fig. 5 shows how the transition from one quark to another can be calculated for the two different cases: d-s and d-s-b mixing.

In Eq. (19), θ_1 is the original Cabibbo angle. As for the others, one can try to calculate the upper limits for θ_2 and θ_3 by using some decay modes or other phenomena involving old quarks (see for example, Harari's review[2]).

a. From $u \to d$, in $n \to p e^- \nu$ one finds
$\cos\theta_1 = 0.974 \pm 0.002$, $\theta_1 = (13.2 \pm 0.5)°$, $(S_1 = 0.227)$.

b. From $u \to s$, in $\Lambda \to p e^- \nu$ or $K \to \pi e \nu$, one finds
$\sin\theta_1 \cos\theta_3 = 0.229 \pm 0.003$,
which gives
$\cos\theta_3 > 0.96$ $\quad \theta_3 \leqslant 16°$, $\quad (S_3 < 0.28)$.

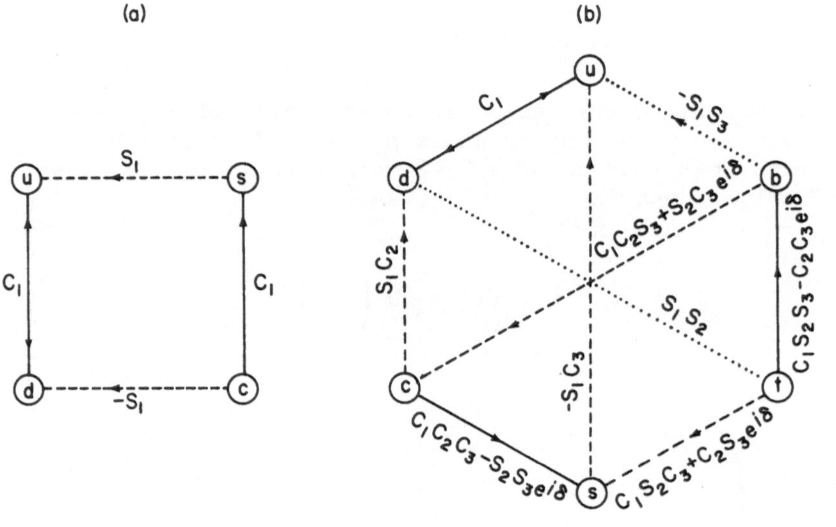

Fig. 5. Graphic representation of transitions between quarks. Solid lines represent the most copious transitions (only $\cos\theta_i$ factors), dashed lines are for transitions with a $\sin\theta_i$ factor, and the dotted lines have two $\sin\theta_i$ factors. (a) is for the four quark case, (b) is for the six quark case.

c. From K_S-K_L mass difference,[32] we get the limit (for m_c = 1.55 GeV, m_t > 5 GeV),

$$\tan^2\theta_2 < m_c/m_t \qquad \text{or} \qquad \theta_2 < 30° \qquad (S_2 < 0.5) .$$

d. Not very much can be said about δ, which is related to CP violation.[33] The only statement one can make at this time from CP violation parameters in $K^0 \to 2\pi$, is that

$$\sin\delta > 5\times10^{-3} \qquad \text{or} \qquad \delta > 0.3° .$$

In conclusion, since θ_2, θ_3 are small, Eq. (19) tells us that the basic content of (15) ($c \to s$ is the favored decay and $c \to d$ is the suppressed decay) still holds for the charmed quark.

3. Nonleptonic Enhancement. With the assumption that the W couples equally to quark or lepton pairs, we would expect, as mentioned in section 1, that the inclusive semi-leptonic decays into e and μ would be about 40%, as from Eq. (18). Does this diagram counting rule work for strange particle decays? The answer is no.

The nonleptonic decay rates of strange particles, both mesons and baryons, are found to be larger than expected. Specifically, the amplitude for the $\Delta I = 1/2$ part of the nonleptonic interaction is enhanced by about a factor of 20 over the $\Delta I = 3/2$ part of the nonleptonic amplitude and the semileptonic amplitude. For example:

$$(K_S \to \pi^+\pi^-) = 0.77 \times 10^{10}\ \text{sec}^{-1}, \quad \Delta I = 1/2,\ 3/2$$

$$(K^+ \to \pi^+\pi^0) = 0.17 \times 10^{8}\ \text{sec}^{-1}, \quad \Delta I = 3/2$$

the $K^+ \to \pi^+\pi^0$ decay rate is much smaller than the first one, as the $\Delta I = 1/2$ enhancement would predict. For a discussion of the experimental evidence for the $\Delta I = 1/2$ rule in K meson and hyperon decays, see the Appendices of the Particle Data Group compilation.[14] Since the $\Delta I = 1/2$ part of the transition appears alone in the octet part of the Hamiltonian, this experimental observation has been called "octet enhancement." As for a dynamical mechanism that would produce such an enhancement, it has been suggested that it could arise from the strong interactions among the constituents at short distance.[34]

What do we expect for charmed particles? Although the mechanism that produces octet enhancement is not fully understood, the same phenomenon has been extended to charmed particle decays by various authors.[35,36] Einhorn and Quigg[37] worked out the necessary group theory for extending the SU(3) phenomenon to SU(4) and concluded that octet enhancement results in 20-plet enhancement for the four-quark case. For SU(3) the weak Hamiltonian reduces to the following representations

$$H_W = \underline{1} \oplus \underline{8} \oplus \underline{27} \tag{20}$$

Here since the $\underline{8}$ representation contains only the $\Delta I = 1/2$ transition, the octet part is found to be the most important one. For the SU(4) case the weak Hamiltonian reduces[37] to

$$H_W = \underline{20} \oplus \underline{84} \tag{21}$$

and by further dividing these two representations into their SU(3) components they conclude that the $\underline{20}$ part will be enhanced.

How large is this enhancement? The predictions differ; for

example, the decay $D^+ \to \bar{K}^0\pi^+$ would be completely forbidden for some authors,[37] allowed for others. Ellis et al[38] predict $\Gamma(D^+ \to \bar{K}^0\pi^+) \sim \Gamma(D^0 \to K^-\pi^+)$. As for the semileptonic decays the predictions vary also, the $D \to e$ inclusive rate could be as low as 3% of the total rate according to Ellis et al.[38] So the extreme case gives

$$\begin{aligned} D^0 &\to K^- + \text{hadrons} && 94\% \\ D^0 &\to K^- + e^+ + \ldots && 3\% \\ D^0 &\to K^- + \mu^+ + \ldots && 3\% \end{aligned} \qquad (22)$$

This has to be compared with the quark diagrams counting case (Eq. 18). It is interesting to compare the experimental results with these two cases. It is clear that only the data will shed some light on the magnitude of the nonleptonic enhancement. We will return to this point in Section IV.D when we will discuss our experimental results.

III. PRODUCTION OF CHARMED PARTICLES

Most of the results discussed in these lectures have been obtained with the SLAC-LBL magnetic detector Mark I. I will refer to two different experiments done with this detector: the earlier SP17 experiment with the detector configuration described in Ref. 39 and the more recent SP26 experiment with a lead-glass array to detect γ and electrons.[40] This last experiment will be called the LGW. The detector in its last configuration is shown in Fig. 6. For details, the reader is referred to Refs. 39 and 40.

In this section we will discuss the evidence for production of charmed mesons and baryons in e^+e^- collisions, as well as the cross sections for production of these states. We will also relate these production rates to the total hadronic cross section.

A. D and D* Mesons

1. Evidence for D and D* Production. The total hadronic cross section normalized to $\sigma_{\mu\mu}$, R in Fig. 2, exhibits a rise with some resonance-like peaks just above the ψ' resonance. The data in the energy region $3.6 < E < 4.6$ GeV are shown in more detail in Fig. 7, taken from the paper of Rapidis et al.[5] The data in this energy region have provided all of the information we have on D and D* mesons: the D and D* have been discovered in the 4.03 and 4.4 GeV regions. Precise masses and D branching ratios have been measured at the 3.772 GeV resonance, the most recently discovered state of $c\bar{c}$.[5]

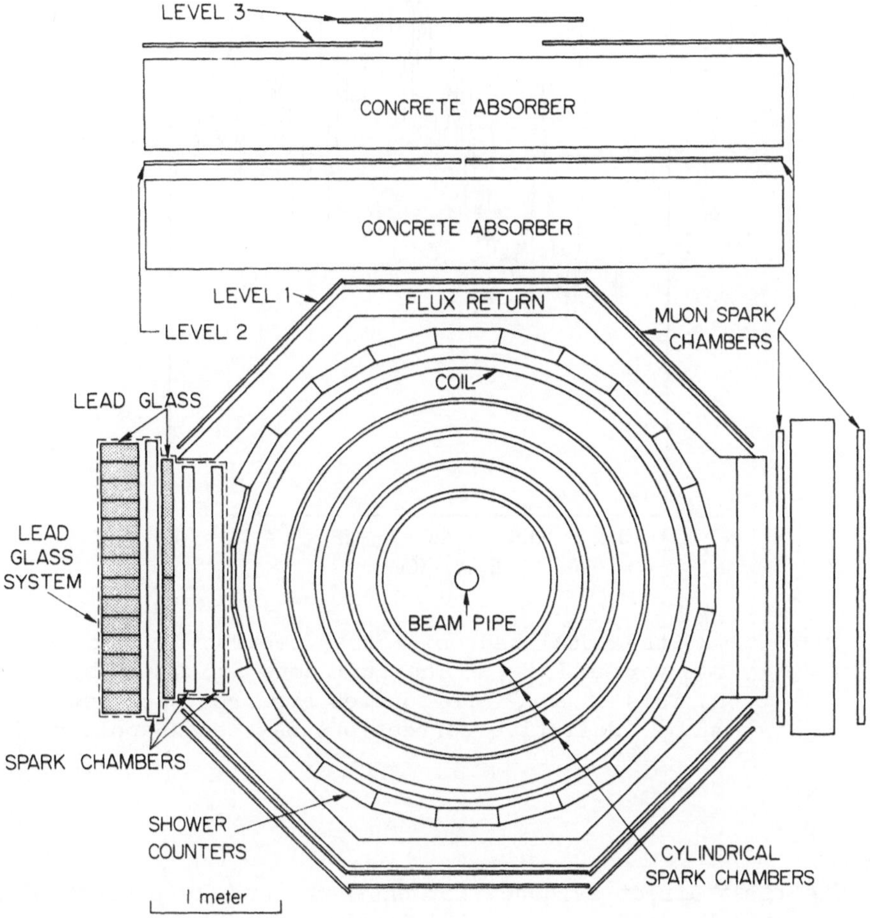

Fig. 6. The SPEAR magnetic detector[39] as seen looking along the beam line. The proportional chambers around the beam pipe and the trigger counters are not shown. The lead glass system (LGW)[40] is shown on the left side of the figure.

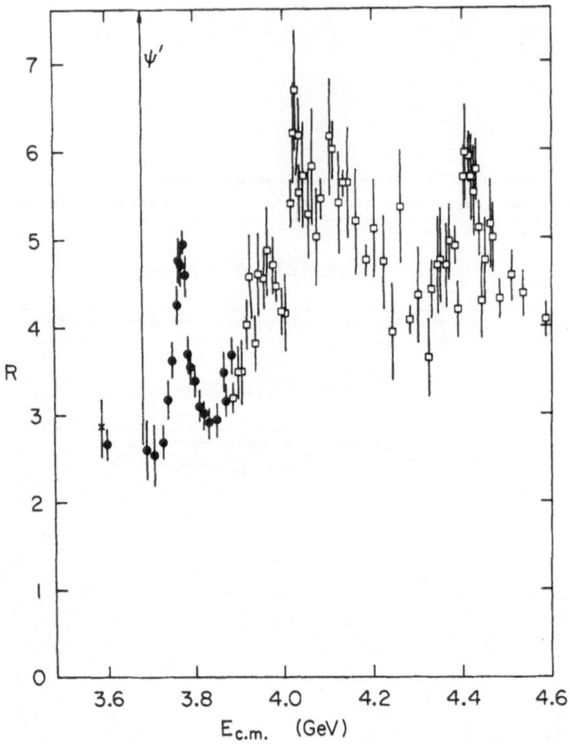

Fig. 7. R, the total hadronic cross section divided by $\sigma_{\mu\mu}$ versus $E_{c.m.}$. The full dots are the data of Rapidis et al.,[5] the squares are from Siegrist et al.[19] Radiative corrections have been applied.

The first direct evidence for charmed particle production has been reported by Goldhaber et al[41] in the invariant mass of the $K^{\mp}\pi^{\pm}$ system (D^{0}, $\bar{D}^{0}$) and shortly after by Peruzzi et al[42] in the invariant mass of the $K^{\mp}\pi^{\pm}\pi^{\pm}$ system ($D^{\pm}$). These results come from the SP17 experiment and are summarized in Fig. 8, taken from Piccolo et al.[43] The charged K's are identified by the time-of-flight measurement for a 1.5 - 2.0 meter flight path in the magnetic detector (the resolution is σ = 0.4 nsec). The neutral kaons are identified by measurement of the dipion mass and by requiring consistency of the dipion vertex with the direction of the kaon momentum.[44]

The following reactions and their charge conjugates are observed:

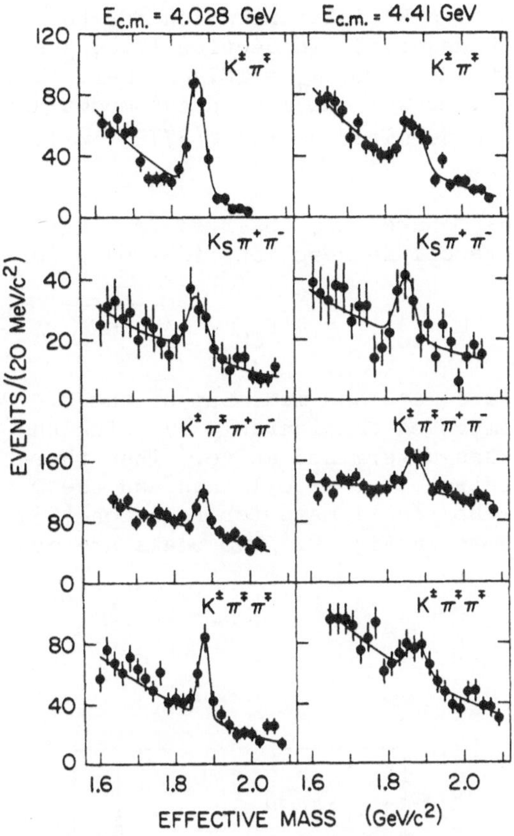

Fig. 8. Invariant mass distribution for various channels showing D^0 and D^+ signals. The data is that of Piccolo et al.[43] at 4.03 and 4.41 GeV total e^+e^- energy.

$$e^+e^- \rightarrow D^0 + \text{recoil}, \quad \text{with} \quad D^0 \rightarrow K^-\pi^+ \tag{23}$$

$$D^0 \rightarrow K_S\pi^+\pi^- \tag{24}$$

$$D^0 \rightarrow K^-\pi^+\pi^+\pi^- \tag{25}$$

$$e^+e^- \rightarrow D^+ + \text{recoil}, \quad \text{with} \quad D^+ \rightarrow K^-\pi^+\pi^+ \tag{26}$$

The widths of the peaks observed are consistent with the experimental mass resolution of 20 MeV for the $K\pi$ system and 15 MeV for the $K^-\pi^+\pi^+$ system. In addition, the $K^-\pi^+\pi^+$ final state is exotic because the overall charge of the state (+) has the opposite sign from the strangeness of the state (-). These facts point clearly toward exclusion of a K* interpretation for this state. As a result of a fit[45] described in Section IV.C, the masses are found to be $M_o = 1863 \pm 3$ MeV and $M_+ = 1874 \pm 5$ MeV. The most recent mass measurements, as well as more decay modes and absolute branching ratios determined at the ψ(3772), will be discussed in Section IV.

The D^{*o} and D^{*+} are also observed in these data as peaks in the mass of the recoil in reactions (23-26). For these reactions,

$$M^2_{recoil} = (E_{CM} - \sqrt{p^2 + M^2})^2 - p^2 \qquad (27)$$

where M and p are mass and momentum of the D. The resolution of M_{recoil} can be improved considerably by using the fixed values for the D^+ or D^o masses determined above. Then there is a one to one correspondence between the recoil mass and the D momentum. The distribution of the recoil mass for reaction (23) at 4.03 GeV total E_{CM} energy is shown in Fig. 9. Two peaks are observed, the lower

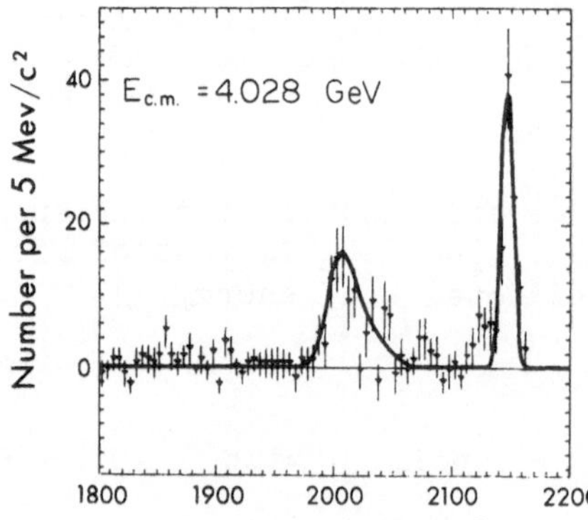

Fig. 9. The mass of the system recoiling against the D^o in e^+e^- interactions at 4.03 GeV. Data are from Goldhaber et al.[45]

one at a mass of 2.01 GeV, the higher one at 2.15 GeV. They are interpreted as due respectively to the reactions

$$e^+e^- \rightarrow D^o \bar{D}^{*o} \text{ (and complex conjugate)} \qquad (28)$$

$$e^+e^- \rightarrow D^{*o} \bar{D}^{*o} \qquad (29)$$

The Q of the first reaction (28) is 159 MeV, which results in a momentum for the directly produced D^o, p_D, of ~580 MeV/c. Because of this large value the uncertainty on p_D contributes considerably to the uncertainty on the recoil mass (see Eq. 27). The widening of the peak is also due to the fact that the observed D^o can be from (a) direct D^o produced, or (b) D^o from the decays $D^{*o} \rightarrow D^o\gamma$ or $D^{*o} \rightarrow D^o\pi^o$, or (c) D^o from the reaction $e^+e^- \rightarrow D^+D^{*-}$ with $D^{*-} \rightarrow D^o\pi^-$. For the second peak, the $D^{*o}\bar{D}^{*o}$ reaction has a small Q (Q = 16 MeV), therefore a smaller p_D that results in a narrower peak. More details on fitting these data to get D^* masses and branching ratios will be given in Section IV.

Before leaving the subject I want to remind you that the only other report of D production for which a peak is seen in an invariant mass distribution is by Baltay et al.[46] This is a Fermilab neutrino experiment in a bubble chamber filled with a heavy mix of hydrogen and neon. The channel observed is $D^o \rightarrow K_S^o\pi^+\pi^-$.

2. Associated Production of D and D*. Above threshold for the reactions

$$e^+e^- \rightarrow D^o\bar{D}^o \qquad (30)$$

$$e^+e^- \rightarrow D^+D^- \qquad (31)$$

associated production of a pair of charmed particles can occur. The $c\bar{c}$ model of Eichten et al[24] predicts resonant states of $c\bar{c}$ above charm threshold, besides the charmonium levels below it. Decays of these states into a pair of charmed particles are allowed by the OZI rule. Therefore, we expect their total widths to be much larger than those of the ψ and ψ' (see Section II.B.2, Table I and Fig. 7).

Eichten et al[47] have extended their $c\bar{c}$ model to these types of decays, and as we will see, can predict some of their properties.

The $\psi(3772)$, ψ'', is the first of such resonances above threshold. Figure 10 shows the detailed shape of R in this energy region as measured by the LGW experiment.[5] Figure 10a shows the raw data, whereas Fig. 10b shows the data after subtraction of the ψ' rapidly descending tail. This is due to the Gaussian resolution of the beam of 1 MeV, much wider than the ψ' width (see Table I) and to radiative effects. Both these corrections have been applied

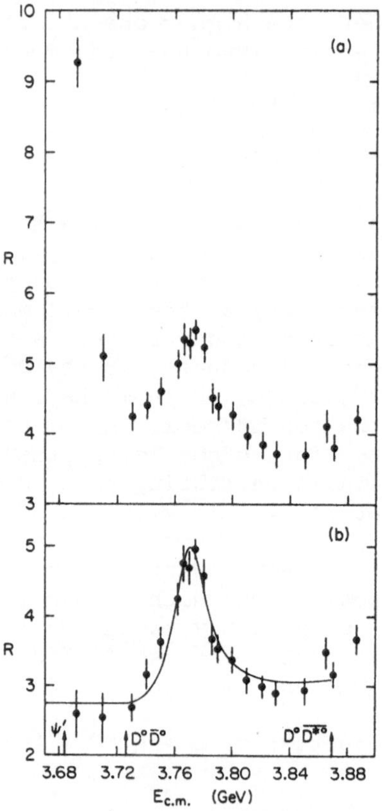

Fig. 10. R versus $E_{c.m.}$ at the ψ(3772). (a) before, and (b) after radiative corrections and ψ' tail subtraction. The curve is a p-wave Breit-Wigner fit to the data.

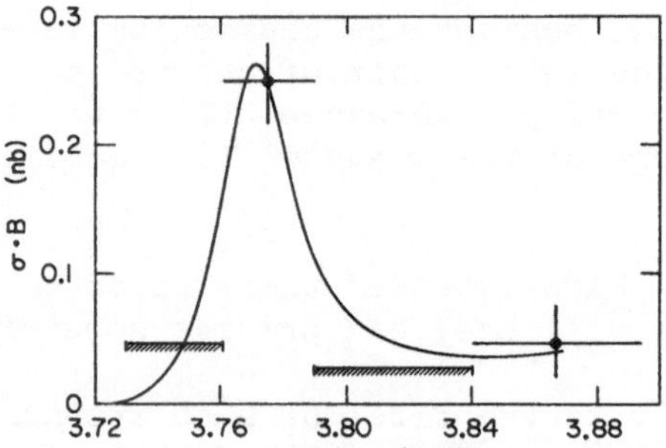

Fig. 11. Data[53] of σ • B versus $E_{c.m.}$ for D^0 and $\bar{D}^0$ decays into $K^{\mp}\pi^{\pm}$. The cross-hatched bars represent upper limits. The curve is the same one shown in Fig. 10 normalized to the 3.77 GeV point.

in Fig. 10b. The curve is a p-wave Breit-Wigner resonant shape. Figure 11 shows evidence that the decay $\psi'' \rightarrow D^0\bar{D}^0$ does in fact take place.[53] The cross section times B, the branching ratio for $D^0 \rightarrow K^-\pi^+$, is plotted versus the e^+e^- total energy. The curve is the same one of Fig. 10, normalized to the highest point.

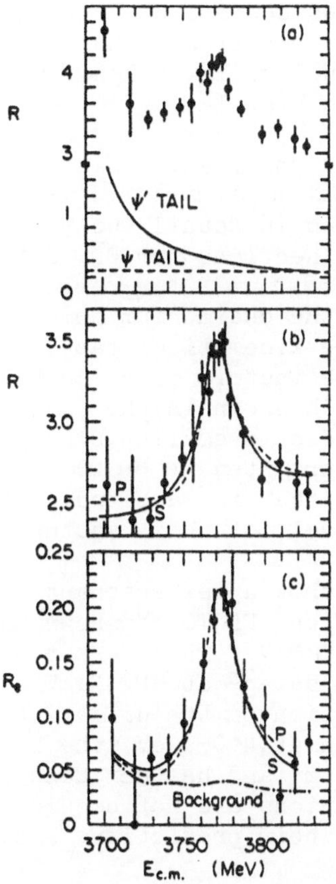

Fig. 12. R as a function of energy as measured by the DELCO experiment[17] at the $\psi(3772)$. (a) is the raw data; (b) the data after radiative correction and subtraction of the ψ' tail; (c) R for events with one electron or more in the final state. The curves are p-wave Breit-Wigner forms.

Figure 12 shows the data of the DELCO experiment[17] in the same energy region. The parameters shown in Table I are derived from Figs. 10 and 12.

The ψ'' is a $^3D_1$ state which is not expected to couple directly to the photon if we use the expression in Section II (Eq. 10) for Γ_e, because for a p-wave the wave function at the origin is zero. However, other effects[82] can produce coupling of the ψ'' to the e^+e^- system, the larger one being mixing with the nearby ψ'. The prediction of the $c\bar{c}$ model[47] for the Γ_e (Γ_e = 150 eV) is closer to the DELCO result[17] (see Table I). The ψ'' is expected to decay almost entirely into $D\bar{D}$, since $D\bar{D}^*$ is not energetically allowed.

No upper limits for the OZI forbidden decay modes are available experimentally.

The 3.95 GeV region. As we go to higher energy in Fig. 7, D and D* associated production is possible. There is a flattening off of the rise in R at about 3.95 GeV, but none of the experiments that measured the total cross section can clearly see a resonant state at this mass. For comparison, the measurements of DASP[18] and PLUTO[50] are shown in Fig. 13, taken from Ref. 18, and the data of DELCO[48] are shown in Fig. 14.

At 4.03 GeV a very striking peak is observed. This is associated with the threshold for $D^*\bar{D}^*$ associated production. As already discussed, Goldhaber et al[45] have studied in detail the composition of this bump and quote a ratio of cross sections for $D\bar{D}$, $D\bar{D}^*$ and $\bar{D}D^*$, and $D^*\bar{D}^*$ associated production. These values are shown in Table II along with the predictions of the charmonium model.[24,47] The $D^*\bar{D}^*$ production is very large if the kinematical factors are taken into account; in fact, phase space factors of p^3 certainly favor $D\bar{D}$ and $D\bar{D}^*$ or $\bar{D}D^*$. The ratios of R shown in the last line of Table II are spin factors, which predict a smaller $D^*\bar{D}^*$ production than the observed one. This fact prompted De Rujula et al[49] to interpret the $\psi(4.03)$ as a molecule, that is, a bound state of $D^*\bar{D}^*$. However, there is no detailed model for this hypothesis.

Above 4.03 GeV the data of Fig. 7 show a new resonant structure at 4.41 GeV, whereas the data of DASP[18] and PLUTO,[50] shown in Fig. 13, have an additional structure at 4.16 GeV. This is not observed clearly in either MARK I data[4] or DELCO data[48] at SPEAR, Fig. 14. At 4.4 GeV the detailed study of production of the different charmed particles is more difficult than at 4.03 GeV (the K identification gets worse with K momentum) and has not been done. As for the $c\bar{c}$ model the present calculations are not considered reliable[25] above 4.1 GeV and therefore their predictions should not be compared with the data.

3. Inclusive D Production Cross Section. The Lead-Glass Wall (LGW) experiment[51,52] has measured the inclusive D production cross sections in the 3.7 to 7.0 GeV energy region. The results are shown in Table III. Note that the D^+ cross section is systematically lower than the D^o cross section. The last column of the table shows $R_{D\bar{D}}$ defined as

$$R_{D\bar{D}} = \frac{\sigma_{D^+} + \sigma_{D^o}}{2\sigma_{\mu\mu}} \tag{32}$$

the factor 2 enters into this expression because it is assumed that a D and a $\bar{D}$ are produced in association, either directly or

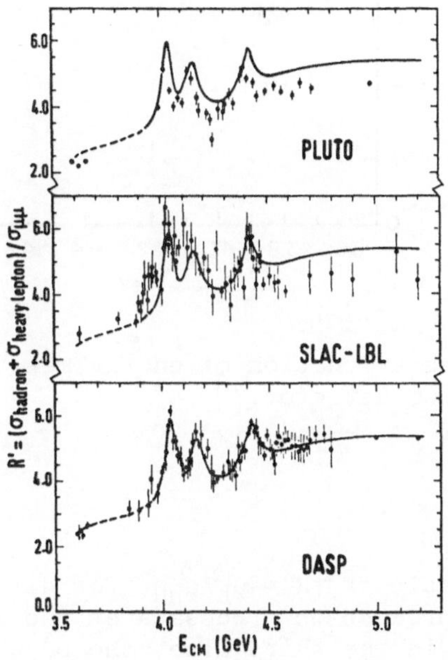

Fig. 13. R as a function of energy taken from the DASP paper of Brandelik et al.[18] The bottom graph shows the DASP data with a fitted curve; the other two graphs show the SLAC-LBL data[4] and the PLUTO data[50] compared with the DASP curve.

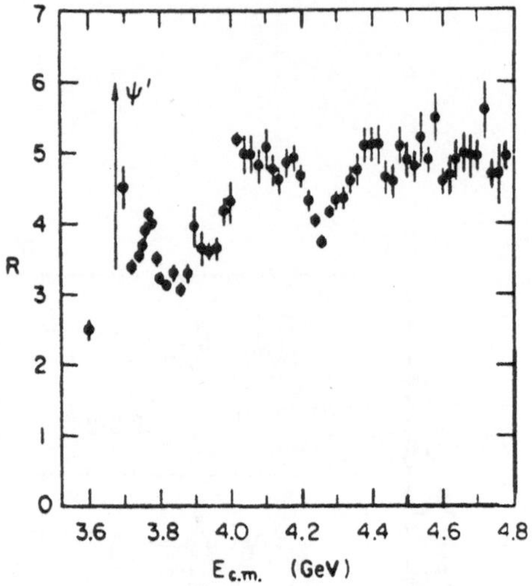

Fig. 14. R as a function of energy measured by the DELCO detector.[48]

TABLE II. Ratio of $R = \sigma_h/\sigma_{\mu\mu}$ for different associated charm production processes at the $\psi(4.03)$. For the values in the third line, the p^3 phase space factors have been explicitly removed.

$R(D\bar{D})$:	$R(D\bar{D}^* + \bar{D}D^*)$:	$R(D^*\bar{D}^*)$	Reference
0.10 ±0.06	0.85 ±0.09	1.00 ±0.10	Goldhaber et al[45]
0.1	4	1	$c\bar{c}$ model, Lane et al[24,47]
0.2 ±0.1	4.0 ± 0.8	128 ±40	Goldhaber et al[45]
1	4	7	spin factors

TABLE III. Cross sections for D^o and D^+ production at different e^+e^- energies.[51] The last column gives the D contribution, $R_{D\bar{D}}$, to the total hadronic cross section expressed as a ratio to $\sigma_{\mu\mu}$.

E Interval (GeV)	$\langle E_{CM} \rangle$ (GeV)	σ_{D^o} (nb)	σ_{D^+} (nb)	$R_{D\bar{D}} = \frac{\sigma_{D^+} + \sigma_{D^o}}{2\sigma_{\mu\mu}}$
3.73 - 3.76	3.74	<1.7	<1.9	<0.29
3.76 - 3.79[a]	3.775	11.5 ±2.5	9.1 ±2.0	1.75 ±0.27
3.79 - 3.84	3.81	<0.7	<0.8	<0.13
3.84 - 3.89	3.87	2.1 ±1.4	1.1 ±1.1	0.28 ±0.16
4.028[b]	4.03	24.2 ±7.0	9.6 ±2.9	3.26 ±0.73
4.0 - 4.2	4.15	16.5 ±5.0	6.2 ±2.5	2.33 ±0.57
4.2 - 4.4	4.28	2.1 ±2.1	6.0 ±2.9	0.88 ±0.40
4.414[b]	4.41	12.6 ±4.2	7.8 ±3.0	2.36 ±0.60
4.4 - 5.0	4.68	9.5 ±3.7	8.9 ±3.1	2.30 ±0.60
5.0 - 5.8	5.36	5.6 ±4.4	2.0 ±2.0	1.30 ±0.83
6.0 - 7.8[c]	7.0	2.3 ±0.8	1.7 ±0.7	1.13 ±0.44

[a]The D cross sections at this energy, measured in Ref. 5, have been reported by I. Peruzzi et al.[53]

[b]These values are calculated by using the σ•B values measured by Piccolo et al[43] and the branching fraction B measured by Peruzzi et al.[53]

[c]From Rapidis et al.[52]

as decay products of D*. The values of $R_{D\bar{D}}$ are plotted in Fig. 15. The measurement of σ has been possible only recently, that is, after the LGW experiment has measured absolute branching ratios[53] for D decays at the ψ" (as discussed in Section IV.D). In fact, when events are observed in a mass plot, as those of Fig. 8, the quantity measured is σ•B, where B is the branching fraction for decay of the D into the final state being considered.

Of course, since D* mesons decay into D mesons, the inclusive D cross sections accounts also for the production of any excited states of the D. In addition to D mesons, F, F* and charmed baryons are expected to contribute to R_{charm} (see Section III.D).

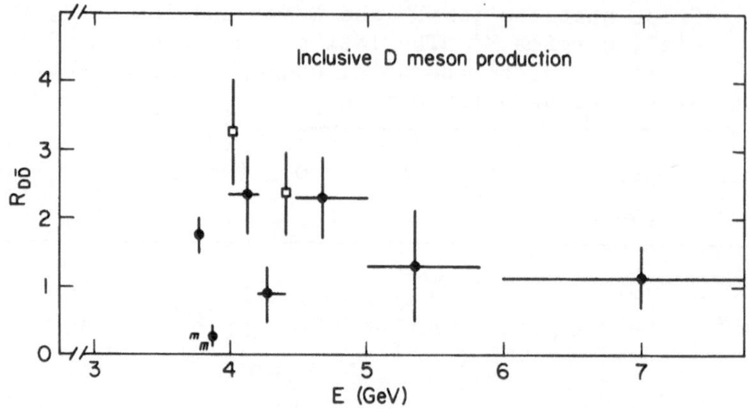

Fig. 15. The cross section for the reaction $e^+e^- \to D\bar{D}$ + anything expressed in units of $\sigma_{\mu\mu}$, $R_{D\bar{D}}$, as a function of energy.[51,52] The solid dots (ɸ) represent the data of the LGW experiment,[51] the squares(ф) are calculated from the σ•B measurements of Piccolo et al.[43]

B. F Meson

The DASP collaboration has reported evidence for F production about a year ago.[54] They have observed five events of the type

$$e^+e^- \to F^+\overline{F^{*-}} \text{ (and c.c.)} \quad \text{with} \quad F^\pm \to \eta\,\pi^\pm,\ \eta \to \gamma\gamma;\quad F^{*\pm} \to F^\pm \gamma_{soft} \tag{33}$$

at the 4.41 GeV resonance. They also observed that the cross section for η production at 4.41 GeV, $\sigma_\eta = 4.1 \pm 0.9$ nb, is much larger than elsewhere. This cross section corresponds to $R_\eta = 0.82$. The masses quoted are: $M_F = 2.03 \pm 0.06$ GeV and $M_{F^*} - M_F = 110 \pm 46$ MeV. As you have heard from Prof. Wolf in his lectures, the same DASP collaboration have now analyzed the data at 4.16 GeV and also observed a large η production cross section:[55] $\sigma_\eta = 1.8 \pm 1.2$ nb. The σ_η as a function of energy is shown in Fig. 16. It is consistent with zero at 4.03, 4.3 and 5.0 GeV, from which the authors infer that the peak at 4.16 GeV is an $F\bar{F}$ state and the 4.41 peak is an $F\overline{F^*}$ state, with the η production showing the same trend. They also find that the fraction of $F \to \eta\pi$ is:

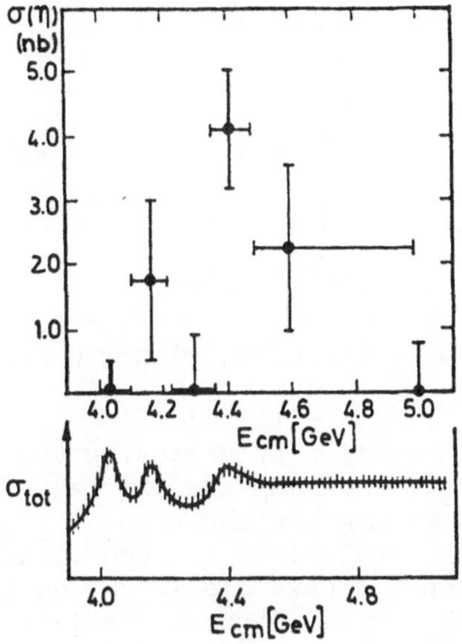

Fig. 16. Inclusive η cross section in e^+e^- annihilation as measured by the DASP experiment.[55] The bottom curve shows the trend of total hadronic cross section as measured by DASP.[18] Plot taken from Flügge.[20]

$$\frac{B(F \to \eta\pi)}{B(F \to \eta + \text{anything})} = 0.09 \pm 0.04 \qquad (34)$$

therefore at 4.16 GeV they find

$$\sigma(e^+e^- \to F\bar{F}) \cdot B(F \to \eta\pi) = 0.08 \pm 0.06 \text{ nb} \qquad (35)$$

where the error is so large because of large uncertainties in the η detection efficiency, which depends not only on the acceptance of the apparatus, but also on the details of the assumed production mechanism.

The lead-glass wall experiment has reported some indication of F production in the $K\bar{K}\pi$ channel.[56] We have reanalyzed that data and will now present the results.

We have studied all channels with a $K\bar{K}$ pair, which should be the other copious decay mode expected for the F. The following final states have been analyzed at 4.16 GeV (the sample had a total integrated luminosity of 940 nb^{-1}):

$$e^+e^- \rightarrow K^+K^-\pi^\pm + X \qquad (36a)$$

$$K^+K^-\pi^+\pi^-\pi^\pm + X \qquad (36b)$$

$$\rightarrow K^+\bar{K}^0 + X \qquad \text{(and c.c.)} \qquad (36c)$$

$$K^+\bar{K}^0\pi^+\pi^- + X \qquad \text{(and c.c.)} \qquad (36d)$$

$$K^0\bar{K}^0\pi^\pm + X \qquad (36e)$$

where X stands for anything else, either charged or neutral particles.

The K^0's were identified by measuring the dipion mass as mentioned in Section III.A.1 and the charged K by time of flight. Since for each track we measure the time of flight, the path length and the momentum we can calculate the confidence level (CL) for it to be a π, K or proton. A particle is chosen to be a K if $(CL)_\pi < (CL)_K > (CL)_p$. This method is reliable up to momenta of 0.9 GeV (σ_{TOF} = 0.4 ns), which is the range of momenta relevant for reactions (36).

The mass resolution was also improved by using the same method applied for the precise mass measurements of the D meson,[53] which will be described in Section IV.A.1. It consists of selecting those events in which the total energy of the $K\bar{K}(n\pi)$ system is within 60 MeV of the beam energy and then replacing the measured total energy of these particles with the beam energy (whose energy resolution is 1 MeV). For these events the likely reaction is a two-body process with equal mass for the two bodies, as expected, if the reaction $e^+e^- \rightarrow F\bar{F}$ were to take place.

A total of 86 events were found with these criteria for the above reactions with $M(K\bar{K}\ n\pi) > 2.0$ GeV. The invariant mass distribution of the $K\bar{K}(n\pi)$ combinations for these events is shown in Fig. 17. The only significant deviation from a flat distribution is found at M = 2040 MeV. Notice that the events are plotted in 2 MeV bins and that the signal is practically all in two bins. This agrees with the expected resolution at this mass. The significance of this signal is not very high. There are 14 events where 4.2 would be expected; this corresponds to a probability of 1.3×10^{-4} for a Poisson distribution. In terms of standard deviations, the significance of the effect is at the 4 standard deviation level. Therefore, we are not prepared to say that we have an F signal. However, since the mass at which we observe this effect is in the general mass region where the DASP collaboration reported an F signal, we can make some comparisons.

Assuming that this 4 standard deviation effect is due to F

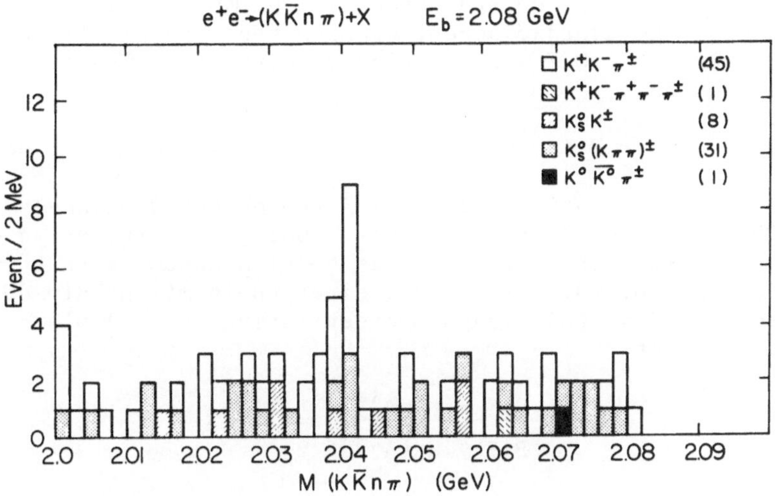

Fig. 17. Invariant mass distribution of the $(K\bar{K}n\pi)$ system as obtained by the LGW experiment. Five different channels contribute, their symbols and number of events are shown in the plot. See text for more details.

production, from the $K^+K^-\pi^\pm$ channel we get

$$\sigma(e^+e^- \to F\bar{F}) \cdot B(F^+ \to K^+K^-\pi^+) = 0.10 \pm 0.05 \text{ nb} \tag{37}$$

where the error includes the contribution from the uncertainty in the detection efficiency for this final state.

Comparing Eqs. (35) and (37) we observe that if the signal in Fig. 17 were due to the F meson

$$r = \frac{B(F^+ \to K^+K^-\pi^+)}{B(F^+ \to \eta\pi^+)} = 1.2 \pm 1.1 \tag{38}$$

in agreement with theoretical expectations. In fact, the predictions of the statistical model of Quigg and Rosner[57] is $r = 1.1$ and the QCD calculations of Cabibbo and Maiani,[58] (to be discussed in more detail in Section IV.D.3 for D decays) give $r = 0.96$.

One further speculation involves the comparison of the overall $K\bar{K}(n\pi)$ rate with the overall $\eta(n\pi)$ rate. This comparison can be done by using the statistical model to calculate the acceptance. The average acceptance for the channels listed in Eqs. (36) is $\varepsilon = 0.027$ with a large uncertainty. The total cross section then is

$$\sigma(e^+e^- \to F\bar{F}) \cdot B(F \to K\bar{K}(n\pi) \text{ of Eqs. (36)}) = 0.23 \pm 0.10 \text{ nb} . \quad (39)$$

According to the statistical model[57] the detected channels constitute 60% of all the $K\bar{K}(n\pi)$ channels. Using this factor we can compare (39) with the total η cross section assuming that it all comes from $F \to \eta(n\pi)$. Of course, since the semileptonic decays are also included, this means overestimating the $F \to \eta(n\pi)$ cross section. We get

$$\frac{B(F \to \eta(n\pi))}{B(F \to K\bar{K}(n\pi))} = 4 \pm 3$$

The statistical model predicts a value of about two for this ratio.

C. Charmed Baryons

There has been no observation of a peak in an invariant mass distribution which could be interpreted as charmed baryon production in e^+e^- collisions.

Cazzoli et al[59] first reported a candidate for a charmed baryon in a neutrino experiment in the BNL hydrogen bubble chamber. They found one event of the type

$$\nu p \to \mu^- \Lambda \pi^+ \pi^+ \pi^+ \pi^- \quad .$$

This reaction violates the $\Delta S = \Delta Q$ rule in weak interactions, but it would be allowed if a charmed baryon ($\Lambda_c^+ \to \Lambda\pi^+\pi^+\pi^-$) were being produced. The mass of the Λ_c^+ using this event was measured to be 2.26 GeV. Subsequently a Fermilab photoproduction experiment[60] observed a peak at the same mass in the $\bar{\Lambda}\pi^-\pi^-\pi^+$ mass spectrum.

The only indication of charmed baryon production in e^+e^- collisions comes from the inclusive baryon cross section measured at SPEAR. The data,[61] a combination of the SP17 and SP26 experiments, are shown in Fig. 18. We have measured the inclusive $\bar{p}$, Λ and $\bar{\Lambda}$ cross sections in the energy region 3.82 to 7.36 GeV. The antiprotons were identified by time-of-flight and momentum measurements, the Λ and $\bar{\Lambda}$ by study of the invariant mass of the $p\pi^-$ and $\bar{p}\pi^+$ combinations, which show peaks at the appropriate mass. Figure 18a shows the ratio $R = \sigma/\sigma_{\mu\mu}$ for production of p and $\bar{p}$.

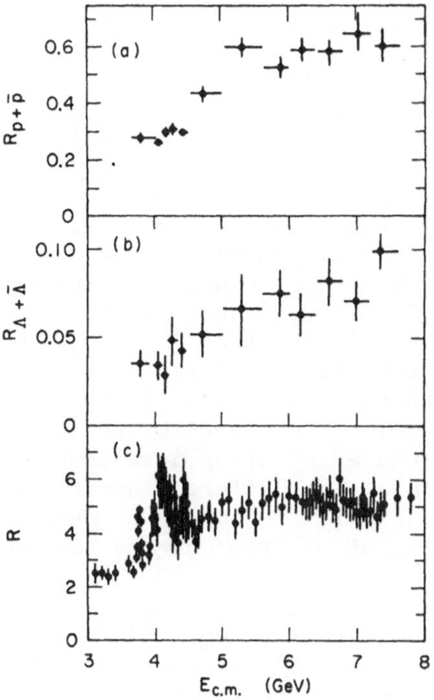

Fig. 18. Inclusive measurement[61] of $R(p+\bar{p})$ shown in graph (a), and $R(\Lambda+\bar{\Lambda})$ shown in graph(b). The measurement of R, the total hadronic production, is shown for comparison in (c) (see Fig. 2 for references).

This is actually $R(p+\bar{p}) = 2R_{\bar{p}}$, since the proton cross section is more difficult to measure because of the large background due to beam-gas interactions. Figure 18b shows the same ratio for $\Lambda+\bar{\Lambda}$ production.

The values of $R(p+\bar{p})$ exhibit a sharp rise of $\Delta R = 0.3$ between 4.4 and 5 GeV. Using the mass for Λ_c mentioned above and the mass formula of De Rujula et al,[62] the thresholds for associated production of charmed baryon pairs ($\Lambda_c^+\bar{\Lambda}_c^-$, $\Sigma_c\bar{\Sigma}_c$, $\Sigma_c^*\bar{\Sigma}_c^*$) are expected to be in this energy region. $R(\Lambda+\bar{\Lambda})$ also shows an increase in the same energy region, although the statistics are not as good and the rise is not as sharp. However, from Fig. 18b we can estimate that the maximum value of $\Delta R(\Lambda+\bar{\Lambda})$ could be ~0.04 or about 10-15% of $\Delta R(p+\bar{p})$. This observation indicates that charmed baryon decays into Λ, Σ^0, and therefore $\Sigma^\pm$ are smaller than decays into strange mesons and nucleons (like $\bar{K}^0p$ or $K^-p\pi^+$).

Prior to the LGW experiment a UCLA group had modified the Mark I magnetic detector in order to identify the $\bar{n}$ and $\bar{p}$ produced. They have measured, as reported by Ferguson et al.,[83] the $\bar{\Sigma}^\pm$ production at 4 and at 7 GeV total energy. They find an increase between these two energies of $\Delta R(\bar{\Sigma}^\pm) = 0.12 \pm 0.05$, somewhat larger than the $\bar{\Lambda}$ result. Although the errors are large, these two results seem to be in disagreement.

D. Summary of Contributions to R

We can now summarize what we have learned so far on the total hadronic cross section, that is, on various contributions to $R = \sigma_h/\sigma_{\mu\mu}$. We have to point out at this point that, since the detectors that have measured R do not cover 100% of the solid angle, all the R measurements depend strongly on Monte Carlo calculations to correct for the events that have not triggered the apparatus. The systematic errors associated with these calculations are estimated to be ±15% for the SPEAR magnetic detector,[4] ±20% for the DELCO detector,[48] ±15% for DASP,[18] and ±11% for PLUTO.[50] The general features in R are similar (see Figs. 2, 13 and 14) but there are differences in the details as already mentioned in discussing the effects in the 3.95 and 4.16 GeV energy regions (see Section III.A.2). The R measured by PLUTO[50] is systematically lower than those measured by other detectors. Because of these problems we will discuss the different contributions to R within the same detector, that is, the SLAC-LBL Mark I detector, since both $R_{D\bar{D}}$ and the contribution from charmed baryons were measured in this detector.

The values of R as measured in the Mark I detector[4,5] are shown in Fig. 2. Since there is a high density of data points in this figure, we have drawn by hand a curve to represent these data so that we can compare it with the sum of the different parts that we have measured and discussed in the previous sections. This curve is shown in Fig. 19. As for the individual contributions to R we can say the following:

1. These data indicate $R = 2.5$ below charm threshold, so we will assume that this is the contribution of the old quarks u, d, and s.
2. The heavy lepton contribution to R can be calculated from QED to be $R_{\tau^+\tau^-} = \beta(3-\beta^2)/2$.
3. The charmed baryon contribution $R_{B\bar{B}}$ has been added in with a value that rises from 0 to ½(0.64) = 0.32 between 4.4 and 5 GeV (the $n+\bar{n}$ contribution is assumed to be the same as that of $p+\bar{p}$). Above 5 GeV it is assumed to be constant as indicated by the data of Fig. 18a.
4. Finally, the $R_{D\bar{D}}$ values of Fig. 15 and Table III have been added to the above as points with error bars.

The sum of these contributions appears to saturate the measured values of R. However, the uncertainties of the $R_{B\bar{B}}$ and $R_{D\bar{D}}$ measurements are such that ½ unit of R of F production or some other process could be easily accommodated.

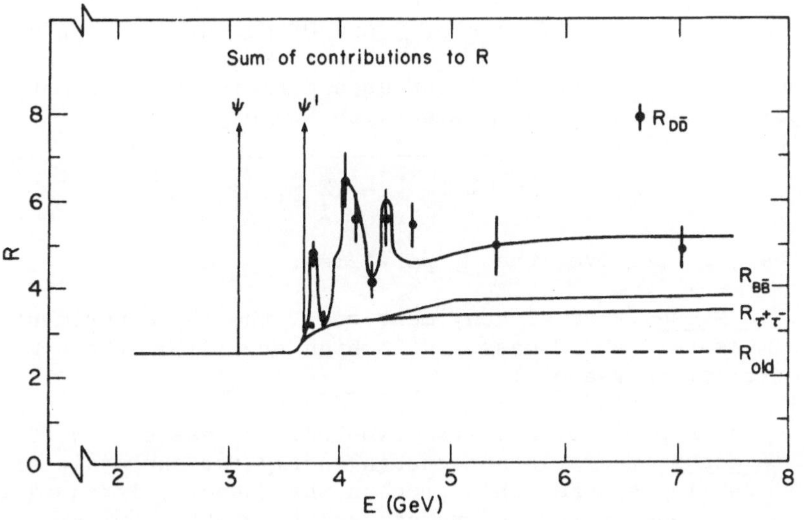

Fig. 19. A composite graph illustrating the various contributions to R, the total hadronic cross section over $\sigma_{\mu\mu}$. The top curve is a sketch of R, hand drawn over the data of Fig. 2. The following contributions are progressively added starting from R = 0: R_{old} is a constant as inferred by the data points below charm threshold; $R_{\tau^+\tau^-}$ is the heavy lepton contribution as calculated from QED; $R_{B\bar{B}}$ is the charmed baryon contribution as inferred by the data[61] of Fig. 18. Finally we add the contribution[51,52] of $R_{D\bar{D}}$ as data points, taken from Fig. 15 and Table III.

IV. PROPERTIES OF CHARMED MESONS

In this section we will review masses and branching ratios for D and D*. In addition, we will review the situation on D^0 and $\bar{D}^0$ mixing. Due to the time available for these lectures, we will not review the spin and parity assignments of the D mesons. Detailed studies made by the SLAC-LBL collaboration[63] find that indeed the D has $J^P = 0^-$.

A. Masses of D and D* Mesons

The mass of a particle found as a peak in an invariant mass of n particles can be calculated with the expression

$$M = \sqrt{(\sum_i E_i)^2 - (\sum_i \vec{p}_i)^2} \tag{40}$$

where the sums are over the n particles.

1. D Masses. As already mentioned, the LGW experiment was able to measure[53] the D masses with high precision at the $\psi(3772)$ for the following reasons:

a. The D production is the two-body process $e^+e^- \to D\bar{D}$ because there is not enough energy for any additional particles. For this process the D energy is equal to the beam energy and we can substitute E_b for $\sum_{i=1}^{n} E_i$ in Eq. (40). This fact improves the resolution considerably because the r.m.s. error[64] of E_b is 1 MeV. Thus E_b is much better determined than the energy obtained by the momentum measurement of the n tracks.

b. The momenta of the secondary particles are low, $p_D \sim 300$ MeV/c. Therefore, the uncertainty on p_D contributes very little to the uncertainty on the mass of the D.

The overall resolution is 3 MeV and the final error on the D mass is dominated by the systematic errors rather than the statistical error.

The invariant mass distributions for the various observed decay channels of the D^+ and D^0 are shown in Fig. 20. Here the charged and neutral kaons have been identified with the same methods described earlier (see Section III.A.1). Among all the hadronic events at the ψ'' peak of Fig. 10, we have chosen those for which $\sum_i E_i$, that is, the measured E_D, is within 50 MeV of the beam energy. For these combinations of n particles

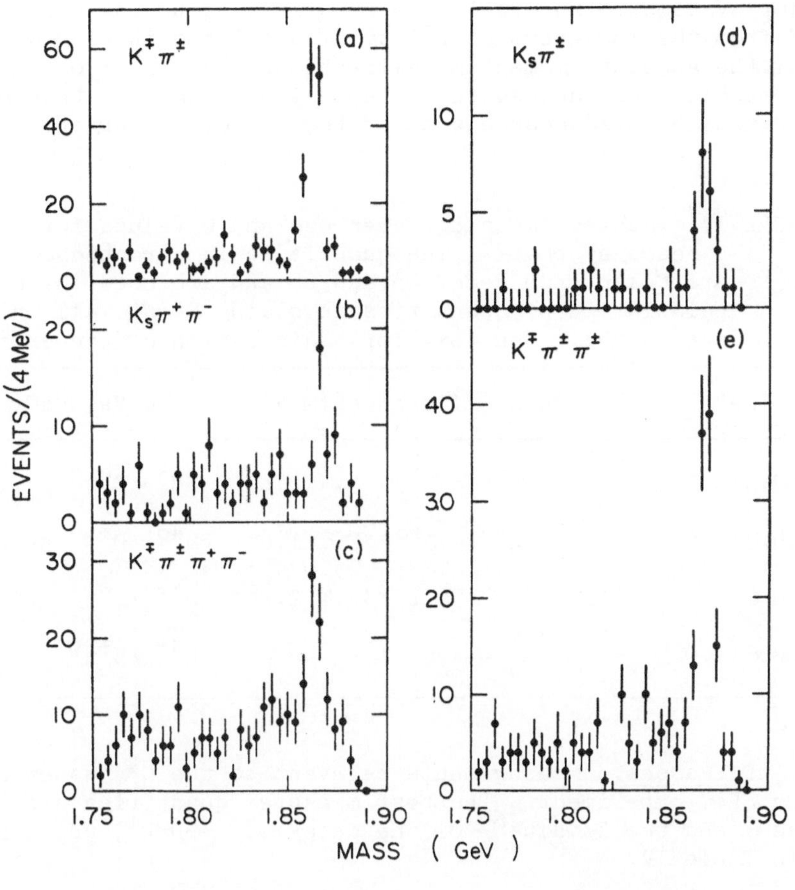

Fig. 20. Invariant mass spectra for various D^0 (on the left) and D^+ (on the right) decay modes. Note that the distributions are plotted in 4 MeV bins.

we use Eq. (40) to calculate the invariant mass with E_b instead of $\sum_i E_i$. Fits to the combined distributions for D^o and D^+ give the mass values

$$M_o = 1863.3 \pm 0.9 \text{ MeV} \tag{41}$$

$$M_+ = 1868.3 \pm 0.9 \text{ MeV} \quad . \tag{42}$$

The major systematic uncertainties contributing to the errors are: 0.5 MeV from the long-term stability of the E_b monitoring and 0.5 MeV from the absolute momentum calibration. The error on the $D^+ - D^o$ mass difference, shown in Table IV, is smaller than either of the D mass errors because some of the systematic errors cancel out.

TABLE IV. Masses, mass differences, and Q values for the D meson system.[53] The quantities in parentheses are taken from Refs. 45 and 65 and are used in the calculation of quantities involving D*'s. All units are in MeV. See text for a discussion of the errors.

Mass (MeV)		Mass Difference (MeV)		Q Values (MeV)	
D^o	1863.3±0.9	$D^+ - D^o$	5.0±0.8	$D^{*o} \to D^o\pi^o$	7.7±1.7
D^+	1868.3±0.9	$D^{*+} - D^{*o}$	2.6±1.8	$D^{*o} \to D^+\pi^-$	-1.9±1.7
D^{*o}	(2006.0±1.5)	$(D^+-D^o)-(D^{*+}-D^{*o})$	2.4±2.4	$D^{*+} \to D^o\pi^+$	(5.7±0.5)
D^{*+}	2008.6±1.0	--	--	$D^{*+} \to D^+\pi^o$	5.3±0.9

2. D* Masses. Measurements relevant to the D* masses come from the SP17 experiment. The best measured quantities are the D^{*o} mass[45] and the Q value[65] of the decay $D^{*+} \to D^o\pi^+$, which are shown in Table IV.

For the D^{*o} they used essentially the same method described for the D. That is, at 4.03 GeV they used the reaction $e^+e^- \to \bar{D}^{*o}D^{*o}$ with $D^{*o} \to D^o\pi^o$. For this two-body reaction the energy of the D^{*o} is, of course, E_b. As for the momentum of the D^{*o}, they assume it to be equal to the measured D^o momentum with a little correction due to the unmeasured π^o. Again the Q of the reaction is small, the momentum of the D^{*o} is small, and its error contributes little to the error on the D^{*o} mass. From Eq. (40) M can be calculated.

The P_D distribution is shown in Fig. 21. The detailed fit[45] of this distribution will be described in Section IV.C. For the

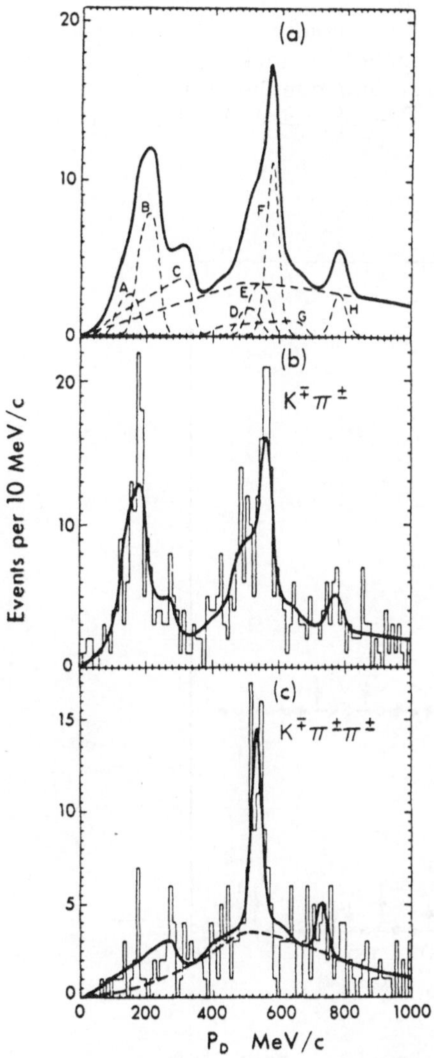

Fig. 21. Studies of p_D at 4.03 GeV.[45]
(a) Contribution to the expected D^0 momentum spectrum from

A $e^+e^- \to D^{*+}D^{*-}$, $D^{*+} \to \pi^+D^0$

B $\to D^{*0}\bar{D}^{*0}$, $D^{*0} \to \pi^0D^0$

C $\to D^{*0}\bar{D}^{*0}$, $D^{*0} \to \gamma D^0$

D $\to D^{*+}D^-$, $D^{*+} \to \pi^+D^0$

E $\to D^{*0}\bar{D}^0$, $D^{*0} \to \pi^0D^0$

F $\to \bar{D}^{*0}D^0$, direct D^0

G $\to D^{*0}\bar{D}^0$, $D^{*0} \to \gamma D^0$

H $\to D^0\bar{D}^0$ direct D^0

(b) $D^0 \to K^-\pi^+$ momentum spectrum, the curve is the result of the fit and (c) $D^+ \to K^-\pi^+\pi^+$ momentum spectrum where the curve is the result of the fit and the dashed line is the background.

D^{*0} mass the main uncertainty comes from the determination of the center of the peak B (see Fig. 21a) due to $D^{*0} \to D^0\pi^0$ in the presence of peak A due to $D^{*+} \to D^0\pi^+$ and peak C due to $D^{*0} \to D^0\gamma$. The fit gives the value

$$M_{D^{*0}} = 2006.0 \pm 1.5 \text{ MeV} \tag{43}$$

The same method has been used for the D^{*+} mass measurement,[45] but due to smaller statistics (Fig. 21c), the errors are twice as large. The best information on the D^{*+} mass comes from direct observation[65] of $D^{*+} \to D^0\pi^+$ at 6.8 GeV where the π^+ momentum is large enough to be measured in the magnetic detector. Again, the Q of the reaction $D^{*+} \to D^0\pi^+$ is small and can be determined accurately. Figure 22a shows the $D^{*+} - D^0$ mass difference which

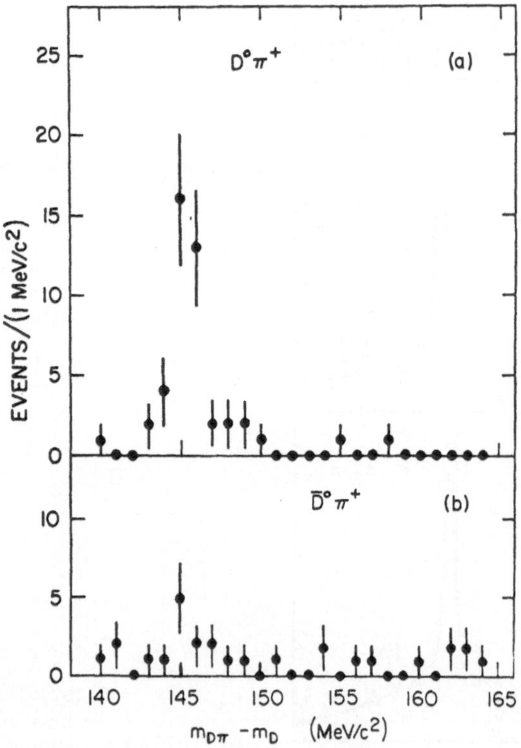

Fig. 22. Study[65] of $D^{*+} \to D^0\pi^+$; the $D\pi$-D mass difference is shown (a) for $D^{*+} \to D^0\pi^+$ with $D^0 \to K^-\pi^+$, and (b) for the sequence $D^{*+} \to \bar{D}^0\pi^+$ with $\bar{D}^0 \to K^+\pi^-$. Events from the charge conjugate reactions are included.

gives

$$Q(D^{*+} \to D^{o}\pi^{+}) \quad = \quad 5.7 \pm 0.5 \text{ MeV} \tag{44}$$

This value is shown in Table IV and in combination with the D^{o} mass, gives the D^{*+} mass shown in the table.

The remaining values in Table IV, essentially the $D^{*+} - D^{*o}$ mass difference and the Q values for D* decays, are quantities derived from the directly measured ones. The quoted errors take into account that some of the systematic errors cancel out in the difference. The Q values for the D* decays are shown in Fig. 23.

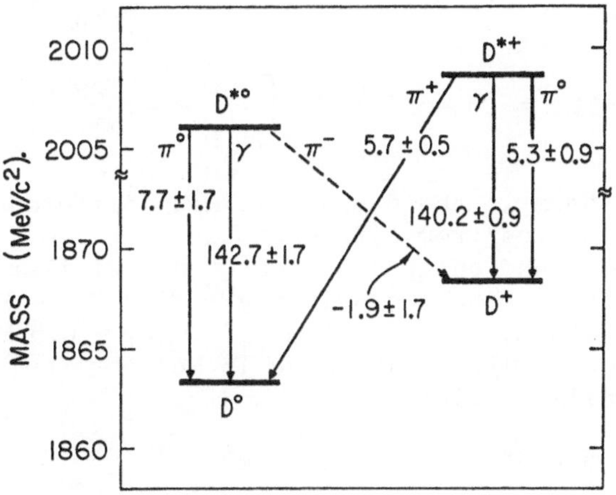

Fig. 23. Mass level diagram for D* and D^{o} states from the measurements shown in Table IV. The arrows represent different decay modes of the D*; the numbers across the lines represent the Q for each decay expressed in MeV. The decay $D^{*o} \to D^{+}\pi^{-}$ cannot take place.

The decay $D^{*o} \to D^{+}\pi^{-}$ is not energetically possible. This observation was already reported before the precise mass measurements of Ref. 53.

3. Charged-Neutral D and D* Mass Differences. Expectation for the masses of charmed particles have been discussed by De Rujula, Georgi, and Glashow[62] (see also Jackson's review[1]). We only mention here the prediction for the mass splitting of members of the same isotopic spin multiplet. The experimental results in

Table IV show that

$$\delta = M_{D^+} - M_{D^o} = 5.0 \pm 0.8 \text{ MeV} \qquad (45)$$

$$\delta^* = M_{D^{*+}} - M_{D^{*o}} = 2.6 \pm 1.8 \text{ MeV} . \qquad (46)$$

For comparison

$$M_{K^+} - M_{K^o} = -4.01 \pm 0.13 \quad . \qquad (47)$$

In the non-relativistic quark model the mass splittings are

$$\delta = M_{D^+} - M_{D^o} = (m_d - m_u) + \frac{2}{3}\alpha\left[<\frac{1}{r_D}> + \frac{2\pi}{m_c m_u}|\psi_D(o)|^2\right] \qquad (48)$$

$$\delta^* = M_{D^{*+}} - M_{D^{*o}} = (m_d - m_u) + \frac{2}{3}\alpha\left[<\frac{1}{r_D}> - \frac{2\pi}{3m_c m_u}|\psi_D(o)|^2\right] \qquad (49)$$

where the first term is the d-u quarks mass difference, the second term is a contribution from single photon exchange. Using current algebra for $m_d - m_u$ and an atomic quark model similar to charmonium for the second term, Lane and Weinberg[67] find $M_{D^+} - M_{D^o} = 7$ MeV to be compared with Eq. (45) and $M_{D^{*+}} - M_{D^{*o}} = 6.5$ MeV. The calculation by De Rujula et al[66] gave 15 MeV for the D mass difference (45). Finally we find

$$\delta - \delta^* = 2.4 \pm 2.4 \text{ MeV} \quad . \qquad (50)$$

This is an electromagnetic hyperfine splitting and is expected to be ~1 MeV in most theoretical models.

B. $D^o - \bar{D}^o$ Mixing

Mixing of the D^o and $\bar{D}^o$ states could arise from $\Delta C = \Delta S$ and $\Delta C = -\Delta S$ transitions of the c quark, as in the two diagrams below

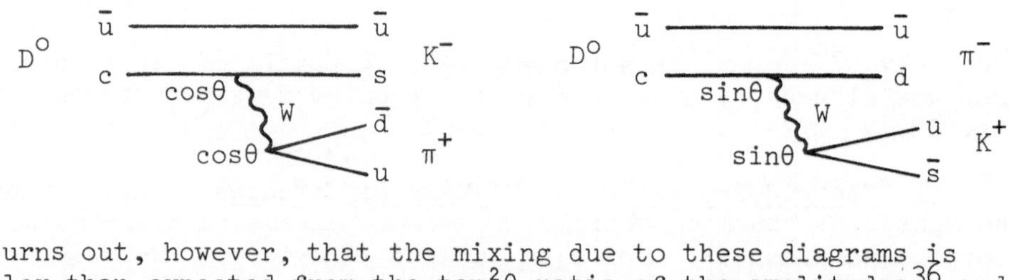

It turns out, however, that the mixing due to these diagrams is smaller than expected from the $\tan^2\theta$ ratio of the amplitudes[36] and is negligible. One other possible source of D^o-$\bar{D}^o$ mixing is the

charm changing neutral current,[69] if it were to exist. In this case one would expect D^o-$\bar{D}^o$ mixing to be complete. Therefore it is very important to check out this hypothesis experimentally.

The only experimental data on D^o-$\bar{D}^o$ mixing was obtained from the SLAC-LBL experiment SP17 in two separate studies. The first[65] was done with D^{*+} produced in the 5 to 7.8 GeV energy region. The reaction studied was

$$e^+e^- \rightarrow D^{*+} + X \qquad \text{(and c.c.)} \tag{51}$$

with

$$D^{*+} \rightarrow D^o\pi^+ \qquad \text{and} \qquad D^o \rightarrow K^-\pi^+ \tag{52a}$$

or

$$D^{*+} \rightarrow \bar{D}^o\pi^+ \qquad \text{and} \qquad \bar{D}^o \rightarrow K^+\pi^- \tag{52b}$$

and their charge conjugates. Here for D^o-$\bar{D}^o$ mixing, one would expect to detect some events with $D^{*+} \rightarrow \bar{D}^o$. The data is shown in Fig. 22b. Five events are observed in the correct mass region compared with the 26 events above background in Fig. 22a. After corrections, these events give an upper limit

$$\frac{N(\text{wrong sign K})}{\text{all } D^{*+} \text{ events}} < 16\% \quad \text{with} \quad 90\% \text{ CL} \quad .$$

The second study[45] was made on the reaction

$$e^+e^- \rightarrow D^o + K^\pm + X \quad ,$$

that is, the sign of the K accompanying the D^o is the signal for D^o-$\bar{D}^o$ mixing. For no D^o-$\bar{D}^o$ mixing one expects strangeness conservation, that is, a K^+ should accompany a D^o. The result of this study is

$$\frac{N(\text{wrong sign K})}{\text{all } D^o \text{ events}} < 18\% \quad \text{with} \quad 90\% \text{ CL} \; .$$

The above results exclude complete D^o - $\bar{D}^o$ mixing.

C. D* Branching Fractions

The measurements of the D* branching fractions have been made by Goldhaber et al[45] at the 4.03 GeV bump in the cross section. As already discussed in Section III.A.2, at this energy there is a large $D^*\bar{D}^*$ production. The measurements are done through the study of the momentum of the D meson detected: the momentum is different depending upon the D* decay mode it comes from, as well as upon the D* or D production reaction. This is illustrated in

Fig. 21a. The curves A through H represent the various possible ways to obtain a D^o either from decays of D^* or directly. Figure 21 shows the result of a simultaneous fit to the D^o and D^+ data. These curves were obtained with the following assumptions:

a. The production of D^o or D^+ processes are the ones shown in Fig. 21a, with the addition of direct D^+D^- production.

b. The decay modes for D^{*o} and D^{*+} are the ones shown in Table V. In fact, $D^{*o} \to D^+\pi^-$ is not energetically possible (see Fig. 23), so the fractions of $D^{*o} \to D^o\pi^o$ and $D^{*o} \to D^o\gamma$ should add to unity.

c. Only three ratios,

$$B(D^{*o} \to \gamma D^o) \tag{53}$$

$$B(D^{*+} \to \pi^+ D^o) \tag{54}$$

and

$$\frac{B(D^+ \to K^-\pi^+\pi^+)}{B(D^o \to K^-\pi^+)} \tag{55}$$

were left free to vary. In order to fit these parameters an isospin-constrained fit was done, so that $D^{*+} \to \pi^+D^o$ and $D^{*+} \to \pi^oD^+$ are related by isospin coefficients.

d. The ratio $\Gamma(D^{*+} \to \gamma D^+)$ over $\Gamma(D^{*o} \to \gamma D^o)$ was assumed from theory[70] to be 1/4. For much smaller values the data with $p_{D^o} \leqslant 300$ MeV/c could not be fitted easily.

The results for the D^* branching fractions are shown in Table V. For a discussion about how sensitive these results are to the assumptions made, the reader is referred to Ref. 45. The quoted errors, however, take into account the uncertainties related to the model dependence of the fit.

Table V also gives the most recent theoretical predictions for these decays, as estimated by Eichten et al.[47] For the $D^* \to D\gamma$ they use the naive quark model formula

$$\Gamma(D^* \to D\gamma) = \frac{4}{3}\alpha\left(\frac{Q_c}{2m_c} + \frac{Q_i}{2m_i}\right)^2 p^3 \tag{56}$$

where $Q_c = 2/3$ and Q_i are the quark charges involved ($Q_u = 2/3$ for D^o and $Q_d = 1/3$ for D^+); $m_c = 1.87$ GeV, as determined in their linear potential calculation using $\Gamma(\psi \to e^+e^-)$ and $M_{\psi'}-M_\psi$ as input

TABLE V. D* branching fractions as measured by Goldhaber et al[45] compared with theoretical expectation.[47]

Mode	Experiment[45] (in %)	Theory[47] (in %)
$D^{*0} \rightarrow D^0\pi^0$	$(45 \pm 15)^a$	53.0
$\rightarrow D^0\gamma$	55 ± 15^a	47.0
$D^{*+} \rightarrow D^+\pi^0$	$(30 \pm 7)^b$	28.4
$D^0\pi^+$	60 ± 15	68.4
$D^+\gamma$	$(10 \pm 17)^c$	3.2

[a] The free parameter in the fit was $D^* \rightarrow D^0\gamma$, the sum of the two decays was constrained to one.

[b] Value derived from $D^{*+} \rightarrow D^0\pi^+$ using isospin factors.

[c] Obtained as difference from unit once the $D^* \rightarrow D^0\pi^+$ is determined.

data; $m_u = m_d = 0.33$ GeV; finally $p = (1/2M_{D*})(M_{D*}^2 - M_D^2)$.

The $D^* \rightarrow D\pi$ width was obtained by assuming a form derived from their $q\bar{q}$ model for higher ψ levels decaying into $D\bar{D}$:

$$\Gamma(D^* \rightarrow D\pi) = \frac{p^3}{72\pi M_{D*}^2} C^2 \left| \sqrt{M_{D*} E_D E_\pi}\, A \right|^2 \tag{57}$$

where E_i is the D or π energy, p their momentum, C a Clebsch-Gordan coefficient, and A is an amplitude depending only on m_u (for $m_c \rightarrow \infty$). A can be estimated from $K^* \rightarrow K\pi$ under the assumption that m_s is very large. This gives

$$A = 47.8 \text{ GeV}^{-3/2} \quad .$$

The estimated branching fractions with these assumptions are shown in Table V. They are in good agreement with the experimental results.

D. Decay Properties of the D Meson

Some hadronic branching fractions of the D have been measured at the $\psi(3772)$ by the LGW experiment.[53] The semileptonic branching fraction has been measured both at SPEAR and at DORIS. The LGW experiment has also measured some inclusive characteristics of the D decays,[71] again at the $\psi(3772)$. A review on all that is known today on D decays follows.

1. Hadronic Decay Modes. As mentioned in Section III.A.1, the SPEAR magnetic detector experiment SP17 has detected a number of D decays[41-43] (see Fig. 8). However, absolute branching fractions were not measured until later, that is, until the D's were copiously produced at the $\psi(3772)$, where it has been possible to measure the cross section for D production.

As discussed in Section III.A.2, one can assume that the ψ'' decays entirely into $D\bar{D}$, therefore the cross section for $D\bar{D}$ production is equal to the resonant cross section,[5] shown in Fig. 10. As for the ratio of D^0 to D^+ production, reactions (30) and (31), it is reasonable to assume[53] that it is given by the ratio of the kinematical and barrier factors present in the p-wave Breit-Wigner formula. These cross sections are shown in Table III.

Figure 20 shows the invariant mass distribution for a number of $K(n\pi)$ mass combinations. These distributions were obtained as explained in Section IV.A.1. If N_i is the number of events found in channel i, we write

$$N_i = 2\sigma(e^+e^- \to D\bar{D})\, B_i\, A_i\, L \qquad (58)$$

where B_i and A_i are the branching fraction and the acceptance of the apparatus for D decaying into that channel, L is the integrated luminosity of the sample analyzed, and the factor 2 is present because either D can decay into that channel. The branching fractions calculated in this way are shown in Table VI. Here the decay $D^0 \to K^-\pi^+\pi^0$, also observed at the $\psi(3772)$ by the LGW experiment,[72] has been added as well as the semileptonic decay fraction measured[73] in the same experiment. For more details on the methods used to measure these branching ratios as well as for a review of LGW results, see Ref. 74.

In Table VI we notice the following:

a. The $D^+ \to \bar{K}^0\pi^+$ decay mode is observed. Comparison with the $D^0 \to K^-\pi^+$ decay mode gives

$$\frac{\Gamma(D^+ \to \bar{K}^0\pi^+)}{\Gamma(D^0 \to K^-\pi^+)} = (0.70 \pm 0.23)\, \frac{\tau_0}{\tau_+} \qquad (59)$$

where τ_o and τ_+ are the lifetimes of the D^o and the D^+. The value (59) shows that if the two lifetimes are not too different, the $D^+ \to \bar{K}^o\pi^+$ decay is of the same order of magnitude as the $D^o \to K^-\pi^+$ decay mode as predicted by Ellis et al[38] (see Section II.C.3). This result is relevant to the understanding of the nonleptonic enhancement and will be discussed in Section IV.D.3 below.

b. We have measured so far only a small fraction of the D decay modes into hadrons:

$$\sum_{\substack{\text{measured}\\\text{modes}}} B_i(D^o \to \text{hadrons}) = (21.4 \pm 6.3)\% \qquad (60)$$

$$\sum_{\substack{\text{measured}\\\text{modes}}} B_i(D^+ \to \text{hadrons}) = (5.4 \pm 1.2)\% \qquad (61)$$

Clearly more data is needed to fully understand the D decay properties.

Table VI shows that the three and four body final states are more copious than the two body final states. It is interesting to find out if there is resonance production in the D decays, that is, if there is evidence for reactions of the type

$$D \to K^*\pi, \quad K\rho, \quad K^*\rho, \quad \text{etc.} \qquad (62)$$

This question has been addressed by Piccolo et al.[43] They find the following:

a. No evidence[75] for K* production in the reaction

$$D^+ \to K^-\pi^+\pi^+$$

b. No evidence[43] for K* production or ρ production in the reaction

$$D^o \to \bar{K}^o\pi^+\pi^-$$

c. Evidence for ρ production in the reaction

$$D^o \to K^-\pi^+\pi^+\pi^-$$

For the last decay they find

Phase Space	$K^-\pi^+\rho^o$	$K^*\pi^+\pi^-$	$K^*\rho^o$
$0.05^{+0.11}_{-0.05}$	$0.85^{+0.11}_{-0.22}$	$0.0^{+0.2}_{-0.0}$	$0.10^{+0.11}_{-0.10}$

TABLE VI. Summary of D decay modes and branching fractions measured by the LGW experiment.[53,72,73]

Mode	B (%)
$D^0 \to K^-\pi^+$	2.2 ± 0.6
$\bar{K}^0\pi^+\pi^-$	4.0 ± 1.3
$K^-\pi^+\pi^-\pi^+$	3.2 ± 1.1
$K^-\pi^+\pi^0$	12 ± 6
$\bar{K}^0\pi^+\pi^-\pi^+\pi^-$	seen
e^+X [a]	7.2 ± 2.6
$D^+ \to \bar{K}^0\pi^+$	1.5 ± 0.6
$K^-\pi^+\pi^+$	3.9 ± 1.0
$\bar{K}^0\pi^+\pi^-\pi^+$	seen
e^+X [a]	7.2 ± 2.6

[a]The quantity measured is an average value for the D^+ and D^0 mesons. Here we assume that the two branching fractions are the same. See Section IV.D.2 for more details.

2. Semileptonic Decays. Evidence for anomalous electron production as a signature for D production and decay into an electron was first reported by the DASP group[76] in events with more than three charged prongs, that is, two charged prongs in addition to the electrons. The semileptonic branching fractions measured by different experiments are summarized in Table VII. Before we discuss the results we point out some characteristics of the events containing decays of D mesons and some difficulties in measuring the branching fractions.

a. The charmed particle decays with an electron are less affected by background in the multiprong events ($n_{ch} \geq 3$). The other source of anomalous electron production at these energies is the τ lepton.[30] The τ is expected to decay about 75% of the time into one charged prong. Therefore it is produced most copiously in 2-prong events, whereas the multiprong events ($n_{ch} \geq 3$) have a smaller contribution from this source. We will see that for events with $n_{ch} \geq 3$, one of which is an electron, the τ background is expected to be ~25% (see for example Fig. 24).

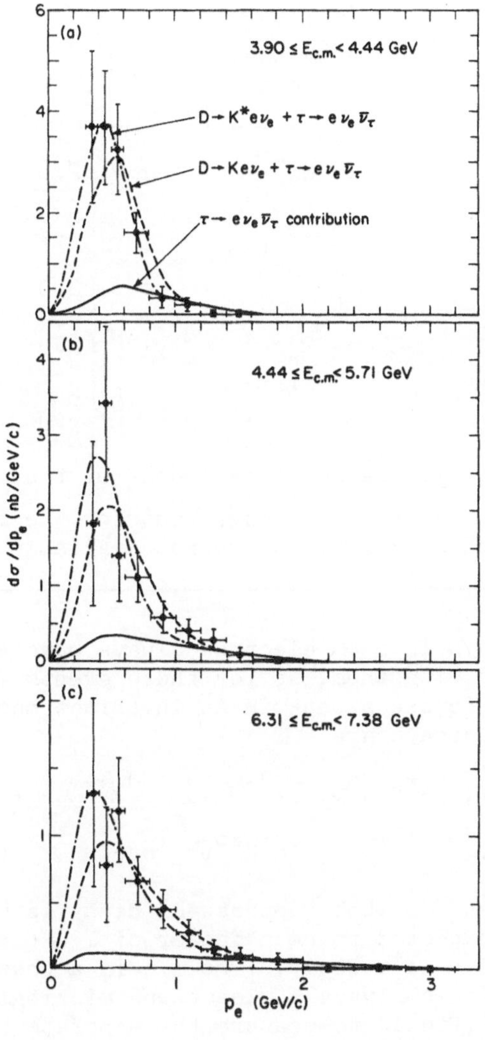

Fig. 24. The momentum spectrum for electrons produced in events with $n_{ch} \geqslant 3$ in three different energy intervals as obtained by the LGW experiment.[77] The curves are labeled in (a) and the energy intervals are indicated in each graph.

TABLE VII. The branching fraction for D semileptonic decay into
electrons as measured by various experiments. For
E > 4.08 GeV other charmed particles may contribute
to the measurement.

E (GeV)	Electron events	Background events	Branching fraction (%)	Reference
3.772	61	25	7.2 ±2.8	LGW[73]
3.90 – 7.38	448	155	8.2 ±1.9	LGW[77]
3.99 – 4.08	$-^a$	$-^a$	8.0 ±2.0	DASP[78]
3.99 – 5.20	182	27	7.2 ±2.0	DASP[78]
3.77	238^b	$-^b$	10 ±2	DELCO[17,48]

[a] This determination is not independent of the following one.

[b] The number of events and backgrounds for the most recent
analysis of this experiment are not available.

b. The D decays into an electron always have a neutrino
associated with them, so for these events it is very
difficult to see a peak in an invariant mass distribution.
The major decays are

$$D^o \rightarrow e^+K^-\nu, \quad e^+K^-\pi^o\nu, \quad e^+\bar{K}^o\pi\nu, \quad e^+K^{*-}\nu, \quad \text{etc.} \qquad (63)$$

$$D^+ \rightarrow e^+\bar{K}^o\nu, \quad e^+K^-\pi^+\nu, \quad e^+\bar{K}^{*o}\nu, \quad \text{etc.} \qquad (64)$$

The largest "Cabibbo-suppressed" decay is $D \rightarrow \pi e^+\nu$,
which is expected to be a factor of $1.6 \tan^2\theta$ smaller
than the $D \rightarrow Ke^+\nu$ decay. The 1.6 is a phase-space factor.
Since there is always at least one missing particle, it
is very difficult to measure the separate branching
fractions for D^o and D^+. This would be possible at the
$\psi(3772)$ for the sequence

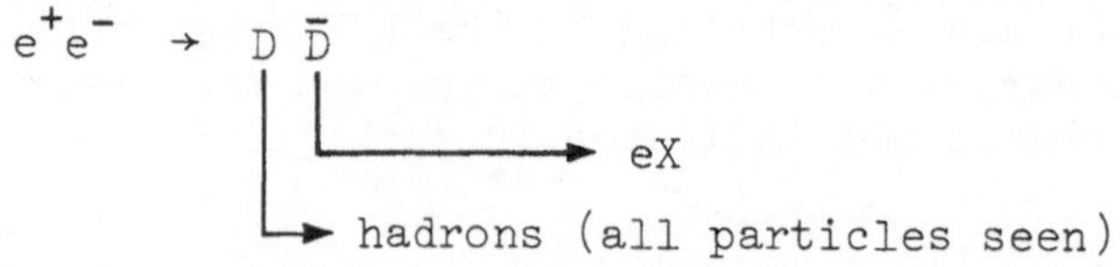

$$e^+e^- \rightarrow D\bar{D}$$
$$\qquad\qquad\qquad \longrightarrow eX \qquad\qquad (65)$$
$$\qquad\qquad \longrightarrow \text{hadrons (all particles seen)}$$

because for these events, tagged D's, the sign and branching fractions of the D decaying into hadrons are known. Using the relation (58) a count of these events for D^0 and D^+ could give us the separate branching fractions. Unfortunately, the statistics[73] collected so far at the ψ'' are not enough to allow such a method. Therefore all the semileptonic branching fractions quoted are averaged over D^+ and D^0.

c. In the experiments done so far only the electron spectrum has been measured. The K or K* decays of Eqs. (63) and (64) predict a different electron spectrum (see for example, Fig. 24). Therefore a large statistics experiment can distinguish among the two and measure each contribution separately. To calculate a branching fraction, B_e, it is necessary to calculate the acceptance (see Eq. (58)) of the apparatus, therefore an assumption has to be made on the relative importance of K and K* final states. The quoted values in Table VII depend on this assumption, although they are not too sensitive to it. The usual assumption is equal contribution from K and K*.

d. To calculate B_e one has to know σ_D, therefore the $\psi(3772)$ or the 4.03 GeV results are more reliable. At higher energies the branching fraction obtained is an average over charm particle semileptonic decays. As discussed in Section III.D, $R_{charm} = R_{D\bar{D}} + R_{F\bar{F}} + R_{B\bar{B}}$. We know $R_{D\bar{D}}$ (see Table III) at some energies, the charmed baryon contribution is at most 0.32 units of R, whereas $R_{F\bar{F}}$ is very uncertain (see Section III.B).

Table VII shows the measured branching fractions for D semileptonic decays. The LGW experiment has made two measurements, in view of (d) above. The first[73] is at the $\psi(3772)$, the second one at higher energies.[77] The electron spectra obtained in three sub-samples of the high energy data are shown in Fig. 24. The contribution of the τ heavy lepton is estimated to be 25%, assuming $B(\tau^- \rightarrow e^- \bar{\nu}_e \nu_\tau) = (18 \pm 2)\%$ and $B(\tau^- \rightarrow \nu_\tau + n_{ch} \geqslant 3) = (25 \pm 10)\%$.

The branching fractions measured at the $\psi(3772)$ and at higher energies agree within errors. Figure 25 shows the branching fractions measured at different energies. They are consistent with a constant value indicating that the contribution from other semileptonic decays of charmed particles is small enough not to alter B_e, or that the B_e for these other decays are not too different.

The DASP data[78] in the 4.03 GeV region are shown in Fig. 26. The branching fractions measured at 4.03 GeV and in the whole energy region are in agreement with the LGW result. Finally, the

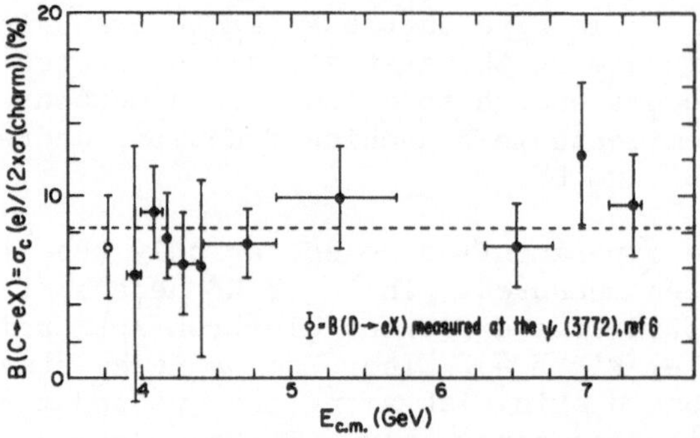

Fig. 25. The branching fraction for charmed particle decay into an electron plus additional particles as a function of energy.[77] The value at the $\psi(3772)$ is from Ref. 73. The dashed line indicates the average value of the ratio for $3.9 < E_{c.m.} < 7.4$ GeV.

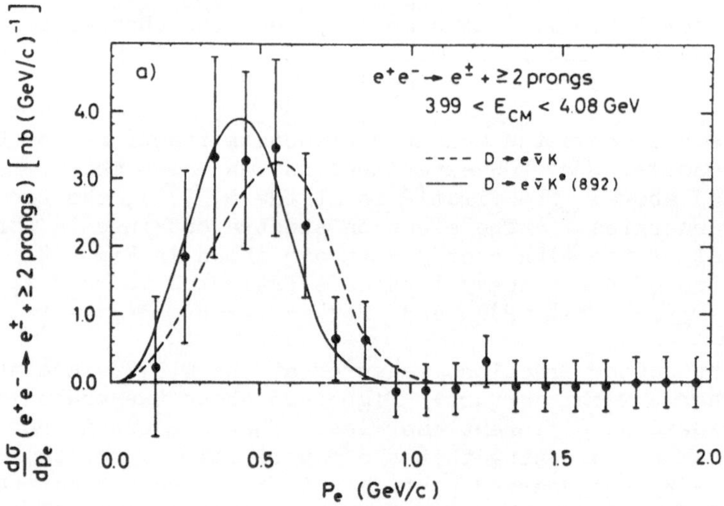

Fig. 26. The electron momentum spectrum for $D \to e\nu X$ as measured by DASP.[78]

DELCO electron spectrum[48] at the $\psi(3772)$ is shown in Fig. 27. The quoted branching fraction is somewhat larger, but not in disagreement with the other determinations.

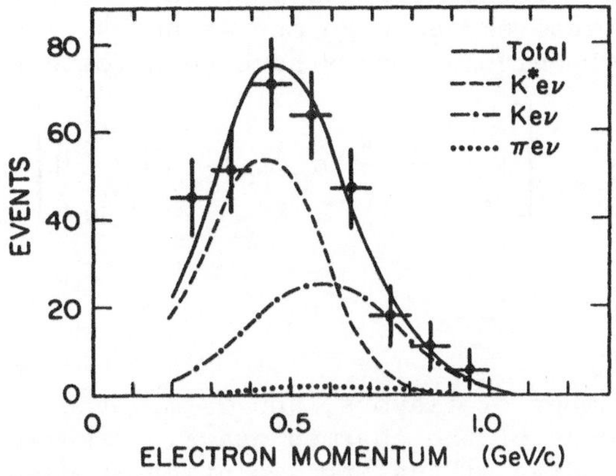

Fig. 27. The electron momentum spectrum from $D \to e\nu X$ as measured by DELCO[48] at the $\psi(3772)$. The curves shown are the result of a fit (see text).

In summary, taking the weighted average of the above results (except for DASP's result at 4.03 GeV), we obtain for the semi-leptonic branching fraction the value

$$B_e = (8.3 \pm 1.1)\% \tag{66}$$

The DELCO experiment has also tried to separate the contributions to B_e from the different semileptonic decay modes. A fit to the electron spectrum shown in Fig. 27 was made to $K^*e\nu$, $Ke\nu$ and $\pi e\nu$. The fraction of $\pi e\nu$ decay was fixed whereas the other two were free to vary. The results are:

$$B(D \to Ke\nu) = (3.7 \pm 2.1)\%$$

$$B(D \to K^*e\nu) = (6.0 \pm 2.3)\%$$

$$B(D \to \pi e\nu) < 2\% \ (90\% \text{ CL})$$

3. <u>The Nonleptonic Enhancement Question</u>. The two results most relevant to nonleptonic enhancement are given in Eqs. (59) and (66). As discussed in Section II.C.3, the semileptonic branching fraction into electrons is expected to be 20% from quark counting and as low as 3% from nonleptonic enhancement calculations.[38] The experimental result is $(8.3 \pm 1.1)\%$.

Ellis et al[38] have calculated the ratio (59); they find

$$r = \frac{\Gamma(D^+ \to \bar{K}^o\pi^+)}{\Gamma(D^o \to K^-\pi^+)} = 4\left(1 + \frac{f_-}{2f_+}\right)^{-2} \tag{67}$$

where f_- and f_+ are the coefficients of the terms in the Lagrangian transforming respectively as a $\underline{20}$ and as an $\underline{84}$. The coefficients f_+ and f_- from QCD calculations[38] have been found to be

$$f_- = \left[1 + \frac{33-2F}{12}\,\alpha_S(m_c)\,\ln\left(\frac{M_W^2}{M_c^2}\right)\right]^{12/(33-2F)} \tag{68}$$

$$f_+ = \frac{1}{\sqrt{f_-}} \tag{69}$$

where F is the number of flavors, and $\alpha_S(m_c)$ the running coupling constant at the mass of the charmed quark. Assuming $\alpha_S(m_c) = 0.7$, as obtained in studies of scaling violation in deep inelastic processes,[79] and $F = 6$ Cabibbo and Maiani[58] have recently calculated r and B_e. They get

$$f_- = 2.15 \qquad \text{and} \qquad f_+ = 0.68 \quad .$$

With these values then

$$r = \frac{B(D^+ \to \bar{K}^o\pi^+)}{B(D^o \to K^-\pi^+)}\,\frac{\tau^o}{\tau^+} = 0.60 \quad . \tag{70}$$

This value is in agreement with the measured value of (0.70 ± 0.23) τ^o/τ^+ if the two lifetimes are not too different. The effect of the $\underline{20}$ enhancement does not result in a large suppression of $D^+ \to \bar{K}^o\pi^+$ which is pure $\underline{84}$ with respect to $D^o \to K^-\pi^+$ because the latter has a small projection in the $\underline{20}$ and a larger one in the $\underline{84}$ representation. In the limit of free quarks, one gets $f_+ = f_- = 1$ and $r = 1.78$.

In the same model these authors[58] have also calculated the semileptonic branching fraction. They get

$$B_e = \frac{1}{2 + 2f_+^2 + f_-^2} \sim 13\% \quad . \tag{71}$$

In the limit of free quarks $B_e = 20\%$. Similar calculations have been done by Fakirov and Stech.[80]

In conclusion, the experimental results indicate that there is nonleptonic enhancement. The magnitude is such that it takes the ratio (59) from r = 1.78 for no enhancement down to the observed value 0.70 and B_e from the expected 20% ($B_e = 1/(1+1+3)$) down to the observed value of 8.3%. This implies a nonleptonic enhancement of about a factor 3. This amount of observed nonleptonic enhancement can be accounted for by QCD calculations.[38,58,80]

4. D Meson Inclusive Decays. The lead-glass wall experiment has reported[71] some inclusive characteristics of D^o and D^+ decays. This has been possible through the use of tagged D's at the $\psi(3772)$. In fact, at this energy if we know that there is a D (or $\bar{D}$) in an event, what recoils against it must be a $\bar{D}$ (or D), since there is not enough energy to produce an additional pion or a D*.

The events used in this study come from the same sample used to measure the masses and branching ratios of D mesons. We have tagged

$$141 \quad D^o \text{ (or } \bar{D}^o) \rightarrow K^{\mp}\pi^{\pm} \tag{72a}$$

$$107 \quad D^+ \text{ (or } D^-) \rightarrow K^{\mp}\pi^{\pm}\pi^{\pm} \tag{72b}$$

by selecting the events in a narrow mass interval around the D. For D^o we use the three highest bins in Fig. 20a; using adjacent bins the background is estimated to be 15.6%. For the D^+ sample we have taken the four highest bins in Fig. 20e. This sample has a background of non D events of 25.2%.

The simplest quantity to measure in this sample is the charged particle multiplicity. Except for background and acceptance correction this measurement requires only counting the number of observed charged particles in the system recoiling against the observed D. Since the solid angle of the detector is only 0.73 of 4π for tracking, that is, for measuring the momentum of a charged particle, a number of charged prongs escape detection. We have calculated by Monte Carlo techniques the efficiency to observe a number of charged prongs, n_{ch}, as a function of the produced number of prongs. We then use these efficiencies to unfold the "true" produced distribution from the observed one. These distributions are shown in Fig. 28. They show that 74% of the events with a D^o have two charged prongs and 15% have four charged prongs in the decay products. For the $D^{\pm}$ we find that 37% have only one charged prong and 59% have three charged prongs in the decay products. The average unfolded multiplicites are:

$$\langle n_{ch} \rangle_{D^o} = 2.3 \pm 0.3 \tag{73a}$$

$$\langle n_{ch} \rangle_{D^+} = 2.3 \pm 0.3 \tag{73b}$$

Next we have measured the K content among these charged prongs. The charged and neutral kaons were identified as mentioned in

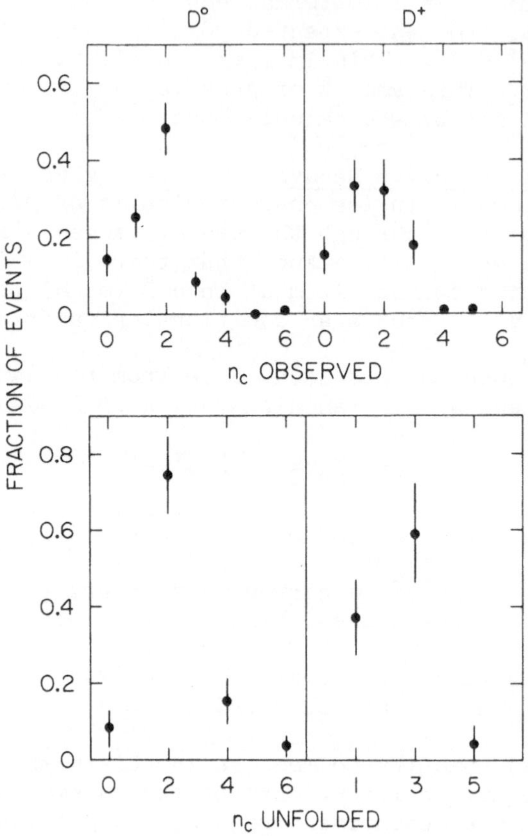

Fig. 28. Charged multiplicity distributions for D^0 and D^+ decays.[71] The data shown in the top graphs are the observed distributions; the bottom graphs have been obtained after corrections for detection efficiency.

Section III.A.1, that is, $K^{\pm}$ by time-of-flight (TOF) measurement, the K^0 by studying the $\pi^+\pi^-$ invariant mass. After correcting the observed $K^{\pm}$ content for decays in flight, TOF efficiency, tracking efficiency, and geometrical acceptance we obtain the fractions of $K^{\pm}$ shown in Table VIII. For K^0 we only detect K_S, so the observed events are corrected for unseen K^0 decays as well as for inefficiency in detecting K_S. Unfortunately, the statistics are very low due to the fact that the detectable $K_S \to \pi^+\pi^-$ are only one-third of all the K^0 produced. In Table VIII we note the following:

a. The total number of K/event are:

$D^o \rightarrow K^{\pm}, K^o$ 0.92 ± 0.28

$D^+ \rightarrow K^{\pm}, K^o$ 0.55 ± 0.30

We expect at least 95% of the events to have a K, since the Cabibbo forbidden decays are ~5%. The observation is in agreement with the expectation, although the D^+ result is a little low.

b. For D^o, the statistical model of Quigg and Rosner[57] (except for a correction due to Cabibbo forbidden decays with no K's in the final state) predicts:

$D^o \rightarrow K^-$ 0.48

$D^o \rightarrow \bar{K}^o$ 0.52

to be compared with 0.34 ± 0.08 and 0.57 ± 0.26 respectively, in fair agreement within the errors.

c. For D^+, the statistical model predicts (with the same small correction mentioned above):

$D^+ \rightarrow K^-$ 0.33

$D^+ \rightarrow K^o$ 0.67

to be compared with the experimental values 0.10 ± 0.07 and 0.39 ± 0.29 respectively. The first value, 0.10 ± 0.07, is therefore not in good agreement with 0.33. The errors are large, therefore at this time there is no cause for alarm, but it is suggestive. An experiment with higher statistics is needed before drawing any conclusions.

TABLE VIII. Fractions of charged and neutral kaons in D^o and D^+ decays.[71]

Mode	Events found	Background events expected	Efficiency	Branching fraction
$D^o \rightarrow K^{\pm}X$	21.2 ± 5.1	2.4 ± 0.6	0.46	0.35 ± 0.10
$D^o \rightarrow K^oX$	7 ± 2.6	1.1 ± 0.8	0.09	0.57 ± 0.26
$D^+ \rightarrow K^-X$	4.8 ± 2.2	1.4 ± 0.5	0.42	0.10 ± 0.07
$D^+ \rightarrow K^+X$	2.8 ± 1.7	1.1 ± 0.4	0.39	0.06 ± 0.06
$D^+ \rightarrow K^oX$	4 ± 2.0	1.3 ± 0.8	0.09	0.39 ± 0.29

Finally, we have measured the average energy going into the different particles in the final state: charged pions, kaons, photons, electrons, and muons. The average values of the energy as a fraction of the D energy are as follows:

	D^o	D^+
$\pi^\pm$	0.53 ± 0.06	0.57 ± 0.08
$K^\pm$	0.15 ± 0.04	0.06 ± 0.04
K^o	0.21 ± 0.11	0.16 ± 0.14
$e^\pm, \mu^\pm$	0.03 ± 0.01	0.03 ± 0.01
γ	0.23 ± 0.10	0.20 ± 0.12
Total	1.15 ± 0.16	1.02 ± 0.21

We expect that some energy will be carried away by the neutrinos associated with the 16% semileptonic decays. This energy will be of the same order of magnitude as that carried by $e^\pm$ and $\mu^\pm$. So we can conclude that within the errors (~20%) all the D decay energy is accounted for.

5. Cabibbo Forbidden Decays. For these decays, as discussed in Section II.C.1, the $c \to d$ transition and $u\bar{s}$ pair creation take place. We expect these rates to be suppressed by at least a $\tan^2\theta = 0.055$ factor. The SPEAR experiment SP17 has searched for these decay modes and found none.[43] Figure 29 shows the invariant mass plots for five of these possible modes; the sixth plot is $D^\pm \to K_S\pi^\pm$, not seen in that experiment, but later detected in the SP26 experiment, as discussed in Section IV.D.1 above.

The upper limits found in this experiment are expressed in terms of $\sigma \cdot B$. Since we now know σ for D^+ and D^o at 4.03 GeV (Table III), we can express the results as a branching fraction upper limit in percent. The results are:

	Mode	Branching fraction
(a)	$D^o \to \pi^+\pi^-$	$< 0.2\%$
(b)	$D^o \to K^+K^-$	$< 0.2\%$
(c)	$D^+ \to K^+K^-\pi^+$	$< 0.6\%$
(d)	$D^+ \to \pi^+\pi^-\pi^+$	$< 0.3\%$
(e)	$D^+ \to K^+\pi^-\pi^+$	$< 0.2\%$

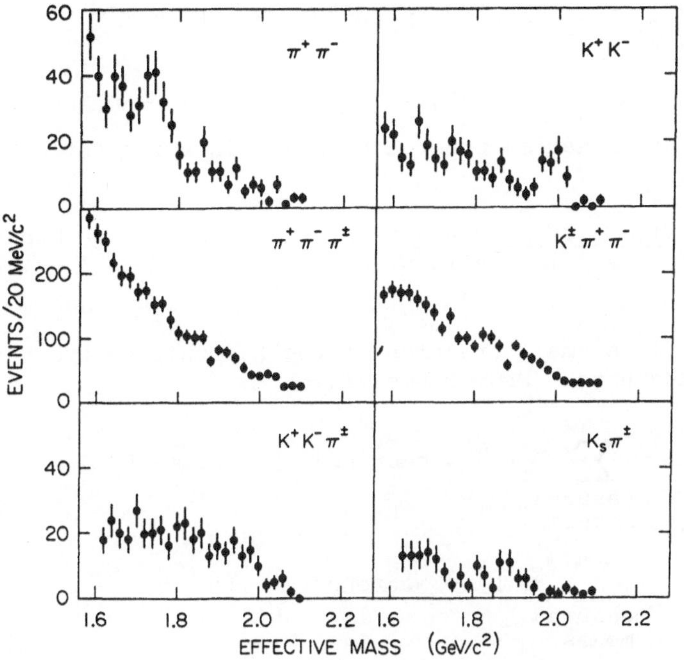

Fig. 29. Search for Cabibbo suppressed decays[43] of the D mesons [$M(D^o) = 1863$ MeV, $M(D^+) = 1868$ MeV]. The decay channel $D^\pm \to K_S\pi^\pm$ is not Cabibbo suppressed; it has been later observed by the LGW experiment[53] (see Fig. 20).

From Fig. 4a and 4e we see that, apart from kinematical factors:

$$\frac{\Gamma(D^o \to \pi^-\pi^+)}{\Gamma(D^o \to K^-\pi^+)} \sim \frac{\sin^2\theta \cos^2\theta}{\cos^4\theta} = \tan^2\theta = 0.055 \qquad (74)$$

that is, according to Eq. (15), each suppressed transition brings a $\sin\theta$ factor in the amplitude and each of the favored transitions brings a $\cos\theta$ factor. From Table VI we expect for $D^o \to \pi^-\pi^+$:

$$B(D^o \to \pi^-\pi^+) \sim (2.2 \times 0.055)\% = 0.12\%$$

smaller than the measured upper limit. For $D^o \to K^+K^-$ the same $\tan^2\theta$ factor is expected.

The D^+ decays (c) and (d) are suppressed by a $\tan^2\theta$ factor over the $D^+ \to K^-\pi^+\pi^+$ decay (see Table VI), whereas the last one,

$D^+ \to K^+\pi^-\pi^-$, is suppressed by $\tan^4\theta$. For $D^+ \to \pi^+\pi^+\pi^-$ we expect, from Table VI

$$B(D^+ \to \pi^+\pi^+\pi^-) \sim (3.9 \times 0.055)\% = 0.21\% \quad ,$$

therefore the measured upper limits are in agreement with expectation.

6. Summary of D Decays. We can summarize the experimental results discussed in sections 1-5 as follows:

a. Only a small fraction of the hadronic decays have been measured. From Table VI,

$$\sum_{\substack{\text{measured}\\ \text{modes}}} B_i(D^0 \to \text{hadrons}) = (21.4 \pm 6.3)\%$$

$$\sum_{\substack{\text{measured}\\ \text{modes}}} B_i(D^+ \to \text{hadrons}) = (5.4 \pm 1.2)\%$$

b. The semileptonic decay branching fraction (Eq. 66) is found to be $B_e = (8.3 \pm 1.1)\%$.

c. The measured branching fraction for $D^+ \to \bar{K}^0\pi^+$ (Table VI) along with that for the semileptonic decay, indicates that there is nonleptonic enhancement in charmed particle decays, in analogy to $\Delta I = 1/2$ or octet enhancement for strange particles. The hadronic decays are enhanced by about a factor 3.

d. Inclusive D decay studies show that

i) $\langle n_{ch} \rangle = 2.3 \pm 0.3$ for D^0 and D^+.

ii) $(D^0, D^+) \to \bar{K}^0$ are more copious than decays into $K^\pm$.

iii) $B(D^+ \to K^-)$ is only $(10 \pm 7)\%$.

iv) There is no energy missing in D decays within the 20% experimental errors.

e. The Cabibbo suppressed decays are not observed; the quoted upper limits are consistent with expectation.

REFERENCES

1. J.D.Jackson, Proceedings of the 1976 Summer Institute on Particle Physics, SLAC Report SLAC-198, p.147 (1976), and LBL Report LBL-5500 (1976).

2. H.Harari, Proceedings of the 1977 Summer Institute on Particle Physics, SLAC Report SLAC-204, p.1 (1977).

3. T.Appelquist, R.M.Barnett and K.Lane, Annual Review of Nucl. and Particle Sci. 28 (1978).

4. R.F.Schwitters, Proc. 1975 Int. Symposium on Lepton Photon Interactions at High Energies, Stanford, California, p.355 (1975).

5. P.A.Rapidis et al., Phys. Rev. Lett. 39, 526 (1977).

6. T.Appelquist and H.Georgi, Phys. Rev. D8, 4000 (1973); A.Zee, Phys. Rev. D8, 4038 (1973).

7. A.De Rujula, S.L.Glashow, Phys. Rev. Lett. 34, 46 (1975).

8. G.Bonneau and F.Martin, Nucl. Phys. B27, 381 (1971); D.R.Yennie, Phys. Rev. Lett. 34, 239 (1975).

9. J.D.Jackson and D.L.Scharre, Nucl. Inst. & Meth. 128, 13 (1975).

10. S.W.Herb et al., Phys. Rev. Lett. 39, 252 (1977).

11. C.Darden et al., Phys. Lett 76B, 246 (1978); Ch.Berger et al., Phys. Lett. 76B, 243 (1978). For Τ see also J.Bienlein et al from the following reference.

12. J.K.Bienlein et al., Phys. Lett. 78B, 360 (1978); C.W.Darden et al., Phys. Lett. 78B, 364 (1978).

13. C.Quigg and J.L.Rosner, Phys. Lett. 72B, 462 (1978); J.L.Rosner, C.Quigg and H.B.Tacker, Phys. Lett. 74B, 350 (1978); see also C.Quigg, "New Quark Flavors," Fermilab-Conf-78/17-THY (1978), talk presented at Orbis Scientiae, 1978.

14. Particle Data Group, C.Bricman et al., Phys. Lett. 75B, 1 (1978)

15. A.M.Boyarski et al., Phys. Rev. Lett. 34, 1357 (1975).

16. V.Lüth et al., Phys. Rev. Lett. 35, 1124 (1975).

17. W.Bacino et al., Phys. Rev. Lett. 40, 671 (1977).

18. R.Brandelik et al., Phys. Lett. 76B, 361 (1978).

19. J.Siegrist et al., Phys. Rev. Lett. 36, 700 (1976).

20. G.Flügge, Proc. of the XIX Int. Conf. on High Energy Physics, 23-30 August 1978, Tokyo, Japan, and DESY Report DESY-78/55.

21. R.Van Royen and V.F.Weisskopf, Nuovo Cimento 50A, 617 (1967), and Nuovo Cimento 51A, 583 (1967).

22. T.Appelquist and H.D.Politzer, Phys. Rev. Lett. 34, 43 (1975).

23. J.E.Augustin et al., Phys. Rev. Lett. 33, 1406 (1974); Aubert et al., Phys. Rev. Lett. 33, 1404 (1974).

24. Eichten et al., Phys. Rev. Lett. 34, 369 (1975); also Phys. Rev. Lett. 36, 500 (1976); Lane et al., Phys. Rev. Lett. 37, 477 (1976); K.Gottfried, Phys. Rev. Lett. 40, 598 (1978).

25. K.Gottfried, Proc. of the 1977 Int. Symposium on Lepton and Photon Interactions at High Energy, August 1977, Hamburg (DESY, Hamburg, 1977), p.667.

26. S.Okubo, Phys. Lett. 5, 105 (1963); G.Zweig, CERN Report TH-401, 412 (1964); J.Iizuka, K.Okada, and O.Shito, Prog. Theor. Phys. 35, 1061 (1966).

27. S.L.Glashow, U.Iliopoulos and L.Maiani, Phys. Rev. D2, 1285 (1970).

28. S.Weinberg, Phys. Rev. Lett. 19, 1264 (1967); A.Salam, in Elementary Particle Physics: Relativistic Groups and Analyticity, edited by N.Svartholm (Almquist and Wiskell, Stockholm, 1968), p.367.

29. N.Cabibbo, Phys. Rev. Lett. 10, 531 (1963).

30. See the most recent review article by M.Perl, "Evidence for and properties of the τ lepton," presented at the Ben Lee Memorial Conf. on Parity Nonconservation, Weak Neutral Currents and Gauge Theories, Batavia, Illinois, October 1977, also published as SLAC-PUB-2055. See also G.Feldman, "Properties of the τ Lepton," in Neutrino-78. also published as SLAC-PUB-2138 (1978).

31. M.Kobayashi and K.Maskawa, Progr. of Theoret. Phys. 49, 652 (1973).

32. J.Ellis et al., Nucl. Phys. B131, 285 (1977).

33. L.Maiani, Phys. Lett. 62B, 186 (1976); S.Pakvasa and H.Sugawara, Phys. Rev. D14, 305 (1976).

34. M.K.Gaillard and B.Lee, Phys. Rev. Lett. 33, 108 (1974); G.Altarelli and L.Maiani, Phys. Lett. 52B, 351 (1974).

35. M.K.Gaillard, B.W.Lee, J.L.Rosner, Rev. Mod. Phys. 47, 277 (1975); G.Altarelli, N.Cabibbo, and L.Maiani, Nucl. Phys. B88, 285 (1975).

36. R.Kingsley et al., Phys. Rev. D11, 1919 (1975).

37. M.B.Einhorn and C.Quigg, Phys. Rev. D12, 2015 (1975).

38. J.Ellis, M.K.Gaillard and D.V.Nanopoulos, Nucl. Phys. B100, 313 (1975).

39. J.E.Augustin et al., Phys. Rev. Lett. 34, 233 (1975).

40. A.Barbaro-Galtieri et al., Phys. Rev. Lett. 39, 526 (1977); also J.Feller et al., IEEE Transactions on Nucl. Sci. NS-25, 304 (1978).

41. G.Goldhaber et al., Phys. Rev. Lett. 37, 255 (1976).

42. I.Peruzzi et al., Phys. Rev. Lett. 37, 569 (1976).

43. M.Piccolo et al., Phys. Letters 70B, 260 (1977).

44. V.Lüth et al., Phys. Lett. 70B, 120 (1977).

45. G.Goldhaber et al., Phys. Lett. 69B, 503 (1977).

46. C.Baltay et al., Phys. Rev. Lett. 41, 73 (1978)

47. Eichten et al., Phys. Rev. D17, 3090 (1978).

48. J.Kirkby, Proc. of the 1977 Int. Symposium on Lepton and Photon Interactions at High Energy, Hamburg, August 1977 (DESY, Hamburg, 1977), p.3. See also J.Kirkby, Proc. of the 1978 Summer Institute on Particle Physics, SLAC, 1978, also published as SLAC-PUB-2231 (1978).

49. A.De Rujula, H.Georgi and S.L.Glashow, Phys. Rev. Lett. 38, 317 (1977).

50. G.Knies, Proc. of the 1977 Int. Symposium on Lepton and Photon Interactions at High Energy, Hamburg, August 1977 (DESY, Hamburg, 1977), p.69.

51. I.Peruzzi et al., "Inclusive K and D production in e^+e^- annihilation," Lawrence Berkeley Laboratory Report LBL-7935, to be published (1978).

52. P.A.Rapidis et al., "Inclusive production of D mesons in e^+e^- annihilation at 7 GeV," to be published (1978).

53. I.Peruzzi et al., Phys. Rev. Lett. 39, 1301 (1977).

54. R.Brandelik et al., Phys. Lett. 70B, 132 (1977). See also S.Yamada, Proceedings of the 1977 Int. Symposium on Lepton and Photon Interactions at High Energy, Hamburg, August 1977, (DESY, Hamburg, 1977), p.47.

55. R.Brandelik et al., "Production characteristics of the F meson," DESY Report (October 1978).

56. D.Lüke, Proceedings of the 1977 Meeting of the Division of Particle and Fields, Argonne 1977, p.441 (1977).

57. C.Quigg and J.L.Rosner, Phys. Rev. D17, 239 (1978).

58. N.Cabibbo and L.Maiani, Phys. Rev. Lett. 73B, 418 (1978).

59. E.G.Cazzoli et al., Phys. Rev. Lett. 34, 1125 (1975).

60. B.Knapp et al., Phys. Rev. Lett. 37, 882 (1976).

61. M.Piccolo et al., Phys. Rev. Lett. 39, 1503 (1977).

62. A.De Rujula, H.Georgi and S.L.Glashow, Phys. Rev. D12, 147 (1975).

63. J.Wiss et al., Phys. Rev. Lett. 37, 1531 (1976); H.K.Nguyen et al., Phys. Rev. Lett. 39, 262 (1977).

64. P.B.Wilson et al., Stanford Linear Accelerator Report SLAC-PUB-1894 (1977).

65. G.J.Feldman et al., Phys. Rev. Lett. 38, 1313 (1977).

66. A.De Rujula, H.Georgi and S.Glashow, Phys. Rev. D12, 147 (1975).

67. K.Lane and S.Weinberg, Phys. Rev. Lett. 37, 717 (1976).

68. See e.g., A.De Rujula, H.Georgi and S.L.Glashow, Phys. Rev. Lett. 35, 69 (1975); R.Kingsley, Phys. Lett. 63B, 329 (1976).

69. S.L.Glashow and S.Weinberg, Phys. Rev. D15, 1958 (1977); E.A.Paschos, Phys. Rev. D15, 1966 (1977).

70. See for example, S.Ono, Phys. Rev. Lett. 37, 655 (1976).

71. V.Vuillemin et al., Phys. Rev. Lett. 41, 1149 (1978).

72. D.L.Scharre et al., Phys. Rev. Lett. 40, 274 (1978).

73. J.M.Feller et al., Phys. Rev. Lett. 40, 274 (1978).

74. A.Barbaro-Galtieri, Proc. of the 1977 Int. Symposium on Lepton and Photon Interactions at High Energy, Hamburg, August 1977 (DESY, Hamburg, 1977), p.21.

75. J.E.Wiss et al., Phys. Rev. Lett. 37, 1531 (1976).

76. W.Braunschweig et al., Phys. Lett. 63B, 471 (1976).

77. J.M.Feller et al., Phys. Rev. Lett. 40, 1677 (1978).

78. R.Brandelik et al., Phys. Lett. 70B, 387 (1978). See updated values of the branching fraction in B.H.Wiik and G.Wolf, "A review of e^+e^- interactions," DESY Report DESY-78/23 (1978).

79. A.De Rujula, H.Georgi and D.Politzer, Ann. Phys. 103, 315 (1977).

80. D.Fakirov and B.Stech, Nucl. Phys. B133, 315 (1978).

81. J.D.Jackson, C.Quigg and J.L.Rosner, Proc. of the XIX Int. Conf. on High Energy Physics, 23-30 August 1978, Tokyo, Japan and LBL Report LBL-7977.

82. J.D.Jackson, Proc. of the European Conf. on Particle Physics, July 1977, Budapest, Hungary, Vol. 1, p.603 (1977).

83. T.Ferguson et al., SLAC Report SLAC-2081r and UCLA Report UCLA-1116r, to be published in Phys. Lett. (1979).

ACKNOWLEDGMENTS

This work was supported by the Physics Division of the U.S. Dept. of Energy under contract N° W-7405-ENG-48.

DISCUSSION

Chairman : A. Barbaro-Galtieri

Scientific Secretaries : J. Bürger, R. M. Morse

DISCUSSION No. 1

- *Whitaker:*

Could you please state the energy and the luminosity of the data from which you studied the F-signal?

- *Barbaro-Galtieri:*

We took some data at 4.11 GeV, the reason being that the SP17 cross section showed a little fluctuation at this energy. We analyzed the data and found no evidence for the F. The cross section from DASP had a bump at 4.16 GeV, so we took data also at 4.16 GeV. At 4.11 GeV we have an integrated luminosity of 0.4 pb^{-1} and at 4.16 we have 0.94 pb^{-1}.

- *Montgomery:*

This morning you suggested that from the ratio of $p\bar{p}$ to $\Lambda\bar{\Lambda}$ production the charmed baryons should decay to kaons. Do you have any information on the associated kaon production?

- *Barbaro-Galtieri*

We have inclusive K^o and K^+ production cross sections. The $K^\pm$ analysis does not go that high in energy because it is more difficult to distinguish π from K (our time-of-flight measurement is not reliable for $p > 0.9$ GeV). The K^o efficiency is not very good (~0.09) and the errors are too large to detect an increase of $\Delta R \lesssim 0.3$.

- *De La Vaissière:*

In the 9 $F \to K\bar{K}\pi$-events, have you seen a ϕ in the $K\bar{K}$ mass?

- *Barbaro-Galtieri:*

We haven't seen $F \to \phi\pi$ with $\phi \to K\bar{K}$ at 4.16 GeV in the F events recorded. Further, the ϕ production cross section, which is very small, shows no detectable increase at 4.16 GeV within the measurement errors.

- *De La Vaissière:*

Have you seen any η in the recoiling system?

- *Barbaro-Galtieri:*

We haven't tried to fit the η in the recoiling system. The efficiency for η detection is very low. The solid angle for γ detection is only 6% of 4π (lead-glass-wall). We are really only sensitive to the $\pi^+\pi^-\pi^0$ decay of the η which is only 24%, therefore, the overall efficiency for η is very low.

- *Kumar:*

What is the K^0 detection efficiency and what is the K^0 decay distance cutoff?

- *Barbaro-Galtieri:*

The K^0 detection efficiency is about 30% of the $K_S \to \pi^+\pi^-$ and we can resolve K^0 decays up to 0.6 cm.

- *Mouzourakis:*

Can you set an upper limit on the channel $F \to \eta h$, which was used at DESY to discover the F? Why is the efficiency for η detection so low in the SLAC-LBL case?

- *Barbaro-Galtieri:*

If the $\eta(n\pi)$ and $K\bar{K}(n\pi)$ decay modes are of the same order of magnitude then we should see one event of the F going into $\eta(n\pi)$. Our η detection is very limited. The solid angle of the lead glass wall is only 6%. For the 2γ decay of the η the acceptance of the LGW is zero below $p_\eta = 1.2$ GeV/c because the $\gamma\gamma$ opening angle is larger than the detector. As for π^0, we can get a reasonably clean sample for $P_{\pi^0} > 400$ MeV/c. So we have a small efficiency for the decay $\eta \to \pi^+\pi^-\pi^0$. With our small η efficiency we have estimated that we expect to see about one η in the LGW. This analysis is not yet complete and we do not have an upper limit to quote at this time.

- *Petersen:*

Can you give the branching fractions of $\psi(4.41)$ into D mesons?

- *Barbaro-Galtieri:*

The data at 4.41 GeV is more difficult to analyze because the K-π ambiguities are larger and because there is more background. The SLAC-LBL collaboration has not given any results on the ratio of production of the various states. From the recoil mass plot they published, it is easy to see that all three processes ($D\bar{D}$, DD^* and $D^*\bar{D}^*$) are present.

- *Petersen:*

Returning to the F, it is my impression that the ratio was determined using the $F \to K\bar{K}\pi$ at 4.16 GeV and the $F \to \eta\pi$ at 4.41 GeV.

- *Barbaro-Galtieri:*

No, both were measured at 4.16 GeV.

- *Motz:*

You mentioned three different theories giving nearly the same result for $BR(F \to K\bar{K}(n\cdot\pi))/BR(F \to \eta(n\pi))$ of about 1/2. If the LGW results and the DASP results are right, we have a value of about 1/20 for this ratio.* What would be the theoretical consequences of this great discrepancy?

- *Barbaro-Galtieri:*

One of the models that I mentioned is a statistical model based on I-spin, which predicts about equal states for $K\bar{K}(n\pi)$, $\eta(n\pi)$ and $\eta'(n\pi)$. The other model is based on QCD calculation and has been used by Cabibbo and Maiani and by Fakirov and Stech.

- *Newman:*

What are the plans for the future of your group? Will you collect more statistics for the F?

- *Barbaro-Galtieri:*

We are presently not running at SPEAR. There is now operating the Mark II detector which, using liquid argon shower detectors, has an acceptance for photons of 60% of 4π. They have taken about twice as much data as that discussed here. The data is being analyzed.

---

*Note added in proof: by this time (12/78) both the LGW and the DASP results have changed. This ratio is now 1/4 (see text for details).

- *De La Vaissière:*

In the diagrams describing the hadronic decays of the charmed mesons, why do you couple predominantly the W^+, which has a spin 1 to a π^+, which has spin 0, and not to a ρ^+ or A_1^+ ?

- *Barbaro-Galtieri:*

This is a somewhat subtle question involving the spin characteristic of vector bosons very far off the mass shell.

- *Isgur:*

I would like to comment here: consider the decay $\pi \to \mu\nu$:

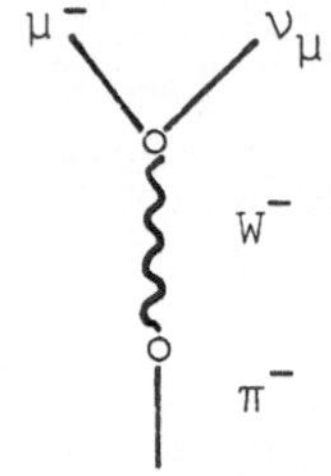

Hopefully this objection is not valid — or the pion couldn't decay to $\mu^-\bar{\nu}_\mu$! The reason the decay goes is that the $W^\mu = (W^0, \vec{W})$ has a spin 1 ($\vec{W}$) and spin zero (W^0) parts; the spin 0 part vanishes at the mass shell (to make the W a spin 1 particle) but in the decay shown, the W is very virtual. If this is unclear, I suggest you examine the W-propagator in detail.

- *Battistoni:*

Is it possible for your apparatus to detect an "all neutral" final state?

- *Barbaro-Galtieri:*

Yes, we had a neutral trigger in the Lead Glass Wall experiment. We have collected data, but we have not yet analyzed them. It's a very difficult analysis; there is a lot of background. So far we have used the $e^+e^- \to \gamma\gamma$ events for calibration purposes.

- *Geshold*

Are there calorimeter measurements of jets at SPEAR?

- *Barbaro-Galtieri:*

Not yet. The SPEAR Mark I detector had only the 6% solid angle of the LGW. There has, however, been extensive study done on jets with the Mark I and one should refer to the reports of G.Hanson and G.Wolf's final lecture.

- *Isgur:*

Prof. Zichichi mentioned this morning that the upper limit on θ_2 is 0.5°; where does that come from?

- *Barbaro-Galtieri:*

I do not know. I think there was a confusion between θ_2 and $\sin\theta_2$. The limit in $\sin\theta_2$ is in fact $S_2 < 0.5$.

- *Kahl:*

What is the nature of the 4.4 resonance and can you comment on the decomposition of the total cross section at this energy?

- *Barbaro-Galtieri:*

There are no relative cross sections quoted because the data has not been completely analyzed. There is an enhancement of D-production in this region. This D can be either produced directly or as a decay product of D*.

- *Kumar:*

Does the contribution to R you described this morning describe the data?

- *Barbaro-Galtieri:*

Yes, you can explain the rise of R from charm threshold up to 7 GeV. However, the errors on R_D are large, so you can still accommodate a half unit of R for F production.

- *Roth:*

Can the R data support another heavy lepton?

- *Barbaro-Galtieri:*

This would add another unit of R away from the threshold for its production. I remind you that the R values mentioned by different groups differ by at least 0.5 among themselves. As for the Mark I results, my previous slide (Fig. 19) showed that within errors we are able to account for R without the necessity of introducing another heavy lepton, although one-half unit of R could be accommodated. One should really look at R in the higher energy region to answer this question.

- *Kahl:*

Can one look for F production at 5 GeV?

- *Barbaro-Galtieri:*

We do not have much data at 5 GeV, in addition, the K identification is more difficult there. The η production reported by DASP is consistent with zero at 5 GeV.

DISCUSSION No. 2

- *Orava:*

You mentioned an upper limit for charm changing neutral currents which was about 16%. This limit is also determined in νN reactions from the wrong sign μ events: $\nu N \to \mu^+ X$. The CDHS collaboration obtains a limit $\sigma(\Delta C=1)/\sigma(\Delta C=0) < 2 \times 10^{-2}$.

- *Barbaro-Galtieri:*

I did not give a limit for charm changing neutral currents. The 16% value I gave this morning is the upper limit for D^0-$\bar{D}^0$ mixing and is measured by looking at the wrong sign K mesons associated with a detected D meson. Such mixing can arise from a charm-changing neutral current. I do not know at the moment how to extract from the 16% upper limit for D^0-$\bar{D}^0$ mixing a ratio of amplitudes for the $\Delta C=1$ and $\Delta C=0$ neutral currents. I am sure some theorist can easily derive the relation between the two.

- *Petersen:*

What did you assume for the ratio of $\pi^{\pm}:\pi^0$ in the Monte Carlo simulation for calculating the total cross section?

- *Barbaro-Galtieri:*

The values of R reported are not too sensitive to this ratio. The calculations have been done using different models for the particle ratios and identification: all pions or a combination of π, K, p and η's. As for the momentum distribution of the particles, a phase space model as well as a jet model have been used. The resulting values of R are not too dependent on which model was used. See, for example, the discussion of V.Lüth in the 1977 SLAC Summer Institute and of G.Hanson in the 1978 Recontre de Moriond.

- *De La Vaissière*

Of the D^+ and D^0 decay modes only ~30% are identified. What are the remaining modes? In the tagged $D\bar{D}$ events could you obtain information about these modes?

- *Barbaro-Galtieri:*

In fact, we have measured only 37% of the D^0 modes and 21% of the D^+ modes. We only have 141 tagged D^0 and 107 tagged D^+. As you can see from Table VI the measured hadronic decays are a few percent for each of the measured modes, so we have only a few events for each of these channels. The branching fractions measured with the tagged events agree with those measured using the entire sample. However, we were unable to find any more modes except for $D^+ \rightarrow \bar{K}^0\pi^+\pi^-\pi^+$ which gave a branching fraction of $(4.8 \pm 2.5)\%$.

- *Kumar:*

In the tagged D events, you can analyze the recoil system on an event-by-event basis to look for the correct assignment of the particle identities which give the $\bar{D}$, since you know that the final state is $D\bar{D}$. Even with a few events you can have a lot of information.

- *Barbaro-Galtieri:*

Yes, as I have shown, we have measured inclusive K content, charge multiplicities, and energy distributions among the different particles. It is very difficult with the small statistics to get more information on the exclusive channels. Also, don't forget that our electron and gamma detector contributes information for only 10% of the events and charged particle are measured only in 70% (4π).

- *Chauveau:*

Of the 107 tagged D^+ events, how many of the recoils have been attributed a definite decay mode?

- *Barbaro-Galtieri:*

For the tagged D^+ we have 8 events completely reconstructed; for the tagged D^0 we have 28 such events.

- *Plothow:*

Could you comment on why the decuplet transition to an octet (as it is for the $\Omega^- \rightarrow \Xi^0 e^- \bar{\nu}$) should have a branching fraction of 10^{-2}, much larger than for the other decays, typically a factor 10 smaller. The CERN hyperon beam experiment has just measured this semileptonic branching ratio to be $\sim 10^{-2}$ (preliminary report at the Tokyo conference).

- *Barbaro-Galtieri:*

I do not know, I would expect that the octet enhancement would produce the same effect on the Ω^- as for the other strange particles, apart from phase space factors.

- *Isgur:*

I would like to comment on the semileptonic enhancement question. The experimental observations $BR(\text{strangeness} \rightarrow X\ell\nu) \simeq 10^{-3}$ and $BR(\text{charm} \rightarrow X\ell\nu) \simeq 10^{-1}$ seem at first sight difficult to reconcile. Naturally one draws the diagrams

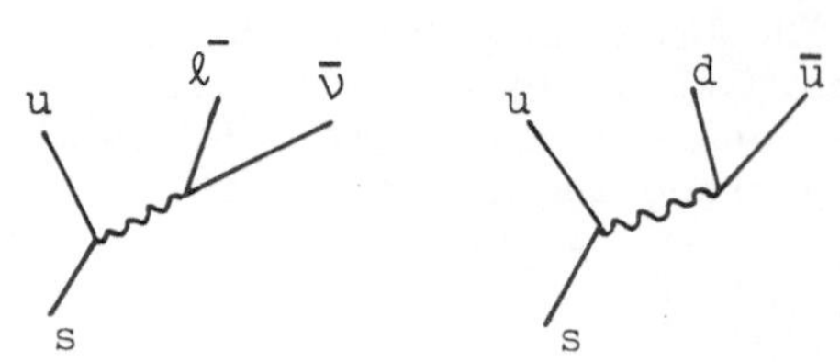

and

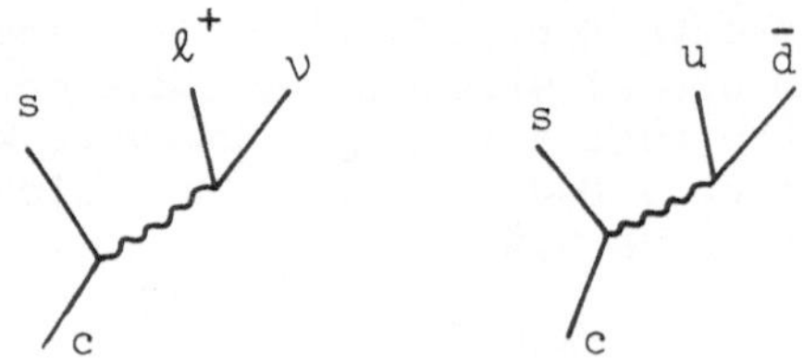

In each case for free (colored) quarks one would naively expect semileptonic branching ratios of 20%, and while this is not far from the charm result, it is completely wrong for strange particle decays. Cabibbo and collaborators have shown that α_s corrections from QCD may even lead to a detailed understanding of the hadronic charm decays; on the other hand, in strange decays, similar considerations do not give a large enough enhancement; although they show that the octet part of the hadronic decay amplitude is enhanced relative to their 27-plet part, as experimentally observed. A possible explanation can be found in terms of diagrams which contain weak interaction loops, as shown below:

which are proportional to $(m_c^2 - m_u^2)$. Such diagrams have pure octet amplitudes and, although no one has been able to calculate their effect convincingly, they may be large. The comparable diagrams for charm decay are suppressed by $\sin\theta$ and by the fact that they are proportional to $(m_s^2 - m_d^2) \ll (m_c^2 - m_u^2)$ and so they can presumably be neglected. Some of the original estimates for charmed particle decays were based on assuming that the octet part of the decay $s \to ud\bar{u}$ was somehow enhanced to the observed value; it was expected that $c \to s\bar{d}u$ would be similarly enhanced.

- De La Vaissière

If the decay branching fraction for $D \to K\pi$ is around 2%, one would expect that the $F \to \eta\pi$ would be of the same order of magnitude. This means that the $F \to K\bar{K}\pi$ is only 0.1%, if the ratio 1/20 for $F \to K\bar{K}X$ to $F \to \eta X$ is correct. Comparing with the D decays into three particle final states, wouldn't this give a discrepancy?

- Barbaro-Galtieri:

Both the DESY result and the SP26 result are preliminary and have large errors. I hope that by the time the final results are presented the two results come closer.* As I mentioned in yesterday's lecture, $F^+ \to K^+K^-\pi^+/F^+ \to \eta\pi^+$ is expected to be about 1.

---

*See footnote to a similar question in the first Discussion, pg.--.

RECENT RESULTS FROM GARGAMELLE

Paul Musset

CERN

Geneva, Switzerland

INTRODUCTION AND OUTLINE

Neutrino experiments have provided three main results[1] in the past few years.

a) <u>the existence of neutral currents</u>:

These reactions are of the type

ν + nucleon $\rightarrow$ ν + hadrons

$\bar{\nu}$ + nucleon $\rightarrow$ $\bar{\nu}$ + hadrons

and are usually compared with the charged current reactions;

ν + nucleon $\rightarrow$ μ^- + hadrons

$\bar{\nu}$ + nucleon $\rightarrow$ μ^+ + hadrons

Neutral current reactions involving only leptons were also observed

ν + electron $\rightarrow$ ν + electron

$\bar{\nu}$ + electron $\rightarrow$ $\bar{\nu}$ + electron

b) <u>the first evidence for charm</u>:

This evidence came from the fast semi-leptonic decays of charmed particles singly produced in neutrino reactions.

$$\nu + \text{nucleon} \rightarrow \mu^- + C + \ldots$$
$$\hookrightarrow \mu^+ + \nu_\mu + \text{hadrons}$$
$$\hookrightarrow e^+ + \text{hadrons} + \nu_e + \text{hadrons}$$

The signature of charm is the presence of two charged leptons in the final state.

c) <u>a confirmation of the quark-parton picture of the nucleon</u>

These studies were made in inclusive reactions

$$\nu + \text{nucleon} \rightarrow \mu^- + \text{hadrons}$$
$$\nu + \text{nucleon} \rightarrow \mu^+ + \text{hadrons}$$

in which only one charged lepton is detected in the final state. They are parallel to the electromagnetic reactions

$$e + \text{nucleon} \rightarrow e + \text{hadrons}$$

The same quark-parton structure of the nucleon is revealed in both the electron and the neutrino reactions.

The first evidence for these new phenomena was followed by a systematic study of their characteristics. These studies took a long time because neutrinos produce rare and complex reactions and because neutrino beams were not known as well as other beams are.

The results recently obtained with Gargamelle at the SPS on these two first topics are presented.

1. EXPERIMENTAL SET-UP

The set-up consists of the large heavy liquid bubble chamber Gargamelle, equipped with a set of surrounding multiwire proportional chambers (fig. 1) and placed in a wide band neutrino beam.

Neutrino events are produced in the bubble chamber, a cylinder of 4.8 m length and 1.9 m diameter, placed in a 2 T magnetic field. The liquid used is a mixture of propane C_3H_8 (90%) and freon CF_3Br (10%) with a radiation length of 60 cm, an interaction length of 1.70 cm and a density of 0.6. These characteristics enable a good detection of electrons, γ-rays, π^o, and neutrons. Tracks of secondary particles are photographed through 8 cameras, scanned and measured for the reactions under study.

A set of counters gives both the position and time of entering or outgoing tracks. It consists, first, of two identical counters at the beginning and the end of the chamber called the veto and the picket fence. Parasitic entering tracks are detected in the veto whereas tracks originating in the chamber pass through the picket fence. Their surface is about $(18 \times 1.2)\ m^2$, and they have three planes of wires with ±20 mm resolution. Also included is an external muon identifier, which is made of two $(6 \times 4)\,m^2$ planes with a resolution ± 16 mm (± 4mm in the central region where the density of tracks is higher). These two planes are separated by an iron absorber with a thickness of about 1 m, in which the pions are absorbed. The secondary muons emitted in charged current neutrino reactions are not absorbed and hit the counters. The counters and their electronics have a recovery time of about 2 μs, i.e. a time largely sufficient to distinguish several events produced during the spill time of 1 ms of the beam.

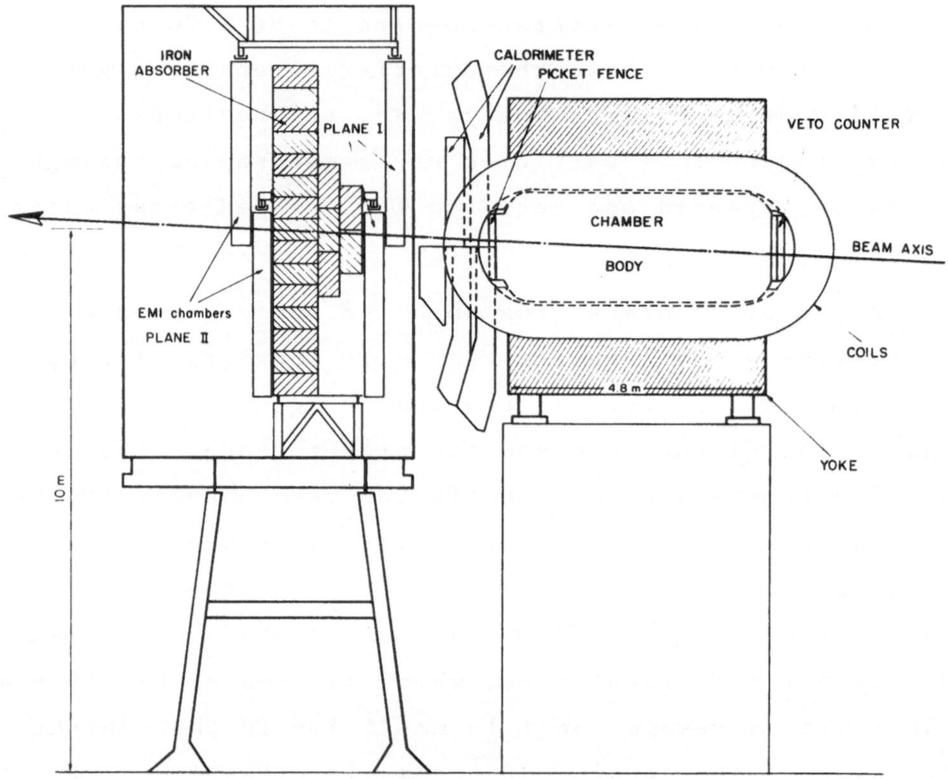

Fig. 1 Experimental Set-up

Data acquired by counters are registered on magnetic tapes. These are used in two ways. Either, it is possible to extrapolate tracks visible in the bubble chamber to search for hits into the counters. Alternatively, it is possible to first search for certain topologies into the counters to select the bubble chamber photographs to be scanned and measured.

The apparatus is placed in the wide band neutrino beam. Pions and kaons produced by the 300 GeV proton beam hitting a metal target are focused by two achromatic magnetic lenses called "horns." These mesons are then left to decay into muons and muon-neutrinos over a length of 400 m. Other decay modes of charged pions and of charged and neutral kaons also produce a small percentage of muon-antineutrino and electron-neutrinos as well as antineutrinos. (see fig. 2). Muons are slowed down by a 180m long iron shielding followed by 300m of earth and other equipment.

The neutrino spectrum is peaked at around 30 GeV, which corresponds to an event rate peaked above 40 GeV, since the neutrino cross section rises approximately linearly with energy.

The muon flux traversing the shielding is continuously monitored in five gaps. The absolute number of muons measured that way then serves as a normalization for the neutrino flux.

2. THE LEPTONIC NEUTRAL CURRENT REACTION

The elastic reaction $\nu_\mu + e^- \to \nu_\mu + e^-$ proceeds only via neutral currents, whereas the reaction $\nu_e + e^- \to \nu_e + e^-$ has both contributions from charged and from neutral currents[1].

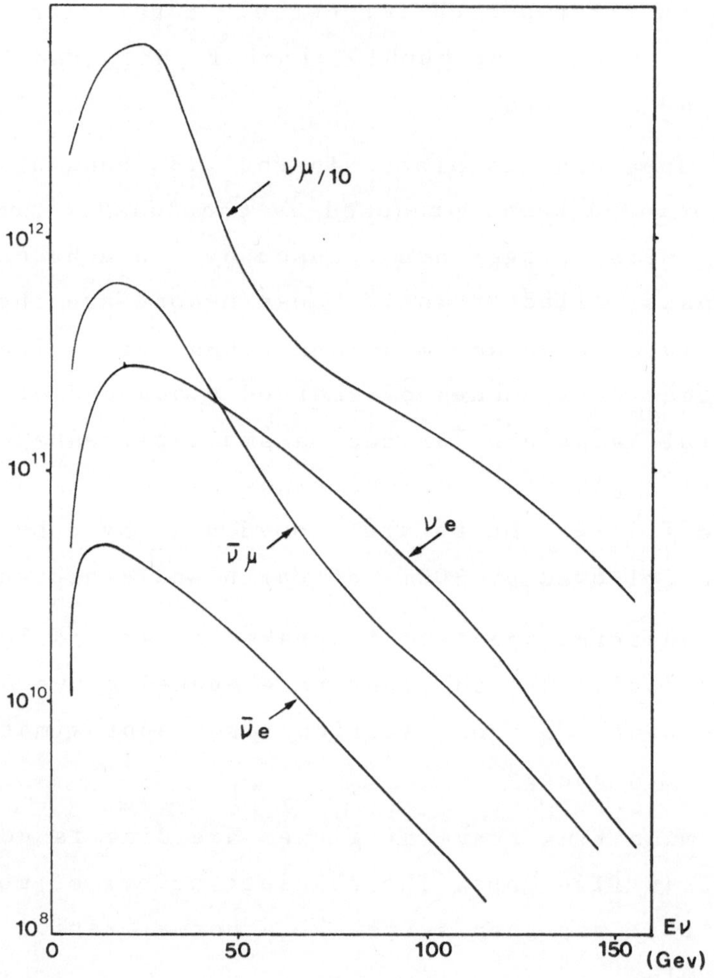

Fig. 2 ν Fluxes for νμ→νμe analysis

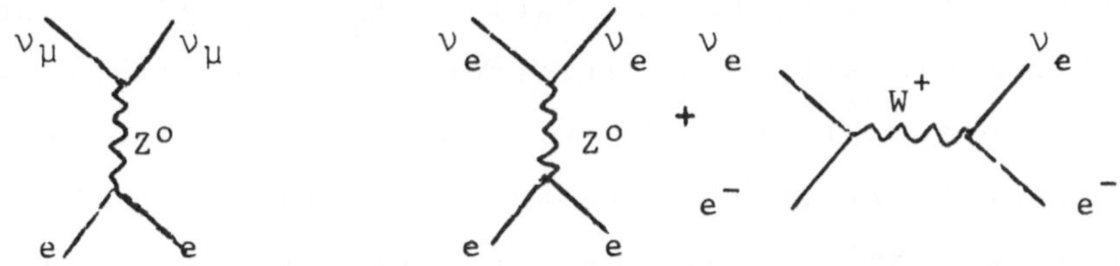

Assuming that only V and A components are present, the cross section σ is given by:

$$\frac{d\sigma}{dE_e} = \frac{G^2}{2\pi} \frac{m_e}{E_\nu^2} \{(g_V + g_A)^2 + (g_V - g_A)^2 (E_\nu - E_e)^2\}$$

where terms in $m_e/E_e \ll 1$ are neglected.
E_ν and E_e are respectively the primary neutrino and secondary electron energies.

For the $\nu_e + e^- \rightarrow \nu_e + e^-$ cross sections, and because of the additional charged current contribution

$$g_{V,A}^{\nu_e} = 1 + g_{V,A}^{\nu_\mu}$$

Note that ν_μ e and ν_ee events are undistinguishable experimentally since only the electron is observable
In the standard SU (2) ⊗ U (1) model, one has

$$g_V = -\frac{1}{2} \cdot (1-4 \sin^2\theta_w), \quad g_A = -\frac{1}{2}$$

2.1 Experimental procedure

The signature for the leptonic neutral current reaction is the observation of a single electron track originating in the liquid. A sample of about 235 K pictures was scanned for spiralizing tracks or electromagnetic showers, with no hadron at the vertex and no visible source in the chamber or in the wall. The fiducial volume is 5.1 m^3 inside a 8 m^3 visible volume.

The electron energy has to satisfy $E_e > 2$ GeV. The events are classified as electrons if there is clearly a single track at the vertex. No e^+ nor e^- above 30 MeV should exist within the first seven centimetres and no e^+ above 2.5 MeV within the first centimetre. Eleven events satisfy this criteria (fig. 3, table 1). Among them, two are ambiguous e/γ which could be due to bremsstrahlung from entering isolated muons. They are removed since e/γ events should in no case be tangent to a muon track. The correction for the losses due to this criteria are estimated to be 2%.

The remaining 9 events satisfy the kinematical condition $\theta_e \leqslant \frac{2\ m_e}{Ee}$

where uncertainties on angle measurements are accounted for. The observed $E\theta^2$ distribution, also characteristic of the reaction, is in good agreement with the predicted curve (fig. 4).

To obtain the rate of events, the signal has to be corrected for the overall detection efficiency:

$$\beta\ (E_e) = 0.88 \times .99 \times (0.92 - 0.037 \log E_{e\ \mathrm{GeV}})$$

where the first term corresponds to scanning efficiency, the second to detection efficiency, the third to the identification criteria (fig. 5).

The signal also has to be corrected for the following background contributions:

a) high energy γ-rays, which by asymmetric e^+e^- pair reaction, compton effect, or annihilation of the e^+ can give rise to isolated electrons. A total of 20 e^+e^- pairs are observed which satisfy the criteria. This

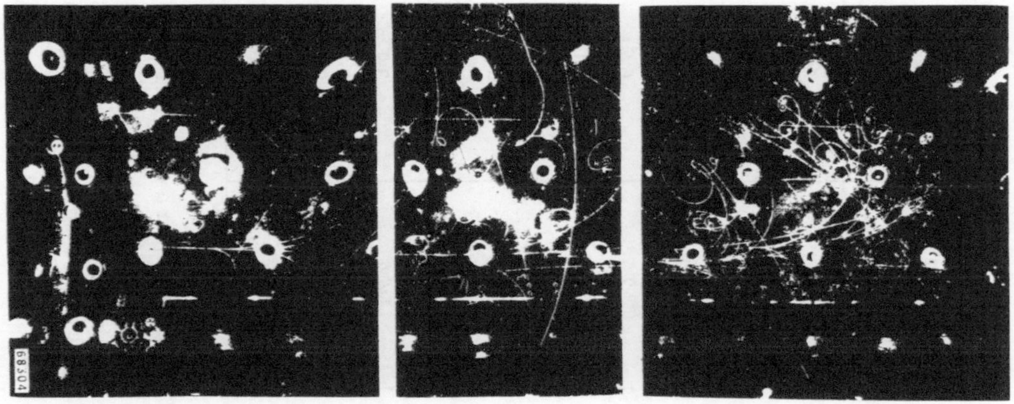

Fig. 3 One of the nine e^- events. The electromagnetic shower development itself through three consecutive views.

TABLE I

Characteristics of the eleven electron candidates

No.	E_{GeV}	θ^2 $mrad^2$	l_{obs} cm	l_{max} cm
1	34±17	296	11	28±7
2	35^{+30}_{-10}	4	50	30^{+10}_{-5}
3	3.3±1.3	324	114	9±2
4	2.2±0.8	225	63	7.5±1
5	67±35	25	64	40±10
6	13.3±1.5	4	101	18±1
7	60^{+60}_{-10}	144	12	38^{+17}_{-4}
8	9±2	3.6	80	16±2
9	3.3±0.3	144	39	9±0.3
10	47±10	4	120	35±3
11	25±10.5	16	30	25±3

Note: If the events were γ-rays, the e^+ and e^- tracks would be separated before a distance L_{max}, calculated from the minimal sagitta of the two tracks in the magnetic field. All the nine events used for the calculation of the cross-section pass this criteria (Events 1 and 7 are accompanied by a muon and not used for that purpose). Bubble counting along the track for these nine events also indicate that they are single (- not double -) tracks.

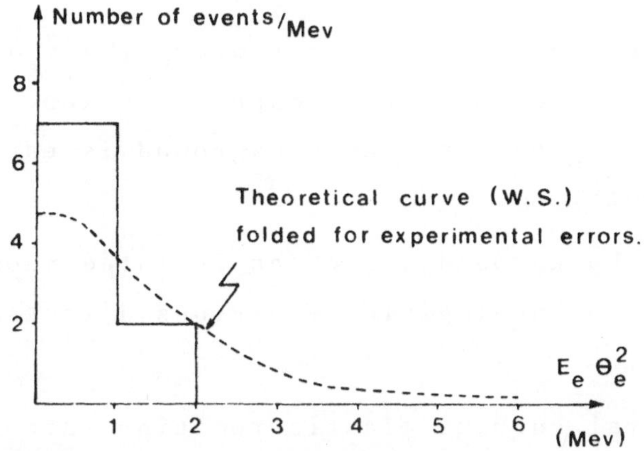

Fig. 4 $E\theta^2$ plot for the eleven e^- candidates. The nine events retained are shown in hatched area. The curve is the theoretical calculation, taking into account experimental errors.

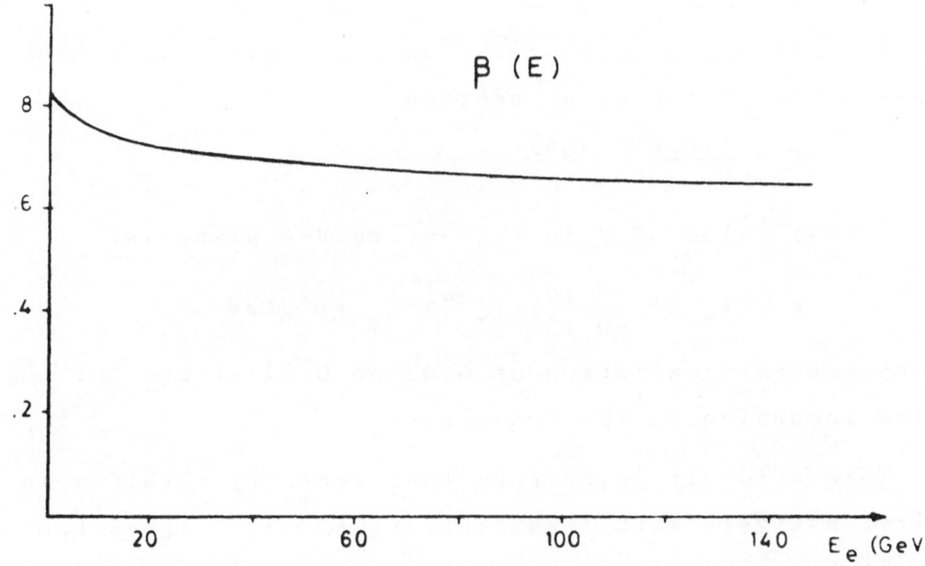

Fig. 5 Detection Efficiency for Electrons

corresponds to a background of <0.2 events;

b) charged current reactions of ν_e (or $\bar{\nu}_e$) reactions with nuclei in which hadrons escape detection. From 4 observed $e^- P_{stop}$ events, the background is estimated to be < 0.35 events.

c) other backgrounds, such as isolated muons or pions are checked to be negligible by various direct measurements in the set-up.

The neutral current elastic reaction rate is then calculated using the ordinary charged-current CC events as a normalization. The number of charged current events is calculated from the flux measurement, and checked with the number of CC events observed in a partial (i.e. $\frac{1}{50}$th) sample of the pictures. It is found to be 42.000 ± 4.000 CC events.

2.2 Results and discussion

The present status of the analysis gives an estimate of the slope of the cross section

$$S = \frac{\sigma(\nu_\mu e \to \nu_\mu e)}{E_\nu}$$

The central value of S in the entire V-A plane is:

$$S \sim (0.41 \, {}^{+0.19}_{-0.15}) \, 10^{-41} \, E_\nu \, cm^2/GeV$$

A supplementary variation of 0.03 on 0.41 allows for the entire incursion in the V-A plane.

This value is lower than that recently obtained in a first analysis with less statistics [2). Nevertheless it is higher than that of the theoretical prediction of the standard SU (2) ⊗ U (1) model, with the value of the mixing parameter $\sin^2\theta_w = 0.25$ [3], i.e.

$$S = 0.14 \cdot 10^{-41} E_\nu \ \mathrm{cm}^2/\mathrm{GeV}.$$

The value obtained in the 15' chamber for the same reaction, i.e. $S = (0.18 \pm 0.008)\ 10^{-41} E_\nu \ \mathrm{cm}^2/\mathrm{GeV}$ [4] is in very good agreement with the SU (2) ⊗ U (1) theory.

Additional statistics will be useful to decide on the reality of a discrepancy between theory and experiment. Such an effect could then possibly be explained by the existence of a new neutral current which is coupled to leptons. Note that atomic experiments also involve the leptonic neutral current. Further measurements on both experiments will be the subject of intense activity this year.

3. PRODUCTION AND DECAY OF CHARMED PARTICLES

Whilst they were intensively searched for in hadron reactions the only charmed particles, directly observed are up to now produced in neutrino reactions [1]. The most characteristic chain of reactions for the identification of a charmed particle C is

$$\nu_\mu + N \rightarrow \mu^- + C + \ldots$$
$$\quad C \rightarrow e^+ + S + \nu_e$$
$$\quad C \rightarrow \mu^+ + S + \nu_\mu$$

where C and S are respectively charmed and strange hadrons. The decay $C \rightarrow e^+ + S + \nu_e$ was already observed a few years ago thanks to the easy electron signature. The equivalent decay $C \rightarrow \mu^+ + S + \nu_\mu$ was to wait for muon identifiers to be isolated.

3.1 Study of decay rates [5)]

The most convenient method to search for such topologies is to select from the counter tape the photographs where two possible muons gave hits in the picket fence and in the two EMI planes, whereas no hit occur in the veto in the same time slot. The tracks leaving the bubble chamber are then measured to check the association with hits in the counters. A χ^2 less than 40 in both EMI planes is required to select the events. With such a procedure 56 $\mu^-\mu^+$ and 17 $\mu^-\mu^-$ are collected in a sample of 19.000 charged current events.

The χ^2 distribution of the μ^- in $\mu^-\mu^+$ events is in agreement with that expected from measurement and multiple scattering errors. On the contrary, μ^+ in the $\mu^-\mu^+$ events and μ^- in the $\mu^-\mu^-$ events have a tail which is to be interpreted as due to background: punch through of hadrons producing hadronic showers and random parasitic hits. To reduce the background, χ^2 is thus required to satisfy $\chi^2 < 10$. The sample of events is then 46 $\mu^-\mu^+$ and 7 $\mu^-\mu^-$ to be compared with a calculated background of 16 $\mu^-\mu^+$ and 10 $\mu^-\mu^-$. The conclusion is that the $\mu^-\mu^-$ signal is compatible with background, and that a genuine $\mu^-\mu^+$ signal of 30 ± 8 events exists.

To compute the rate of dilepton events, a correction has to be applied for the loss of muons below about 2.5 GeV. Assuming the simple charm production model, the dimuon detection efficiency is found to be 27%, and the relative rate to ordinary charged-current events is:

$$\rho = \frac{\nu N \to \mu^-\mu^+ \text{ hadrons}}{\nu N \to \mu^- \text{ hadrons}} = (0.66 \pm 0.20)\ 10^{-2}$$

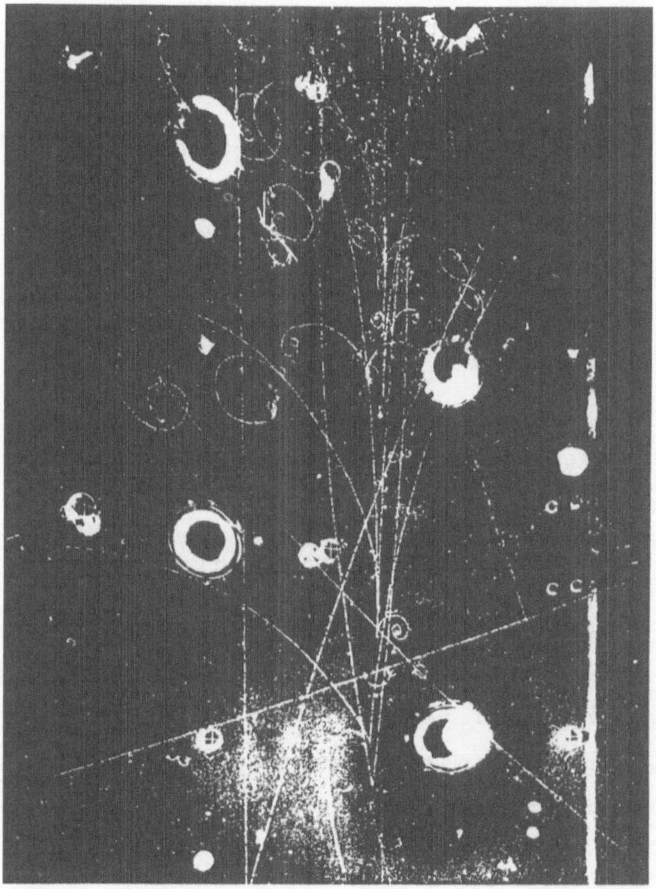

Fig. 6 A dimuon event. An identified K_o also points to the main vertex.

This is in agreement with the rate measured at the same energies for the corresponding decay mode into electrons.

An important way to test the charm hypothesis is the measurement of the fraction f of events in which strange neutral particles are observed. Two previous measurements lead to some inconsistency

i.e. $f = 1.84 {}^{+.6}_{-.5}$ and $f = 0.6 \pm 0.2$ [6].

Correcting for losses, in the present experiment, we found

$$f = \frac{\mu^- \mu^+ V^o}{\mu^- \mu^+} = 0.55 {}^{+0.30}_{-0.25}$$

in good agreement with what is expected from the current charm picture.

3.2 Limit on the charmed particles lifetime

It is possible to search for a visible (short) decay of the type $D^+ \rightarrow \mu^+$ + neutrals, in the experiment. This is done by carefully inspecting the vicinity of the vertex, using also the extrapolation of the tracks from the surrounding region, to exclude any possible decay beyond some distance L_{max} to the vertex. No candidate for such a decay was observed. A likelihood function is then calculated taking into account the background and the probability of observing such a decay within the length L_{max}. The fractional momentum taken by the μ^+ in the D^+ decay was taken to be 0.3 ∿ 0.5. The D^+ decay branching fraction into μ^+ is estimated to be $\leqslant 0.4$. This leads to an upper limit $\tau \lesssim 10^{-12}$ seconds as a preliminary result.

3.3 Charm production in proton-nucleus interactions

Experimental searches for charmed particles produced in hadronic reactions are numerous. Nonetheless only indirect evidence was found, by the observation of "prompt" leptons, supposedly produced by the decay of charmed particles. The difficulty of observing hadron-produced charmed particles is mainly due to the possibly very small cross section.

As the result of recent beam dump experiments [7)] done at CERN, a new indirect evidence for the semi-leptonic decay of short lived particles has been obtained, through the observation of an excess of neutrino events of the ν_e type. The value obtained for the cross-sections in these experiments is higher than that expected for charm and this question will be discussed in conjunction with data on charm production coming from other experiments.

The aim of the experiment is to look for neutral penetrating particles produced directly at the target. The proton beam is dumped onto a thick copper target of ϕ = 0.27 m and L = 2 m, followed by a 0.4 x 0.4 x 0.8 m^3 iron block, in order to absorb all hadrons. In the Gargamelle experiment 70 K pictures were taken, corresponding to a total of 3.5 10^{17} protons on the target. Events were divided into three categories: events having a muon identified in the EMI (category ν_μ CC); events having an electron at the vertex (category ν_eCC); events without a charged lepton at the vertex (category NC).

After corrections for EMI efficiency and for background 18 events fall into the ν_μ category, 9 in the ν_e category, 5 into the NC category.

The visible energy distribution of these three classes of events is shown in fig. 7.

The number of NC events is consistent with the observed number of ν_μ CC and ν_e CC events and the well-known NC/CC ratios. On the contrary the number of ν_e events is surprisingly high.

If ν_e induced events were produced by pions, kaons and hyperons, a relative rate of 6% to ν_μ induced events would be expected. The production of heavy leptons pairs $\tau\bar{\tau}$ is expected to be too low to account for the ν_e events.

The hypothesis of the production of charmed D mesons to explain the data has been investigated using a Monte Carlo to generate production and decay of D's. The production was assumed to follow Bourquin Gaillard [8) and the decay is according to the electron spectrum measured at SPEAR[9)].

A more recent evaluation gives

$$\sigma_D \sim (150 \pm 50)\ \mu b$$

corresponding to a D production $\sim (1-x)^5$, where x is the Feynman variable, and an average energy of the events of

$$E_{\nu_e} \sim 75\ GeV$$

A preliminary study of the possible D reinteraction inside the nuclei lead to (another source of uncertainty)

$$\sigma_D \rightarrow \sigma_D \times 2.5.$$

Such high values for σ_D have to be compared to the very low limit $\sigma_D \lesssim 1\ \mu b$ deduced from an emulsion experiment with protons at 300 GeV.

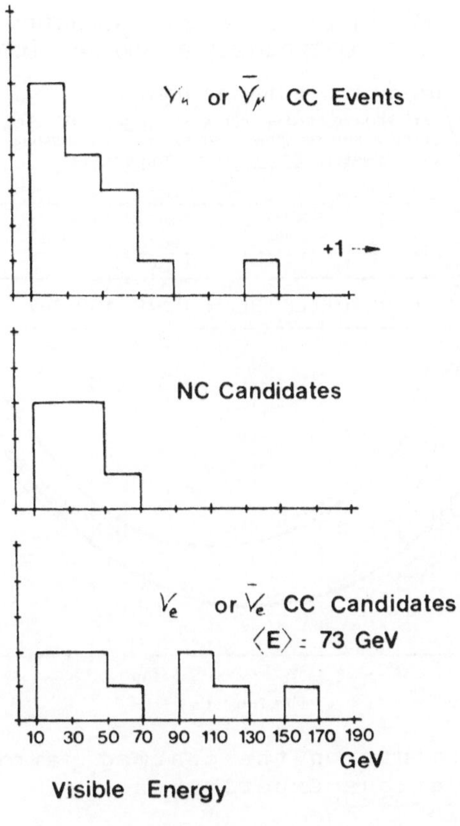

Fig. 7 Visible energy distribution of three classes of events.

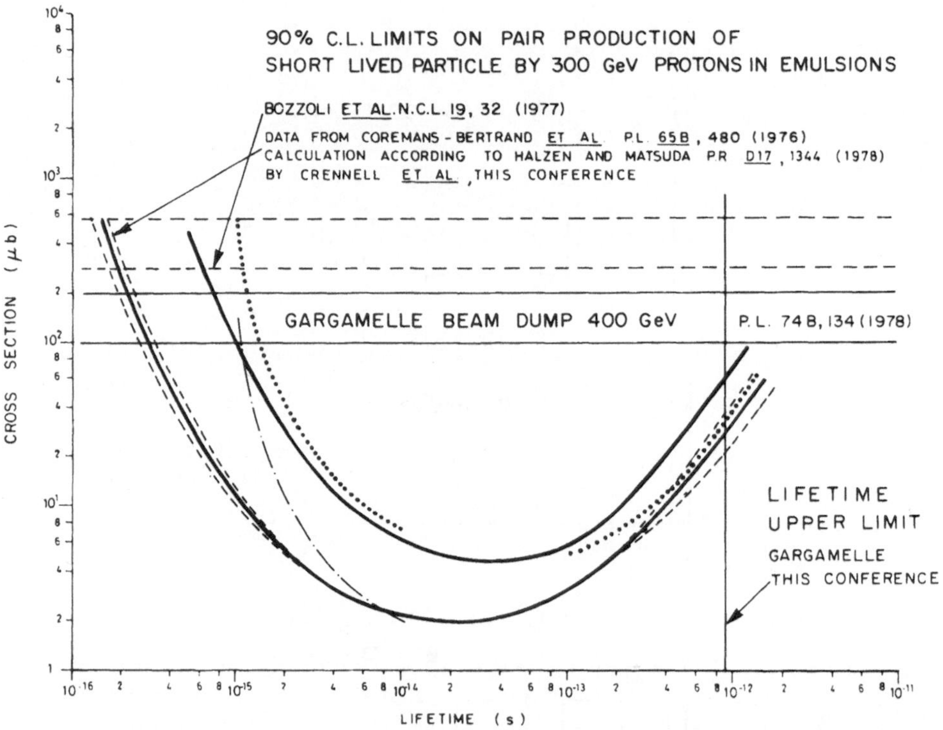

Fig. 8 Constraints on the charmed particle lifetimes from various experiments.

In fig. 8, the results of two emulsion experiments have been reproduced[10]. Effects of varying the production index n in $(1-x)^n$ from 2 to 4 are shown in broken lines, whereas reinteractions inside the nucleus are indicated by dotted lines. The beam-dump experiment in Gargamelle and the dimuon result also constrain charm production. Taken at face value, the results of all these experiments would indicate that the lifetime of charmed particles (D) would be less than about 10^{-15} s. Nevertheless, it should be stressed that the values obtained in all the experiments strongly depend on the models assumed. In that case, a small region around 10^{-12} seconds would still be allowed for the charmed particle lifetime. The possibility that the beam dump observations are not due entirely to charm, but to other fast decaying particles is not excluded, but there is no compelling reason why these new particles should be produced with such large cross sections.

Experiments of these three types: beam-dump, emulsion and dimuon are in progress and will certainly answer the question of the charmed particle lifetime in the near future.

REFERENCES

1) For a review, see for example P. Musset and J.P. Vialle Physics reports, Vol. 39C, Number 1 (1978).

2) P. Alibran et al., Phys. Let. B 74B (1978), 422.

3) For a compilation of neutral current coupling constants, see for example P. Musset. Proceedings of the 1977 International Symposium on Lepton and Photon Interactions at High Energies, Hamburg (1977).

4) A.M. Cnops et al., Submitted to Physics Rev. Let.

5) A. Blondel, Gargamelle Collaboration. Topical Conference on Neutrino Physics at Accelerators, Oxford (1979).

6) C. Baltay, Topical Conference on Neutrino Physics at Accelerators, Oxford (1978).

7) P. Alibran et al., Phys. Let. B 74B (1978) 134;
T. Hans et al., Phys. Let. B 74B (1978) 139;
P. Bosetti et al., Phys. Let. B 74B (1978) 143.

8) M. Bourquin and J.M. Gaillard, Phys. Let. 59B (1975) 191.

9) J.M. Feller et al., Phys. Rev. Let. 40 (1978) 274.

10) G. Conforto, Topical Conference on Neutrino Physics at Accelerators, Oxford (1978).

DISCUSSION

CHAIRMAN: P. Musset

Scientific Secretaries: G. Anastaze and Y. Ha

DISCUSSION

- MERRITT:

The Delco experiment at SPEAR has obtained a value of the D lifetime by measuring the branching ratio $D \to \kappa e \nu$. The result is

$$5\% \leqslant \frac{D \to \kappa e \nu}{D \to \text{All}} \leqslant 10\%$$

This, combined with the (fairly reliable) theoretical calculation of the width Γ $(D \to \kappa e \nu)$, gives

$$3.5 \times 10^{-13} \leqslant \tau_D \leqslant 7 \times 10^{-13} \text{ sec.}$$

This appears to be consistent with the CERN beam dump experiment, the STANFORD-CALTECH direct experiment at FERMILAB and the emulsion experiment.

- MUSSET:

This is a very interesting comment. This new result could contribute appreciably in settling the question of the charmed particle lifetime.

- JACOB:

You said that $(1-x)^5$ is consistent with the energy distribution. However, if one assumes that charmed baryons dominate on the ground that one biases oneself in favour of large x particles, can one argue that a different Q value would make the distribution compatible with a weaker x dependence and hence a lower cross section?

- MUSSET:

It is true that the result is strongly dependent on the x-dependence of the cross section σ. Strictly speaking one is not measuring a total cross section, but merely exploring a very small part of the cross section in the beam dump experiments. Even without invoking charmed baryon, the "result" on the total cross section σ varies by about 25% for every unit of n in the $(1-x)^n$ dependence. This is one of the main reasons why the comparison of the different experiments on cross sections and lifetimes of charmed particles has to be taken as only roughly indicative.

- MONTGOMERY:

Does the recent result from SLAC with polarized electrons not exclude anomalous neutral current couplings for the electron?

- MUSSET:

In the standard SU(2) x U(1) model, there is only one free parameter, $\sin^2\theta_W$. Hence, any experimental result on a neutral current effect will completely predict all the others. Nevertheless, it is always possible to envisage models with more parameters, and test the universality of the neutral current in different experiments.

- TYNDEL:

You have estimated the background due to γ to be 0.2 event? How do you reconcile this with the 2 events which were removed because they were associated with muons?

- MUSSET:

The background due to γ is 0.2 from asymmetric γ materialization, Compton and e^+ annihilation close to the vertex. Candidates electrons and γ rays associated to muons are all removed: consequently there is no corresponding background.

- BHANOT:

How sensitive is the number you quoted for S, ($S=.14 \times 10^{-41} cm^2/GeV$ at $\sin^2\theta_W=.25$), to change in θ_W? What value of $\sin^2\theta_W$ do you need to get your experimental answer of $S=.41 \times 10^{-41} cm^2/GeV$?

- MUSSET:

In the SU(2) x U(1) model of Weinberg-Salam, S is given by

$$S = \frac{G^2}{2\pi} m_e \left\{ (1-2\sin^2\theta_W)^2 + \frac{4}{3} \sin^4\theta_W \right.$$

- ORAVA:

What is the γ momentum spectrum in the events scanned for $\nu_\mu e \to \nu_\mu e$ events and the electron momentum spectrum in the $\nu_\mu e \to \nu_\mu e$ candidates?

- MUSSET:

The γ spectrum is of lower energy than the electron spectrum.

- ORAVA:

How well is the sign of high energy e^- defined?

- MUSSET:

We have 8 e^-, 1 $e^\pm$, 4 e^-p, no e^+p.

- ORAVA:

What is the efficiency of the External Muon Identifier coupled to Gargamelle?

- MUSSET:

The electronic efficiency is > 99%. The geometrical efficiency is $\sim$ 100% for μ tracks along the beam. It is $\sim$ 30% for muons of the dimuons events which are peaked at low energy.

- ORAVA:

Atomic physics results are contradictory; isn't this due to the difficulties in the theoretical treatment of the various correction terms?

- MUSSET:

According to experts, uncertainties up to a factor of 2 in the calculation are possible. The observed numbers in the null experiments rule out an effect more than $\sim$ 1/5 the calculated value. On the other hand, the SEATTLE and NOVOSIBIRSK experiments measure the same quantity and should have the same result independently from calculations. However, they don't.

WHAT CAN A PARTICLE PHYSICIST LEARN FROM SUPERLIQUID ^{3}He?

H. Kleinert

Institut für Theorie der Elementarteilchen

D-1000 Berlin 33, Arnimallee 3

ABSTRACT

The superliquid ^{3}He is shown to be a wide field of application for many recently developed methods in particle physics and field theory:

1) There is a transformation from fundamental to collective fields via path integral techniques just as exists from Thirring to Sine-Gordon fields and as one would like to find from quarks to hadrons.
2) There are many classical field configurations, monopoles, strings, and solitons, which all can be produced and investigated in the laboratory.
3) Topological quantum numbers are helpful in classifying the stable field configurations.

Such applications to realistic situations in other branches of physics, apart from being useful in themselves, may extend our intuition and help us finding new methods and approximation procedures.

Work supported in part by Deutsche Forschungsgemeinschaft/Kl256/6..

Many arguments used recently in the attempts to deduce the confinement of quarks in Yang-Mills theories [1] rely heavily upon our intuition about flux loops derived from what we know about superconductors and superliquid He-II [2]. I would like to draw your attention to another superliquid, ^3He, whose collective description has a non-trivial topological structure and may therefore be a complimentary or even more inspiring source of information. Instead of a pure phase, $e^{i\phi}$, the order parameter describing the condensate of Cooper pairs at each point is given by a complex 3x3 matrix of the specific form

$$A_{a,i} \propto d_a(\underset{\sim}{\phi}^{(1)} + i\underset{\sim}{\phi}^{(2)})_i \tag{1}$$

Here $\underset{\sim}{\phi}^1, \underset{\sim}{\phi}^2$ are orthogonal unit vectors which can be thought of as the two axes of a dreibein (see Fig.I) with $\underset{\sim}{\ell} \equiv \underset{\sim}{\phi}^{(1)} \times \underset{\sim}{\phi}^{(2)}$ as the third axis while d_a is another

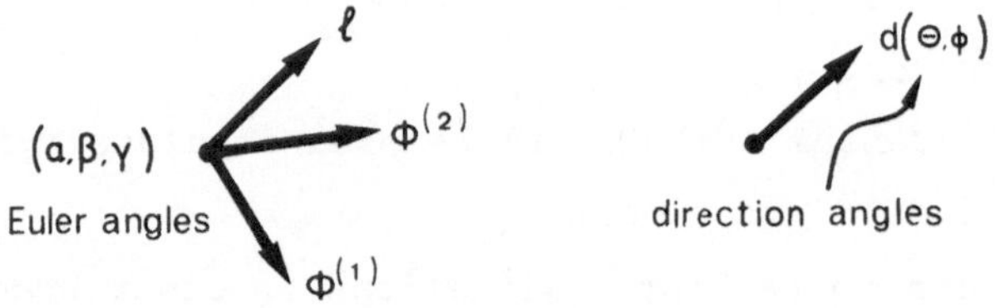

Fig. I

unit vector. Physically, $\underset{\sim}{\ell}$ points in the direction of the angular momentum of the Cooper pairs while $\underset{\sim}{d}$ denotes the axis along which the total spin has a vanishing component. The parameter space of this field is topologically equivalent to $S^2 x SO_3/Z_2$ where Z_2 corresponds to the simultaneous reflection of $\underset{\sim}{\phi}^1, \underset{\sim}{\phi}^2$, and $\underset{\sim}{d}$ (with $\underset{\sim}{\ell}$ staying fixed to preserve the orientation of the dreibein).

Due to this feature there exist many non-trivial topologically inequivalent field configurations very similar to those in gauge theories. Topology is very useful in classifying the different solutions. Contrary to gauge theories, however, many of these fields can be prepared in the laboratory by an appropriate choice of the container walls, external fields, and currents. It is possible to bring the liquid into states which are separated from the ground state by a potential barrier and study the "fate of the false vacuum" [3).].

Finally, the very derivation of the collective field theory from that of the fundamental ^{3}He action is completely analogous to what has recently become popular in 1+1 dimensional theories: The transition from the Thirring model to the Sine-Gordon equation [4)]. Since it is quite plausible that eventually a similar transition will be found from quantum chromodynamics of quarks and gluons to a dual theory of hadrons [5)], it may be an inspiring experience to learn how well a liquid containing many strongly interacting fermions can be described in terms of a few Bose fields [6)]. Also, Landau's way of arguing for an approximation of the original theory by another one formulated in terms of weakly interacting quasiparticles may give related insights into the connection between current and constituent quarks [7].

The fundamental action of ^{3}He is

$$\mathcal{A} = \int d^4x\psi^{+\alpha}(x)(i\partial_t+\frac{\nabla^2}{2m}+\mu)\psi_\alpha(x)$$
$$-\frac{1}{2}\int d^4xd^4x'\psi^{+\alpha}(x)\psi^{+\beta}(x')V(x-x')\psi_\beta(x')\psi_\alpha(x) \quad (2)$$

In addition, there is a weak hyperfine interaction between the nuclear dipole moments which we shall at first neglect. The potential V(r) is displayed on Fig.II.

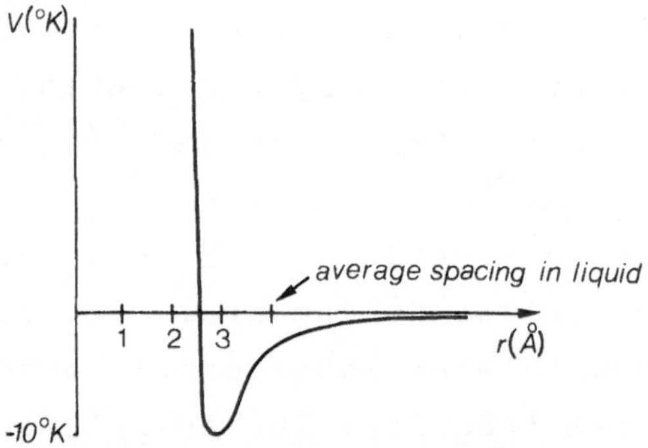

Fig. II

If one calculates the average distance between two atoms one finds $\langle r\rangle \approx 3.5$ Å. Thus the atoms attract each other all the time and one is confronted with a strong-coupling problem. It would be hopeless to attempt a perturbative solution were it not for Landau's important observation:

Instantaneous screening effects dress the particle, moving through the liquid, with a small cloud and generate a quasiparticle. The interaction between the quasiparticles is quite weak. In order to test this idea one may calculate specific heat, susceptibility, and compressibility for a free system of quasiparticles. As the temperature varies, these behave as T, constant, T^{-2}, with coefficients depending on the mass. It turns out that all formulas agree very well with experiment if one uses the quasiparticle mass $m_{Qp} \approx 3m_{^3He}$. Thus Landau's idea of almost free quasiparticles seems to be correct in ^{3}He with the screening cloud consisting, on the average, of two neighbouring atoms. If one now writes an <u>effective</u> action for the quasiparticles again in the form (2), the potential is expected to be strongly screened and of short range. Only the hard core will definitely stick out. Now there is hope of Feynman graphs leading to realistic answers.

Notice that since the fields are now not completely local but describe a time and space average over a quasiparticle size (≈ 3 Å) one cannot use the effective action to answer questions concerning distances of this order. We shall therefore restrict the following discussion to much larger distances, say starting with 100 Å.

In quark physics, something similar and much more drastic seems to happen: Fundamental quarks are strongly interacting objects of small mass ($M_u^o \approx 15$ MeV) [9]. When moving through the quark matter inside a hadron they seem to become heavy ($M_u \approx 310$ MeV) and almost free. Thus the current versus constituent quark picture is in complete analogy with Landau's idea of Fermi liquid effects.

The quasiparticle action of the form (2) can most efficiently be treated by approximating the potential with another one of the same characteristic shape but with a squeezed radius

$$V(x) = V_o \delta(x) - \frac{3g}{4p_F^2} \nabla^2 \delta(x) \qquad (3)$$

Here $V_o > 0$ accounts for the repulsive core and $g > 0$ for the attractive well around the core. If one neglects V_o (which can be taken care of afterwards in form of what is called paramagnon correction) the interaction can be reordered and written as

$$\mathcal{A}_{int} = \frac{1}{2} \frac{3g}{4p_F^2} \int dx \psi^{+\alpha} i\overleftrightarrow{\nabla}_i \psi^{+\beta} \psi_\beta i\overleftrightarrow{\nabla}_i \psi_\alpha \qquad (4)$$

We have left out a term proportional to $\underset{\sim}{\nabla}(\psi^{+\alpha}\psi_\rho)\underset{\sim}{\nabla}(\psi^{+\beta}\psi_\sigma)$ since for phenomena with wavelength much larger than 3 Å this is neglegible compared to (4): There the derivative $\overleftrightarrow{\nabla}$ stands between the fields which gives at the Fermi surface a momentum $2p_F$. Since $p_F \approx 1/(1.1\ \text{Å})$ this is indeed large compared with $\underset{\sim}{\nabla}(\psi^{+\alpha}\psi_\beta)\ \underset{\sim}{\nabla}(\psi^{+\beta}\psi_\alpha)$ which goes with the total momentum of the phenomena we want to study (< 1/(100 Å)).

With the interaction (4) we can write the generating functional of the quantum field theory of ^{3}He as the path integral

$$\mathcal{Z} = \int \mathcal{D}\psi \mathcal{D}\psi^+ e^{i\int dx \mathcal{L}(x)} \qquad (5)$$

with

$$\mathcal{L}(x) = \psi^{+}(i\partial_t - \xi(-i\nabla))\psi + \frac{3g}{4p_F^2}\psi^{+} i\overleftrightarrow{\nabla}_i \frac{\sigma_a}{2} C^{+}\psi^{+}\psi i\overleftrightarrow{\nabla}_i C\frac{\sigma_a}{2}\psi$$

Notice that we have inserted $C\sigma_a$ matrices ($C \equiv i\sigma^2$) between the fields in the interaction term without changing the expression (4) because of the anticommutativity of the fields.

The letter ξ stands short for the energy measured from the chemical potential:

$$\xi(-i\nabla) \equiv -\frac{\nabla^2}{2m} - \mu. \tag{6}$$

In order to allow for temperature ensembles the fields are periodic in Euclidean time such that the energy integrals of the Feynman rules are sums over Matsubara frequencies

$$i\int \frac{dk^4}{2\pi} \longrightarrow T \sum_{k^0 = i\left\{\begin{matrix}2n+1\\ 2n\end{matrix}\right\}\pi T} \quad \text{for } \left\{\begin{matrix}\text{fermions}\\ \text{bosons}\end{matrix}\right\} \tag{7}$$

We now transform (5) into Bose form by using the standard trick of multiplying Z with a constant gaussian functional integral [10,5,6)]

$$1 \equiv \int \mathcal{D}A\,\mathcal{D}A^{+}\, e^{-\frac{i}{3g^2}\int dx \left| A_{a,i} - \frac{3g}{2p_F}\psi i\overleftrightarrow{\nabla}_i C\frac{\sigma_a}{2}\psi\right|^2} \tag{8}$$

Now the exponent of (5) times (8) can be written as

$$\mathcal{L}(x) = \frac{1}{2} f^+(x) \begin{pmatrix} i\partial_t - \xi & i\overleftrightarrow{\nabla}_i \sigma_a A_{ai} \\ i\overleftrightarrow{\nabla}\sigma_a A_{ai}^+ & i\partial_t + \xi \end{pmatrix} f(x) - \frac{1}{3g} A_{ai}^+(x) A_{ai}(x) \tag{9}$$

where $f(x) \equiv \begin{pmatrix} \psi(x) \\ c^+\psi^+(x) \end{pmatrix}$. The fields f can be integrated out (for details see Ref. 6)) and the functional becomes a pure integral over A fields

$$Z = \int \mathcal{D}A \, \mathcal{D}A^+ e^{i\mathcal{A}_{coll}[A]} \tag{10}$$

with the collective action

$$\mathcal{A}_{coll}[A] = -\frac{i}{2} \operatorname{tr} \log \begin{pmatrix} i\partial_t - \xi & i\overleftrightarrow{\nabla}\sigma_a A_{ai} \\ i\overleftrightarrow{\nabla}\sigma_a A_{ai}^+ & i\partial_t - \xi \end{pmatrix} - \frac{1}{3g} \int dx A_{ai}^+(x) A_{ai}(x) \tag{11}$$

This depends only on the complex 3x3 Bose field A_{ai} and describes completely the ^{3}He liquid for phenomena varying over distances > 100 Å, say. In regions of small A one can expand (11) in a power series. The lowest terms are for time independent fields

$$\begin{aligned} \mathcal{A} = \int dx \, \{ & \tfrac{1}{3}(1-\tfrac{T}{T_c}) A_{ai}^+ A_{ai} \\ & - \xi_o^2 (\partial_i A_{ai}^+ \partial_j A_{aj} + \partial_i A_{aj}^+ \partial_j A_{ai} + \partial_i A_{ai}^+ \partial_j A_{aj}) \\ & - \tfrac{2}{5} \xi_o^2 \big[\beta_1 A_{ai}^+ A_{bj} A_{ai}^+ A_{bj} + \beta_2 (A_{ai}^+ A_{ai})^2 \\ & + \beta_3 A_{ai}^+ A_{aj} A_{bi}^+ A_{bj} + \beta_4 (A_{ai}^+ A_{bi} A_{bj}^+ A_{aj}) \\ & + \beta_5 A_{ai}^+ A_{bi} A^+_{aj} A_{bj} \big] \} \end{aligned} \tag{12}$$

Here we have

1) divided out an overall factor

$$N(0) \equiv \frac{mp_F}{2\pi^2} = \frac{3}{4}\frac{\rho}{m}\frac{1}{T_F} \tag{13}$$

which is the density of states at the surface of the Fermi sphere,

2) introduced the critical temperature T_c as a solution of the gap equation

$$T_c \equiv \omega_{cutoff}\; 2\,\frac{e^{\gamma}}{\pi}\, e^{-1/g\,N(0)} \tag{14}$$

with the parameter ω_{cutoff} which limits the energy integration. This cutoff is determined by the frequency at which the quasiparticle approximation breaks down (say $\frac{1}{10}$ MHz),

3) used the length parameter +)

$$\xi_0 = \sqrt{\frac{7\zeta(3)}{48\pi^2}}\;\frac{v_F}{T_c} \approx .134\,\frac{v_F}{T_c} \approx 150\ \text{Å} \tag{15}$$

called coherence length for reasons to be seen shortly

4) set $v_F \approx 3\times10^3$ cm $\equiv 1$ ++) thus converting freely energy into length^{-1} units ($\hbar = K_{Boltzmann} \equiv 1$ from the beginning). If m$\mathring{\text{K}}$ is used as energy unit, the frequency and length units become 131.6 MHz and 305 Å, respectively.

+) The constants are $\zeta(3) \equiv \sum_1^{\infty} \frac{1}{n^3} \approx 1.202$, $v_F \equiv \frac{p_F}{m}$

++) This is at melting pressure. At zero pressure $v_F \approx 5.5\times10^3$ cm/sec.

The coefficients of the quartic term are actually determined by the calculation as

$$-2\beta_1=\beta_2=\beta_3=\beta_4=-\beta_5=1 \tag{16}$$

It turns out, though, that the effect of the repulsive core and, to a smaller extent, also of the remaining part of the neglected quasiparticle interaction do cause some changes in β (up to 30%). Therefore we keep β as parameters. The same corrections also modify the coefficients of the derivative terms but only by a few percent such that they can be neglected.

Notice that higher derivatives as well as higher powers in A_{ai} carry, for dimensional reasons, one more power of ξ_o each. Thus for wavelength $> \xi_o$ higher derivatives can be neglected. How about higher powers in A? Looking at the potential we see that there is a phase transition as $T < T_c$ since then the mass term picks up the wrong sign. The fields A will fluctuate around a new minimum determined by the quartic interaction. Its size is of the order of

$$|A| \sim \frac{1}{\xi} \equiv \frac{1}{\xi_o} \sqrt{1 - \frac{T}{T_c}} \tag{17}$$

for dimensional reasons. Thus, as one is close enough to T_c, also higher powers in A can be neglected. The action (12) is complete as far as the long wavelength limiting behaviour of the system is concerned. In the modern jargon of critical phenomena, it contains all infrared relevant terms [11].

What information can be extracted from the collective

field theory (12) of ^{3}He? Obviously , one is confronted with an $SU_2 \times SU_2 \times U(1)$ σ type of model whose critical phenomena have been studied in the literature [11]. For brevity, we shall focus here only on the classical aspects:

As $T < T_c$ the vacuum settles at a new minimum determined by β_i. There is no complete analysis of the potential in the 18 parameter space of A_{ai}.[12] However, there are two minima which describe very well all properties of the two main phases of ^{3}He, A and B (see Fig.III). These minima

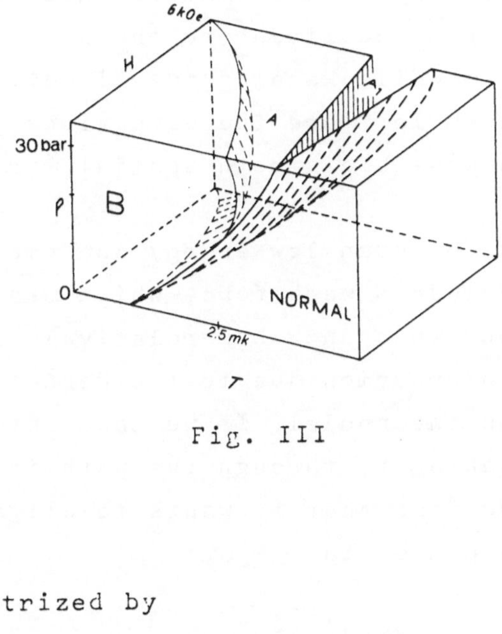

Fig. III

can be parametrized by

$$A^{o}_{a,i} = \frac{1}{2}\sqrt{\frac{5}{6}}\ \frac{1}{\xi}\ d_a(\phi^{(1)}+i\,\phi^{(2)})_i \qquad (18)$$

and

$$A^{o}_{a,i} = \frac{1}{2}\frac{1}{\sqrt{3}}\frac{1}{\xi} R_{a,i}(\mathbf{n},\theta)e^{i\phi} \qquad (19)$$

The first minimum has a degeneracy corresponding to a space $SO_3 \times S^2/Z_2$ as discussed in the beginning. Similarly, the second space is described by $SO_3 \times U(1)$.

Great simplicity is gained by restricting oneself to phenomena with wavelength $\gg \xi_o$. Then the energy density stays much smaller than $1/\xi_o$ and the field is pinned down tightly at the minimum of the potential valley. The size of A is fixed and the energy depends only on the changes in the directions of the field vectors. It is therefore often referred to as bending energy. In the standard field theory of the σ model this corresponds to the transition from the linear to the non-linear σ model by letting $m_\sigma \to \infty$ while keeping $\langle\sigma\rangle$ fixed. Now the only degrees of freedom left are the directions of the different vacua with possible smooth spatial variations.

By concentrating on such low-energy phenomena it becomes important to include a weak force which was left out in the beginning but which now has relatively large effects: The hyperfine interaction due to the magnetic dipole-dipole forces in the nuclei. If we calculate its collective form (by taking it through the path integral transformation [6]) we find that it wants to align $\mathbf{d}$ and $\boldsymbol{\ell}$ vector in the A phase via

$$\mathcal{A}_d = \frac{1}{6}\frac{\xi_o^2}{\xi^2}\frac{1}{\xi_d^2}\int dx(\mathbf{d}\cdot\boldsymbol{\ell})^2 \qquad (20)$$

where the length parameter ξ_d characterizes the strength of the interaction. The microscopic calculation of ξ_d [6]

agrees with the experimental determination

$$\xi_d \approx 10^{-3} \text{ cm} \approx 300\ \xi_o \tag{21}$$

Thus the dipole interaction drives $\underset{\sim}{d}$ parallel to $\underset{\sim}{\ell}$ and becomes important at length scales 10μ. This alignment force is easily understood. In the Cooper pair the configuration $S \parallel L$ is clearly of higher energy than $S \perp L$ since equal magnetic poles are always adjacent to each other while in $S \perp L$ they are in line for half the orbit (see Fig.IV). But $S \perp L$ means $d \parallel \ell$ since $\underset{\sim}{d}$ is the axis

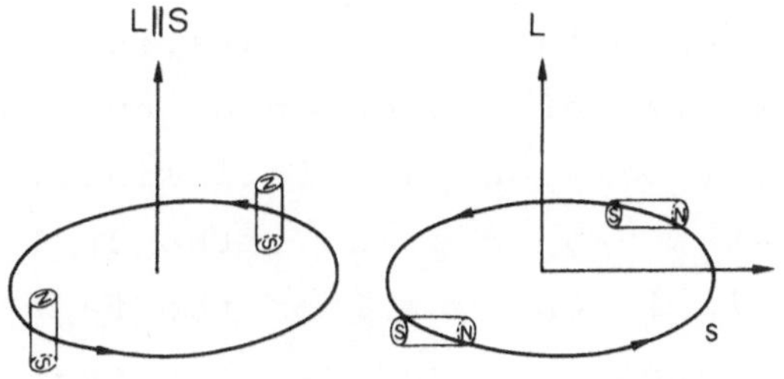

Fig. IV

along which $S_3 = 0$.

The collective free energy can now be written (dropping an overall factor $\frac{1}{6}\frac{\xi_o^2}{\xi^2}$ and surface terms) as

$$f = \frac{1}{2}\left(|\partial\Phi|^2 + \frac{1}{2}|\partial_i \phi_j|^2 + |\underset{\sim}{\phi}\underset{\sim}{\partial} d_a|^2 + (\partial_i d_a)^2\right) - \frac{1}{\xi_d^2}(\underset{\sim}{\ell}\cdot\underset{\sim}{d})^2 \tag{22}$$

Similarly one can write f for the B phase in terms of the fields $\underset{\sim}{n}$, θ, and ϕ and a dipole energy which drives the angle θ to $\theta_d \approx \cos(-\frac{1}{4})$ with a length scale of the same order of magnitude.

This non-linear σ model (22) (and a corresponding version for the B phase) can now be investigated with popular methods of field theory: It contains the Sine-Gordon equation for particular field configurations thereby giving rise to solitons. The topology of the parameter space allows for the existence of non-trivial field configurations for the ground state.

Before we present a few of the phenomena we have to realize that the superliquid is always in a finite container, usually with a size of the order of cm. This imposes boundary conditions upon the field lines. It can easily be derived that $\underset{\sim}{\ell}$ has to stand orthogonal to the walls. Physically, this is obvious from the fact that the size of the Cooper pairs is given by the coherence length and is therefore a few hundred times larger than the atomic distance. Thus if the liquid is to stay 'super' up to the walls, the orbital planes of the pairs have to be parallel to the walls in order to avoid break-up (see Fig.V).

Another important external effect is given by a magnetic field. Since this introduces a quantization axis for the spin the vector d along which $S_3 = 0$ is forced orthogonal to H (see Fig.VI).

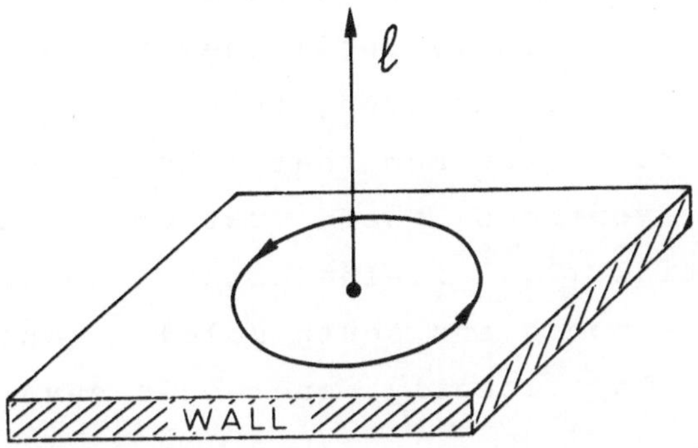

Fig. V

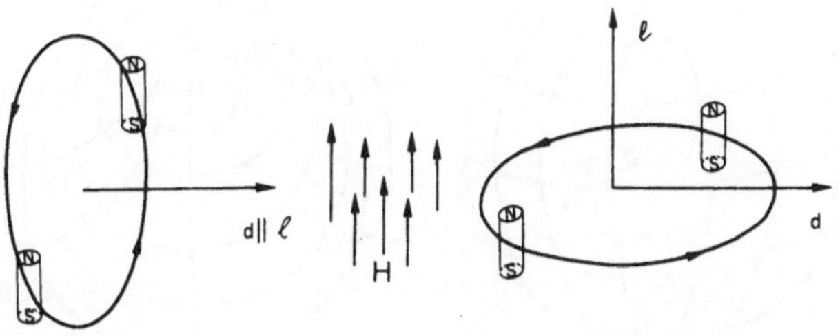

Fig. VI

With this preparation we can now present a few interesting classical field configurations.

1) Monopoles

Cooling a sphere smoothly through T_c plants the $\underset{\sim}{\ell}$ vectors orthogonal to the walls with $\boldsymbol{\phi}^1, \boldsymbol{\phi}^2$ covering them like orthonormal coordinates on a globe. With the liquid becoming "super" more and more inwards one may think that the $\boldsymbol{\ell}$ lines grow radially until they hit a point singularity at the center. However, this is not true: Because of the mathematical theorem that a hedgehog cannot be combed without vortices, there must be singularities in the tangential $\boldsymbol{\phi}^1, \boldsymbol{\phi}^2$ fields either two of flux one (for example at north and south pole) or one of flux two (see Fig.VII). Thus ideally, one would have field lines

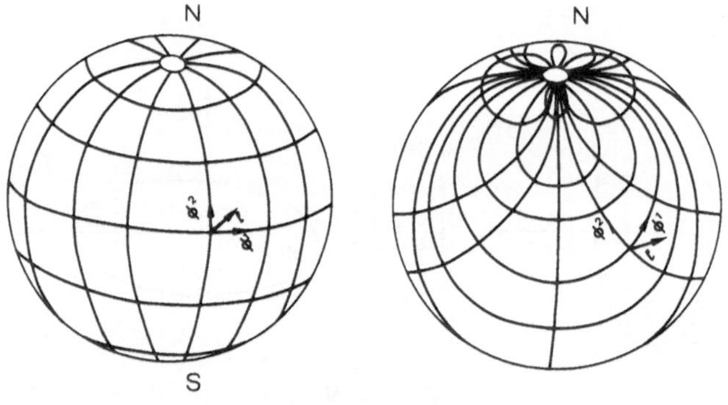

Fig. VII

growing as shown in Fig. VIIIa,b where the curly lines trace the singularities of the $\boldsymbol{\phi}^1, \boldsymbol{\phi}^2$ fields for the two possible extremes. But along these singular vortex lines the liquid must be normal since only if the field A_{ai} vanishes can the direction of $\boldsymbol{\phi}^1, \boldsymbol{\phi}^2$ be ill-defined. Thus the vortex lines form strings with a thickness equal to

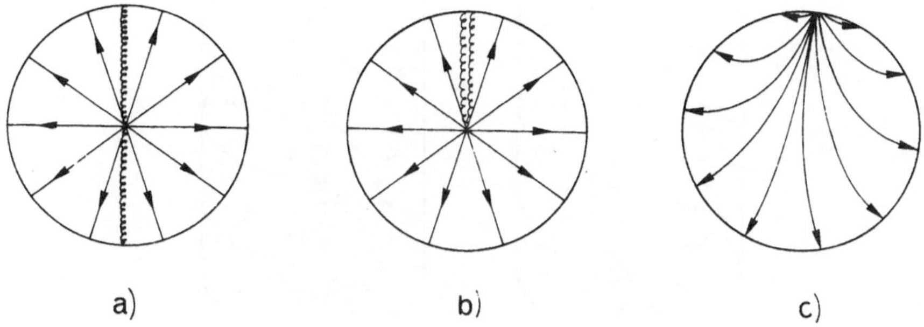

Fig. VIII

the coherence length inside of which there is an accumulation of condensation energy (the difference between f(A=0) and f(A = equ.(18),(19)). Thus the liquid likes to keep these strings as short as possible and pulls the point singularity to the wall with an LlogL potential energy resulting in the flower-like field configuration VIIIc (called boojum). This object has an intrinsic angular momentum and and should be detectable by observing the rotation of little ^{3}He droplets as they are cooled through T_c into the superliquid phase.

2) Line Singularities

In a cylinder the $\hat{\ell}$ lines will develop inwards until they form a singular line at the axis (see the left of Fig.IX). But this again contains the large condensation energy and the field lines prefer avoiding it by flaring upwards like flames in a chimney [14] (see the right of Fig.IX).

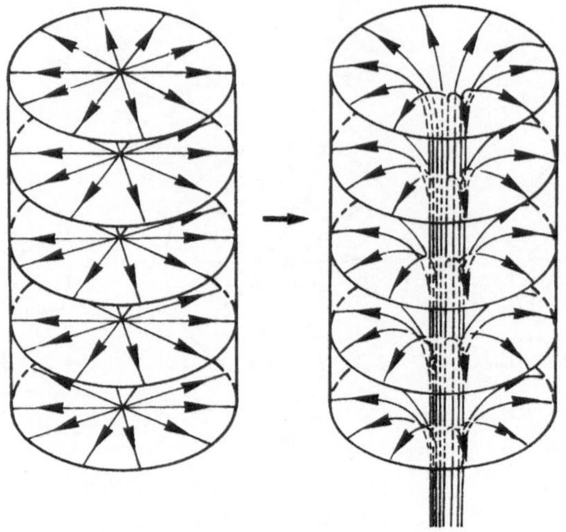

Fig. IX

3) Planar Singularities (Solitons)

In a magnetic field along the z axis the $\underset{\sim}{d}$ vectors will be forced in the xy plane, say

$$\underset{\sim}{d} = \sin\psi\, \hat{x} + \cos\psi\, \hat{y} \qquad (23)$$

Due to the dipole force, also ℓ wants to have this direction. If the superliquid could develop smoothly, one would obtain a completely uniform $d \parallel \ell$ field. However, small perturbations will create defects with $d \parallel \ell$ in some direction and antiparallel in others. The size of such domain walls will be determined by the dipole length ξ_d. In order to study a particular (twist) domain wall assume $\underset{\sim}{\ell}$ to have the form

$$\underset{\sim}{\ell} = \sin\chi\, \hat{x} + \cos\chi\, \hat{y} \qquad (24)$$

which is compatable with a Φ vector of the form

$$\Phi = e^{i\phi}\,(-\cos\chi\,\hat{x} + \sin\chi\,\hat{y} + iz) \qquad (25)$$

The free energy reads

$$f = \{(\partial\psi)^2 + (\partial\phi)^2 + \tfrac{1}{2}(\partial\chi)^2\} + \frac{1}{\xi_d^2}\sin^2(\chi-\psi) \qquad (26)$$

This can be diagonalized by

$$v = \chi - \psi \qquad (27)$$

$$u = \chi + 4\psi$$

as

$$f = \phi_z^2 + \frac{1}{20}u_z^2 + \frac{1}{5}v_z^2 + \frac{1}{\xi_d^{\,2}}\sin^2 v \qquad (28)$$

The classical extrema are ϕ = const., u = const. with a soliton in the v variable [15] (see Fig. X)

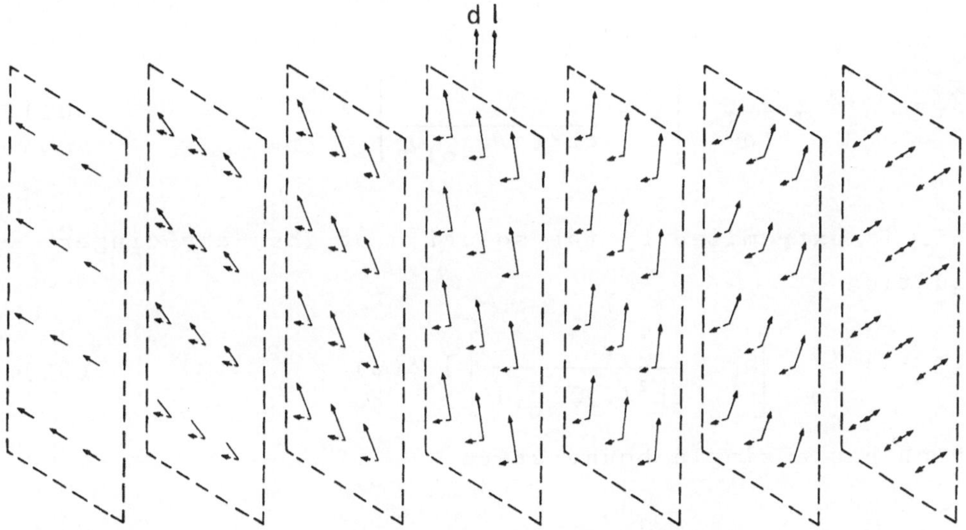

Fig. X

$$\mathrm{tg}\,\frac{v_{sol}}{2} = e^{\pm z/\xi_{sol}} \quad , \sin v_{sol} = \mathrm{ch}^{-1}(z/\xi_{sol}) \qquad (29)$$

of size

$$\xi_{sol} = \frac{1}{\sqrt{5}}\xi_d$$

The detection of all these classical field configurations proceeds most simply via nuclear magnetic resonance experiments. A vibrating $\underset{\sim}{H}_{\omega}$ field causes the $\underset{\sim}{d}$ vector to oscillate around the equilibrium position generating spin waves in the system. For these any spatial inhomogeneity acts as a potential wall (or mountain) which can trap (or repell) them. For example, the soliton just found will catch spin waves in a bound state. Consider small deviations

$$\psi = \psi_{sol} + \delta \qquad (30)$$

Then the energy fluctuates as

$$\delta^2 f = \delta_z{}^2 + \frac{1}{\xi_d{}^2}\left[1 - \frac{2}{\mathrm{ch}^2(z/\xi_{sol})}\right]\delta^2 \qquad (31)$$

This is extremized by the solution of the Schrödinger equation

$$\{-\partial_z{}^2 + \frac{1}{\xi_d{}^2}\left[1 - \frac{2}{\mathrm{ch}^2(z/\xi_{sol})}\right]\}\,\delta(z) = \omega^2\,\delta(z) \qquad (32)$$

which has a single bound state

$$\delta(z) \propto \frac{1}{[\mathrm{ch}(z/\xi_{sol})]^s} \qquad (33)$$

with

$$s = \frac{1}{2}\left[-1+\sqrt{1+4\frac{2}{\xi_d^2}\xi_{sol}^2}\right] = \frac{1}{2}\left[-1+\sqrt{\frac{13}{5}}\right] \approx .306 \qquad (34)$$

The energy is

$$\omega^2 = \frac{1}{2}(\sqrt{65}-7) = \frac{1}{\xi_d^2} \qquad (35)$$

while the continuum has a spectrum

$$\omega^2 = k^2 + \frac{1}{\xi_d^2} \qquad (36)$$

This bound state can be seen as a satellite frequency in NMR experiments shifted down by a factor of $\sqrt{\frac{1}{2}(\sqrt{65}-7)} \approx .728$ with respect to the normal line in complete agreement with the data [16].

Finally, let us give an example of the use of topology. Since the dipole force aligns $\underset{\sim}{d}$ and $\underset{\sim}{\ell}$, the parameter space of ^{3}He-A is for extended objects ($\gg \xi_d$) SO_3 only. Its homotopy groups are

$$\pi_1 = Z_2 \quad , \quad \pi_2 = 0. \qquad (37)$$

Notice that the first statement implies that ^{3}He-A is not really a superliquid at all since it is not able to pile up a large number of flux quanta in a torus (for this π_1 has to be equal to Z).[17]

In the B phase, θ is pinched to $\theta_d = \arccos(-\frac{1}{4}) \approx 104^\circ$ such that the remaining parameter space is $S^2 \times S^1$ (S^1 from the phase and S^2 from the direction vector $\underset{\sim}{n}$). Now the homotopy groups are

$$\pi_1 = Z \quad , \quad \pi_2 = Z \tag{38}$$

such that there are infinitely many line and point singularities. ^{3}He-B is a superliquid just as He-II.

For a more complete discussion of the non-trivial field configurations you are referred to Ref. 18).

Apart from these applications of recently popular methods of particle physics and field theory, ^{3}He may also serve directly high-energy experiments whenever a coherent accumulation of small effects must be detected. An example could be the electric dipole moments of the Cooper pairs caused by P violating (T conserving) neutral currents. These would line up with LxS and could pile up to a macroscopic effect in the condensate [19)].

With ^{3}He physics being a rapidly expanding field of research, I have been able to present only a glimpse of the many interesting phenomena to be explained by theory. Hopefully, the similarity of the problems as well as the methods of their solution may provide particle physicists with the same degree of inspiration as has been derived in the past from superconductors and superliquid ^{4}He.

References

1) G. 't Hooft, Utrecht preprint, Dec.1977.
H.B. Nielsen and P. Olesen, Nucl. Phys. B61, 1973.
2) See the lecture by K. Huang at this school.

3) S. Coleman, Phys. Rev. D15, 2929 (1977).
P.H. Frampton, Phys. Rev. D15 2922 (1977).
C. Callan and S. Coleman, Phys. Rev. D16, 1762 (1977).
4) S. Coleman, Phys. Rev. D11, 2088 (1975).
5) H. Kleinert, Hadronization of Quark Theories, Erice Lectures 1976, Plenum Press (in print).
6) H. Kleinert, Collective Quantum Fields, Lectures presented at the I International School on Low Temperature Physics, Erice 1977 and Fortschr. Physik 26, Nb. 8/9 (1978).
7) M. Gell-Mann, Strong and Weak Interactions, Present Problems, Proc. of 1966 Erice School, ed. by A. Zichichi (Acad. Press NY 1967), p.202.
H.J. Melosh, Phys. Rev. D9, 1095 (1974).
F. Buccella, H. Kleinert, C.A. Savoy, E. Celeghini, and E. Sorace, Nuovo Cimento 69A, 133 (1970).
8) See D.Pines and P. Nozières, Theory of Quantum Liquids, V.I. Benjamin, New York 1966.
9) H. Kleinert, Phys. Lett. 62B, 77 (1976).
10) P.T. Mathews and A. Salam, Nuovo Cimento 12, 563 (1954); 2, 120 (1955).
11) H.E. Stanley, Phase Transitions and Critical Phenomena, Clarendon Press, Oxford (1971).
F.J. Wegener, Physe Transitions and Critical Phenomena, ed. by C. Domb and M.S. Green, p.7, Academic Press (1976).
E. Brezin, J.C. Le Guillou, and J. Zinn-Justin, ibid. p.125.
See also L. Kadanoff, Rev. Mod. Phys. 49, 267 (1977).
12) N.D. Mermin, Lectures presented at the Erice Summer School on Low Temperature Physics, June 1977.

N.D. Mermin, Remarks prepared for the Sanibel Symposium on Quantum Fluids and Solids, 1977, and Physica 90B, 1 (1977).

13) G. Barton and M.A. Moore, Journal of Physics C, 7, 4220 (1974).

14) N.D. Mermin and T.L. Ho, Phys. Rev. Lett. 36, 594 (1976).

15) K. Maki and P. Kumar, Phys. Rev. Lett. 38, 557 (1977).

16) C.M. Gould and D.M. Lee, Phys. Rev. Lett. 37, 1223 (1976).

17) For a discussion of this subject see H. Kleinert, Berlin preprint, Sept.1978.

18) H. Kleinert, Berlin preprint, Sept.1978.

19) A. Legget, Sussex preprint 1977.

IS CONFINEMENT THE ULTIMATE TRUTH?

Walter Thirring

Institut für Theoretische Physik der Universität Wien

Boltzmanngasse 5, A-1010 Vienna, Austria

1. INTRODUCTION

In the past years evidence has accumulated as to there existing quarks with rest energies easily available. However, we do not see them, because they are confined by an increasing potential. The latter seems to contradict the Källen-Lehman representation according to which a potential in field theory has the general form

$$V(r) = \int_0^\infty dm\rho(m)e^{-mr}/r \ .$$

If the theory has no ghosts, i.e. if all energies and probabilities are ≥ 0, then $\rho(m) \geq 0$ and hence $\frac{\partial V}{\partial r} \leq 0$.

In this seminar I shall discuss a field theory which leads to a r-potential and therefore to confinement. Such a potential is the static Green-function of the equation $\Box\Box\phi=0$. Equations of this type were proposed in the 50's [1] and have recently been revived because of various reasons.

a) A propagator $(K^2)^{-2}$ improves the infrared divergence.
b) The r-potential is a good candidate for the gluon potential in hadrons [5].
c) In gauge theories quadratic in the curvature an equation of this type will arise [4].
d) If coupled to the energy-momentum tensor it leads to a dimensionless coupling constant [8].

The main trouble with these theories is the ghost problem [2]. Hence I shall concentrate on this aspect following [7]. The conclusion will be that the ghost difficulty corresponds classically to an instability due to a resonance phenomenon. For some couplings this does not occur and the theory behaves reasonably. Also the introduction of an indefinite metric [3] in Hilbert space is artificial and unnecessary.

2. THE FREE TIME EVOLUTION

The Lagrangian for two scalar fields $A_{o,1}$

$$L = \dot{A}_o \dot{A}_1 - \vec{\nabla} A_o \vec{\nabla} A_1 - \tfrac{1}{2} A_o^2 \tag{1}$$

leads to the field equation $(\Box = \Delta - \frac{\partial^2}{\partial t^2})$

$$\Box A_1 = A_o, \quad \Box A_o = 0 \rightarrow \quad \Box\,\Box A_1 = 0. \tag{2}$$

They lead to the time evolution τ_t^o :

$$\begin{aligned} \tau_t^o A_o := A_o(\vec{x},t) = \int d^3x' \Big[& \dot{D}(\vec{x}-\vec{x}',t)\, A_o(\vec{x}',0) + \\ & D(\vec{x}-\vec{x}',t)\, \dot{A}_o(\vec{x}',0) \Big] \end{aligned} \tag{3}$$

$$\begin{aligned} \tau_t^o A_1 := A_1(\vec{x},t) = \int d^3x' \Big[& \dot{D}(\vec{x}-\vec{x}',t)\, A_1(\vec{x}',0) + \\ & D(\vec{x}-\vec{x}',t)\, \dot{A}_1(\vec{x}',0) + \dot{D}(\vec{x}-\vec{x}',t)\, A_o(\vec{x}',0) + \\ & D(\vec{x}-\vec{x}',t)\, \dot{A}_o(\vec{x}',0) \Big] \end{aligned}$$

with

$$D(\vec{x},t) = \varepsilon(t)\ \delta(\vec{x}^2-\vec{t}^2)/2\pi\ ,\ D(\vec{x},t) = \varepsilon(t)\ \theta(\vec{x}^2-\vec{t}^2)/8\pi,$$

$$(\Box D = D).$$

The canonical commutation relations

$$[A_o(\vec{x}), \dot{A}_1(\vec{x}')] = i\delta^3(\vec{x}-\vec{x}') = [A_1(\vec{x}), \dot{A}_o(\vec{x}')] \tag{4}$$

can be satisfied if we represent the Fourier transforms by creation and destruction operators

$$\tilde{A}_o(k) = \{\alpha_1(k) + \alpha_1^\dagger(k) - i(\alpha_2(k) - \alpha_2^\dagger(k))\}\ /2\sqrt{k}$$

$$\tilde{A}_1(k) = \{\alpha_2(k) + \alpha_2^\dagger(k) - i(\alpha_1(k) - \alpha_1^\dagger(k))\}\ /2\sqrt{k}$$

$$\dot{\tilde{A}}_0(k) = - \{\alpha_1(k) + \alpha_1^\dagger(k) + i(\alpha_2(k) - \alpha_2^\dagger(k))\} \sqrt{k}/2$$

$$\dot{\tilde{A}}_1(k) = - \{\alpha_2(k) + \alpha_2^\dagger(k) + i(\alpha_1(k) - \alpha_1^\dagger(k))\} \sqrt{k}/2$$

$$[\alpha_i(k), \alpha_j^\dagger(k')] = \delta_{ij}\delta^3(k-k'). \tag{5}$$

For the α's the time evolution becomes

$$\tau_t^o\alpha_1(k) = \alpha_1(k)\ (1 + \frac{it}{4k})\ \cos kt + \alpha_1^\dagger(k)\ \frac{it}{4k}\ \cos kt +$$
$$\alpha_2(k)\ (\frac{t}{4k} + \frac{1}{4k^2} - i)\ \sin kt - \alpha_2^\dagger(k)\ (\frac{t}{4k} + \frac{1}{4k^2})\ \sin kt$$

$$\tau_t^o\alpha_2(k) = \alpha_2(k)\ (1 + \frac{it}{4k})\ \cos kt - \alpha_2^\dagger(k)\ \frac{it}{k}\ \cos kt +$$
$$\alpha_1(k)\ (\frac{t}{4k} - \frac{1}{4k^2} - i)\ \sin kt + \alpha_1^\dagger(k)\ (\frac{t}{4k} - \frac{1}{4k^2})\ \sin kt. \tag{6}$$

Remarks (7)

1. The linear increase with t arises because D is a constant in the light cone and is to be expected from a relativistic theory which leads to a potential $\sim r$. It can be viewed as a resonance phenomenon since the field equations in Fourier space

$$\ddot{\tilde{A}}_0 = - k^2\ \tilde{A}_0,$$

$$\ddot{\tilde{A}}_1 = - k^2\ \tilde{A}_1 - \tilde{A}_0$$

correspond to one oscillator driven by another with the same frequency.

2. The time evolution is a Bogoliubov transformation and is not unitarily implementable in a Fock representation $\alpha_i(k)|\ \rangle = 0$. ($\alpha(k) \to K_1(k,k')\ \alpha(k') + K_2(k,k')\ \alpha^\dagger(k)$ is unitarily implemented iff $\mathrm{Tr}\ K_2^\dagger\ K_2 < \infty$.) Correspondingly the formal Hamiltonian

$$\text{"H"} = \sum_k (\dot{\tilde{A}}_1(k)\ \dot{\tilde{A}}_0(k) + k^2\tilde{A}_1(k)\ \tilde{A}_0(k) + \frac{1}{2}\ \tilde{A}_0(k)^2) =$$

$$\sum_k |\vec{k}|\ (\alpha_1^\dagger(k)\ \alpha_2(k) + \alpha_2^\dagger(k)\ \alpha_1(k) + \frac{1}{2}\ \sum_k \tilde{A}_0(k)^2$$

is not an operator but only a (non-definite) quadratic form.

3. The Fock vacuum is not a time invariant vector and in fact one can prove that there is no time invariant state at all. Denoting

$$\langle\alpha_j^\dagger(k)\ \alpha_j(k)\rangle = \rho_j(k),\quad \langle\alpha_j(k)\ \alpha_j(k)\rangle = \langle\alpha_j^\dagger(k)\ \alpha_j^\dagger(k)\rangle = \sigma_j(k)$$

we find

$$\langle\tau_t^o\ \alpha_1^\dagger(k)\ \alpha_1(k)\rangle = \rho_1\ \cos^2 kt + \rho_2 \sin^2 kt + (2\rho_1+1-2\sigma_1)\ \frac{t^2}{k^2}\cos^2 kt$$
$$+ (2\rho_2+1-\sigma_2)\ \left(\frac{t}{k} + \frac{1}{k^2}\right)^2 \sin^2 kt +$$

terms with sin kt cos kt and t sin kt cos kt.

This is time independent only if $\rho_1=\rho_2$, $\sigma_1=\sigma_2=\rho_1+1/2$.

However, $\langle(\alpha^\dagger + c\alpha)\ (\alpha + c\alpha^\dagger)\rangle \geq 0\ \forall\ c \in R$ implies $|\sigma_j| \leq \sqrt{\rho_j(\rho_j+1)}$ and thus this condition cannot be met. The absence of a time invariant state poses no problem for the interpretation of the theory, it does not exist for one free nonrelativistic particle ($H=p^2/2m$) either. Only if there were a time invariant state, its G.N.S. representation would furnish a unitary representation for the time evolution.

3. FIELD WITH AN EXTERNAL SOURCE

Adding $-j(\vec{x},t)\ A_1(\vec{x},t)$ to the Lagrangian the field equations become

$$\Box A_1 = A_o,\quad \Box A_o = j \rightarrow \Box\Box A_1 = j. \tag{8}$$

This changes the time evolution to

$$\begin{aligned}\tau_t A_o(\vec{x}) &= \tau_t^o A_o(\vec{x}) - \int_o^t dt'\ dx'\ D(\vec{x}-\vec{x}')\ j(\vec{x}')\\ \tau_t A_1(\vec{x}) &= \tau_t^o A_1(\vec{x}) - \int_o^t dt'\ dx'\ D(\vec{x}-\vec{x}')\ j(\vec{x}').\end{aligned} \tag{9}$$

This is again not a unitary transformation, but relative to the free evolution it means

$$\tau_t\ \tau_{-t}^o\ \alpha_1(k) = \alpha_1(k) + \frac{1}{2\sqrt{k}} \int_o^t dt'\ \tilde{j}(k,t')\ \left[(\sin kt' - \cos kt')\left(\frac{it'}{2k} - 1\right) - \frac{i}{2k^2}\sin kt'\right] \tag{10}$$

$$\tau_t \tau^o_{-t} \alpha_2(k) = \alpha_2(k) + \frac{1}{2i\sqrt{k}} \int_0^t dt' \, \tilde{j}(k,t') \left[(\sin kt' - \cos kt') \left(\frac{it'}{2k} + 1\right) + \frac{i}{2k^2} \sin kt'\right].$$

Remarks (11)

1. Since $\alpha(k) \to \alpha(k) + c_t(k)$ is a unitary transformation if the c-number $c_t(k) \in L_2(d^3k)$, we see that for reasonable currents there may be a unitary wave operator Ω: $\lim_{t\to\infty} \tau_t \tau^o_{-t} \alpha(k) = \Omega^\dagger \alpha(k) \Omega$. For this to happen the limit of $c_t(k)$ has to exist and to be square integrable. Thus if $\tilde{j}$ goes sufficiently to zero for $k \to 0$ (zero total charge) and for $k \to \infty$ there will be a unitary S-matrix.

2. Although the Hamiltonian does not exist as an operator, its expectation value exists for a dense set of states and in these one can talk about the energy loss to the field. In the Fock state $\alpha| > = 0$ it is calculated to

$$\Delta E = \int_{-\infty}^{\infty} dt \left\langle \frac{\partial H}{\partial t} \right\rangle = \int_{-\infty}^{\infty} dt \left\langle A_1(\vec{x},t) \frac{\partial j}{\partial t} \right\rangle d^3x =$$

$$\int \frac{d_\omega \, d^3k}{(2\pi)^3} |\tilde{j}(\omega,k)|^2 |\omega| \, \delta'(\omega'^2 - k^2).$$

It can have either sign and is positive if $|\tilde{j}(\omega,k)|^2\omega$ is monotonically increasing with k.

4. INTERACTION WITH AN OSCILLATOR

Our findings so far raise the question as to whether such a field in interaction with matter will lead to dynamical instabilities. For a first exploration we represent matter by a harmonic oscillator and study the Lagrangian

$$L = - A_{o,\mu} A_1^{\,\mu} - \frac{1}{2} A_o^2 + \frac{\dot{q}^2}{2} - \omega_o^2 \frac{\dot{q}^2}{2} - q \int \beta(x) \, A_1(\vec{x}) \, d^3x.$$

To solve the classical initial value problem of the Euler equations

$$\begin{aligned} \Box A_o(x,t) &= - \beta(x) \, q(t) \\ \Box A_1 &= A_o \\ \ddot{q}(t) &= - \omega_o^2 \, q(t) - \int d^3x \, \beta(x) \, A_1(x,t) \end{aligned} \qquad (12)$$

we make the usual restriction of fields for t=0 to be $\varepsilon\ L_1(d^3\vec{x})$ so that the Fourier transform $\tilde{A}$ of the field exists. In $\vec{K}$-space the linear equations (12) can be considered as ordinary first order differential equations in t for the quantity $(q,\ \dot{q},\ \tilde{A}_0(k),\ \dot{\tilde{A}}_0(k),\ \tilde{A}_1(k),\ \dot{\tilde{A}}_1(k)) =: A$

$$\dot{A} = S\ A \tag{13}$$

where S is the generator of a sympletic transformation. For the solution of the initial value problem

$$A(t) = \frac{1}{2\pi} \int_C d\omega \frac{e^{i\omega t}}{(i\omega - S)} A(0) \tag{14}$$

(C encloses the spectrum of S) we have to calculate the inverse of

$$i\omega - S = \begin{vmatrix} i\omega & -1 & 0 & 0 & 0 & 0 & 0 & 0 & . \\ \omega_0^2 & i\omega & 0 & 0 & \tilde{\beta}(k_1) & 0 & 0 & 0 & . \\ 0 & 0 & i\omega & -1 & 0 & 0 & 0 & 0 & . \\ -\tilde{\beta}(k_1) & 0 & k_1^2 & i\omega & 0 & 0 & 0 & 0 & . \\ 0 & 0 & 0 & 0 & i\omega & -1 & 0 & 0 & . \\ 0 & 0 & 1 & 0 & k_1^2 & i\omega & 0 & 0 & . \\ 0 & 0 & 0 & 0 & 0 & 0 & i\omega & -1 & . \\ -\tilde{\beta}(k_2) & 0 & 0 & 0 & 0 & 0 & -k_2^2 & i\omega & . \\ . & . & . & . & . & . & . & . & . \end{vmatrix} \tag{15}$$

Direct computation of the determinants involved leads to the following expressions for the motion of the oscillator

$$q(t) = \frac{1}{2\pi} \int_{-\infty}^{\infty} \frac{d\omega}{D(\omega)} e^{i\omega t} \{i\omega q(0) + \dot{q}(0) + \int \frac{d^3k}{(2\pi)^3} \tilde{\beta}(k)$$

$$\left(\frac{\tilde{A}_0(k,0) i\omega}{(k^2-\omega^2)^2} + \frac{\dot{\tilde{A}}_0(k,0)}{(k^2-\omega^2)^2} - \frac{i\omega \tilde{A}_1(k,0)}{(k^2-\omega^2)} - \frac{\dot{\tilde{A}}_1(k,0)}{(k^2-\omega^2)} \right)\}$$

$$D(\omega) = \omega_0^2 - \omega^2 - \int \frac{d^3k\ |\tilde{\beta}(k^2)|^2}{(k^2-\omega^2)^2}\ . \tag{16}$$

All depends now on whether $D(\omega)$ has zeros for complex ω or not. If so we get contributions increasing exponentially with time since with ω_o also $-\omega_o$, ω_o^*, $-\omega_o^*$ are zeros of $D(\omega)$. (This reflects the symplectic nature of e^{St}.) For second order field equations complex zeros occur only if oscillator and field have opposite sign of the energy [6].

Let us choose

$$\mathcal{B}^2(k) = \frac{k^2}{k^2+M^2}\ \frac{8\beta^2}{\pi} \tag{17}$$

so that the integral does not become singular for $\omega \to 0$. We calculate for Im $\omega > 0$, $M > 0$,

$$\int \frac{d^3k}{(2\pi)^3}\ \frac{\mathcal{B}^2(k)}{(k^2-\omega^2)^2} = \frac{2}{\pi}\int_{-\infty}^{\infty} \frac{dk\ k^4\ \beta^2}{(k+\omega)^2(k-\omega)^2(k+iM)(k-iM)}$$

$$= \beta^2\ \frac{i\omega-2M}{(\omega+iM)^2}\ . \tag{18}$$

To find possible zeros of $\omega^2 - \omega_o^2 + \beta^2\ \frac{i\omega-2M}{(\omega+iM)^2}$ in the upper half plane we set $\omega = \mu+i\nu$, $\nu > 0$ and calculate the imaginary part of the above expression

$$\text{Im}\ \{(\mu+i\nu)^2 - \omega_o^2 + \beta^2\ \frac{i\mu-\nu-2M}{(\mu+i(\nu+M))^2}\} = \mu\{2\nu + \frac{\beta^2(\mu^2+4\nu M+3M^2+\gamma^2)}{[\mu^2+(\nu+M)^2]^2}$$

For $\nu > 0$ we have { } > 0 and thus there is no zero for $\mu \neq 0$. On the imaginary axis D becomes $-\nu^2 - \omega_o^2 + \beta^2(\nu+2M)(\nu+M)^{-2}$. As long as $\omega_o^2 > 2\beta^2/M$ this is negative for $\nu > 0$. Thus we conclude that for some types of coupling there is no instability irrespective of the initial conditions.

Remarks (19)

1. Since the field energy is not definite, no sign of the field Lagrangian is a priori preferred. We have chosen the opposite sign to (1), since the other sign would reverse the sign of β^2 and produce complex zeros.

2. If the oscillator interacts only with one mode, $\mathcal{B}^2(k) = \beta^2\delta(k-k_o)$, we have

$$D(\omega) = \omega_o^2 - \omega^2 - \frac{\beta^2}{(k_o^2-\omega^2)^2}\ .$$

This has only real frequencies iff $k_o^2<\omega_o^2$ and $\beta< \frac{2}{3\sqrt{3}}(\omega_o^2-k_o^2)^{3/2}$.

3. More generally one can argue for small β as follows. Let $\tilde{\beta}(k)$ go to zero for $k \to 0$ (zero total charge) such that $\int d^3k\, \tilde{\beta}^2(k)/(k^2-\omega^2)^2$ is bound in the upper half plane. The zeros of $D(\omega)$ will depend continuously on β and for $\beta = 0$ they are at $\omega = \pm\omega_0$. It now depends whether they move for $\beta > 0$ into the physical or the unphysical ω-sheet. Thus iff $\mathrm{Im} \int d^3k\tilde{\beta}^2(k)/(k^2-\omega^2)^2 = -\int d^3k\tilde{\beta}^2(k)\, \delta'(k^2-\omega^2)$ is positive as ω approaches ω_0 from above the zeros will move into the unphysical sheet and there will be no complex poles. This condition resembles the requirement of a positive energy loss (11,2).

4. Certainly there are solutions of the equation (12) where $A \sim e^x$ and they will always lead to exponentially increasing amplitudes of q. They have been excluded by our requirement $A \in L_1$. Our result shows that for some interactions the oscillator does not increase its amplitude and gains the energy from the emission of quanta of negative energy.

5. If there are no complex poles and $\mathrm{Im}\,(S-i\omega)^{-1}$ approaches an integrable function for $\omega \to$ real axis then $Q(t) \to 0$ as $t \to \pm\infty$. This follows from the Riemann-Lebesque lemma. Physically this means that the A-field produces a radiation force similar to the electromagnetic field.

Open Questions (20)

1. Are there some representations where the time evolution is unitarily implementable?

2. Can one find couplings of the A-field to other forms of matter (i.e. a Dirac field) such that the negative energies do not cause trouble?

3. Since the A-quanta have zero mass, they do not exist in reality. Can one make a gauge theory such that the A-quanta also have A-charge such that their emission is forbidden by the requirement of A-charge neutrality?

REFERENCES

[1] H. Bhabha, Phys. Rev. 77, 665 (1950)
W. Thirring, Phys. Rev. 77, 570 (1950)

[2] A. Pais, G.E. Uhlenbeck, Phys. Rev. 79, 145 (1950)

[3] W. Heisenberg, Nucl. Phys. 4, 532 (1957)
M. Froissart, Nuovo Cim. Suppl. XIV, 197 (1959)

[4] P. Havas, GRG 8, 631 (1977)
E. Pechlaner, R. Sexl, Comm. Math. Phys. 2, 165 (1966)
S. Ferrara, B. Zumino, Structure of Conformal Supergravity, CERN-preprint (1977)

[5] S.K. Kaufmann, Nucl. Phys. B87, 133 (1975)
S. Blaha, Phys. Rev. D10/12, 4268 (1974)

[6] F. Schwabl, W. Thirring, Ergebnisse der exakten Naturwissenschaften, Springer 1964, p. 219

[7] H. Narnhofer, W. Thirring, Phys. Lett 76B, 428 (1978)

[8] Tonin and co-author, Padova Preprint (1978)

DISCUSSION

CHAIRMAN: W. Thirring

Scientific Secretaries: K. Lackner, H.R. Gerhold

DISCUSSION No. 1

- BHANOT:

The motion of a mechanical system in phase space is either closed or will come arbitrarily close to its starting point in a finite time. Does this contradict ergodicity?

- THIRRING:

If a trajectory of a mechanical system in phase space does not escape to infinity, it will come arbitrarily close to its starting point in a finite time. This theorem is due to Poincaré. However, this mathematically correct statement is absolutely irrelevant to statistical mechanics. The reason is that the time which is needed to come back within a reasonable neighborhood of the starting point grows exponentially with the number of degrees of freedom. For 10^{24} particles this time is much larger than the lifetime of the universe. This problem was long disputed by Boltzmann and Zermelo. Boltzmann correctly realized Poincaré's theorem to be totally irrelevant, because the times involved are far too long.

- SURSOCK:

In statistical mechanics what breaks the symmetry of time in the fundamental equations of motion?

- THIRRING:

Starting from first principles we shall take it as granted that the equations of motion are time-reversible. So the only possibility left is that certain initial conditions are more likely than others. This is not only true in statistical mechanics but also in classical electrodynamics where one has a radiative reaction force which is not invariant under time reversal. It distinguishes a certain time direction. Clearly for any solution with the usual sign of the reactive force there is also another solution with the opposite sign, but we believe the initial condition of the latter solution to be extremely unlikely. Therefore we get a definite sign of the reactive forces.

In statistical mechanics one can get time irreversibility in a similar manner.

- LACKNER:

In the case of the harmonic oscillator the motion proceeds on a torus in phase-space. By disturbing the oscillator slightly one can destroy the torus, but if the frequencies are reasonably irrational, the torus will only be deformed. So you still have a variable which is constant during all time. Does that mean, that the system is still integrable? Can one estimate how strong the deformation is?

- THIRRING:

The constants of motion are completely destroyed no matter how small the perturbation might be. The reason is that for every point where one still has a surface to which the motion is confined there, arbitrarily close, is another point in which the motion is completely unconfined. So the system is not integrable even if it behaves in some cases very similarly to an integrable one.

- KENNEDY:

How useful are existence theorems such as those of Penrose for the existence of singularities in General Relativity, as they seem to be very sensitive to small changes in the choice of the theory? (e.g. taking Einstein - Cartan theory).

- THIRRING:

It is true that one can change the equations of motion so that the singularity theorem does not hold anymore. An example is the introduction of torsion by Trautmann. If E + 3P will not be positive in a given theory there is no basis for the singularity theorem anymore. Nevertheless I think the theorem is of interest, because singularities might arise from situations which are well explored and where Einstein's equations hold very well. The critical density for developing into a black hole of an object of galactic mass is the density of air. There we are pretty sure that the laws of nature hold as we know them. Even if the equations of motion do not lead to a shrinking into a point, matter should certainly fall below the Schwartzchild radius. But looking from outside we do not care so much what happens inside. The essential point is that the object shrinks below the event-horizon from where nothing can come back. There are good reasons to believe that these things actually do happen.

- WOLF:

In calculating the energy of a N-particle ensemble an important ingredient for stability was the Fermi nature of particles. What happens with bosons?

- THIRRING:

It can be shown that a mixture of bosons and fermions is also stable provided the bosons all carry the same charge. An example of this is deuterium. If there are also bosons of opposite charge the system becomes instable. If there were charged stable bosons which only interact electromagnetically, the universe could not be stable, its energy would be unbound from below. Since the energy needed for pair production is proportional to N, it would eventually be overwhelmed by the binding energy which is at least proportional to $N^{7/5}$.

DISCUSSION No. 2 (Scientific Secretaries: H. Grosse, K. Lackner)

- MOTZ:

Is the Hamiltonian you used an operator in Hilbert space when you did the quantum theory?

- THIRRING:

We want a theory which is time translation invariant. That means there is to be a time automorphism τ_t which transforms each field operator α into another field operator $\alpha_t = \tau_t\alpha$ in such a way that the commutator relations remain invariant. We would like to define a selfadjoint Hamiltonian H as the infinitesimal generator of that time evolution; or in other words we would like to have a unitary transformation $U_t = \exp(iHt)$ so that $\alpha_t = U_t \alpha U_t^{-1}$ For a finite number of degrees of freedom such a U always exists; because there is this famous theorem by John von Neumann that all representations of the CCR for bose operators (this is also working for fermion operators) are unitarily equivalent. For infinitely many degrees of freedom no such U is generally in existence and therefore also no Hamiltonian. So the question arises: what is the meaning of the formally derived Hamiltonian? It is now not a generator which you can apply to all vectors (it makes a vector of infinite length out of any vector in Hilbert space), but nevertheless expectation values of H for a dense set of vectors are perfectly well defined (one can even scale it to zero for the vacuum). Therefore you can even calculate the energy loss of a system coupled to something else, if the energy is not conserved.

- BLASI:

In what sense is the S-matrix defined? Are there |in) and |out) states?

- THIRRING:

First you take the product of the free time evolution times the full time evolution τ_t (in inverse order), then there exists a unitary operator Ω_t which does the following

$$\alpha_t = \tau_t \tau_{-t}^c \alpha = \Omega_t \alpha \Omega_t^{-1}$$

The S-matrix is then defined as the following limit

$$S = \lim_{t\to\infty} \Omega_{-t}^{-1} \Omega_t$$

and is a unitary operator mapping the algebra at $t=-\infty$ onto the algebra at $t=+\infty$.

$$S \alpha_{-\infty} S^{-1} = \alpha_{+\infty}$$

The |in) and |out) states on the other hand could be defined in principle by applying $\alpha_{-\infty}^{+}$ to the vacuum state $|0)$. Since $\alpha_{-\infty}|0) = 0$ gives zero and since S in the first example does not mix creation and annihilation operators, also $\alpha_{+\infty}|0) = 0$. But then you can build up your |in) and |out) states, although they would not have the usual physical significance, because this vacuum is not a time invariant state(no time invariant state even exists at all) and we have to remember that $\Omega_{+\infty}$ is not the limit of the time evolution but the limit of the free time evolution times the full one in reversed order.

- BLASI:

Are these states now particles?

- THIRRING:

One can now ask how one could see these states for instance in a bubble chamber. I guess they have some momentum and energy so you could see recoil electrons, on the other hand they are a sort of continuous background.

- BHANOT:

Is it true that a propagator like $^1/q^4$ makes the theory noncausal?

- THIRRING:

I cannot agree, because the Green function has support in the past light cone and is zero otherwise; the so called Velo-Zwanziger phenomenon of the light cone being deformed and something being able to go faster than light does not occur for a coupling as I wrote it down. On the other hand, one sometimes says that ghosts are non-causal.

One should distinguish between different versions of causality: one is that the light cone is deformed and something may go with a velocity faster than the light velocity. But you have to consider that also the light cone of the photon might be deformed.

Then it may happen that you used the quantization with indefinite metric and you may get negative probabilities which would certainly not be acceptable. If you avoid these negative probabilities you can get something like the preacceleration as is found in classical electrodynamics where you have Dirac's run-away solutions, sometimes called acausal. Here this may also happen if you have poles in the complex plane for the D-function. In electrodynamics you start with a negative infinite mass of the electron in order to compensate the positive infinite self energy so as to get a finite value. Here it is the reverse in a sense: you have an oscillator with potential energy coupled to the field with positive and negative energy.

- NEWMAN:

Can these confinement theories say anything about the $J/\psi-\psi'$ and $\gamma-\gamma'$ mass differences?

- ISGUR:

Charmonium models which used a combination of a linear and a Coulomb potential underestimate the $\Upsilon-\Upsilon'$ splitting by about 25%; this failure is sensitive to the exact way the potential makes the transition from the $^1/r$ to the confinement potential and should not, in my opinion, be taken seriously.

- THIRRING:

There is one feature of the linear potential that the wave functions at the origin for all S-states are equal, but then you naturally have corrections.

- ISGUR:

The corrections due to the Coulomb potential then go into the right direction.

- KLEINERT:

How about asymptotic freedom in you model?

- THIRRING:

In a way it is even better than asymptotic freedom, because you have this dipole type of propagation where the ultraviolet divergencies are damped very strongly. That actually is the reason, why these theories have been discussed before.

- ISGUR:

What are the renormalization properties of a quantum field theory based on $\Box\Box\phi = \rho$?

- THIRRING:

All these renormalization arguments work only in perturbation theory, where you start with a single quark emitting a single gluon. I tried to convince you that a single quark in the world is a completely impossible situation, because it is not colour-neutral. So you should talk only about confined neutral objects. One should invent a perturbation theory where one talks only about colour-neutral entities. Another perturbation theory may give you infrared troubles which are physically not present, because there should not be any single quark state within the asymptotic states. Power counting in our model in the usual perturbation theory is very good; there is only one vacuum diagram which is divergent.

- HOFFMAN:

How general is the result of colour neutrality you obtained, if one considers the potentials of the r^{α} type?

- THIRRING:

Intuitively I would say that if you have a potential which keeps going on to infinity, the theory will make sense only if you have colour neutralization.

- BHANOT:

What happens if somebody sees a quark tomorrow? Can you modify your theory to take free charges into account?

- THIRRING:

I can tell you what is bad, if you don't have total charge zero. I first solved for you the initial value problem for a certain time interval $[t_0, t]$ and this has a perfect solution even if the total charge is not zero. What one wants to have afterwards is the connection between $t = -\infty$ and zero, or zero and $t = +\infty$. Now this limit does not exist unless the total charge is zero. This gives an indication that for large times there will be troubles related to the linear rise of the potential.

- BLASI:

There is a model called the Froissart model which takes into account multiple ghosts. The conclusion therein is that even without any interactions you will not have any $|in)$ and $|out)$ states.

SUCCESSES AND FAILURES IN MATHEMATICAL PHYSICS

Walter Thirring

Institut für Theoretische Physik der
Universität Wien
Boltzmanngasse 5
A-1010 VIENNA, Austria

1. INTRODUCTION

Let me first explain what I consider as the task of mathematical physics. One sometimes gets the impression that it tries to cleanse physics from earthly blemishes by purifying abstractions. I see the problems somewhere else. At present we know very well the laws of matter surrounding us, and only that which happens in the very large or the very small is unknown. Saying we know the laws means that we can write some equations on a piece of paper but they are of little service, if we cannot deduce consequences from them for the reason that they are beyond human calculational power. Thus one proceeds semi-empirically and makes approximations guided or seduced by intuition. In this way the logical unity of physics gets lost and it decays into specialized disciplines. These all have their own folklore and their own experts, which through mutual agreement decide what the truth is. However, mathematics has also progressed and sometimes can help us with problems where our intuition is helpless, since it has not yet been trained in this direction. Thus one sometimes can answer some fundamental qualitative questions and close gaps in the structure of physics. I shall illustrate this by three examples[1] from the most difficult problems which have been solved in mathematical physics in the last decades.

1) The three examples are treated in detail in Volumes I, II and IV of my lectures in mathematical physics (German edition Springer Wien 1977, English edition Springer New York 1978). They contain references to the original literature.

2. THE KAM-THEOREM

Most examples found in textbooks on classical mechanics are so-called integrable systems. Such a system of n degrees of freedom has n angle variables and n action variables which are constant. A prototype is a system of n harmonic oscillators:

$$H = \sum_{i=1}^{n} (\tfrac{1}{2})(p_i^2 + \omega_i^2 q_i 2) \tag{1}$$

It has the n constants $p_i^2 + \omega_i^2 q_i^2$ and the orbit does not fill the 2n-1-dimensional energy shell but only an n-dimensional torus. Integrable systems are of this form with the ω's depending generally on the action variables. For almost a hundred years the belief was cherished that (1) is a degenerate case and an arbitrarily small perturbation would render the system ergodic so as to have the orbit filling the energy shell densely. Fermi still thought that he could prove this mathematically. On the contrary, it was shown by Kolmogorow, Arnold and Moser that under a small perturbation most of the tori are only slightly deformed and the system behaves practically like an integrable one. Let me briefly discuss why Fermi's brilliant intuition failed here.

Periodic motions average out the effect of perturbations unless they are amplified by resonance phenomena. This becomes apparent already if one adds to (1) with n=2 a perturbation $H' = \lambda\, q_1 q_2$. The new frequencies

$$\omega_{\pm}^2 = \tfrac{1}{2}(\omega_1^2 + \omega_2^2) \pm \sqrt{\left(\frac{\omega_1^2 - \omega_2^2}{2}\right)^2 + \lambda^2} \tag{2}$$

can be expanded in λ^2 if $|\omega_1^2 - \omega_2^2| > 2|\lambda|$, but are of the order of λ if $\omega_1 = \omega_2$. Thus the effect of the perturbations is small except when the frequencies are degenerate. The KAM-theorem generalizes this to the statement: the effect of a general perturbation is small, if the frequencies are sufficiently rationally independent. One may wonder what this means, since the rational numbers are dense and one is always close to rationally related frequencies. The exact condition is that resonance denominators $(\sum_{i=1}^{n} \omega_i g_i)^{-1}$, $g_i =$ integer have to remain bound by polynomials in the g's.

$$\left|\sum_{i=1}^{n} \omega_i g_i\right|^{-1} \leq c \left|\sum_{i=1}^{n} |g_i|\right|^{n} \tag{3}$$

In this case the perturbation can be eliminated by a convergent iteration procedure. Eq. (3) allows the resonance denominator to become small for big g_i, i.e. if the ω's are only remotely rationally similar. To a musician this is plausible, because he is not disturbed by the dissonant far-away overtunes these being only faintly excited. However, the ω's violating (3) and thus destroying the invariant tori form an open and dense set, since they contain all rational points. Thus one wonders where Eq. (3) can be satisfied and is surprised to learn a lot to be left over. It turns out that the ω's satisfying Eq. (3) form one of these strange sets which are dense nowhere and yet have, for large c, a large measure. In one dimension such a set S can be constructed by excluding the rational numbers r_n, $n = 1, 2 \ldots \infty$ by open intervals of shrinking size

$$S = R \setminus \bigcup_{n=1}^{\infty} (r_n - 2^{-n} \varepsilon , r_n + 2^{-n} \varepsilon)$$

The measure of the excluded part is $< \sum_{n=1}^{\infty} 2 \varepsilon 2^{-n} = 2 \varepsilon$ and goes with ε to zero. It is clear that the physicists of the twentieth century were not prepared to distinguish between these topological and measure-theoretical features. Since the latter determine the probabilities, they are the ones, which are relevant for physics and tell us that for weak perturbations it is very unlikely that the system will behave non-integrably. The weakness of the result is that the estimates for c in Eq. (3) are so poor for many degrees of freedom that the theorem is proven to apply only for ridiculously small perturbations.

3. GRAVITATIONAL COLLAPSE AND SINGULARITY THEOREMS

At the beginning of World War II Robert Oppenheimer and various collaborators published solutions of Einstein's equations describing the collapse of a star. It contracts in finite time to a point and only a singularity is left over of the star. This perhaps is not so surprising, since in their radially symmetric solution everything falls straight into the center. The only new feature is that even pressure cannot stop it, because in Einstein's theory it produces even more gravity. But it was frequently believed that this is a pathological feature of radially symmetric solutions and would not happen in general. However, R. Penrose and other relativists from the school of D. Sciama succeeded in showing that under fairly general circumstances space develops some singularities in Einstein's theory.

The proof of the existence of singularities starts with the obvious fact that gravitation only strengthens the convergence of the world lines of freely falling particles. This reflects the attractive nature of gravity and holds also true in Newton's theory if all masses are positive. In reasonable theories, where energy and pressure are positive, it carries over to Einstein's theory. Now one shows that the geodesics do not only converge but cross the neighboring geodesics. This in itself is no disaster but geodesics which have crossed themselves are no longer extremal lines, namely one can do better by rounding off the corner. However, this can lead to a contradiction, since one can show that for an initial surface intersected by all timelike geodesics there is a line of longer proper time to every point. If all orthogonal geodesics through the initial surface converge and cross after a finite proper time, they cannot be these extremal lines. On the other hand, the latter have to be geodesics and the only way out is the conclusion that singularities prevent the continuation of the geodesics beyond the crossing point. One sees the weakness of these geometrical arguments, since they do not tell us what happens at the singularity. Presumably the singularity is felt by infinite tidal forces, but so far the physical nature of the singularity is known only in special examples.

4. STABILITY OF MATTER

In thermodynamics one postulates that the energy is an extensive quantity and thus the energy per particle should tend to a limit for an infinite system. Since ordinary matter consists of electrons and nuclei, the laws of which are well-known, one should be able to deduce this hypothesis. More specifically, the Hamiltonian for a Coulomb system

$$H_N = \sum_{i=1}^{N} \left(p_i^2/2m_i + \sum_{j<i} e_i e_j / | x_i - x_j | \right) \tag{4}$$

should give an inequality

$$E_N > -A \cdot N, \quad \forall N \tag{5}$$

where A is a constant of the order of a Rydberg. Although this is a question of localizing the lowest eigenvalue of the Schrödinger equation, Eq. (5) was proved for the first time in 1965 by Dyson and Lenard. The obvious difficulty in proving Eq. (5) is the fact that Eq.(4) contains a double sum with $\sim N^2$ terms and hence a large amount of cancellation is necessary to acquire something $\sim N$. In particular, if all terms were negative, as it happens for gravitational interactions, E_N goes $\sim N^{7/3}$ for fermions and $\sim N^3$ for bosons. But even if one takes the Coulomb potential, Eq. (5) does not hold generally as is indicated by the following naive argument with the uncertainty relation. If the system is in a volume $\sim R^3$, the kinetic energy (for bosons) is of the order of $N\hbar^2/2mR^2$. The potential energy will be $-Ne^2$/(distance to the next neighbour) $\sim -N^{4/3}e^2/R$, since the field of the farther-away charges will be shielded. The minimum of the total energy is reached for $R\sim N^{-1/3}$ and goes $\sim -N^{5/3}$. Only for fermions, where the kinetic energy goes $\sim N^{5/3}\hbar^2/2mR^2$ does the minimum have the extensive properties $R\sim N^{1/3}$, $E_N \sim -N$. Of course, this rough argument is not a proof, but it points to the following features of the problem:

a) In the instable system R decreases with increasing N and thus it is not the long range property of the Coulomb force which is responsible for the instability in the boson case. In fact, the proof of stability is just as hard for the $e^{-\mu r}/r$-potential as for the $1/r$-potential.

b) By removing the $1/r$-singularity by a nuclear form factor at 10^{-13}cm one gets stability. However, as this gives an A~MeV, this certainly is not the correct explanation for it.

c) As the reliable molecular calculations deal with systems of one or two electrons, they cannot teach us anything about stability, because they do not depend on Fermi statistics.

Dyson and Lenard proved Eq. (5) by a heroic attack digging out the kinetic energy of the electrons from the zeros of the wave function Ψ generated by its antisymmetry. Since their analysis became somewhat lengthy, they had to pay rather a high price, namely $A\sim 10^{14}$ Rydberg. Ten years later E. Lieb and myself got a reasonable A through formulating the naive argument in rigorous mathematical inequalities. The main point is that the kinetic energy is bound by the electron density, $E_{kin} > cN^{5/3}/R^2$, or more precisely

$$E_{kin} > c \int d^3x \rho^{5/3}(x). \qquad (6)$$

Such an expression is used in the Thomas-Fermi theory and presumably the inequality is correct with c from the Thomas-Fermi theory. We did not succeed in proving the inequality with this constant, but only with a c smaller by 4π. Using some results from the Thomas-Fermi theory one thus will get a simple proof of Equation (5). Therefore in this case Fermi's intuition got to the heart of the matter and turned out to be more valuable than the knowledge of mathematicians.

SUPERSYMMETRY AND $SU(2)_L \times U(1)_{L+R}$

CLOSING LECTURE*

A. Zichichi

CERN

Geneva, Switzerland

In the opening lecture, which is based on last year's "Whys"[1], the status of the most outstanding problems in subnuclear physics has been sketched. This closing lecture will be devoted mainly to two very important areas of research which have attracted my attention; namely "supersymmetry" and the "unified theory of weak and electromagnetic interactions".

1. One of the most striking relations of supersymmetry is that the anticommutator of two spinorial charges Q_α and Q_β gives a space-time translation P_μ, i.e.

$$\{Q_\alpha, Q_\beta\} = -2\gamma^\mu_{\alpha\beta} P_\mu \ .$$

This means that the spinorial charge Q_α is essentially the square root of a translation. In a world governed by supersymmetry, translation is not the primary motion. There are more elementary motions than the space-time displacements in Minkowski space, and these motions occur in a superspace where points are labelled by a set of bosonic coordinates x_μ and a set of fermionic coordinates θ_α.

The structure of space-time is modified in an intrinsic way: each space-time point x_μ has an additional spinorial coordinate θ_α. This allows the concepts of mass and spin to be put on an equal

---

* Scientific secretary: Y. Ha

1 A. Zichichi, The Whys of Subnuclear Physics, in Vol. 15 of the Subnuclear Series (Plenum Press, 1979).

footing. In supergravity, the curvature is related to mass density and the torsion to spin density. Thus, not only the mass but also the spin are derived from the structure of space-time. And this was one of the two basic dreams of Einstein.

An immediate application of "extended supergravity" theories (i.e. supergravity with internal degrees of freedom such as flavours and colours) is to examine whether the phenomenology of these theories is compatible with the observed properties of elementary particles and their symmetries. For illustration, we may naively consider, following Gell-Mann, the phenomenology of an SO(8) theory, even though the group SO(8) does not contain the basic strong, weak, and electromagnetic group of $SU^c(3) \times SU(2)_L \times U(1)_{L,R}$. For the spin 1 sector, the pattern of $SU^c(3)$ content is

Colour multiplet states =	$8 \oplus$	$1 \oplus$	$1 \oplus$	$3 \oplus$	$3 \oplus$	$3 \oplus$	$\bar{3} \oplus$	$\bar{3} \oplus$	$\bar{3}$	$\equiv 28$
Electric charges =	0	0	0	$-1/3$	$-1/3$	$2/3$	$1/3$	$1/3$	$-2/3$	
	coloured gluons	photon	Z^0	superheavy vector bosons						

We notice that in this scheme there is an octet of coloured gluons, the colour singlet photon, the colour singlet neutral weak boson, and many colour triplets charged vector bosons with fractional charges necessary in some unification models of the strong, weak, and electromagnetic interactions. However, the charged vector bosons W^+ and W^- are absent, and if one wants to accept this scheme one would have to assume either that these W's are composite ones or that the charged current weak interactions do not arise from spontaneous breakdown of gauge invariance, but possibly from contact terms of the SO(8) supergravity Lagrangian.

The spin ½ sector in this model is also interesting in that the 56 Majorana spinors break up into complex Dirac spinors according to:

Colour multiplet states =	$3 \oplus$	$3 \oplus$	$3 \oplus$	$3 \oplus$	$6 \oplus$	$8 \oplus$	1
Electric charges =	$2/3$	$-1/3$	$1/3$	$2/3$	$-1/3$	0	-1
	u	d	s	c	fifth flavour	neutral octet	electron

and two neutral Majorana spinors which are SU(3) colour singlets and may be identified with the neutrino. However, the muon and its neutrino, and the τ particle are missing. Notice that in this scheme the fifth flavour should be a colour sextet, not a colour triplet as are u, d, s, and c.

These are a few examples. Naturally, one would consider more realistic supersymmetric models and confront their predictions with experimental evidence. A genuine unification based on local supersymmetry and super-Higgs mechanism would include all interactions, including gravitation. Let me just mention an interesting possible consequence of supersymmetry. If the gluon exists, a spin-½ gluino must exist. This implies, for example, the existence of baryons with bosonic properties and mesons with fermionic ones.

2. Now we turn to the unified theory of electromagnetic and weak interactions. Let me recall some basic steps.

Start with the "electromagnetic current" due to the existence of "electrons" only. Notice that e denotes the spinor, which consists of a "left" and a "right" component:

$$J_\mu^{em} = -\bar{e}\gamma_\mu e = -\bar{e}_L\gamma_\mu e_L - \bar{e}_R\gamma_\mu e_R ,$$

with

$$e_L = \tfrac{1}{2}(1 + \gamma_5)\psi_e , \quad e_R = \tfrac{1}{2}(1 - \gamma_5)\psi_e ,$$

and

$$\bar{e}_L\gamma_\mu e_L = \underbrace{\bar{\psi}_e[(1 - \gamma_5)/2]}_{\bar{e}_L}\ \gamma_\mu\ \underbrace{[(1 + \gamma_5)/2]\psi_e}_{e_L} ,$$

etc.

If we add and subtract $\tfrac{1}{2}(\nu_L\gamma_\mu\nu_L)$, with some grouping of left and right terms, we get

$$J_\mu^{em} = \underbrace{\tfrac{1}{2}[\bar{\nu}_L\gamma_\mu\nu_L - \bar{e}_L\gamma_\mu e_L]}_{J^3_{\mu(L)}} - \underbrace{\tfrac{1}{2}[\bar{e}_L\gamma_\mu e_L + \bar{\nu}_L\gamma_\mu\nu_L] - \bar{e}_R\gamma_\mu e_R}_{\frac{1}{2}J^Y_{\mu(L,R)}} . \quad (1)$$

Notice that $J^3_{\mu(L)}$ consists only of "left"-handed terms, while $J^Y_{\mu(L,R)}$ is made up with left and right terms; $J^3_{\mu(L)}$ is the electroweak isospin current, while $J^Y_{\mu(L,R)}$ is the electroweak hypercharge current. From the coefficients in front of each term which makes up J_μ^{em}, we immediately deduce the values of $T^3(\nu)_L = +\frac{1}{2}$, $T^3(e)_L = -\frac{1}{2}$, $T^3(e)_R = 0$, $Y(\nu)_L = -1$, $Y(e)_L = -1$, $Y(e)_R = -2$.

Repeating the same exercise for all known leptons and quarks, the electroweak quantum numbers of Table 1 are obtained. Notice that the leptons are colour singlets, while the quarks exist in three colours. The equality between "lepton" multiplets and "quark" multiplets is a condition dictated by the famous ABJ (Adler, Bell, Jackiw) anomaly. In order to avoid that the axial vector currents become divergent, it is necessary that the sum of the electric charges of all leptons and all quarks be zero. The charges of the known leptons add up to -3. The sum of the charges of all known quarks (ud), (cs), (tb) is +1. Taking into account that quarks come in three colours, we get +3, which cancels the -3 of leptons. Notice the need for three colours in this argument.

Table 1

The electroweak quantum numbers of the point-like particles

				T^3	Y	Q^{em}	
Leptons	$\begin{pmatrix}\nu_e\\ e^-\end{pmatrix}_L$	$\begin{pmatrix}\nu_\mu\\ \mu^-\end{pmatrix}_L$	$\begin{pmatrix}\nu_\tau\\ \tau^-\end{pmatrix}_L$	$+\frac{1}{2}$ $-\frac{1}{2}$	-1 -1	0 -1	Left states
	$(e^-)_R$	$(\mu^-)_R$	$(\tau^-)_R$	0	-2	-1	Right states
	no neutrinos (R) because m_ν = all zero						
Quarks × 3 colours	$\begin{pmatrix}u\\ d_C\end{pmatrix}_L$	$\begin{pmatrix}c\\ s_C\end{pmatrix}_L$	$\begin{pmatrix}t\\ b_C\end{pmatrix}_L$	$+\frac{1}{2}$ $-\frac{1}{2}$	$+\frac{1}{3}$ $+\frac{1}{3}$	$+\frac{2}{3}$ $-\frac{1}{3}$	Left states
	$(d_C)_R$	$(s_C)_R$	$(b_C)_R$	0	$-\frac{2}{3}$	$-\frac{1}{3}$	Right states
	$(u)_R$	$(c)_R$	$(t)_R$	0	$\frac{4}{3}$	$+\frac{2}{3}$	
T = isospin*) Y = hypercharge	$[Q^{em} = T^3 + Y/2]$						

*) This electroweak isospin is not to be confused with the usual strong one because: first, it is only left-handed, in contrast to T^3 strong, which is (L + R); secondly, because this $(T^3)_L$ refers to Cabibbo mixed particles; thirdly, it is defined for leptons also, which the strong isospin is not.

Formula (1) can be synthetically rewritten as

$$J^{em}_{L+R} = J^3_L + \tfrac{1}{2} J^Y_{L+R} \tag{2}$$

to emphasize how J^{em} can be constructed to contain a purely "left" current, which is also a third component of an isospin. This current (the electromagnetic) is the one that everybody was thinking of when trying to put some order in the "weak interactions".

With reference to Table 1, notice that

i) all right-handed states are isosinglets;

ii) there are no right-handed neutrinos because all neutrino masses are taken to be zero;

iii) no experimental evidence has yet been obtained for the "top" quark, but it should be there as the electroweak companion of "bottom"; this is a prediction of the theory;

iv) the subscript C in the quark states means "Cabibbo mixed states".

The other basic equation, which relates the neutral weak current to $J^{em}_{L,R}$ and to J^3_L, is

$$J^{NC}_{L,R} = J^3_L - \sin^2\theta \cdot J^{em}_{L+R} . \tag{3}$$

This formula tells us that, in order to know the "neutral" weak coupling g, all we need to know are the values of the electroweak isospin T^3_L and of the electric charge of a given particle (leptons or quarks) as given by Table 1.

Remember that the electroweak isospin is only left; it contributes only to the "left" neutral weak coupling g_L. The electric charge is left and right; therefore, it contributes to the "left" as well as to the "right" neutral weak couplings g_L, g_R. All this is shown below:

$$\begin{aligned} J^{NC}_{L,R} &= J^3_L - \sin^2\theta \cdot J^{em}_{L+R} \\ \rightarrow g_L &= T^3_L - \sin^2\theta \cdot Q^{em}_L \leftarrow \\ \rightarrow g_R &= \text{zero} - \sin^2\theta \cdot Q^{em}_R \leftarrow \end{aligned} \tag{4}$$

For example, take the "up" quark. The electroweak isospin third component is $T_L^3(\text{up}) = +\frac{1}{2}$, while the electric charge is $+\frac{2}{3}$; the result is

$$g(u)_L = +\tfrac{1}{2} - \sin^2\theta \cdot \tfrac{2}{3} .$$

If $\sin^2\theta = \frac{1}{4}$, we have $g(u)_L = +\frac{1}{3}$.

The values of the weak neutral coupling constants for all known leptons and quarks are given in Table 2.

Table 2

Neutral weak coupling constants of leptons and quarks as predicted by the $SU(2)_L \times U(1)_{L+R}$ standard theory

	Spinors	T_L^3	Q_L^{em}	Q_R^{em}	g_L	g_R	g_V	g_A
Leptons	ν_e, ν_μ, ν_τ	$+\frac{1}{2}$	0	0	$\frac{1}{2}$	0	$\frac{1}{4}$	$-\frac{1}{4}$
	e^-, μ^-, τ^-	$-\frac{1}{2}$	-1	-1	$-\frac{1}{2} + \sin^2\theta$	$\sin^2\theta$	$-\frac{1}{4} + \sin^2\theta$	$+\frac{1}{4}$
Quarks	u, c, t	$+\frac{1}{2}$	$+\frac{2}{3}$	$+\frac{2}{3}$	$\frac{1}{2} - \frac{2}{3}\sin^2\theta$	$-\frac{2}{3}\sin^2\theta$	$\frac{1}{4} - \frac{2}{3}\sin^2\theta$	$-\frac{1}{4}$
	d_C, s_C, b_C	$-\frac{1}{2}$	$-\frac{1}{3}$	$-\frac{1}{3}$	$-\frac{1}{2} + \frac{1}{3}\sin^2\theta$	$+\frac{1}{3}\sin^2\theta$	$-\frac{1}{4} + \frac{1}{3}\sin^2\theta$	$+\frac{1}{4}$
	If $\sin^2\theta = \frac{1}{4}$:							
Leptons	neutrinos (ν_e, ν_μ, ν_τ)				$+\frac{1}{2}$	0	$+\frac{1}{4}$	$-\frac{1}{4}$
	charged leptons (e^-, μ^-, τ^-)				$-\frac{1}{4}$	$+\frac{1}{4}$	0	$+\frac{1}{4}$
Quarks	up-like quarks (u, c, t)				$+\frac{1}{3}$	$-\frac{2}{12}$	$+\frac{1}{12}$	$-\frac{1}{4}$
	down-like quarks (d_C, s_C, b_C)				$-\frac{5}{12}$	$+\frac{1}{12}$	$-\frac{2}{12}$	$+\frac{1}{4}$

Notice that in $SU(2)_L \times U(1)_{L+R}$, $T_R^3 = 0$ for all quarks and leptons. Therefore g_R, the "right" neutral weak coupling, can be $\neq 0$ only for particles with $Q^{em} \neq 0$. In other words, in the "standard" $SU(2)_L \times U(1)_{L+R}$ theory the "right" coupling is coming from the existence of "electrically" charged spinors. Otherwise the weak neutral coupling would be "left-handed" only.

Notice also that the "vector" (g_V) and "axial" (g_A) neutral weak couplings can be worked out in terms of the "chiral" neutral weak couplings (g_L, g_R) by writing J^{NC} explicitly in terms of the basic spinors and γ-matrices:

$$J^{NC}_{L+R} = J^3_L - \sin^2\theta \cdot J^{em}_{L+R} \equiv \bar{\psi}\gamma_\mu\left[\frac{1-\gamma_5}{2} g_L + \frac{1+\gamma_5}{2} g_R\right]\psi \equiv$$

$$\equiv \underbrace{\tfrac{1}{2}(g_L + g_R)}_{g_V}\bar{\psi}\gamma_\mu\psi + \underbrace{\tfrac{1}{2}(-g_L + g_R)}_{g_A}\bar{\psi}\gamma_\mu\gamma_5\psi \ . \qquad (5)$$

In fact,

$$\bar{\psi}\gamma_\mu\left(\frac{1-\gamma_5}{2}\right)\psi \equiv \text{left-handed current,}$$

$$\bar{\psi}\gamma_\mu\left(\frac{1+\gamma_5}{2}\right)\psi \equiv \text{right-handed current,}$$

$$\bar{\psi}\gamma_\mu\psi \equiv \text{vector current,}$$

$$\bar{\psi}\gamma_\mu\gamma_5\psi \equiv \text{axial current,}$$

where

$\psi \equiv$ Dirac spinor for leptons and quarks,

$\gamma_\mu (\mu = 1, 2, 3, 4)$ and $\gamma_5 \equiv$ Dirac matrices.

Combining formulae (4) and (5), we get

$$g_V = \tfrac{1}{2}(g_L + g_R) = \tfrac{1}{2}T^3_L - \sin^2\theta \cdot Q^{em} \ ,$$

$$g_A = \tfrac{1}{2}(g_R - g_L) = -\tfrac{1}{2}T^3_L \ .$$

All this explains how the weak neutral coupling constants of quarks and leptons, in terms of the "chiral" (g_L, g_R) or of the "vector" (g_V) and "axial" (g_A), are related. Table 2 is an example of the predictive power of the theory. The comparison with experimental data, as reported in K. Winter's lecture, is shown in Table 3.

Table 3

Comparison of the $SU(2)_L \times U(1)_{L+R}$ weak neutral coupling constant with experiments

Glashow-Salam-Weinberg predictions	Taking $\sin^2\theta = 1/4$	Experimental
$g(u)_L = 1/2 - 2/3 \sin^2\theta$	+0.33	+0.35 ± 0.07
$g(d)_L(*) = -1/2 + 1/3 \sin^2\theta$	-0.42	-0.40 ± 0.07
$g(u)_R = -2/3 \sin^2\theta$	-0.17	-0.19 ± 0.06
$g(d)_R(*) = +1/3 \sin^2\theta$	+0.08	0.00 ± 0.11

*) The Cabibbo angles are neglected here. The exact formulae should read $g(d)_L$ = (same)$\cdot\cos\theta_C$, $g(d)_R$ = (same)$\cdot\cos\theta_C$. These effects are too small, compared with the experimental uncertainties.

What have we learnt?

The knowledge of the following five quantities:

i) α, the fine structure constant,

ii) θ, the mixing angle between the two gauge groups $SU(2)_L$ and $U(1)_{L+R}$,

iii) the Clebsch-Gordan coefficients of the $SU(2)_L \times U(1)_{L+R}$ symmetry groups,

iv) the Fermi coupling constant, G_F, or one mass, $m_{W^\pm}$ or m_{Z^0},

v) the generalized Cabibbo angles,

is all that is needed to describe weak and electromagnetic processes, in the framework of a theory which is renormalizable. The old times when weak processes needed a cut-off are over.

Let me say a few words on the simple Spontaneous Symmetry Breaking (SSB).

Here comes an impressive experimental check. In a weak interaction theory, with the Higgs mechanism unknown, there are two unknown parameters: the famous mixing angle θ; and the ratio ρ of neutral to charged currents, introduced in order to keep free the masses of the intermediate bosons:

$$\rho^2 = \frac{\text{rate of neutral currents}}{\text{rate of charged currents}} = \frac{g_{NC}^4}{g_{CC}^4} \cdot \frac{(1/m_{Z^0}^2)^2}{(1/m_{W^\pm}^2)^2} \quad . \tag{6}$$

If SSB really takes place, as suggested by Salam-Weinberg, i.e. via the simplest Higgs mechanism, the masses of the charged and neutral intermediate bosons are related:

$$m_{W^\pm}/m_{Z^0} = \cos\theta \ . \tag{7}$$

In this case, the damping of the neutral currents, with respect to the charged ones, is compensated exactly by the ratio of the coupling constants: $g_{CC}^2/g_{NC}^2 = \cos^2\theta$; in fact, as shown in Fig. 1, $g_{CC} = g^\pm = g \cdot \cos\theta$, and $g_{NC} = g$. Therefore

$$\rho^2 = \frac{m_{W^\pm}^4}{m_{Z^0}^4} \cdot \frac{1}{\cos^4\theta} = 1 \ . \tag{8}$$

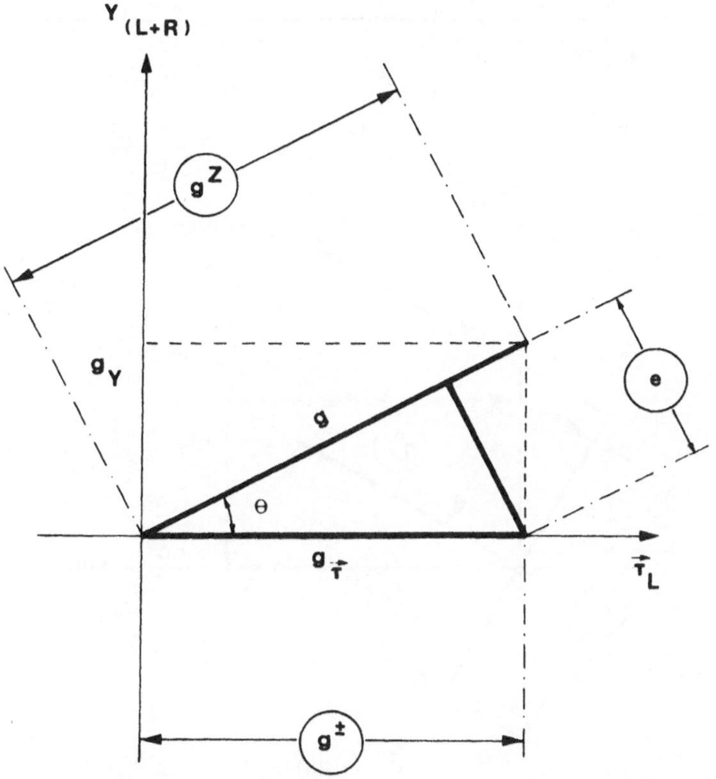

Fig. 1

The experimental result, as reported in the lecture by Dr. K. Winter, is

$$\rho = 0.98 \pm 0.05 \ .$$

In terms of the basic fields, let me show a synthesis of the Glashow-Salam-Weinberg theory. In Fig. 2a, W_μ^Y is the electroweak

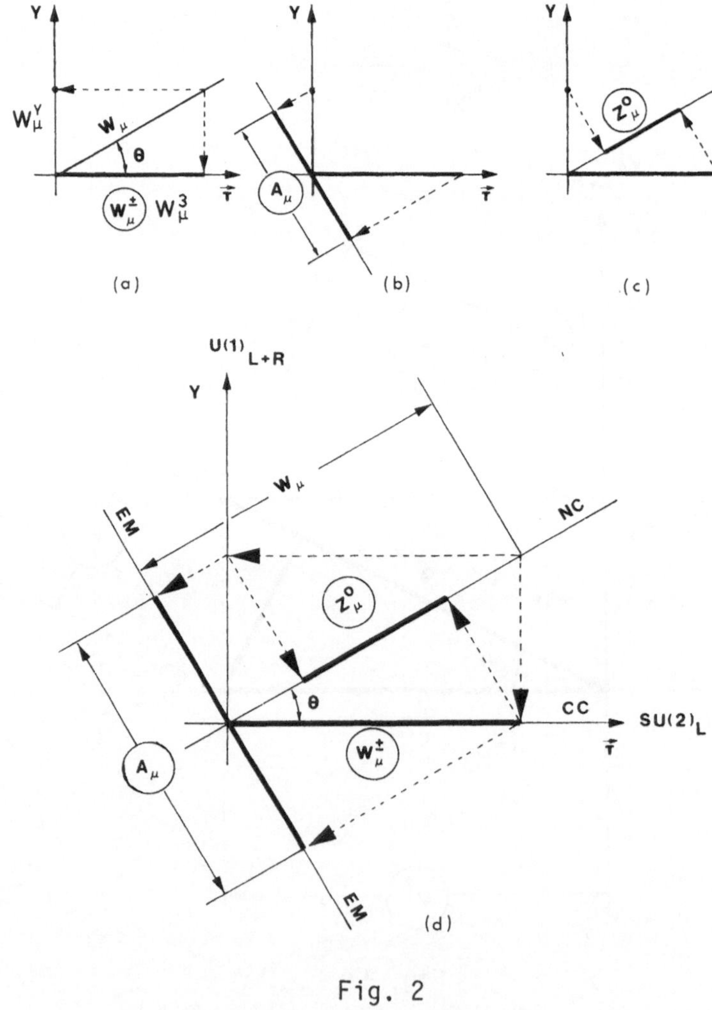

Fig. 2

hypercharge field and W_μ^3 the third component of the electroweak isospin field. In terms of these vector fields, the physical fields are:

$$A_\mu = W_\mu^Y \cos\theta + W_\mu^3 \sin\theta \text{ (Fig. 2b)}$$

$$Z_\mu^0 = -W_\mu^Y \sin\theta + W_\mu^3 \cos\theta \text{ (Fig. 2c)}.$$

The summary of all this is shown in Fig. 2d. Notice that circled quantities indicate the fields whose quanta are observable. Thus $W_\mu^\pm$ correspond to the charged weak bosons; W_μ^3 and W_μ^Y do not have observable quanta. Their mixing produces A_μ and Z_μ, whose quanta are the photon and the neutral weak boson. In Figs. 2a-d the thick lines indicate where the observable quantities come from.

Before the $SU(2)_L \times U(1)_{L+R}$ electroweak theory, our knowledge was all along the $\vec{\tau}$ axis, where we knew only the so-called charged currents (more correctly these currents should be called "electric charge-changing currents"). Nobody had suspected the existence of τ_3. We knew that A_μ exist but we did not know of the existence of the neutral-current (NC) axis, nor of the intimate connection between A_μ and Z_μ^0.

Note that $\vec{\tau}$ indicates the existence of the three components (τ^+, τ^-, τ^3). These are the generators of the group $SU(2)_L$ while Y is the generator of the group $U(1)_{L,R}$. The electroweak angle θ determines the relative weight of these two basic gauge groups, whose merging generates the electromagnetic and the weak interactions.

We have learned that the e.m. field is not a fundamental field; it is made up of two other fields, W_μ^3 and W_μ^Y. Their mixing generates A_μ and Z_μ^0; the quanta of these fields are the observable quantities. The Z_μ^0 is associated with a particle whose mass is expected to be near 85 GeV or so. Wait and see.

Let me call your attention to the following three features of the GSW theory:

i) the existence of the two quantum numbers, the electroweak isospin and the electroweak hypercharge, shared by quarks and leptons;

ii) the discovery of a new law which relates the weak neutral current to the electromagnetic current:

$$J_{L,R}^{NC} = J_L^3 - \sin^2\theta \cdot J_{L+R}^{em} ;$$

iii) the strength of neutral to charged current effects, i.e. $\rho = 1$, which implies that the simplest SSB is at work.

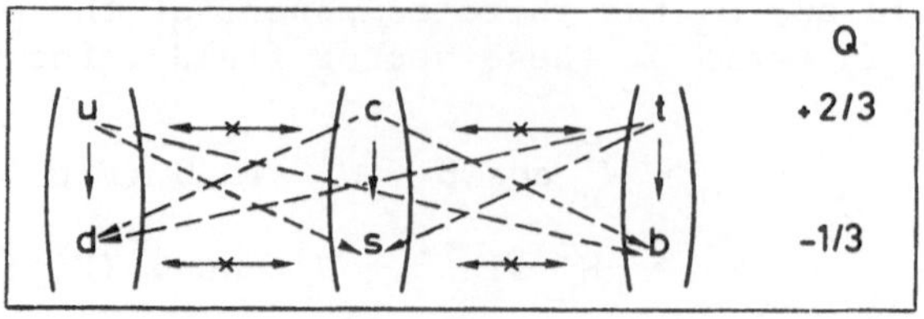

Fig. 3

3. Within the framework of the $SU(2)_L \times U(1)_{L+R}$ gauge theory of electroweak interactions, another important point is the possibility to incorporate PC and T violations. This can be done if the number of quarks is at least six, as shown in Fig. 3. Notice that the allowed neutral currents are $\bar{u}u$, $\bar{c}c$, $\bar{t}t$, $\bar{d}d$, $\bar{s}s$, $\bar{b}b$. Generalized Cabibbo mixing opens the dashed channels. Without it, the dashed transitions would not exist. The only allowed transitions would be those inside a quark doublet, indicated by the full arrows. Like the introduction of charm, which naturally forbids any charm-changing neutral currents and strangeness-changing neutral currents, the new quarks will similarly forbid the existence of the new types of neutral currents in new decay processes, as illustrated in Fig. 3 by the "crossed" arrows. The way in which the generalized Cabibbo angles come in the various decay processes is illustrated in Fig. 4. The quantities needed are a phase δ and three generalized Cabibbo mixing angles θ_i.

$S_i = \sin \theta_i$
$C_i = \cos \theta_i$

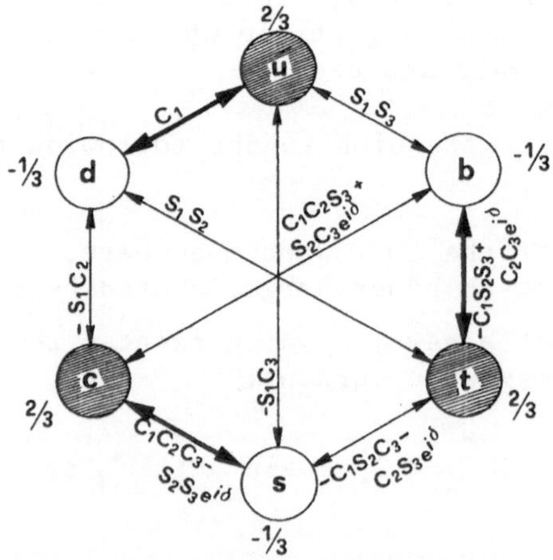

Fig. 4 Generalized Cabibbo mixing with six quarks

4. Let me close by quoting two points in the present phenomenology of charmed particle decays. Once again, new discoveries bring with them new problems. It is not true that the phenomenology of the new particles follows simple expectations. We have learnt from Lina Barbaro-Galtieri that, for example, the ratio $F \to \eta / F \to K\bar{K}$ is 25 rather than the expected value of 1; and that charmed baryons like to decay preferentially into the NK rather than the Λ particles.

* * *

The topics mentioned are just a few examples of the over-all effort to understand the subnuclear world.

D I S C U S S I O N

CHAIRMAN: A. Zichichi

Scientific Secretaries: Y. Ha and R. Orava

DISCUSSION

- SURSOCK:

Why has Nature chosen three colours rather than two, four or any other number?

- ZICHICHI:

If one assumes only two colours it is impossible to form any colour singlet state out of a three quark system (e.g. Δ^{++}). On the other hand, assuming more than three colours the ratio

$$R = \frac{e^+ e^- \to \text{hadrons}}{e^+ e^- \to \mu^+ \mu^-}$$

would become too large and the lifetime of the π^0 would become too short. For more details on the problem "why colour" see my closing lecture of last year's School.

- THIRRING:

How well does a factor of 9 due to colour account for the lifetime of the $\pi^0 \to 2\gamma$ decay? Is the agreement as good as that for the $\eta \to 2\gamma$ case?

- ZICHICHI:

The most recent results do agree within errors with the factor of 9 in the η decay.

- SURSOCK:

Are there any theoretical arguments other than the experimental evidence for three colours?

- ISGUR:

If one considers a Weinberg-Salam type model and demands it to be a consistent field theory, then in order to cancel the Adler-Bell-Jackiw anomalies in this model one has to associate with each lepton doublet a doublet of quarks with charges 2/3 and -1/3 and in three colours.

- ZICHICHI:

What this means is that in order to cancel the anomalies in a model of this type the total charge of the quarks must cancel the total charge of the leptons. In a model with three lepton doublets, namely, e, μ, τ and their respective neutrinos ν_e, ν_μ, ν_τ, and three doublets of quarks (u,d), (s,c), (t,b), the total charge for the leptons is -3, whereas the total charge for the quarks is thrice (2/3 - 1/3) and hence 1. In order to balance the charge of the leptons we must bring in three colours for each quark flavour. However, I don't consider this argument more important than the evidence for the ratio R or the π^0 lifetime.

- HA:

Three colours are necessary to make the occurrence of symmetric SU(6) representations for low lying baryons compatible with the fermionic character of quarks, and this was the original motivation for the introduction of colour. The only candidates for a simple Lie group of colour are SU(3) and SO(3). With SO(3) one loses the desired property of asymptotic freedom in QCD when the number of flavours exceeds 2, while for SU(3) one can have up to 16 flavours. Since four and possibly five flavours are now known, SU(3) is the only viable possibility.

- MERRITT:

Regarding the J/ψ production in proton-proton and proton-antiproton reactions I believe that the difference between the two production processes is not due to the difference in the "ocean" quark-antiquark contribution in proton and antiproton but due to a combination of gluon fusion and valence quark-antiquark annihilation in the process $q\bar{q} \to \chi \to J/\psi$ for the $p\bar{p}$ interactions

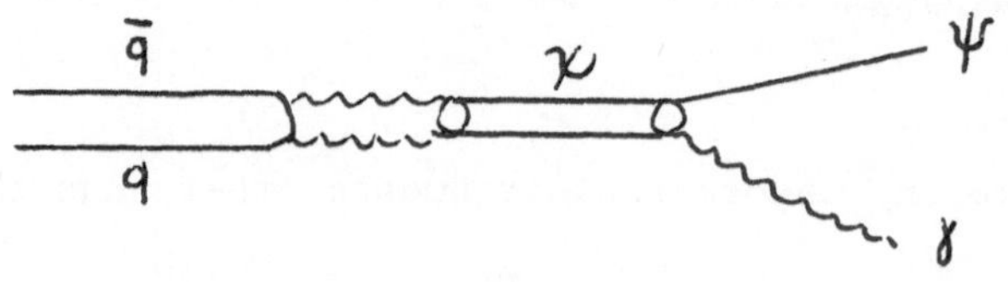

- ZICHICHI:

I agree with you. However, as I said in the lecture, the point beyond any doubt is the exclusion of those models predicting equal rates for p and $\bar{p}$ production.

- JACOB:

There are several processes which seem to be all relevant at moderate energies. The two gluon exchange process for $q\bar{q} \to \chi \to J/\psi$ and the three gluon exchange one for $q\bar{q} \to J/\psi$ could distinguish pp and $p\bar{p}$ induced production, whereas the gluon gluon processes and the charm sea annihilation processes will not. As the energy increases, the charm-anticharm annihilation process may become very important. It is, however, not yet the case at 40 GeV where data have been obtained.

- MUSSET:

In Nature, there seems to be an equal number of quark flavours and a number of different leptons. Is there any theoretical attempt to explain this coincidence or the fact that these spectra may arise from an internal degree of freedom?

- ZICHICHI:

There are models in which one can calculate the maximum number of flavours allowed in QCD, however, the connection between the number of quarks and the number of leptons is not established and is only a phenomenological observation. Concerning the number of flavours the only theoretical argument we know is due to QCD, which fixes their maximum number up to 16, if asymptotic freedom has to be kept valid. The number of leptons has no theoretical predictions. The equality between quark flavours and lepton states is needed as I have said before, in order to cancel the famous Adler, Bell, Jackiw anomaly. I know no theoretical attempt able to produce the observed quarks and lepton spectra.

CLOSING CEREMONY

The Closing Ceremony took place on Thursday, 10 August 1978. The Director of the School presented the prizes and scholarships to the winners as specified below.

PRIZES AND SCHOLARSHIPS

Prize for *Best Student* - awarded to Dr. Frank S. MERRITT, SLAC, Stanford University, Stanford, CA, USA.

Prize for *Best Scientific Secretary* - awarded to Mr. Yuan K. HA, Yale University, New Haven, CT, USA.

Twelve scholarships were open for competition among the participants - they were awarded as follows:

Patrick M.S. Blackett Scholarship - awarded to Mr. Anthony D. KENNEDY, University of Sussex, Brighton, Sussex, UK.

James Chadwick Scholarship - awarded to Dr. Harald GROSSE, CERN, Genève, Switzerland.

Amos De-Shalit Scholarship - awarded to Dr. Robert ROTH, Weizmann Institute of Science, Rehovot, Israel.

Gunnar Källen Scholarship - awarded to Mr. Guy ANASTAZE, Centre de Recherches Nucléaires, Strasbourg, France.

André Lagarrigue Scholarship - awarded to Ms. Catherine B. NEWMAN, University of Chicago, Chicago, IL, USA.

Giulio Racah Scholarship - awarded to Mr. Gyan BHANOT, Cornell University, Ithaca, NY, USA.

Giorgio Ghigo Scholarship - awarded to Dr. Hans Reinhard GERHOLD, Österreichische Akademie der Wissenschaften, Wien, Austria.

Enrico Persico Scholarship - awarded to Dr. Jean-Pierre SURSOCK, University of California, Berkeley, CA, USA.

Peter Preiswerk Scholarship - awarded to Dr. Gertraud HANSL, CERN, Genève, Switzerland.

Gianni Quareni Scholarship - awarded to Dr. J. Scott WHITAKER, CERN, Genève, Switzerland.

Antonio Stanghellini Scholarship - awarded to Dr. Alberto BLASI, Università di Genova, Italy.

Alberto Tomasini Scholarship - awarded to Dr. Bhaskar Ravindranath KUMAR, CERN, Genève, Switzerland.

The following Students received *honorary mentions* for their contributions to the activity of the School:

Dr. Luuk H. KARSTEN, Universiteit van Amsterdam, Amsterdam, The Netherlands.

Dr. Klaus LACKNER, Universität Heidelberg, Heidelberg, FRG.

Dr. Hugh MONTGOMERY, CERN, Genève, Switzerland.

Mr. Risto O. ORAVA, CERN, Genève, Switzerland.

The following participants gave their collaboration in the scientific secretarial work:

Guy ANASTAZE
Luciano BARONE
Giuseppe BATTISTONI
Gyan BHANOT
K. Wyatt BROWN
Jochen BÜRGER
Ivan DADIĆ
Christian DE LA VAISSIÈRE
Hans Reinhard GERHOLD
Karl-Ludwig GIBONI
Harald GROSSE
Yuan K. HA
Gertraud HANSL
Hanna JACOB-HOFFMAN
Luuk H. KARSTEN
Anthony D. KENNEDY
Klaus LACKNER
Ilkka LIEDE
Juha LINDFORS
Robert M. MORSE
Catherine B. NEWMAN
Risto O. ORAVA
Joseph ROSEN
Robert ROTH
Jean-Pierre SURSOCK
Stephen TEMPLETON

PARTICIPANTS

Guy ANASTAZE	Centre de Recherches Nucléaires Division des Hautes Energies F-67037 STRASBOURG CEDEX, France
Lina BARBARO GALTIERI	University of California Lawrence Berkeley Laboratory BERKELEY, CA 94720, USA
Antonio BARONCELLI	Istituto Superiore di Sanità Viale Regina Margherita, 299 I-00161 ROMA, Italy
Luciano BARONE	Istituto di Fisica dell'Università Piazzale delle Scienze, 5 I-00185 ROMA, Italy
Giuseppe BATTISTONI	Laboratori Nazionali - INFN Casella Postale 13 I-00044 FRASCATI (Roma), Italy
Hans-Uno BENGTSSON	Institutionen för Teoretisk Fysik Sölvegatan 14 A S-223 62 LUND, Sweden
Raimondo BERTINI	CERN - EP Division CH-1211 GENEVE 23, Switzerland
Gyan BHANOT	Cornell University Laboratory of Nuclear Physics ITHACA, NY 14853, USA
Renato BIRSA	Istituto di Fisica dell'Università Via A. Valerio, 2 I-34100 TRIESTE, Italy

Alberto BLASI	Istituto di Scienze Fisiche dell'Università Viale Benedetto XV, 5 I-16132 GENOVA, Italy
Kors BOS	CERN - EP Division CH-1211 GENEVE 23, Switzerland
Wyatt BROWN	California Institute of Technology Charles C. Lauritsen Laboratory of High Energy Physics PASADENA, CA 91125, USA
Jochen BÜRGER	Gesamthochschule Siegen Fachbereich für Physik Hölderlinstrasse 3 Postfach 21 02 09 D-5900 SIEGEN 21, FRG
Antonio CAPONE	Istituto di Fisica dell'Università Piazzale delle Scienze, 5 I-00185 ROMA, Italy
Jacques CHAUVEAU	CERN - EP Division CH-1211 GENEVE 23, Switzerland
Ivan DADIC	Institute "Ruder Boskovic" P.O. Box 1016 41001 ZAGREB, Yugoslavia
Christian DE LA VAISSIERE	Laboratoire de Physique Nucléaire et de Hautes Energies Université de Paris VI Tour 32 4, Place Jussieu F-75230 PARIS CEDEX 5, France
Marianna DI IANNI	Istituto di Fisica dell'Università Piazza Torricelli I-56100 PISA, Italy
Werner DORTH	Institut für Hochenergiephysik Albert-Überle Strasse 2 D-6900 HEIDELBERG 1, FRG
John C. ECCLES	"Ca' a la Gra'" CH-6611 CONTRA (Locarno) Switzerland

Glennys R. FARRAR — Lauritsen Laboratory of Physics
California Institute of Technology
PASADENA, CA 91125, USA

Joël FELTESSE — CERN - EP Division
CH-1211 GENEVE 23, Switzerland

Sergio FERRARA — CERN - Th Division
CH-1211 GENEVE 23, Switzerland

Hans Reinhard GERHOLD — Österreichische Akademie
der Wissenschaften
Institut für Hochenergiephysik
Nikolsdorfergasse 18
A-1050 WIEN, Austria

Karl-Ludwig GIBONI — III Physikalisches Institut
Physikzentrum
D-5100 AACHEN, Germany

T.P. GRAY — Cavendish Laboratory
University of Cambridge
CAMBRIDGE, UK

Harald GROSSE — CERN - Th Division
CH-1211 GENEVE 23, Switzerland

Yuan K. HA — Yale University
Physics Department
217 Prospect Street
NEW HAVEN, CT 06520, USA

Gertraud HANSL — CERN - EP Division
CH-1211 GENEVE 23, Switzerland

Kerson HUANG — Massachusetts Institute of
Technology
77 Massachusetts Avenue
CAMBRIDGE, MA 02139, USA

Nathan ISGUR — Department of Physics
University of Toronto
TORONTO, ONTARIO M5S 1A7, Canada

Maurice JACOB — CERN - Th Division
CH-1211 GENEVE 23, Switzerland

Hanna JACOB-HOFFMAN — Rutherford High Energy Laboratory
CHILTON, DIDCOT, Berks., UK

Thomas KAHL — Max-Planck-Institut für Physik
und Astrophysik
Föhringer Ring 6
D-8000 MÜNCHEN 40, FRG

Luuk H. KARSTEN — Instituut voor Theoretische Fysica
Valckenierstraat 65
AMSTERDAM 1004, The Netherlands

Anthony D. KENNEDY — University of Sussex
School of Mathematical &
Physical Sciences
FALMER, BRIGHTON, Sussex, UK

Hagen KLEINERT — Freie Universität Berlin
Institut für Theoretische Physik
der Elementarteilchen
Fachbereich Physik
Arnimallee 3
D-1000 BERLIN 33, FRG

Mehmet KOCA — Cukurova Universitesi
Temel Bilimler Faküttesi
Köpruköy
ADANA, Turkey

Bhaskar Ravindranath KUMAR — CERN - EP Division
CH-1211 GENEVE 23, Switzerland

Klaus LACKNER — Institut für Theoretische Physik
Philosophenweg 16
D-6900 HEIDELBERG, FRG

Morten LAURSEN — Niels Bohr Institute
Blegdamsvej 17
DK-2100 KØBENHAVN Ø, Denmark

Ilkka LIEDE — Helsinki University
Research Institute of Theoretical
Physics
Siltavuorenpenger 20 B
SF-00170 HELSINKI 17, Finland

Giuseppe LIGUORI
Istituto di Fisica Nucleare
dell'Università
Via A. Bassi, 6
I-27100 PAVIA, Italy

Juha LINDFORS
Helsinki University
Research Institute of Theoretical Physics
Siltavuorenpenger 20 B
SF-00170 HELSINKI 17, Finland

Michele LIVAN
Istituto di Fisica Nucleare
dell'Università
Via A. Bassi, 6
I-27100 PAVIA, Italy

Bruno MACHET
Laboratoire de Physique Théorique
et Hautes Energies
Bat. 211 - Centre d'Orsay
F-91405 ORSAY, France

Daniela MAURIZIO
Istituto di Fisica
Corso Massimo d'Azeglio, 46
I-10125 TORINO, Italy

Emanuela MERONI
Istituto di Fisica
Via Celoria, 16
I-20133 MILANO, Italy

Frank S. MERRITT
Stanford Linear Accelerator Center
P.O. Box 4349
STANFORD, CA 94305, USA

William R. MOLZON
University of Chicago
The Enrico Fermi Institute
5630 Ellis Avenue
CHICAGO, IL 60637, USA

Hugh MONTGOMERY
CERN - EP Division
CH-1211 GENEVE 23, Switzerland

Robert M. MORSE
University of Wisconsin
Department of Physics
1150 University Avenue
MADISON, WI 53706, USA

G. MOTZ	CERN - EP Division CH-1211 GENEVE 23, Switzerland
Panayiotis MOUZOURAKIS	CERN - EP Division CH-1211 GENEVE 23, Switzerland
Paul MUSSET	CERN - EP Division CH-1211 GENEVE 23, Switzerland
Catherine B. NEWMAN	University of Chicago The Enrico Fermi Institute 5630 Ellis Avenue CHICAGO, IL 60637, USA
Bengt NILSSON	Institute of Theoretical Physics GÖTEBORG 5, Sweden
Risto O. ORAVA	CERN - EP Division CH-1211 GENEVE 23, Switzerland
Fernanda PASTORE	Istituto di Fisica Nucleare dell'Università Via A. Bassi, 6 I-27100 PAVIA, Italy
Alfred PETERSEN	DESY Notkestr. 85 D-2000 HAMBURG 52, FRG
Hartmuthe PLOTHOW	CERN - EP Division CH-1211 GENEVE 23, Switzerland
Joseph ROSEN	Tel-Aviv University Department of Physics & Astronomy RAMAT-AVIV, TEL-AVIV, Israel
Michael RÖSSLER	Gesamthochschule Wuppertal Theoretische Physik Fachbereich 8 - Naturwissenschaften 1 Gaubstr. 20 D-5600 WUPPERTAL 1, FRG
Robert ROTH	The Weizmann Institute of Science Department of Nuclear Physics REHOVOT, Israel

A.N.J.J. SCHELLEKENS	Catholic University Institute for Theoretical Physics Toernooiveld NIJMEGEN, The Netherlands
Leslie SHORT	University of Glasgow Natural Philosophy Department GLASGOW G12 8QQ, UK
Roberto SOLDATI	Dipartimento di Matematica e Fisica della Libera Università I-38050 POVO (Trento), Italy
E. Van Der SPUY	Atomic Energy Board Private Bag X256 PRETORIA 0001, South Africa
Robert A. STERN	Imperial College The Blackett Laboratory High Energy Nuclear Physics Group Prince Consort Road LONDON SW7 2AZ, UK
Jean-Pierre SURSOCK	University of California Lawrence Berkeley Laboratory Theoretical Group BERKELEY, CA 94720, USA
Stephen TEMPLETON	University of Durham Department of Mathematics Science Laboratories, South Road DURHAM DH1 3LE, UK
Walter THIRRING	Institut für Theoretische Physik der Universität Wien Boltzmanngasse 5 A-1010 WIEN, Austria
Johannes J.M. TIMMERMANS	CERN - EP Division CH-1211 GENEVE 23, Switzerland
Michael TYNDEL	Rutherford Laboratory Science Research Council CHILTON, DIDCOT, Berks., UK

Marcel VIVARGENT	Laboratoire de Physique des Particules IUT (IN2P3) Chemin de Bellevue B.P. 909 F-74019 ANNECY-LE-VIEUX CEDEX France
Rudiger VOSS	CERN - EP Division CH-1211 GENEVE 23, Switzerland
Wolfgang WEINZIERL	Max-Planck-Institut für Physik und Astrophysik Föhringer Ring 6 Postfach 40 12 12 D-8000 MÜNCHEN 40, FRG
J. Scott WHITAKER	CERN - EP Division CH-1211 GENEVE 23, Switzerland
Klaus WINTER	CERN - EP Division CH-1211 GENEVE 23, Switzerland
Günter WOLF	DESY Notkestrasse 85 D-2000 HAMBURG 52, FRG

INDEX

R-hadrons, 61, 72-73